Chronology
of Feedback

1868

Maxwell, Flyball stability analysis

1927

al analyzer

1932

stability criterion

1938

uency response methods

1942

ner, Optimal filter design

1947

ns; Nichols, Nichols chart

1948

Evans, Root locus

1950

nberger, Nonlinear analysis

1956

Pontryagin, Maximum principle

1957

Bellman, Dynamic programming

1960

Draper, Inertial navigation; Kalman, Optimal estimation

1969

Hoff, Microprocessor

Second Edition

Feedback Control of

Dynamic Systems

Gene F. Franklin
Stanford University

J. David Powell
Stanford University

Abbas Emami-Naeini
Systems Control Technology, Inc.

 ADDISON-WESLEY PUBLISHING COMPANY

Reading, Massachusetts • Menlo Park, California • New York
Don Mills, Ontario • Wokingham, England • Amsterdam • Bonn
Sydney • Singapore • Tokyo • Madrid • San Juan • Milan • Paris

Sponsoring Editor: Thomas Robbins
Production Supervisor: Karen Myer
Art Editor: Dick Morton
Production Coordinator: Patsy DuMoulin
Manufacturing Supervisor: Hugh Crawford

Many of the designations used by manufacturers and sellers to distinguish their products are claimed as trademarks. Where those designations appear in this book, and Addison-Wesley was aware of a trademark claim, the designations have been printed in initial caps or all caps.

The programs and applications presented in this book have been included for their instructional value. They have been tested with care, but are not guaranteed for any particular purpose. The publisher does not offer any warranties or representations, nor does it accept any liabilities with respect to the programs or applications.

Library of Congress Cataloging-in-Publication Data

Franklin, Gene F.
  Feedback control of dynamic systems / Gene F. Franklin, J. David
Powell, Abbas Emami-Naeini. —2nd ed.
    p.   cm.
  Includes bibliographical references and index.
  ISBN 0-201-50862-1
  1. Feedback control systems.   I. Powell, J. David, 1938– .
II. Emami-Naeini, Abbas. III. Title.
TJ216.F723   1991
629.8'3—dc20   90–1193
                                                                          CIP

Reprinted with corrections April, 1991

23456789-HA-9321

This book is in the **Addison-Wesley Series in Electrical and Computer Engineering: Control Engineering**

Adaptive Control, 09720
Karl J. Åström and Björn Wittenmark

Introduction to Robotics, Second Edition, 09528
John J. Craig

Modern Control Systems, Fifth Edition, 14278
Richard C. Dorf

Digital Control of Dynamic Systems, Second Edition, 11938
Gene F. Franklin, J. David Powell, and Michael L. Workman

Computer Control of Machines and Processes, 10645
John G. Bollinger and Neil A. Duffie

Feedback Control of Dynamic Systems, Second Edition, 50862
Gene F. Franklin, J. David Powell, and Abbas Emami-Naeini

Adaptive Control of Mechanical Manipulators, 10490
John J. Craig

Modern Control System Theory and Application, Second Edition, 07494
Stanley M. Shinners

To

Gertrude, David, Carole
Valerie, Daisy, Annika, Dave
Malahat, Sheila

Preface

Feedback Control of Dynamic Systems is for a first course in feedback control at the senior level for all engineers. The book includes a review of dynamic response and the Laplace transform but assumes that the student is familiar with the essential ideas, including writing equations of motion and using the Laplace transform to solve linear constant ordinary differential equations. The emphasis of the book is on controller design, and the intention is to introduce the root locus, frequency response, and state feedback design methods as alternative approaches to the same goal: designing controllers that will process sensor measurements so as to cause a dynamic system to meet tracking accuracy and disturbance-rejection specifications in spite of model errors.

The material in the book is used at Stanford University as the basis for two ten-week courses taken by seniors and first-year graduate students, especially in the departments of Aeronautics and Astronautics, Electrical Engineering, and Mechanical Engineering. The courses complement a graduate course in linear systems and are prerequisites to courses in digital control, optimal control, flight control, and smart product design. The book can also be used to support a one-semester (15 week) course or a single one-quarter course. An outline of how the material might be used is given following an overview of the philosophy and contents of the book.

In this second edition, we have retained the basic philosophy and content of the first edition while being responsive to the suggestions of students, colleagues, and reviewers. Throughout the book we have rewritten material to improve clarity and to generally help the reader understand the material better. Specifically, we have added material on the Laplace transform, time response, and block-diagram manipulation in Chapter 2. We have reorganized the presentation and expanded the treatment of system Type and Routh's stability criterion in Chapter 3.

In the treatment of stability margins and sensitivity in Chapter 5, we have begun to tie in the single-input, single-output concepts with the more general methods necessary for the multivariable case. In Chapter 6 we introduce the idea of a tradeoff curve between control effort and tracking accuracy in the context of the linear quadratic regulator design. As background for dynamic-response and frequency-response material, we have added an appendix on complex analysis. We have also thoroughly reworked the problems in each chapter to be sure that the statements are clear and unambiguous and that in each case the questions are appropriate to the section to which the problem applies.

We believe that this book is distinguished by its emphasis on controller design, by a uniform level of treatment of the root locus, frequency response, and state-feedback ideas, and by the extent to which the several design methods are integrated into a comprehensive set of tools to serve the control designer. As a consequence of these features, the book is a superior foundation for further study of systems, feedback, and advanced control. The emphasis on design begins in Chapter 3 following the review of dynamic response. Before any sophisticated analysis method is presented, the influence of feedback on disturbance rejection, tracking accuracy, and model errors is introduced. Here, in the context of a first-order model for speed control, the concepts of proportional, integral, and derivative (PID) control are introduced. In this way, the student gets the idea of what control is all about *before* the tedious rules of the root locus or of the Nyquist Stability Criterion are developed. By this approach, the central issues of control *design* are brought forward and can be kept in the foreground during the necessary analysis that goes with construction of sophisticated design tools.

The level of treatment of the root locus, frequency response, and state-feedback techniques is that of a senior engineering student who has had a course in circuits or dynamic response that uses the Laplace transform to solve linear constant differential equations. The advanced algebra essential to using state-variable methods on multivariable systems is omitted, to be treated in a subsequent course. Our objective in the first course is to teach the student to design single-input, single-output systems by each of the methods and to experience the unique insights given by each of them on the design problem. Also emphasized is the fact that it is mainly *insight* that the techniques provide; the details of numerical calculations of specific parameter values is expected to be done by computer. The unity of the methods is illustrated by the use of similar examples during the individual presentations and by the collection in Chapter 7 of several case studies where all three methods are used to contribute to the designs. We suggest that instructors should consider a look-ahead to these examples and perhaps select one to be considered as a continuing illustration of each technique as it is introduced.

The contents of the book are organized into eight chapters and three appendixes. The chapters include some sections of advanced or enrichment material marked with □ that can be omitted without interfering with the flow of the material. The appendixes include background and reference material such as matrix and complex variable analysis that support the book but are not part of it. In Chapter 1, a brief history of control is traced from the ancient beginnings of process control to the contributions of flight control and electronic feedback amplifiers. Our intent is to help the student appreciate where this field came from and to introduce the names of those pioneers who contributed to getting it to the present state of development. Chapter 2 is a review of writing equations of motion for dynamic systems and solving for the dynamic response. Emphasis is on mechanical, electrical, and electromechanical devices. This material can be omitted if the students are well prepared or it can be used for review to smooth out the usual nonuniform preparation of students in the course. At a minimum, the material on the effects of extra zeros and poles on the transient response should be covered. Chapter 3 introduces feedback in an elementary setting to permit concentration on the essential ideas of using feedback to reduce tracking errors, improve disturbance rejection, and reduce sensitivity to model errors. PID controllers, anti-windup circuits, and the idea of system type are treated here.

Following the overview of feedback, the core of the book presents the design methods based on root locus, frequency response, and state feedback in Chapters 4, 5, and 6 respectively. In Chapter 7, the three approaches are integrated in several case studies and a framework for design is described that includes a touch of the real-world context of practical control design. The final Chapter 8 is a brief summary of the main issues in the design of digital controllers and is a transition to this important branch of control design.

In a ten-week quarter, it is possible to cover the nonboxed sections of Chapters 3, 4, 5, and 6 but the pace must be brisk. For those on the semester calendar, several lectures can be devoted to review of dynamic response and some of the boxed sections can be treated, such as the material on analysis of zero-memory nonlinearities, by describing functions and the root locus. It should also be possible to include treatment of one of the case studies and to introduce a computer-aided design tool. Finally, a two-quarter course can cover the first five chapters in the first quarter and Chapter 6 (state-space methods) and Chapter 7 (case studies) in the second quarter with computer-aided analysis and design throughout. This is how we use the book at Stanford; we do not cover Chapter 8 as we have later courses in digital control for which another book is used.

A word is in order respecting computer aids. Many of the original root locus plots in the book were generated with programs written in BASIC at Stanford some time ago. In recent years we have used and made available to the students PC-MATLAB and the Control Toolbox from The Mathworks. We also used CTRL-C, a product of Systems Control Technology, Inc. for many of the calculations and plots in the book. The nonlinear simulations were done with SIMNON, a product of Prof. K. J. Åström's Department of Control at Lund Institute of Technology. Many other computer-aided control system design packages exist, including MATRIX$_x$ from Integrated Systems, Inc. and CC from Systems Technology, Inc. It is our assumption that any engineer doing control design today will have access to one or more of these tools or their equivalent but it is also our firm belief that to use these tools effectively the engineer must understand the details of the methods being used so that the results from the computer can be evaluated and checked for reasonableness by independent analysis. The First Law of Computers for engineers remains "Garbage In, Garbage Out."

Finally, we wish to acknowledge our great debt to all those who have contributed to the development of feedback control into the exciting field it is today and specifically to the considerable help and education we have received from our students and our colleagues. In particular we have benefitted in this effort by many discussions with others who have taught introductory control at Stanford: Profs. A. E. Bryson, Jr., R. H. Cannon, Jr., D. B. Debra, and S. Rock. We also appreciate the comments and help of Dr. M. Anderson, Dr. T. Trankle, T. Iwata, H. Aghajan, Prof. M. Rabins, Prof. S. Desa, Prof. J. Chiasson, David M. Ward, Peter Sherman, and H. Lipkin. Smooth production of the book was again managed by Tom Robbins, Karen Myer, and the Addison-Wesley staff.

Stanford, California   G. F. F.
                                                                J. D. P.
                                                                A. E.-N.

Contents

1

An Overview and Brief History of Feedback Control 1

2

Dynamic Models and Dynamic Response 17

3

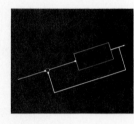

Basic Principles of Feedback 107

4

The Root-Locus Design Method 155

5

The Frequency-Response Design Method 243

6

State-Space Design 361

7

Control-System Design: Principles and Case Studies 491

Digital Control 585

1

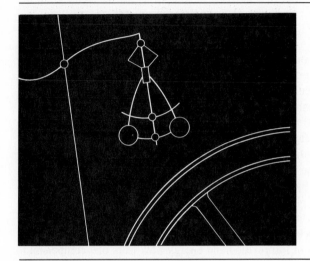

An Overview

and Brief History

of Feedback

Control

1.1 Introduction

Control is a very common concept. The term is used to refer to certain specific human-machine interactions, as in driving an automobile, where it is necessary to control the vehicle if one is to arrive safely at a planned destination. Such systems are called manual controls. Automatic control involves machines only, as in room-temperature control, where a furnace is turned on and off depending on a thermostat reading to control the temperature in winter, and a similarly controlled air conditioner is used to control the temperature in summer. An extensive body of knowledge common to both manual and automatic control has evolved into the discipline of control systems design, the subject of this book. The list of variables subject to control is vast, being limited virtually only by one's imagination. In mechanisms such as robots, control has been applied to position, speed, and force, for example. In the chemical industry, control is applied to fluid flow and liquid level, to gas flow and gas pressure, to chemical concentrations, and to many other variables. Within the human body, blood pressure, blood sugar, cell carbon dioxide, and eye-pupil diameter are only a few of the many variables controlled by biological

1

FIGURE 1.1
Component block
diagram of a
room-temperature
control system.

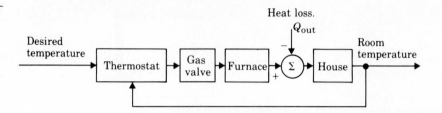

mechanisms that are equivalent to automatic control and can be studied by the methods of feedback control. The occurrences of control are very widespread, both in nature and as a result of engineering design, and the methods of analysis and design required of control engineers can be very useful indeed.

Especially important is the special class of control systems composed of those that use feedback. Characteristic of this class, the variable being controlled— whether temperature, speed, or whatever—is measured by a sensor, and the information is fed back to the process to influence the controlled variable. The principle is readily illustrated by a very common system, the household furnace controlled by a thermostat. A picture of the components is suggested by the sketch in Fig. 1.1. We call such a picture a *component block diagram*. It identifies the major components as blocks in the system, omits details, and, without math, shows the major directions of information and energy flow from one component to another.

A qualitative analysis of the operation of this system is readily done. Suppose that when the power is applied, both the temperature in the home where the thermostat is located and the outside temperature are significantly below the desired temperature. The thermostat will be *on*, and the control logic will transmit power to the furnace gas valve, which will open, causing the furnace to fire, the blower to run, and heat to be supplied to the house. If the furnace is properly designed, the input heat Q_{in} will be much larger than the heat loss Q_{out}, and the room temperature will rise until it exceeds the thermostat setting by a small amount. At this time, the furnace will be turned off, and room temperature will start to fall toward the outside value. When it falls a degree or so below the set point of the thermostat, the thermostat comes on again, and the cycle repeats. From this example, we can identify the generic components of the elementary feedback control system. These are shown in Fig. 1.2.

The central component in Fig. 1.2 is the *process* or plant,* one of whose variables (temperature) is to be controlled. In our illustrative example, the "process" is the house. The output is the house temperature, and the disturbance signal is the flow of heat from the house due to conduction through the walls, roof, windows, and

---

*As many control problems are involved in factories, or "plants," the term is often used as a generic name of the object to which control is applied.

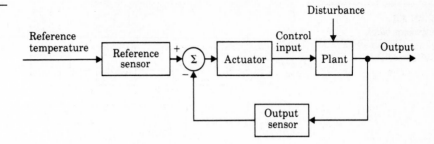

FIGURE 1.2
Component block
diagram of elementary
feedback control.

doors to the lower outside temperature. (The outward flow of heat also depends on wind, open doors, and so on.) The *actuator* is the device that can influence the controlled variable of the process; in our case, the actuator is the gas furnace. Actually, the furnace has a pilot light, which usually involves feedback; a gas valve, which also involves feedback; and a blower fan with various nonfeedback controls to time its on and off cycles, based on efficient operation of the system. These details are mentioned to illustrate that many feedback systems contain components that themselves form other feedback systems.* The component we labeled "thermostat" in Fig. 1.1 has been divided into three parts in Fig. 1.2. These are the *reference* and *output sensors* and the *comparator* (the summation symbol). For purposes of feedback control, one needs to measure the output variable (house temperature), to sense the reference variable (desired temperature), and to compare the two.† Our purpose in this book is to present methods for the analysis of feedback control systems and their components and to present the most important classes of design techniques the engineer can use with confidence in applying feedback to solve problems of control. We will also study the specific advantages of feedback that compensate for the additional complexity it demands.

1.2 A First Analysis of Feedback

The value of feedback can readily be demonstrated by quantitative analysis of a simplified model of a familiar system, the cruise control of an automobile. A component block diagram of the system is drawn in Fig. 1.3. In order to analyze

*Jonathan Swift (1733) said it this way: "So, Naturalists observe, a flea Hath smaller fleas that on him prey; And these have smaller still to bite 'em; And so proceed, ad infinitum."

†It is sometimes not possible for the controlled variable and the sensed variable to be the same. For example, while we wish to control the house temperature, the thermostat is in one particular room that may or may not be the same temperature as the rest of the house. If you work in the study, while the thermostat is in the living room near a roaring fireplace, you can freeze with the desired temperature set at 75°!

FIGURE 1.3
Component block
diagram of automobile
cruise control.

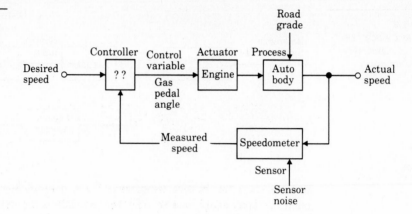

this system, we need a set of mathematical relations among the variables of our system, a set that is called a model of the system. For this example, we will ignore the dynamic response of the car and consider only the static behavior; dynamics will play a major role in our later studies. Furthermore, we will assume that for the range of speeds to be used by the system, the relations may be approximated as linear. For the vehicle, we measure speed on a level road at 55 miles per hour (mph) and find that a unit change in our control, which is the gas pedal angle, causes a 10 mph change in speed. When the grade changes by 1%, we measure a speed change of 5 mph. The speedometer is found to be accurate to a fraction

FIGURE 1.4
Functional block diagram
of cruise control.

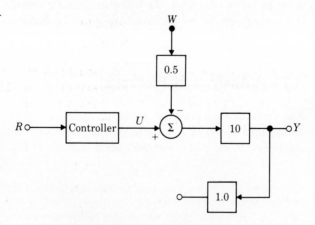

R=reference speed, miles per hour (mph)
U=gas pedal angle, degrees
Y=actual speed, miles per hour
W=road grade, percent

FIGURE 1.5
Open-loop cruise control.

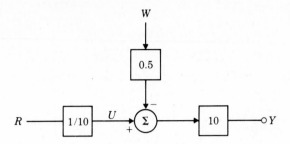

of 1 mph and can be considered exact. With these relations, we can draw the *functional block diagram* shown in Fig. 1.4. The idea of this block diagram is to represent specific mathematical relationships. The lines are like wires that carry signals, and the block is like an amplifier that multiplies the signal at its input by the number or gain marked in the block to give the output signal. To sum two or more signals, we show lines for the signals coming into a circle with a Σ inside and algebraic signs beside each to show if it is added with a plus or a minus sign to give the output of the summer. In Fig. 1.4, no relation is given for the controller; we wish to compare the effects of a 1% grade when the speed is set for 55 with and without feedback in the controller. In the first case, the controller does not use the speedometer reading and sets $U = R/10$, as shown in Fig. 1.5. This type of control is called "open-loop" because there is no closed path or loop around which signals can go in the topology of the block diagram. In this case, the output speed is given by

$$Y_{ol} = 10\,(U - 0.5W)$$
$$= 10\,[(R/10) - 0.5W]$$
$$= R - 5W.$$

If $W = 0$, a level road, and $R = 55$, then the speed will be 55, and there is no error. However, if $W = 1$, a 1% grade, then the speed will be 50, and we have a 5 mph error in the speed. Let's compare this with a feedback scheme, as shown in Fig. 1.6. In this case, the equations are

$$Y_{cl} = 10U - 5W$$
$$U = R - 0.9Y_{cl}$$
$$Y_{cl} = 10R - 9Y_{cl} - 5W$$
$$Y_{cl} = R - W/2.$$

Thus, we see that in this case, if the input set-point $R = 55$, and the grade $W = 1$, then the speed will be 54.5, only a 0.5 mph error. The effect of feedback has been to reduce the speed error by a factor of 10!

FIGURE 1.6
Closed-loop cruise
control.

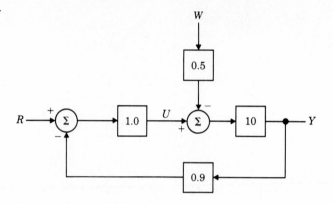

If we had included a gain factor in the controller greater than 1, the improvement in the error would have been even greater. Unfortunately, there are limits to
how high this gain can be made, both because of limits in the power of the engine
and, more importantly, because when dynamics are introduced, the feedback can
make the response worse than before. The dilemma is illustrated by a familiar
situation where it is easy to change a loop gain. If one tries to raise the gain of
a public-address amplifier too much, the sound system will squeal in a most un-
pleasant way. The gain in the feedback loop from the speakers to the microphone
through the amplifier and back to the speakers is too much. Trying to get the gain as
large as possible, to reduce the errors without making the system become unstable
and squeal, is what feedback control is all about.

1.3 A Brief History

An interesting history of early work on feedback control has been written by Mayr
(1970). He traces the control of mechanisms to antiquity and describes some of
the earliest examples—the control of flow rate to regulate a water clock, and the
control of liquid level in a wine vessel, which is thereby kept full regardless of how
many cups are dipped from it. Actually, the control of fluid flow rate is reduced
to the control of fluid level, since a small orifice will produce constant flow if the
pressure is constant, which is the case if the level of the liquid above the orifice
is constant. The mechanism of the liquid-level control invented in antiquity, and
still used for level control, is a float valve similar to that in the water tank of the
ordinary flush toilet. The float is arranged so that, as the water level falls, the flow
into the tank increases; and as the level rises, the flow is reduced and, if necessary,
cut off. A sketch of such a device is seen in Fig. 1.7. Notice that here sensor and
actuator are combined in one device, the carefully shaped float and supply-tube
combination.

FIGURE 1.7
Control of liquid level
and flow as practiced by
the ancients.

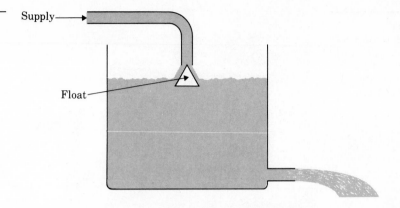

A more recent invention described by Mayr that uses feedback control is the temperature regulator of a furnace used to heat an incubator, a system designed by Drebbel about 1620. A sketch is shown in Fig. 1.8. The furnace consists of a box to contain the fire, with a flue at the top, fitted with a damper. Inside the fire box is the double-walled incubator box, the hollow walls filled with water to transfer the heat evenly to the incubator. The temperature sensor is a glass vessel filled with alcohol and mercury and placed in the water jacket around the incubator box. As the fire heats the box and the water, the alcohol expands and the riser floats up, lowering the damper on the flue. If the box is too cold, the alcohol contracts, the damper is opened, and the fire burns hotter. The desired temperature is set by the length of the riser, which sets the opening of the damper for a given expansion of the alcohol.

A famous problem in the chronicles of automatic control was the search for a means to control the speed of rotation of a shaft. Much early work (Fuller, 1976) seems to have been motivated by the desire to control automatically the speed of

FIGURE 1.8
Sketch of Drebbel's
incubator for hatching
chicken eggs. *(Adapted
from Mayr, 1970.)*

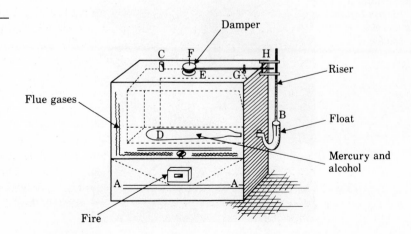

FIGURE 1.9
Photograph of a steam
engine from the shop of
J. Watt. *(British Crown
Copyright, Science
Museum, London.)*

the grinding stone in a wind-driven flour mill. Of various methods attempted, the
one with the most promise turned out to be one using a conical pendulum, or
flying-ball governor. This device was used to measure the speed of the mill; the
sails of the driving windmill were rolled up or let out via ropes and pulleys, much
like a window shade, to maintain fixed speed. But it was not the windmill that
made the flying-ball governor famous; it was its adaptation to the steam engine in
the laboratories of James Watt around 1788. A photograph of an early engine is
shown in Fig. 1.9, a close-up of a flying-ball governor in Fig. 1.10, and a sketch
showing the operation of the device in Fig. 1.11.

FIGURE 1.10
Watt steam engine (1789-
1800) with centrifugal
governor. *(British Crown
Copyright, Science
Museum, London.)*

FIGURE 1.11
Sketch of the flying-ball governor

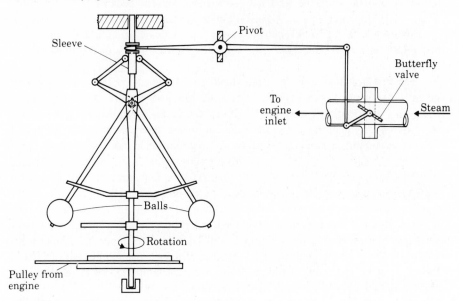

The action of the centrifugal governor is simple to describe. Suppose the engine is operating in equilibrium when a load is suddenly applied. At this point, engine speed will slow, and the balls of the governor will drop to a smaller cone. Thus, the ball angle is used to sense the output speed. This action, through the levers, will open the main valve to the steam chest (the actuator) and admit more steam to the engine, restoring most of the lost speed. To hold the steam valve at a new position, it is necessary for the fly balls to rotate at a different angle, implying that the speed under load is not exactly the same as before. To recover exactly the same speed in the system, it would be necessary to reset the desired speed setting by changing the length of the rod from the lever to the valve. Other inventors introduced mechanisms that integrated the speed error and thus provided automatic reset. An analysis of these systems is presented in Chapter 3, where it will be shown that integral action can cause feedback systems to have zero steady-state error to constant disturbances.

Watt was a practical man, like the millwrights before him, and he did not engage in theoretical analysis of the governor. The early development of control theory has been traced by Fuller (1976) from studies of Huygens in 1673 to Maxwell in 1868. Fuller gives particular credit to the contributions of G. B. Airy, who was professor of mathematics and astronomy at Cambridge University from 1826 to 1835 and Astronomer Royal at Greenwich Observatory from 1835 to 1881. Airy was concerned with speed control; if his telescopes could be rotated counter to the rotation of the earth, a fixed star could be observed for extended periods. He used the centrifugal-pendulum governor and discovered that it was capable of unstable motion—"and the machine (if I may so express myself) became perfectly wild" (Airy, 1840; quoted in Fuller, 1976, 1). Fuller attributes to Airy the first discussion of instability in a feedback control system, the analysis of such a system using differential equations, and hence the beginnings of the study of the dynamics of feedback control.

The first systematic study of the stability of feedback control was apparently given in the paper "On Governors" by J. C. Maxwell (1868). An exposition of Maxwell's contribution is given in Fuller (1976). In this paper, Maxwell developed the differential equations of the governor, linearized them about equilibrium, and stated that stability depends on the roots of a certain (characteristic) equation having negative real parts. Maxwell attempted to derive conditions on the coefficients of a polynomial that held if all the roots had negative real parts. He was successful only for second- and third-order cases. The problem of determining criteria for stability was set as the problem for the Adams Prize of 1877, which was won by E. J. Routh.* His criterion, developed in his prize-winning essay, remains of sufficient interest that control engineers still learn how to apply his simple technique. Analysis of the characteristic equation remained the foundation of control theory until the invention of the electronic feedback amplifier by H. S. Black in 1927 at Bell Telephone Laboratories. As an aside, we should note here that shortly after the publication of Routh's work, the Russian mathematician A. M. Lyapunov (1892) began studying the question of stability of motion. His studies, based on the nonlinear differential equations of motion, also included results for linear equations that are equivalent to Routh's criterion. His work was fundamental to what is now called the state-variable approach to control theory, but it was not introduced into the control literature until about 1958.

The development of the feedback amplifier is briefly described in an interesting article based on a talk by H. W. Bode (1960), reproduced in Bellman and Kalaba (1964). With the introduction of electronic amplifiers, long-distance telephoning became possible in the decades following World War I. However, as distances increased, so did the loss of electrical energy in spite of the use of large-diameter wire, and more and more amplifiers were needed to replace what was lost. Unfortunately, with so many amplifiers came much distortion, as the small nonlinearities of the vacuum tubes then used in electronic amplifiers were multiplied over and over. As a solution to this problem of distortion reduction, Black proposed the feedback amplifier. As we will see in Chapter 3, it turns out that the more we wish to reduce such distortion, the more feedback we need to apply. The loop gain from actuator to plant to sensor to actuator must be made very large. As we

---

*E. J. Routh was first academically in his class at Cambridge University in 1854. J. C. Maxwell was second. In 1877, Maxwell was on the Adams Prize Committee that set the problem of stability as the topic for the year.

mentioned earlier in connection with the automobile cruise control, with high gain the feedback loop begins to sing and is unstable. Here was Maxwell's and Routh's stability problem in a different technology, where the dynamics were so complex (differential equations of order 50 being common) that Routh's criterion was not very helpful. As communications engineers familiar with the ideas of frequency response and with the mathematics of complex variables, the workers at Bell Labs turned to complex analysis. In 1932, H. Nyquist published a paper describing how to determine stability from a graphical plot of the loop frequency response. From this theory, there developed an extensive methodology of feedback-amplifier design described by Bode (1945) and still used widely in the design of feedback controls.

Simultaneously with the development of the feedback amplifier, feedback control of industrial processes was becoming standard. In this field, characterized by processes that are not only highly complex but also nonlinear and subject to relatively long time delays between actuator and sensor, there developed the practice of proportional-plus-integral-plus-derivative control—the PID controller, as described by Callender, Hartree, and Porter (1936). This technology, based on extensive experimental work and simple linearized approximations to the system dynamics, led to standard experiments suitable to application in the field and eventually to satisfactory "tuning" of the coefficients of the PID controller. Also under development at this time were devices for the guidance and control of aircraft; especially important was the development of sensors suitable for measuring aircraft altitude and speed. An interesting account of this branch of control is given in McRuer (1973).

An enormous impulse was given to feedback control during World War II. In the United States, engineers and mathematicians at the MIT Radiation Laboratory combined their knowledge to bring together not only the feedback amplifier theory of Bode and the PID control of processes, but also the theory of stochastic processes developed by N. Wiener (1930). The result was the development of a comprehensive set of techniques for the design of control mechanisms, or *servomechanisms*, as they came to be called. Much of this work was collected and published in the records of the Radiation Laboratory by James, Nichols, and Phillips (1947).

Another approach to control-system design was introduced in 1948 by W. R. Evans, who was working in the field of guidance and control of aircraft. Many of his problems had unstable or neutrally stable dynamics, which make the frequency methods difficult, and he suggested a return to the study of the characteristic equation that had been the basis of the work of Maxwell and Routh nearly 70 years earlier. However, Evans developed techniques and rules allowing one to follow graphically the paths of the roots of the characteristic equation as a parameter was changed. His method, the root locus, is suitable for design as well as for stability analysis and remains an important technique today.

During the 1950s, several authors, including Bellman and Kalman in the United States and Pontryagin in the U.S.S.R., began again to consider the ordinary differential equation (ODE) as a model for control systems. Much of this work was stimulated by the new field of control of artificial earth satellites (where the ODE is a natural form for writing the model) and was supported by digital computers, which could be used to carry out calculations that would have been unthinkable 10 years earlier. The work of Lyapunov was translated into the language of control at about this time, and the study of optimal controls, begun in the work of Wiener and Phillips during the war, was extended to optimization of trajectories of nonlinear systems based on the calculus of variations. Much of this work was presented at the first conference of the newly formed International Federation of Automatic Control, held in Moscow in 1960.* This work did not use the frequency response or the characteristic equation but rather worked directly with the ordinary differential equation in "normal" or "state" form and typically called for extensive use of computers. The methods so introduced are now often called *modern control* as opposed to the complex variable methods of Bode et al., which are termed *classical*. Of course, Maxwell, Routh, and Lyapunov had laid the foundations of the study of the differential equation in the late nineteenth century. Finally, beginning during the decade of the 1970s and continuing into the 1990s, we find a growing body of work that seeks to use the best features of each of the available techniques (Doyle and Stein, 1981). We might call these methods eclectic control.

Thus, we come to the current state of affairs, where we feel that the well-prepared engineer should understand many techniques, be able to choose the method best suited to the problem at hand, and be able to use that method to guide and verify automatic computer calculations. For more background on the history of control, a set of survey papers appears in the *IEEE Control Systems Magazine,* November 1984.

1.4   An Overview of the Book

The central purpose of this book is to provide an introduction to the most important techniques for control-system design. First, we must review (or briefly introduce) analysis techniques necessary to obtain models and compare design alternatives. In Chapter 2, we present such an introduction to model making for mechanical, electric, electromechanical, and a few other physical systems. Analysis of dynamic response is described by Laplace transforms, and the relations between time response and poles and zeros are discussed. Also described are the linearization of nonlinear models and the construction of models from experimental data.

In Chapter 3, the basic features of feedback are described. An analysis of the effects of feedback on disturbance rejection, tracking accuracy, and dynamic

---

*Optimal control gained a large boost when Bryson and Denham (1962) showed that the path of a supersonic aircraft should actually dive at one point in order to reach a given altitude in minimum time. This nonintuitive result was later demonstrated to skeptical fighter pilots in flight tests.

response is given. In the most elementary of situations, we present the essential problems of control and how they can be addressed. In Chapters 4, 5, and 6, we introduce the techniques for realizing, in more complex dynamic systems, the objectives identified in Chapter 3. These are the methods of root locus, frequency response, and state-variable pole placement. The methods are not independent; rather, they are alternatives to the same end, with differing advantages and disadvantages. The methods are fundamentally complementary, and each needs to be understood to achieve the most effective control-system design. Application of these techniques to problems of substantial complexity is discussed in Chapter 7. There the student will find all methods brought to bear simultaneously on a specific problem. Finally, in Chapter 8, we briefly introduce the problems associated with digital control, how sampling requires another analysis tool—the z transform—and how each of the design methods can, by suitable modification, be applied to design of computer-based control.

Needless to say, many topics are *not* treated here. We do not extend the methods to systems with more than one input or output, so-called multivariable control. Nor is optimal control treated. Most lamentable is the omission of real consideration of nonlinear control. However, we believe that mastery of the material here will provide a foundation of understanding upon which the student can build knowledge of these more advanced topics—a foundation strong enough to allow you to build your *own* design method in the spirit of all those who toiled to give us what we study here.

Problems

1.1 Several common feedback control systems are listed below.

a) The manual steering system of an automobile
b) The incubator of Drebbel
c) The water level controlled by a float and valve
d) Watt's steam engine with flying-ball governor
e) The automatic steering of an ocean-going ship
f) A public address system

You are to draw the component block diagram for these and identify the following for each:

- the process
- the actuator
- the sensor
- the reference input
- the controlled output
- the actuator output
- the sensor output

1.2 The thermostat in your home or office could be one of several different kinds—it might use a tube filled with liquid, a bimetallic strip, spring bellows, or an electronic controller. Describe how it works.

FIGURE 1.12
Sketch of a papermaking machine. *(From Åström, 1970, p. 192. Reprinted with permission.)*

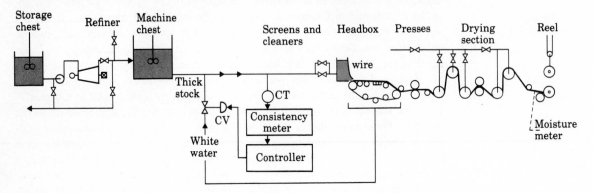

1.3 A machine for making paper is diagrammed in Fig. 1.12. There are two main parameters under feedback control; the density of fibers, as controlled by the consistency of the "thick stock" that flows from the headbox onto the wire, and the moisture content of the final product that comes out of the dryers. Stock from the machine chest is diluted by white water returning from under the "wire" as controlled by a control valve (CV). A meter supplies a reading of the consistency. At the "dry end" of the machine, there is a moisture sensor. Draw a component block diagram and identify the seven items listed in Problem 1.1 for

   a) control of consistency
   b) control of moisture

1.4 Many variables in the human body are under feedback control. For each of the following controlled variables, draw the component block diagram showing the process being controlled, the sensor that measures the variable, the actuator that causes it to increase and/or decrease, the information path that completes the feedback path, and the disturbances that upset the variable. (You may need to consult an encyclopedia or textbook on human physiology for information about this problem.)

   a) Blood pressure
   b) Blood sugar concentration
   c) Heart rate
   d) Eye pointing angle
   e) Eye-pupil diameter
   f) Blood calcium level

1.5 Draw the component block diagram for temperature control in a refrigerator.

1.6 Draw the component block diagram for an elevator-position control. Indicate how you would measure the position of the elevator car. Consider a combined "coarse" and "fine" measurement system. What accuracies do you suggest for each sensor?

1.7 For feedback control, it is necessary to sense the variable being controlled. Because of the ease with which electrical signals are transmitted, amplified, and generally processed, it is often the case that we want our sensor to deliver as output a voltage proportional to the variable being measured. Devise the principles of operation, and sketch a diagram suitable to explain the operation of a sensor that would measure

a) Temperature
b) Pressure
c) Liquid level
d) Flow of liquid along a pipe (or blood along an artery)
e) Linear position
f) Rotational position
g) Linear velocity
h) Rotational speed
i) Translational acceleration
j) Force
k) Torque

In each case, comment on the expected dynamic range, the linearity, and the signal/noise ratio of your device.

1.8 Each of the variables listed in Problem 1.7 can be brought under feedback control. Describe alternative actuators that can influence each variable. Comment on the range of power levels available in each case.

2

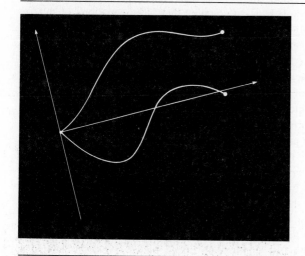

Dynamic

Models

and Dynamic

Response

2.1 Introduction

The goal of feedback control is to use the principle of feedback to cause the output variable of a dynamic process to follow a desired reference variable accurately in spite of the path the reference variable may take, the external disturbances, and any changes in the dynamics of the process. Before design of feedback can begin, however, the control engineer must construct a mathematical model of the system to be controlled and must understand the dynamic responses expected from the system to be able to see how a trial design might match the desired performance. Thus, two of the foundation skills of feedback-control engineering are the ability to model dynamic systems and the ability to obtain a quick approximate analysis of dynamic responses.

In many cases, the modeling of complex systems is difficult, expensive, and time consuming, especially when the important steps of experimental verification are included. In this introductory text, we review the most basic principles of modeling for the most common physical systems and refer the engineer to more comprehensive sources such as Cannon (1967) and to specialized texts such as

Elgerd (1982), Etkin (1959), and Buckley (1964) for further details. For our purposes, we will introduce model making by writing equations of motion for simple mechanical, electrical, and electromechanical systems. We will also present results from a few other fields of interest in control, including fluid flow and heat exchange. As a final topic in modeling, we will discuss some of the techniques used to build a model based on experimental data. Experimental techniques are especially useful when the process is so complex that basic equations of motion are too complicated or simply not available. Experimental methods are also useful in checking the validity of models obtained by direct means.

Once a model is obtained in the form of dynamic equations, usually ordinary differential equations, the next step is to solve the equations to learn the nature of the response of the system. In general, solving dynamic equations is very difficult because most realistic models are time-varying, nonlinear, and of high order. However, in many control problems we are often concerned with keeping the system at an equilibrium point and can thus restrict our study to the response of the system to small signals, all near this equilibrium. For this purpose, we can approximate the nonlinear behavior by a linear model for small signals around the selected equilibrium. Furthermore, the time variations are often very slow when compared with the speed of the signals of interest, and we can consider the system parameters to be constant for the initial analysis. For these reasons, we devote our attention in this first course to the study of the response of systems that are linear and time-invariant. In addition to small-signal linear models, we show how it is sometimes possible to obtain a linear model for design purposes by using part of the control effort to counteract the nonlinear terms in the model. This approach is widely used in the field of rigid-link robotics, where the technique is called "computed torque."

We study dynamic response within three domains. These three are the s plane, the frequency response, and the state-space domains. The basic mathematical tool of the s plane is the Laplace transform. A summary of this theory is given in Section 2.7, and a useful table of Laplace transforms and properties are on the inside back cover to this book. The main role of the s plane and the Laplace transform is to transform the linear constant ordinary differential equations of our process model to algebraic equations, in the complex variable s. From the solution to these algebraic equations, we can find the transform of the signals of interest and with the transform we can compute poles and zeros that are associated with the time-domain solution to the differential equations. By study of special cases, we develop insight and intuition as to how these poles and zeros correspond to the time responses of interest, and, thus, guide the design in the algebraic domain of the s plane rather than always requiring the complicated time-domain solutions. The basic s plane design technique is the root locus. The frequency domain is a closely related idea but associated with sinusoidal signals only. The amplification and phase shift that a system imparts to an input signal that is sinusoidal are easily measured in the laboratory and are often more easily visualized as a physical concept when compared to the more abstract complex s plane representation. The state-space domain follows from a standard construction of the ordinary differential equations as a simultaneous set of first-order equations. This form permits introduction of matrix notation and, for linear models, the extensive algebra of linear matrix equations can be used to help understand the dynamic response.* The state domain is particularly well suited to computer-aided analysis and design and for study of systems with more than one input or more than one output. We feel that these three domains — s plane, frequency response, and state-space — are three different directions from which to view, to understand, and to design control systems. The well-prepared control engineer needs to be familiar with each of them.

2.2 Dynamics of Mechanical Systems

The cornerstone for obtaining a mathematical model, or the *equations of motion*, for any mechanical system is Newton's law,

$$\mathbf{F} = m\mathbf{a}, \tag{2.1}$$

where

$\mathbf{F}$ = vector sum of all forces applied to each body in a system, Newtons (N) or pounds (lb),

$\mathbf{a}$ = vector acceleration of each body with respect to an "inertial" reference frame (i.e., one that is not accelerating or rotating), m/s^2 or ft/s^2,

m = mass of the body, kg or slug.[†]

Application of this law typically involves defining convenient coordinates to account for the body's motion — position, velocity, and acceleration — determining the forces on the body using a free-body diagram, and then writing the equations of motion from Eq. (2.1). The procedure is simplest when the coordinates chosen express the position with respect to an inertial frame because, in this case, the accelerations needed for Newton's law are simply the second derivatives of the position coordinates.

An example should help clarify ideas. Suppose we have a cart, as shown in Figure 2.1. For simplicity, we assume that the rotational inertia of the wheels is negligible and that there is friction retarding the motion of the cart, friction that is proportional to the cart's speed. The cart can then be approximated for modeling purposes using the free-body diagram seen in Fig. 2.2, which defines coordinates, shows all forces on the body (heavy lines), and indicates the acceleration (dashed

---

*A summary of background results in matrix linear algebra is given in Appendix C.

[†] In English units, we commonly refer to the "weight" of an object in units of pounds. To obtain the mass in slugs, divide the weight by g (i.e., 32.2 ft/s^2). Therefore, an object weighing 1 lb has a mass of 1/32.2 slugs. A slug has dimensions lb · s^2/ft. Unfortunately, in English units, we use pounds both for force and for mass so there is frequent confusion. In MKS units, one Newton will impart an acceleration of 1 m/s^2 to 1 kilogram.

FIGURE 2.1
Example system.

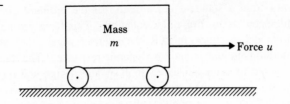

lines). The coordinate of the cart's position, x, is the distance from the reference line shown and chosen to be positive to the right. Note that the inertial acceleration is simply the second derivative of x ($\mathbf{a} = \ddot{x}$) in this case because the cart position is measured with respect to an inertial reference. The equation of motion is found using Eq. (2.1). The friction force acts opposite the direction of motion and, therefore, is drawn opposite the direction of positive motion and entered as a negative force in Eq. (2.1). The result is

$$u - b\dot{x} = m\ddot{x}, \qquad (2.2)$$

or

$$\ddot{x} + \frac{b}{m}\dot{x} = \frac{u}{m}.$$

Another example is shown in Fig. 2.3. It consists of two masses connected by a spring and a damper. The force from the spring acts on both masses in proportion to their relative displacement, while the damper yields a force on each mass proportional to its relative velocity. Figure 2.4 shows the free-body diagram of each mass. Note that the forces from the spring on the two masses are equal in magnitude but occur in the opposite direction, and likewise for the damper. Note that a positive displacement of mass m_2 will result in a force from the spring in the directions shown. However, as indicated by the minus sign in the spring force term, a positive displacement of mass m_1 will result in a force from the spring opposite the direction shown. Applying Eq. (2.1) to each mass and noting that the forces on the right mass are in the negative direction yields

$$u + b(\dot{y} - \dot{x}) + k(y - x) = m_1\ddot{x}$$
$$-k(y - x) - b(\dot{y} - \dot{x}) = m_2\ddot{y}, \qquad (2.3)$$

FIGURE 2.2
Free-body diagram
and position coordinate.

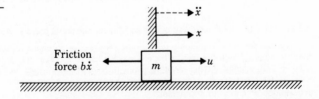

FIGURE 2.3
Resonant system
example.

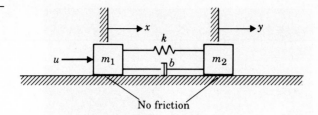

No friction

or, with some rearranging,

$$\ddot{x} + \frac{b}{m_1}(\dot{x} - \dot{y}) + \frac{k}{m_1}(x - y) = \frac{u}{m_1}$$

$$\ddot{y} + \frac{b}{m_2}(\dot{y} - \dot{x}) + \frac{k}{m_2}(y - x) = 0. \tag{2.4}$$

Application of Newton's law to one-dimensional rotational systems requires that Eq. (2.1) be modified to

$$M = I\alpha, \tag{2.5}$$

where

M = sum of all moments about the center of mass of a body, N · m or lb · ft,

I = body's moment of inertia about its center of mass, kg · m² or slug · ft²,

α = angular acceleration of the body, rad/s².

In the special case where a point in the body is fixed with respect to an inertial reference, as is the case with a pendulum, Eq. (2.5) can be applied with M being the sum of all moments about the *fixed* point and I being the moment of inertia about the fixed point.

An example of the application of these equations is found in the model of attitude motion of a satellite, such as that shown in Fig. 2.5. Satellites usually require attitude control so that antennas, sensors, and solar panels are properly oriented. For example, antennas are usually pointed toward a particular location on earth while solar panels need to be oriented toward the sun for maximum power generation.

FIGURE 2.4
Forces on each mass
for resonant system.

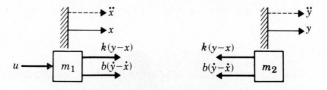

FIGURE 2.5
Communications
satellite. *(Courtesy
Ford Aerospace
& Communications
Corp.)*

To gain insight into the full three-axis attitude-control system, we often consider one axis at a time. Figure 2.6 depicts this case, where motion is only allowed about the axis perpendicular to the page. The angle θ that describes the satellite orientation must be measured with respect to an "inertial" reference, that is, a reference that has no angular acceleration. The control force comes from reaction jets that produce a moment ($F_c d$) about the mass center. There may also be small disturbance moments on the satellite (M_D), primarily arising from solar pressure acting on any asymmetry in the solar panels. Applying Eq. (2.5) yields the equation of motion,

$$F_c d + M_D = I\ddot{\theta}. \tag{2.6}$$

If this system is controlled by measuring θ from the antenna signal and that signal is used to drive the control jets, there is the possibility of some flexibility in the structure between the measured θ and the applied control force. Schematically, the dynamic model for this common situation is as shown in Fig. 2.7. It is the rotational version of the resonant system shown in Fig. 2.3 and results in equations of motion identical in form to Eq. (2.4). The moments on each body are shown in the free-body diagrams in Fig. 2.8.* When summed according to Eq. (2.5), the result is

$$I_1\ddot{\theta}_1 + b(\dot{\theta}_1 - \dot{\theta}_2) + k(\theta_1 - \theta_2) = M_c + M_D$$
$$I_2\ddot{\theta}_2 + b(\dot{\theta}_2 - \dot{\theta}_1) + k(\theta_2 - \theta_1) = 0. \tag{2.7}$$

A two-dimensional mechanical system that consists of both translational and rotational portions is the hanging crane shown in Fig. 2.9. The free-body diagrams

---

*Note that the moment from the spring is drawn in the same direction on each inertia but the moments are labeled to have opposite signs. This is equivalent to the situation in Fig. 2.4, where the forces were drawn in opposite directions and labeled to have equal signs.

FIGURE 2.6
Satellite control
schematic.

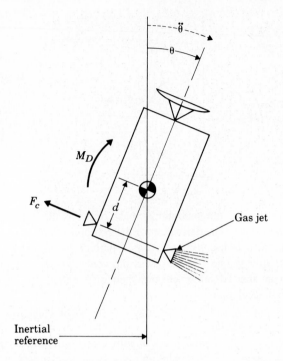

are shown in Fig. 2.10. In the case of the pendulum, the forces are shown with bold lines while the components of the inertial acceleration of its center of mass are shown with dashed lines. The inertial acceleration needs to be determined because the **a** in Eq. (2.1) is with respect to an inertial reference. The total inertial

FIGURE 2.7
Rotational resonant
system.

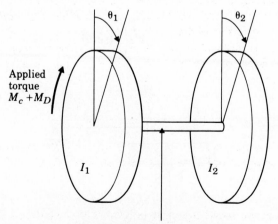

Rotational spring with constant k and
rotational damper with constant b

FIGURE 2.8
Free-body diagrams of
the rotational resonant
system.

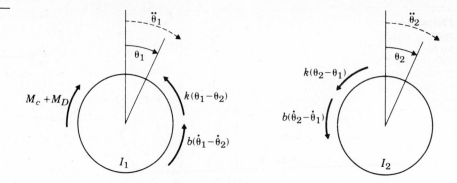

acceleration of the pendulum's mass center is the vector sum of the three dashed arrows shown. The derivation of the components of an object's acceleration, called *kinematics*, is usually studied as a prelude to the application of Newton's Laws. Alternatively, one could express the center of mass of the pendulum as a vector from an inertial reference and then differentiate that vector twice to obtain an inertial acceleration.

Using the results of a kinematic study, we find that the component of acceleration along the pendulum is $l\dot{\theta}^2$ and is called the "centripetal" acceleration. It is present for any object whose velocity is changing direction. The $\ddot{x}$ component of acceleration is a consequence of the pendulum pivot point accelerating at the trolley's acceleration and will always have the same direction and magnitude as the trolley. The $l\ddot{\theta}$ component is a result of angular acceleration of the pendulum and is always perpendicular to the pendulum. The total inertial acceleration will automatically be the vector sum of the three acceleration components shown in Fig. 2.10(b).

In order to understand the $l\ddot{\theta}$ and $l\dot{\theta}^2$ terms better, consider the situation in Fig. 2.10(c), where the $\hat{\mathbf{i}}$ and $\hat{\mathbf{j}}$ axes are inertially fixed. A vector $\mathbf{r}$ describing the position of the pendulum center of mass can be expressed as

$$\mathbf{r} = l(\sin\theta\hat{\mathbf{i}} - \cos\theta\hat{\mathbf{j}}).$$

FIGURE 2.9
Crane with hanging load.

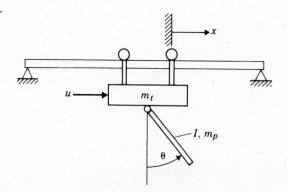

FIGURE 2.10
Hanging crane:
(a) trolley free-body
diagram; (b) pendulum
free-body diagram;
(c) pendulum position
vector.

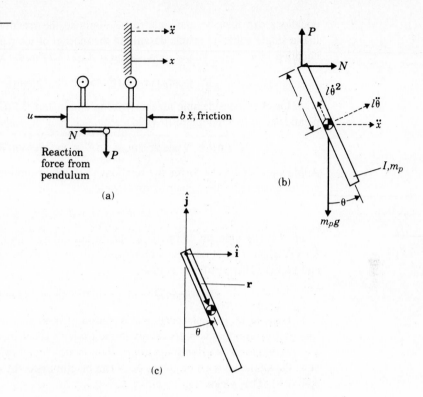

The first derivative of $\mathbf{r}$ is

$$\dot{\mathbf{r}} = l\dot{\theta}(\cos\theta\hat{\mathbf{i}} + \sin\theta\hat{\mathbf{j}}).$$

Likewise, the second derivative of $\mathbf{r}$ is

$$\ddot{\mathbf{r}} = l\ddot{\theta}(\cos\theta\hat{\mathbf{i}} + \sin\theta\hat{\mathbf{j}}) - l\dot{\theta}^2(\sin\theta\hat{\mathbf{i}} - \cos\theta\hat{\mathbf{j}}).$$

Note that the $l\dot{\theta}^2$ term is aligned along the pendulum pointing toward the axis of rotation and that the $l\ddot{\theta}$ term is aligned perpendicular to the pendulum pointing in the direction of a positive rotation.

Having all the forces and accelerations for the two bodies, we proceed to apply Eq. (2.1). In the case of the trolley, Fig. 2.10(a), we see that it is constrained by the tracks to move only in the x direction; therefore, application of Eq. (2.1) in this direction yields

$$m_t\ddot{x} + b\dot{x} = u - N. \tag{2.8}$$

Conceptually, Eq. (2.1) can be applied to the pendulum of Fig. 2.10(b) in the vertical and horizontal directions, and Eq. (2.5) can be applied for rotational motion to yield three equations in the three unknowns: N, P, and θ. These three equations can then be manipulated to eliminate the reaction forces N and P so that a single equation results describing the motion of the pendulum, i.e., a single equation in θ. For example, application of Eq. (2.1) in the x direction yields

$$N = m_p\ddot{x} + m_p l\ddot{\theta}\cos\theta - m_p l\dot{\theta}^2\sin\theta. \qquad (2.9)$$

However, considerable algebra will be eliminated if Eq. (2.1) is applied perpendicular to the pendulum to yield

$$P\sin\theta + N\cos\theta - m_p g\sin\theta = m_p l\ddot{\theta} + m_p\ddot{x}\cos\theta. \qquad (2.10)$$

Application of Eq. (2.5) for the rotational pendulum motion where the moments are summed about the center of mass yields

$$-Pl\sin\theta - Nl\cos\theta = I\ddot{\theta}, \qquad (2.11)$$

where I is the moment of inertia about the pendulum's mass center. The reaction forces, N and P, can now be eliminated relatively easily by combining Eqs. (2.10) and (2.11). This yields the equation

$$(I + m_p l^2)\ddot{\theta} + m_p gl\sin\theta = -m_p l\ddot{x}\cos\theta. \qquad (2.12)$$

It is identical to a simple pendulum equation of motion, except that it contains a forcing function that is proportional to the trolley's acceleration.

An equation describing the trolley motion was found in Eq. (2.8), but it contains the unknown reaction force, N. N can be eliminated by combining Eqs. (2.9) and (2.8). This yields

$$(m_t + m_p)\ddot{x} + b\dot{x} + m_p l\ddot{\theta}\cos\theta - m_p l\dot{\theta}^2\sin\theta = u. \qquad (2.13)$$

Equations (2.12) and (2.13) constitute the differential equations that describe the motion of the crane with its hanging load. For an accurate calculation of the motion of the system, these nonlinear equations need to be solved. However, nonlinear equations are much more difficult to solve than linear ones and the kinds of possible motions resulting from a nonlinear model are much more difficult to categorize than those resulting from a linear model. It is therefore useful to linearize models in order to gain access to linear analysis methods. It may be that the linear models and linear analysis are used only for the design of the control system (whose function may be to maintain the system in the linear region). Once a control system is synthesized and shown to have desirable performance based on linear analysis, it is then prudent to carry out an accurate analysis of the system with all the nonlinearities in order to validate that performance.

To linearize the equations of motion of the hanging crane, we could assume the pendulum has small motions about the vertical where $\cos\theta \cong 1$, $\sin\theta \cong \theta$, and $\dot{\theta}^2 \cong 0$; thus, the equations may be approximated by

$$(I + m_p l^2)\ddot{\theta} + m_p gl\theta = -m_p l\ddot{x}$$
$$(m_t + m_p)\ddot{x} + b\dot{x} + m_p l\ddot{\theta} = u. \qquad (2.14)$$

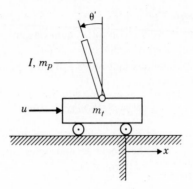

FIGURE 2.11
Inverted pendulum.

As a second case, we could assume the pendulum has small motion about a vertical *upward* direction, as is the case for the inverted pendulum shown in Fig. 2.11, in which case $\theta = \pi + \theta'$. In this case, $\cos\theta \cong -1$ and $\sin\theta \cong -\theta'$ in Eqs. (2.12) and (2.13), and they become*

$$(I + m_p l^2)\ddot{\theta}' - m_p g l \theta' = m_p l \ddot{x}$$
$$(m_t + m_p)\ddot{x} + b\dot{x} - m_p l \ddot{\theta}' = u. \tag{2.15}$$

To summarize, the physics necessary to write the equations of motion of a rigid body is given entirely by Newton's law of motion. The method for systems such as the inverted pendulum (Fig. 2.11) is as follows:

1. Assign variables such as x and θ necessary and sufficient to describe an *arbitrary* position of the object.
2. Draw a free-body diagram of each component and indicate *all* forces acting on each body and the accelerations of the center of mass with respect to an inertial reference.
3. Apply Newton's law in translation (Eq. 2.1) and rotation (Eq. 2.5) form.
4. Combine the equations to eliminate internal forces.

The combination of the equations to eliminate internal forces is sometimes expedited by an intelligent choice of directions for the application of Eq. (2.1), as illustrated by the preceding example. It often is useful to try alternate directions and to evaluate the algebra required to eliminate the internal reaction forces. Application of Newton's laws in different directions sometimes yields what appear to be quite different sets of equations; however, after manipulations, the equations can always be shown to be equivalent.

---

*The inverted pendulum is often described with the angle of the pendulum being positive for *clockwise* motion. In this case, the sign on all terms in Eq. (2.15) in θ' or $\ddot{\theta}'$ need to be reversed.

State Variable Form

In Chapter 6 we will consider control system design using matrix notation. For that approach, we will need to arrange the differential equations that describe the plant into a set of simultaneous first-order equations known as state variable form. For example, Eq. (2.2) is

$$\ddot{x} = -(b/m)\dot{x} + u/m.$$

If we define the position and the velocity as the state variables x_1 and x_2, so that $x_1 = x$ and $x_2 = \dot{x}_1$, then the single second-order equation can be written in an equivalent way as two first-order equations

$$\dot{x}_1 = x_2$$
$$\dot{x}_2 = -(b/m)x_2 + (1/m)u.$$

For more complex systems of rigid bodies, selecting the positions and velocities of the separate bodies as the state variables is usually adequate, but for any but the simplest systems, a more advanced study of dynamics is necessary (see Kane, 1985).

All the preceding examples contained one or more rigid bodies, although some were connected to others by springs. Actual structures—for example, satellite solar panels, airplane wings, or robot arms—usually bend; Figure 2.12(a) shows a very simple flexible beam. The equation describing its motion is a fourth-order partial differential equation:

$$EI\frac{\partial^4 w}{\partial x^4} + \rho\frac{\partial^2 w}{\partial t^2} = 0,$$

where

   $\mathbf{E}$ = Young's modulus
   $\mathbf{I}$ = beam cross-sectional moment of inertia
   ρ = beam density
   $\mathbf{w}$ = beam deflection.

The exact solution to this equation is too cumbersome for use in the design of control systems, but it is often important to account for the gross effects of bending in control-system design.

The continuous beam in Fig. 2.12(b) has an infinite number of vibration mode shapes, all with different frequencies. Typically, the lowest-frequency modes have the largest amplitude and are the most important to approximate well. The simplified model in Fig. 2.12(c) can be made to duplicate the essential behavior of the first bending mode shape and frequency and would usually be adequate for controller design. If frequencies higher than the first bending mode are anticipated in the control-system operation, it may be necessary to model the beam as shown in Fig. 2.12(d), which could be made to approximate the first two bending modes and

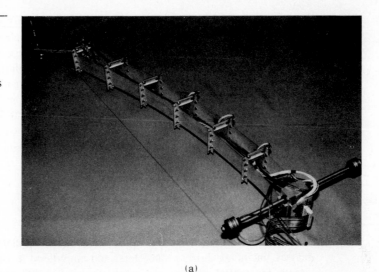

(a)

(b)   (c)   (d)

frequencies. Likewise, higher-order models could be used if such accuracy and complexity were deemed necessary (Thomson, 1981, and Schmitz, 1985).

2.3   Models of Electric Circuits

Electric circuits consist of interconnections of sources of electric voltage and current with passive elements such as resistors, capacitors, and inductors, and with active electronic elements such as transistors and especially operational amplifiers. Electric circuits are frequently used components of feedback-control systems because of the enormous flexibility they give the designer to modify and to process electric signals. Operational amplifiers are in themselves examples of complex feedback systems, and some of the most important methods of feedback system design were developed by the designers of high-gain feedback amplifiers, mainly at the Bell Telephone Laboratories between 1925 and 1940. Electric and electronic components also play a central role in electromechanical energy conversion devices such as electric motors and generators and electrical sensors. In this brief survey, we cannot derive the physics of electricity or give a comprehensive review of all the important analysis techniques. We will define the variables, describe the relations imposed on them by typical elements and circuits, and describe a few of the most effective methods available for solving the resulting equations.

Symbols for some linear circuit elements and their current-voltage relations are given in Fig. 2.13. Passive circuits consist of interconnections of resistors, capacitors, and inductors.

With electronics, we increase the set of electrical elements by adding active devices, including diodes, transistors, and amplifiers. In signal processing and in circuits for control systems, the most common amplification device is the operational amplifier, or op amp.* The symbol of the op amp is shown in Fig. 2.14. In typical amplifiers, the zero frequency or DC gain A_0 varies from 10^5 to 10^7. We can take A_0 to be infinity with satisfactory accuracy for most control uses.

The basic equations of electric circuits are called Kirchhoff's laws, which are

Kirchhoff's current law (KCL): *The algebraic sum of all currents leaving a junction or node in a circuit is zero.*

Kirchhoff's voltage law (KVL): *The algebraic sum of all voltages taken around a closed path in a circuit is zero.*

With complex circuits of many elements, it is essential that the writing of the equations be well organized and carefully done. Of the numerous methods for doing this, we select for description and illustration the popular and powerful scheme known as *node analysis*. We consider the circuit shown in Fig. 2.15 for illustration. The first idea of node analysis is to select one node as a reference and to select the voltages of all other nodes as unknowns. The choice of reference is arbitrary as far as the method is concerned, but in active electronic circuits the common, or ground, terminal for signals is the obvious and standard choice. In the example circuit, we select node ④ as the reference and the voltages $v_①$, $v_②$, and $v_③$ as the unknowns. The second idea of node analysis is to write equations for the selected unknowns using the current law (KCL) at each node. We express these currents in terms of the selected unknowns by using the element equations.

*Heaviside introduced the mathematical operation p to signify differentiation so that pv was to be taken as dv/dt. The Laplace transform incorporates this idea, using the complex variable s. Ragazzini, Randall, and Russell (1947) demonstrated that an ideal, high-gain electronic amplifier permitted one to realize arbitrary "operations" in s, and they named such an amplifier the "operational amplifier."

FIGURE 2.13
Elements of electric
circuits.

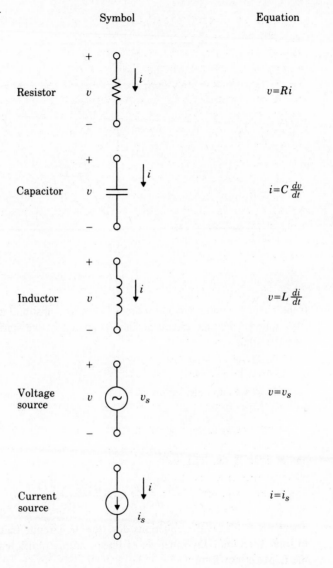

Symbol	Equation

Resistor $v=Ri$

Capacitor $i=C\frac{dv}{dt}$

Inductor $v=L\frac{di}{dt}$

Voltage source $v=v_s$

Current source $i=i_s$

FIGURE 2.14
The operational amplifier.

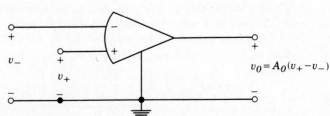

$v_O = A_O(v_+ - v_-)$

FIGURE 2.15
An example circuit,
the bridged tee.

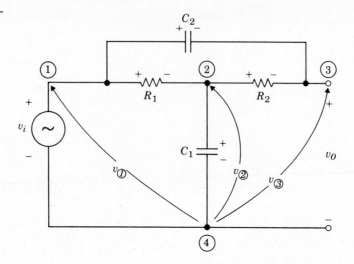

If the circuit contains voltage sources, we must substitute a voltage law (KVL) for such sources. For the circuit at hand (Fig. 2.15), we start with a trivial KVL to the effect that

$$v_① = v_i. \tag{2.16a}$$

At node ② we have the equation

$$\frac{v_② - v_①}{R_1} + \frac{v_② - v_③}{R_2} + C_1\frac{dv_②}{dt} = 0, \tag{2.16b}$$

and at node ③ the KCL is

$$\frac{v_③ - v_②}{R_2} + C_2\frac{d(v_③ - v_①)}{dt} = 0. \tag{2.16c}$$

These three differential equations describe the circuit. Later, we will discuss how to solve these and any other set of linear, ordinary differential equations by using the Laplace transform.

   As a supplementary exercise, we can again write the equations as a set of simultaneous first-order differential equations by selecting the capacitor voltages v_{C_1} and v_{C_2} as unknowns. Here $v_{C_1} = v_②$, $v_{C_2} = v_① - v_③$, and still $v_① = v_i$. Thus, $v_① = v_i$, $v_② = v_{C_1}$, and $v_③ = v_i - v_{C_2}$. In terms of v_{C_1} and v_{C_2}, Eq. (2.16b) is

$$\frac{v_{C_1} - v_i}{R_1} + \frac{v_{C_1} - (v_i - v_{C_2})}{R_2} + C_1\frac{dv_{C_1}}{dt} = 0.$$

Rearranging this equation into standard form,

$$\frac{dv_{C_1}}{dt} = -\frac{1}{C_1}\left(\frac{1}{R_1} + \frac{1}{R_2}\right)v_{C_1} - \frac{1}{C_1}\left(\frac{1}{R_2}\right)v_{C_2} + \frac{1}{C_1}\left(\frac{1}{R_1} + \frac{1}{R_2}\right)v_i. \qquad (2.17a)$$

In terms of v_{C_1} and v_{C_2}, Eq. (2.16c) is

$$\frac{v_i - v_{C_2} - v_{C_1}}{R_2} + C_2\frac{d}{dt}(v_i - v_{C_2} - v_i) = 0.$$

Again, in normal form, the equation is

$$\frac{dv_{C_2}}{dt} = -\frac{v_{C_1}}{C_2R_2} - \frac{1}{C_2R_2}v_{C_2} + \frac{v_i}{C_2R_2}. \qquad (2.17b)$$

Equations (2.16a, b, and c) are entirely equivalent to Eqs. (2.17a and b) in describing the circuit. Equation (2.17) gives the equations in state-variable form.

As another simple example of node analysis, we consider the standard op amp circuit shown in Fig. 2.16(a). Node analysis will give us an equation from KCL at node ①. This equation is

$$\frac{v_① - v_i}{R_1} + \frac{v_① - v_o}{R_2} + C\frac{d}{dt}(v_① - v_o) = 0.$$

The law of the amplifier as given in Fig. 2.14 is in this case

$$v_o = -A_0 v_①.$$

If we let $A_0 \rightarrow \infty$, then $v_① = 0$, and the equation reduces to

$$-\frac{v_i}{R_1} - \frac{v_o}{R_2} - \frac{C\,dv_o}{dt} = 0.$$

Finally, if we let $R_2 \rightarrow \infty$ (an open circuit), then we have the circuit of Fig. 2.16(b) and the equation

$$-\frac{C\,dv_o}{dt} = \frac{v_i}{R_1},$$

or

$$v_o = -\frac{1}{R_1C}\int_0^t v_i(\tau)\,d\tau + v_o(0). \qquad (2.18)$$

Thus, the ideal op amp in this circuit performs the operation of integration.

The equations of much more complicated circuits can be established by nodal analysis. However, when the equations are linear, it is more common to use the Laplace transform directly on the element equations, reducing them to algebraic form, and to solve the resulting linear algebraic equations.

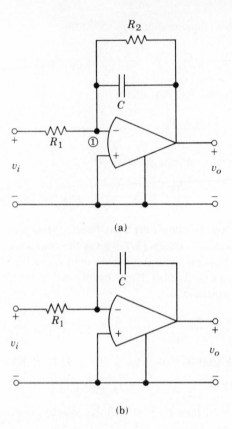

(a)

(b)

2.4 Models of Electromechanical Systems

Electric current and magnetic fields interact in many ways, three of which are important to an understanding of the operation of most electromechanical energy conversion devices, such as linear and rotary motors and electric sensors of motion. In this section, we will state these relations and sketch a few typical configurations used in control devices. For greater detail respecting electromagnetic principles as discovered by researchers such as Ampere, Faraday, Weber, Tesla, and many others, a book on physics, such as Halliday and Resnick (1970), should be consulted. For a general introduction to the application of these principles to electromechanical energy conversion devices, see Smith (1980); for a thorough treatment of DC motors, see Electrocraft (1980).

The first important relation we note between current and magnetism is that an

FIGURE 2.17
Toroid wrapped with N
turns.

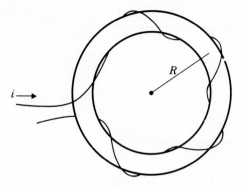

FIGURE 2.17
Toroid wrapped with N
turns.

electric current establishes a magnetic field. The strength of the field at a particular point depends on the intensity of the current, the materials, and the geometry of the situation. A typical idealized case is sketched in Fig. 2.17, where a toroid of radius R made of material with permeability μ is wrapped with N turns of wire carrying i amperes. The field strength at the center of the toroid is given by

$$B = \frac{\mu}{2\pi R} N i. \tag{2.19}$$

The important fact is that the field strength is proportional to the current strength and to the number of turns of wire around the toroid. The permeability μ is a property of the material and represents the extent to which the material aids the current in establishing the field. For air, $\mu_o = 4\pi \times 10^{-7}$ Wb/A·m. For a ferromagnetic material such as iron under typical nonsaturated conditions, the permeability can be several thousand times μ_o. The net effect of this relationship between current and magnetic field is that a strong field can be established by sending current through many turns of wire wrapped around a ferromagnetic material. If a small gap is cut in the toroid core of the scheme in Fig. 2.17, the field will be slightly reduced but will be accessible in the gap to interact with current-carrying conductors, which can be placed there. Such a device is an electromagnet, so called because the magnetic field is formed and controlled by the electric current in the wire. Some materials, known since prehistoric times, can establish a magnetic field because of the electrical properties of their constituent molecules. These are called *permanent magnets,* since the field does not come and go with an external current. In many cases, such as in small motors and generators used in many control applications, a permanent magnet may be entirely satisfactory for establishing the required magnetic field.

The second electromagnetic effect of interest to us is the fact that a charge q moving with vector velocity $\mathbf{v}$ in a magnetic field of intensity $\mathbf{B}$ experiences a force $\mathbf{F}$ given by the vector cross-product $\mathbf{F} = q\mathbf{v} \times \mathbf{B}$. If the moving charge is composed of a current of i amperes in a conductor of length l meters arranged at right angles to the field strength of B tesla, then the force at right angles to the

FIGURE 2.18
Geometry of a
loudspeaker; a form
of motor for linear
motion.

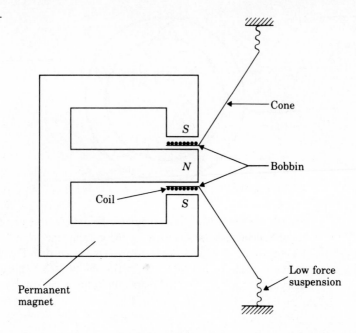

plane of i and B has the magnitude

$$F = Bli \text{ newton.} \qquad (2.20)$$

This equation is the basis of using a magnetic field to convert electric energy from
the source of current i into mechanical energy by arranging for the force F to do
work on some mechanical object. Such devices are called *motors,* and Eq. (2.20) is
the *law of motors.* For example, a loudspeaker for reproducing sound from electric
signals contains such a motor. A typical geometry is sketched in Fig. 2.18. The
permanent magnet establishes a radial field in the cylindrical gap between the poles
of the magnet. When a current is caused to flow in the wire wound on the bobbin,
the force causes the bobbin to move left or right and the cone to move against the
air. As the cone moves back and forth, the fluctuations of air pressure propagate
and are heard as sound.

   If we approximate the effects of the air as if the cone had equivalent mass M
and viscous damping coefficient d, we can write the equations of motion of the
device. Suppose the magnet establishes a uniform field of 0.5 T, and the bobbin
has 20 turns at 2-cm diameter. The current is at right angles to the field, and the
force of interest is at right angles to the plane of i and B, so Eq. (2.20) applies.
In this case, the parameter values are $B = 0.5$ T and

$$l = 20 \times \frac{2\pi}{100} m = 1.26m.$$

Thus,

$$F = 0.63i \text{ newton.}$$

The mechanical equation, as developed in Section 2.2, is

$$M\ddot{x} + d\dot{x} = 0.63i. \tag{2.21}$$

Equation (2.20) also can be used in finding the force on the rotor of a motor and in many other useful applications.

In addition to Eq. (2.20), which expresses force (a mechanical variable) in terms of current (an electrical variable), there is also a relation giving the effect of mechanical motion on electricity. The basic fact is that if a charge is moving in a magnetic field and is forced along a conductor, an electric voltage is established between the ends of the conductor. The relation is that if a conductor (which is filled with charged particles) of length l meters is moved at a velocity v meters per second through a constant field of B teslas at right angles to the direction of the field, then the voltage between the ends of the conductor is given by

$$e(t) = Blv \text{ volts.} \tag{2.22}$$

This expression is called the *law of the generator,* for it is the basis for generating electric power by moving a conductor in a field and letting e cause current to flow in an external circuit.

Considering again the situation of the loudspeaker (Fig. 2.18), we see that if motion results according to Eq. (2.21), the voltage across the coil is given by Eq. (2.22), where the velocity is $\dot{x}$. The result is

$$e_{\text{coil}} = 0.63\dot{x}. \tag{2.23}$$

In the circuit of Fig. 2.19, these two effects are both important. Assume the amplifier output resistance is R and the speaker equivalent inductance is L. The equations of motion are, for the mechanical parts,

$$M\ddot{x} + d\dot{x} = 0.63i, \tag{2.24a}$$

FIGURE 2.19
A loudspeaker showing the electric circuit.

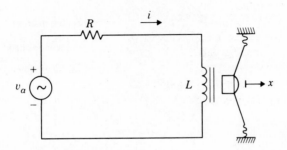

and for the electrical parts,

$$L\frac{di}{dt} + Ri = v_a - 0.63\dot{x}. \tag{2.24b}$$

For control of rotary motion, a common device is the direct-current (DC) motor. A sketch of the basic components is given in Fig. 2.20. In addition to housing and bearings, the nonturning part (stator) has magnets, which establish a field across the turning part (rotor). The magnets may be electromagnets or, for small motors, permanent magnets. The rotor is wound with wire through which current is forced via the brushes. The (rotating) commutator causes the current always to be sent through the armature, a collection of conductor windings, so that it will produce the maximum torque in the desired direction. If the direction of the current is reversed, the direction of the torque is reversed.

The underlying principles are given by the motor law (*Bli*) and the generator law (*Blv*). However, rather than give the field strength and winding geometry, it is standard to express the torque developed in the rotor in terms of the armature current and a torque constant K_t as

$$T = K_t i_a, \tag{2.25}$$

and to express the voltage generated as a result of rotation* in terms of the shaft rotational velocity $\dot{\theta}$ and an electric or emf constant K_e as

$$e = K_e \dot{\theta}. \tag{2.26}$$

In consistent units, it is true that K_t equals K_e, but in some cases a torque constant will be given in other units, such as ounce-inches per ampere, and the electric constant may be expressed in units of volts per 1000 rpm. In such cases, the engineer must make the necessary translations to be certain the equations are correct.

Using these relations, a set of equations for the DC motor with armature driven

---

*Because the generated electromotive force (emf) works against the applied armature voltage, we call it the *back emf*.

FIGURE 2.20
Sketch of a DC motor.

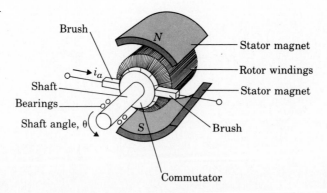

by an amplifier can be written by considering the network with back emf included for the electrical side and a free-body diagram of the rotor with motor torque included for the mechanical side. A typical electric circuit is shown in Fig. 2.21(a) and the corresponding free-body diagram in Fig. 2.21(b). In the mechanical diagram, we have assumed significant compliance in the motor shaft, which connects the motor rotor, with inertia J_m at angle θ_m, to the load, with inertia J_l at angle θ_l. We assume the twisting of the shaft can be modeled adequately as a spring of displacement torque constant k and viscous torque constant b. The electrical equation is

$$L_a \frac{d i_a}{dt} + R_a i_a = v_a - K_e \dot{\theta}, \tag{2.27}$$

and the mechanical equations are (see Eq. 2.7)

$$J_m \ddot{\theta}_m + b(\dot{\theta}_m - \dot{\theta}_l) + k(\theta_m - \theta_l) = K_t i_a$$
$$J_l \ddot{\theta}_l + b(\dot{\theta}_l - \dot{\theta}_m) + k(\theta_l - \theta_m) = 0. \tag{2.28}$$

These are the equations we will need to use when studying feedback control of DC motors.

Another device used for electromechanical energy conversion is the AC induc-

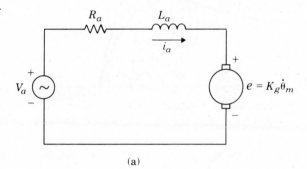

(a)

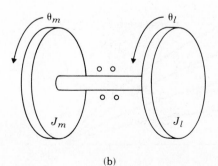

(b)

tion motor invented by N. Tesla and based on the following concept. Suppose we construct a rotor of conductors with no commutator and no brushes. Suppose the stator magnetic field can be made to rotate at high speed (perhaps 1800 rpm). The moving field will induce currents (the principle of induction) in the rotor, which will, in turn, be subject to forces according to the motor law. These forces will cause the rotor to turn so that it chases the rotating magnetic field.

Elementary analysis of the AC motor is more complex than is that of the DC motor so, rather than give the details in terms of the motor geometry, we present the experimental facts in the form of a set of curves of speed versus torque for fixed frequency and varying amplitude of applied voltage. A typical set appears in Fig. 2.22(a) for a low-rotor-resistance machine used for power applications and in Fig. 2.22(b) for a high-resistance machine typical of servomotor applications.

For analysis of a control problem involving an AC motor, such as that described by Fig. 2.22(b), we make a linear approximation to the curves for speed near zero

FIGURE 2.22
Torque-speed curves for two typical induction motors: (a) low-rotor-resistance machine; (b) high-rotor-resistance machine, with four values of armature voltage given.

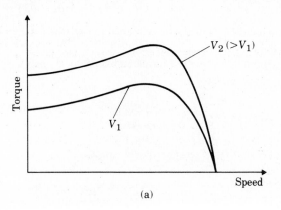

(a)

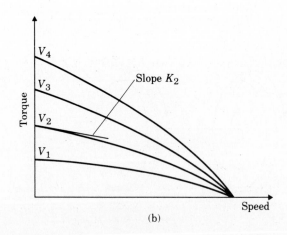

(b)

and voltage at midrange to obtain the expression

$$T = K_1 v_a - K_2 \dot{\theta}. \tag{2.29}$$

For the electrical side, experiments are performed to measure the armature resistance and inductance for inclusion in the analysis. Given K_1, K_2, R_a, and L_a, the analysis proceeds as in the case of the DC motor.

In addition to the DC and AC motors mentioned here, control systems use brushless DC motors (Electrocraft, 1980), stepping motors (Kuo, 1982), and variable reluctance motors (Kuo, 1982). The development of models for these machines is given in the works cited, and the models do not differ in principle from the motors considered above. In general, the analysis, supported by experiment, develops a function of the torque in terms of voltage and speed similar to the AC-motor torque-speed curves given in Fig. 2.22. From these curves, a linearized formula such as Eq. (2.29) is obtained for use on the mechanical side, and an equivalent circuit consisting of a resistance and a reactance is obtained for use on the electrical side.

2.5   Elementary Aspects of Other Dynamic Systems

2.5.1 Heat Flow

Some control systems involve regulation of temperature for portions of the system. The dynamic models of temperature-control systems involve the flow and storage of heat energy. Heat energy flows through substances at a rate proportional to the temperature difference across the substance; that is,

$$q = \frac{1}{R}(T_1 - T_2), \tag{2.30}$$

where

$\quad q$ = heat energy flow, J/s,
$\quad R$ = thermal resistance, $^\circ$C/J/s,
$\quad T$ = temperature, $^\circ$C.

The net heat-energy flow into a substance affects the temperature of the substance according to

$$\dot{T} = \frac{1}{C}q, \tag{2.31}$$

where

$\quad C$ = thermal capacity, J/$^\circ$C.

Typically, there would be several paths for heat to flow into or out of a substance, and the q in Eq. (2.31) would be the sum of heat flows obeying Eq. (2.30). For example, a box with all but two sides insulated ($1/R = 0$) is shown in Fig. 2.23.

FIGURE 2.23
Thermal dynamics
example.

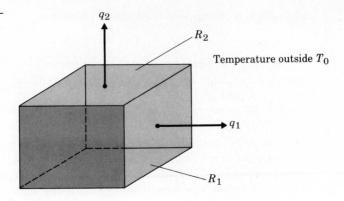

Application of Eqs. (2.30) and (2.31) yields

$$\dot{T}_I = \frac{1}{C_I}\left(\frac{1}{R_1} + \frac{1}{R_2}\right)(T_0 - T_1),$$

where

C_1 = thermal capacitance of substance within box,
T_0 = temperature outside,
T_1 = temperature inside,
R_2 = thermal resistance of box top,
R_1 = thermal resistance of box side.

Normally, the material properties are given in tables as

1. The specific heat at constant volume c_v, which is then converted by

$$C = mc_v,$$

where m is the mass of the substance;

2. The thermal conductivity* k, which is converted to R by

$$\frac{1}{R} = \frac{kA}{l},$$

where A is the cross-sectional area and l is the length of the heat-flow path.

For a more complete discussion of dynamic models for temperature-control systems, see Cannon (1967) or textbooks on heat transfer.

2.5.2 Incompressible Fluid Flow

Fluid flows are common in many control-system components, one of the most common being the hydraulic actuators used to move the control surfaces on airplanes.

---

*In the case of insulation for houses, resistance is quoted as "R" values, where R-11, for example, refers to a substance that has resistance to heat flow equal to 11 in. of solid wood.

FIGURE 2.24
Water tank example.

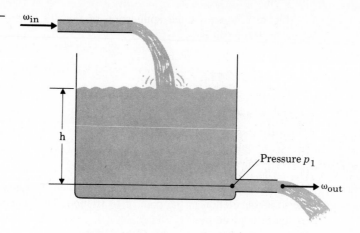

The physical relations governing fluid flow are continuity, force equilibrium, and resistance. The continuity relation is simply a statement of the conservation of matter:

$$\dot{m} = \omega_{in} - \omega_{out},$$ (2.32)

where

$m =$ fluid mass within a prescribed portion of the system,
$\omega_{in} =$ mass flow rate into the prescribed portion of the system,
$\omega_{out} =$ mass flow rate out of the prescribed portion of the system.

Application of this relation to the water tank shown in Fig. 2.24 yields

$$\dot{h} = \frac{1}{A\rho}(\omega_{in} - \omega_{out}),$$ (2.33)

where

$A =$ area of tank,
$\rho =$ density of water,
$h = m/Arho =$ height of water,
$m =$ mass of water in tank.

Force equilibrium must apply exactly as described by Eq. (2.1) for mechanical systems. However, for fluid-flow systems, some forces may be the result of a fluid pressure acting on a piston. In this case, the force from the fluid would be

$$f = pA,$$ (2.34)

where

$f =$ force,
$p =$ pressure in fluid,
$A =$ area on which fluid acts.

Figure 2.24 shows the application of Eqs. (2.1) and (2.5) to a piston-type actuator. The result is

$$Ap - F_D = M\ddot{x},$$

where

A = area of piston,
p = pressure in chamber,
M = mass of piston,
x = position of piston.

In many cases of fluid-flow problems, there is some restriction of the flow due to a constriction in the path or to friction. The general form of the effect of resistance is given by

$$\omega = \frac{1}{R}(p_1 - p_2)^{1/\alpha}, \tag{2.35}$$

where

ω = mass flow rate,
p_1, p_2 = pressures at ends of path through which flow is occurring,
R, α = constants whose values depend on the type of restriction.

The constant α takes on values between 1 and 2, the most common being approximately 2 for high flow rates—having a Reynolds number $(R_e) > 10^5$—through pipes or through short constrictions or nozzles. For very slow flows through long pipes or porous plugs where the flow remains laminar $(R_e < 1100)$, $\alpha = 1$. Flow rates in between these extremes can yield intermediate values of α.

Applying this relation to the water tank example in Fig. 2.25 would relate the outflow to the water level. The relatively short restriction at the outlet would

FIGURE 2.25
Force balance example.

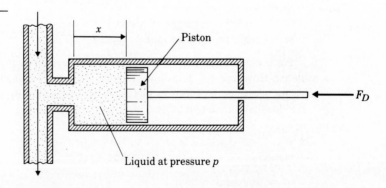

typically yield $\alpha = 2$; therefore

$$\omega_{\text{out}} = \frac{1}{R}(p_1 - p_a)^{1/2}, \tag{2.36}$$

where $p_1 = \rho g h$, the hydrostatic pressure, and p_a is the ambient pressure outside the restriction.

Note that $\alpha = 2$ indicates that the flow is proportional to the square root of the pressure difference and therefore produces a nonlinear equation of motion. For the initial stages of control-system analysis and design purposes, it is typically very useful to linearize these equations so that the design techniques described in this book can be applied. Linearization involves the selection of an "operating point"—for example, h_0 and p_0 in the water tank example—and expanding the nonlinear term according to

$$(1 + \varepsilon)^{\beta} \simeq 1 + \beta \varepsilon, \tag{2.37}$$

where $\varepsilon \ll 1$. Equation (2.36) can thus be written as

$$\begin{aligned}
\omega_{\text{out}} &= \frac{\sqrt{p_0 - p_a}}{R} \cdot \sqrt{1 + \frac{\Delta p}{p_0 - p_a}} \\
&\simeq \frac{\sqrt{p_0 - p_a}}{R}\left(1 + \frac{1}{2}\frac{\Delta p}{p_0 - p_a}\right),
\end{aligned}$$

where the p_1 in Eq. (2.36) has been expressed as $p_0 + \Delta p$. The linearizing approximation made in Eq. (2.38) is valid as long as $\Delta p \ll p_0 - p_a$, that is, as long as the system pressure excursions from the chosen operating point are relatively small.

Combining Eqs. (2.33) and (2.38) yields the linearized equations of motion for the water tank level:

$$\Delta \dot{h} = \frac{1}{A\rho}\left[\omega_{\text{in}} - \frac{\sqrt{p_0 - p_a}}{R}\left(1 + \frac{1}{2}\frac{\Delta p}{p_0 - p_a}\right)\right].$$

But since $p = \rho g h$, the equation reduces to

$$\Delta \dot{h} = -\frac{g}{2AR\sqrt{p_0 - p_a}}\Delta h + \frac{\omega_{\text{in}}}{A\rho} - \frac{\sqrt{p_0 - p_a}}{\rho AR},$$

where $p_0 = \rho g h$.

Hydraulic actuators obey the same relations as the water tank. For example, consider the system shown in Fig. 2.26. When the pilot valve is at $x = 0$, both passages are closed and no motion results. When $x > 0$ as shown, the piston will be forced to the left; when $x < 0$, the high-pressure oil will enter the left side of the chamber and push the piston to the right.

We assume the flow through the orifice formed by the pilot valve is proportional

FIGURE 2.26
Hydraulic actuator.

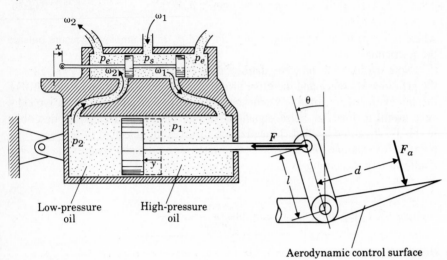

Low-pressure oil   High-pressure oil

Aerodynamic control surface

to x; that is,

$$\omega_1 = \frac{1}{R_1}(p_s - p_1)^{1/2}x. \tag{2.39}$$

Similarly,

$$\omega_2 = \frac{1}{R_2}(p_2 - p_e)^{1/2}x. \tag{2.40}$$

The continuity relation yields

$$\rho A\dot{y} = \omega_1 = \omega_2, \tag{2.41}$$

where ρ is the fluid density, and A is the piston area.

The force balance on the piston yields

$$A(p_1 - p_2) - F = m\ddot{y}, \tag{2.42}$$

where m is the mass of the piston and the attached rod, and F is the force applied by the piston rod to the control surface attachment point. Furthermore, the control surface moment balance yields

$$I\ddot{\theta} = Fl - F_a d, \tag{2.43}$$

where I is the moment of inertia of the control surface and attachment about the hinge, and F_a is the applied aerodynamic load.

To solve this set of five equations, we can combine Eqs. (2.42) and (2.43) by noting that the kinematic relation between θ and y is $y = l\theta$. Furthermore, if the pilot valve exposes the two passages equally and $R_1 = R_2$, Eqs. (2.41), (2.39), and (2.40) imply that

$$p_s - p_1 = p_2 - p_e. \tag{2.44}$$

Because of the nonlinearity in Eqs. (2.39) and (2.40), a reference operating point needs to be selected, and small perturbations about that point can then be described by linear equations amenable to analysis. One possibility is for $\dot{y}$ = constant and no applied load; that is, $F = 0$. In this case, Eqs. (2.42) and (2.44) indicate that

$$p_1 = p_2 = \frac{p_s + p_e}{2}, \tag{2.45}$$

and therefore

$$\dot{\theta} = \frac{\sqrt{p_s - p_e}}{\sqrt{2A\rho R l}} x. \tag{2.46}$$

If $\dot{y}$ = constant but $F \neq 0$, then Eqs. (2.42) and (2.44) indicate that

$$p_1 = \frac{p_s + p_e + F/A}{2}$$

and

$$\dot{\theta} = \frac{\sqrt{p_s - p_e - F/A}}{\sqrt{2A\rho R l}} x. \tag{2.47}$$

As long as the commanded values of x produce θ motion that has a sufficiently small value of $\ddot{\theta}$, the approximation given by Eqs. (2.46) and (2.47) are valid and no other linearized dynamic relations are necessary. However, as soon as the commanded values of x produce $\ddot{\theta}s$ where the inertial forces ($m\ddot{y}$ and the reaction to $I\ddot{\theta}$) are a significant fraction of $p_s - p_e$, the approximations are no longer valid, and one must incorporate these forces into the equations, thus obtaining a dynamic relation between x and θ that is more involved than the pure integration implied by Eqs. (2.46) and (2.47).

☐ 2.6 Linearization, Amplitude Scaling, and Time Scaling

The differential equations of motion for almost all processes selected for control are nonlinear. On the other hand, as we will soon see, both analysis and control design are far easier for linear than for non-linear models. Fortunately, as Lyapunov proved over 100 years ago, if the small-signal linear model valid near an equilibrium is stable, there is a region (which may be small, of course) containing the equilibrium within which the nonlinear system is stable.* With Lyapunov's result, we can safely make a linear model and design a linear control for it, and at least in the neighborhood of the equilibrium, our design will be stable. Since a very important role of feedback control is to maintain the process variables near equilibrium, such small-signal linear models are a frequent starting point for control models. An alternative approach to obtain a linear model for use as the basis of control system design is to use part of the control effort to cancel the nonlinear terms and to design the remainder of the control based on linear theory. This approach is popular in the field of robotics and there is called the "method of computed torque." This method is also a topic of research for control of aircraft. We will illustrate these two techniques by simple examples and by general formulas.

2.6.1 Small-Signal Linearization

Figure 2.27 shows a device for the magnetic levitation of a ball. The physical arrangement of the levitator is depicted in Fig. 2.28. The equation of motion for the ball from Newton's equations (2.1) is

$$m\ddot{x} = f(x, i) - mg, \qquad (2.48)$$

where the force $f(x, i)$ is caused by the field of the electromagnet and described by the experimental curves of Fig. 2.29 for a ball with 1-cm diameter and a mass of 8.4×10^{-3} kg. At current $i_2 = 600$ mA and displacement x_1, the magnet force f will just cancel the gravity force mg of 82×10^{-3} N (the mass of the ball

---

*In 1949, the Russian scientist Aizerman conjectured that if a certain class of systems is stable with any linear gain between two limits, the nonlinear system with a nonlinear gain characteristic kept between the same limits would also be stable. This conjecture is not true, and that's a shame!

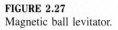

FIGURE 2.27
Magnetic ball levitator.

FIGURE 2.28
Model for ball levitation.

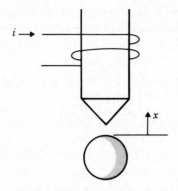

is 8.4×10^{-3} kg and the acceleration of gravity is 9.8 m/s^2); therefore this point represents an equilibrium, and a linearized model can be constructed about the point x_1, i_2. The force is written in terms of the deviations from the equilibrium values as the expansion

$$f(x_1 + \delta x, \, i_2 + \delta i) = f(x_1, \, i_2) + K_x \delta x + K_i \delta i + \cdots. \qquad (2.49)$$

The linear gains are found as follows: K_x is the slope of the force versus x along the curve for $i = i_2$. This is found to be about 0.14 N/cm, or 14 N/m; K_i is the change of force with current for fixed $x = x_1$. Here it is found that for $i = i_1 = 700$ mA at $x = x_1$, the force is about 122×10^{-3} N; and at $i = i_3 = 500$ mA at $x = x_1$, the force is about 42×10^{-3} N. Thus

$$K_i \cong \frac{122 \times 10^{-3} - 42 \times 10^{-3}}{700 - 500} = \frac{80 \times 10^{-3} \text{ N}}{200 \text{ mA}}$$

$$\cong 400 \times 10^{-3} \text{ N/A}$$

$$\cong 0.4 \text{ N/A}.$$

FIGURE 2.29
Experimentally
determined force
curves.

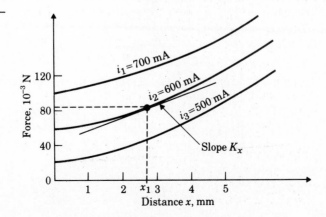

Substituting these values into Eq. (2.49) leads to the linear approximation for the force in the neighborhood of equilibrium as

$$f \cong 82 \times 10^{-3} + 14 \delta x + 0.4 \delta i.$$

And, with this expression in Eq. (2.48), the linearized model is (using the numerical value for the mass and for the gravity force)

$$(8.4 \times 10^{-3}) \ddot{x} = 82 \times 10^{-3} + 14 \delta x + 0.4 \delta i - 82 \times 10^{-3}.$$

Now, since $x = x_1 + \delta x$ and $\ddot{x} = \delta \ddot{x}$, the equation in terms of δx is

$$(8.4 \times 10^{-3}) \delta \ddot{x} = 14 \delta x + 0.4 \delta i$$

$$\delta \ddot{x} = 1667 \delta x + 47.6 \delta i \qquad (2.50)$$

For such a simple system, the magnitudes of the parameters are not too bad, but even so it is good practice to try to bring constants into the range between 0.01 and 100 or even to the range between 0.1 and 10, if possible, by careful selection of scale factors. The scale factors of variables may be viewed as the selection of appropriate units in which to measure the variables. The process is called *amplitude scaling*. In this case, the displacement is measured in meters in Eq. (2.50), whereas centimeters or even millimeters are more reasonable in view of the size of a typical motion. An equivalent process is associated with selection of appropriate units for time, called *time scaling*. As a general guideline, amplitude scale factors should be selected so that the expected maximum deviation of the (scaled) variables is ± 1.0, and time scaled so the expected transient time is about 1 unit.

To consider the particular case for the moment, normalized variables are defined as follows:

$$y = \frac{\delta x}{x_0}; \quad u = \frac{\delta i}{i_0}; \quad \tau = \omega_0 t. \qquad (2.51)$$

The effect of the time scaling is to change the differentiation so that, with time scaling,

$$\frac{d \delta x}{dt} = \frac{d \delta x}{d(\tau/\omega_0)} = \omega_0 \frac{d \delta x}{d \tau}.$$

Using this and substituting Eq. (2.51) into Eq. (2.50) results in

$$\frac{\omega_0^2 x_0 d^2 y}{d\tau^2} = (1667 x_0) y + (47.6)(i_0) u.$$

Dividing by $\omega_0^2 x_0$ gives

$$\frac{d^2 y}{d\tau^2} = \frac{1667 y}{\omega_0^2} + \frac{47.6 i_0}{\omega_0^2 x_0} u. \qquad (2.52)$$

The coefficient of y can be set equal to 1 by taking $\omega_0^2 = 1667$ or $\omega_0 = 40.83$.*
With this constant, the unit of time is $\approx 1/40$ s. For the scale factors in δx and i,
it appears reasonable that position be measured in centimeters, so $x_0 = 10^{-2}$, and
current in amperes, so $i_0 = 1$. With these values, the scaled, linearized model of
the ball balancer is

$$\frac{d^2 y}{d\tau^2} = y + 2.86u. \tag{2.53}$$

Clearly, we could have selected the current scale factor i_0 so as to make the
coefficient of u have a magnitude of 1.0 as well.

Models with simple even numbers for coefficients will frequently be used in
later examples, where it is understood that these could have been obtained by
similar amplitude and time scaling from real physical situations.

2.6.2 Linearization by Feedback

To illustrate linearization by feedback, we consider the equation of a simple pendulum, which is readily obtained from Eq 2.12 with $\ddot{x}$ set to 0 and an applied
torque τ added. The equation is

$$(I + m_p l^2)\ddot{\theta} + m_p g l \sin(\theta) = \tau \tag{2.54}$$

If we compute the torque to be

$$\tau = m_p g l \sin(\theta) + u \tag{2.55}$$

then the motion is described by

$$(I + m_p l^2)\ddot{\theta} = u, \tag{2.56}$$

and this is a linear equation *no matter how large the angle θ becomes*. For robots
with two or three rigid links, this computed torque approach leads to effective
control.

☐ 2.6.3 General Formulas for Linearization and Scaling

The steps described above can be expressed in greatly reduced form if one is willing
to use matrix notation. The basic facts about matrices are collected in Appendix
C. For small-signal linearization we begin with the original model in matrix form as

$$\dot{\mathbf{x}} = \mathbf{f}(\mathbf{x}, \mathbf{u}), \tag{2.57}$$

*In view of the probable error in the plots of Fig. 2.29, a reasonable choice would be to let $\omega_0 = 40$
and approximate 1667 with 1600 to make the constant easy to remember.

where $\mathbf{x} = [x_1 x_2 \cdots x_n]^T$ is the state vector, and $\mathbf{f} = [f_1 f_2 \cdots f_n]^T$ is the vector of functions that make up the equations. We assume that the equilibrium solution is for the values $\mathbf{x}_0$, $\mathbf{u}_0$, so that $\dot{\mathbf{x}}_0 = 0 = \mathbf{f}(\mathbf{x}_0, \mathbf{u}_0)$. The matrix of first derivatives $\mathbf{F}$ is defined as the matrix with element in row i and column j as $[\delta f_i/\delta x_j]_{x_0, u_0}$, where $\mathbf{x}_0$ and $\mathbf{u}_0$ are the equilibrium values. The matrix $\mathbf{G}$ has elements $[\delta f_i/\delta u_j]_{x_0, u_0}$. With these partial derivatives, Eq. (2.54) can be written for deviations about $\mathbf{x}_0$, $\mathbf{u}_0$ as

$$\dot{\mathbf{x}}_0 + \delta\dot{\mathbf{x}} + \mathbf{F}(\mathbf{x}_0, \mathbf{u}_0) + \mathbf{F}(t)\delta\mathbf{x} + \mathbf{G}(t)\delta\mathbf{u} + \text{higher-order terms.}$$

Thus, the small-signal linearized equations are

$$\delta\dot{\mathbf{x}} = \mathbf{F}\delta\mathbf{x} + \mathbf{G}\delta\mathbf{u}. \tag{2.58}$$

Time scaling with $\tau = \omega_0 t$ replaces Eq. (2.58) with

$$\delta\dot{\mathbf{x}} = \frac{1}{\omega_0}\mathbf{F}\delta\mathbf{x} + \frac{1}{\omega_0}\mathbf{G}\delta\mathbf{u}.$$

Amplitude scaling of the state corresponds to replacing $\delta\mathbf{x}$ with $\mathbf{z} = \mathbf{D}_x^{-1}\delta\mathbf{x}$, where $\mathbf{D}_x$ is a diagonal matrix of scale factors. Input scaling corresponds to replacing $\delta\mathbf{u}$ with $\mathbf{v} = D_u^{-1}\delta\mathbf{u}$. With these substitutions,

$$\mathbf{D}_x\dot{\mathbf{z}} = \frac{1}{\omega_0}\mathbf{F}\mathbf{D}_x\mathbf{z} + \frac{1}{\omega_0}\mathbf{G}\mathbf{D}_u\mathbf{v}$$

$$\dot{\mathbf{z}} = \frac{1}{\omega_0}\mathbf{D}_x^{-1}\mathbf{F}\mathbf{D}_x\mathbf{z} + \frac{1}{\omega_0}\mathbf{D}_x^{-1}\mathbf{G}\mathbf{D}_u\mathbf{v}. \tag{2.59}$$

Equation (2.59) compactly expresses the time- and amplitude-scaling operations. Regrettably it does not relieve the engineer of the responsibility of actually thinking of good scale factors so that scaled equations are in good shape.

2.7 A Review of Dynamic Response

We have seen that models of dynamic components can frequently be found in the form of an ordinary differential equation or a set of ordinary differential equations. Furthermore, near an equilibrium the equations are well approximated as linear and time-invariant. Once such model equations are found, attention is turned to their solution and, in control engineering, even more to the analysis of their solutions so that one can estimate major features of the dynamic response and can also visualize how the system might be changed to modify the response in some desirable direction. A computer program suitable for solving ordinary differential equations is sufficient to obtain a specific solution. To gain insight into *why* the solution has certain features—such as transient rise time, overshoot of final value, or transient duration (settling time)—requires a more complete and comprehensive analysis technique.

There are two equivalent approaches to presenting model information so that the required analysis can be done; one of these is via pole-zero patterns and the other is by steady-state frequency response. We therefore begin with a review of the Laplace transform and frequency response analysis. Tables of properties and corresponding time-transform pairs are presented. Using knowledge of the effects of poles and zeros, simple formulas are developed relating time-domain features to the transforms of signals.

2.7.1 Introduction to the Laplace Transform

There are two basic facts about linear constant systems that form the basis for almost all analysis techniques for these systems. These are

A linear system response obeys the principle of superposition.

The response of a linear *constant* system is the convolution of the input with the system unit impulse response.

We will show that from the second of these properties, it follows immediately that the response of a linear constant system to an exponential input is also exponential, and from this flows the usefulness of Fourier and Laplace transforms in the study of linear constant systems.

The principle of superposition states that if the system has an input which can be expressed as a sum of signals, the response of the system can be expressed as the same sum of the individual responses to the respective signals. We can express this principle mathematically as follows. Consider the system to have input u and output y, and suppose further that, with the system at rest (zero initial conditions), we apply the input $u_1(t)$ and observe the output $y_1(t)$. We restore the system to rest and reset time and apply a second input $u_2(t)$ and again observe the output, which we call $y_2(t)$. Finally, we form the composite input $u(t) = \alpha_1 u_1(t) + \alpha_2 u_2(t)$. If superposition applies, which is true if and only if the system is linear, then the response will be $y(t) = \alpha_1 y_1(t) + \alpha_2 y_2(t)$.

EXAMPLE 2.1   Consider the system modeled by the first-order linear differential equation

$$\dot{y} + ky = u.$$

Let $u = \alpha_1 u_1 + \alpha_2 u_2$, and *assume* that $y = \alpha_1 y_1 + \alpha_2 y_2$, then $\dot{y} = \alpha_1 \dot{y}_1 + \alpha_2 \dot{y}_2$. If we substitute these expressions into the system equation, we obtain

$$\alpha_1 \dot{y}_1 + \alpha_2 \dot{y}_2 + k(\alpha_1 y_1 + \alpha_2 y_2) = \alpha_1 u_1 + \alpha_2 u_2.$$

From this, it follows that

$$\alpha_1(\dot{y}_1 + ky_1 - u_1) + \alpha_2(\dot{y}_2 + ky_2 - u_2) = 0. \tag{2.60}$$

If y_1 is the solution with input u_1, and y_2 is the solution with input u_2, then Eq. (2.60) is satisfied, and our assumption is correct: The response is the sum of the individual responses and superposition holds.   ∎

Using the principle of superposition, we can solve for the system responses to a set of elementary signals and are then able to solve for the response to a general signal simply by decomposing the given signal into a sum of the elementary components and, by superposition, the response to the general signal is the sum of the responses to the elementary signals. In order for this process to work, the elementary signals need to be such that responses to them are easy to find and they need to be sufficiently "rich" that any reasonable signal can be expressed as a sum of them. The most common candidates for elementary signals for use in linear systems are the impulse and the exponential.

The idea for the impulse comes from dynamics. Suppose we wish to study the motion of a baseball hit by a bat. The details of the collision between the bat and ball can be very complex as the ball deforms and the bat bends; however, for purposes of computing the path of the ball, the effect of the collision can be summarized by the net velocity change given to the ball in a very short time. It is as if the ball is subjected to a very intense force for a very short time: an impulse. The physicist Dirac suggested that such forces could be represented by a mathematical impulse, $\delta(t)$, which has the property that if $f(t)$ is continuous at $t = \tau$, then

$$\int_{-\infty}^{+\infty} f(\tau)\delta(t - \tau)d\tau = f(t). \qquad (2.61)$$

In other words, the δ is so short and so intense that no value of f matters, except at the point where the δ occurs. If we replace f by u, then Eq. (2.61) represents an input $u(t)$ as the integration of impulses of intensity $u(t - \tau)$. To find the response to an arbitrary input by the principle of superposition, we need to find the response to a unit impulse. For a general linear system, we can express the impulse response as $h(t, \tau)$, which is the response at time t to an input applied at time τ. The total response is then the integral of these with intensity u, which is

$$y(t) = \int_{-\infty}^{\infty} u(\tau)h(t, \tau)d\tau.$$

This is the superposition integral. Some linear systems are constant and some are time-varying. Consider a cart to be pulled along a road. If the cart is loaded with melting ice, the mass will vary with time and the system will not be constant. A system described by a linear differential equation with constant coefficients is constant. If a system is constant, the response depends on what input is applied but does not depend on the time at which the input is applied. If the system is linear and constant, the impulse response is given by $h(t - \tau)$, since the response at t to an input applied at τ depends only on the time difference between when the impulse is applied and the time we are observing the response. Constant systems are called "shift invariant" for this reason. For constant systems, the superposition integral then takes the special form

$$y(t) = \int_{-\infty}^{\infty} u(\tau)h(t - \tau)d\tau$$

$$= \int_{-\infty}^{\infty} u(t - \tau)h(\tau)d\tau. \tag{2.62}$$

Equation (2.62) is called the *convolution integral*.

EXAMPLE 2.2   We can illustrate convolution with a simple system. Consider the impulse response for the system described by the differential equation

$$\dot{y} + ky = u = \delta(t).$$

Because $\delta(t)$ only has effect near $t = 0$, we can integrate this equation from just before zero to just after zero with the result

$$\int_{0^-}^{0^+} \dot{y}dt + k \int_{0^-}^{0^+} ydt = \int_{0^-}^{0^+} \delta dt.$$

The integral of $\dot{y}$ is just y, the integral of y over so small a range is zero, and the integral of the impulse over the same range is unity. The result of the integration is

$$y(0^+) - y(0^-) = 1.$$

Since the system was at rest before application of the impulse, $y(0^-) = 0$, and we are left with the effect of the impulse that $y(0^+) = 1$. For positive time, we have the differential equation

$$\dot{y} + ky = 0 \qquad y(0^+) = 1.$$

If we assume a solution $y = Ae^{st}$, then $\dot{y} = Ase^{st}$, and the equation is

$$Ase^{st} + kAe^{st} = 0$$
$$s + k = 0$$
$$s = -k.$$

Also, to match $y(0^+) = 1$, it is necessary that $A = 1$. Thus, the solution for the impulse response is $y(t) = h(t) = e^{-kt}$ for $t > 0$. To take care of the fact that $h(t) = 0$ for negative time, we define the unit step function

$$1(t) = 0 \qquad t < 0$$
$$= 1 \qquad t \geq 0.$$

With this definition, the impulse response of the first-order system is

$$h(t) = e^{-kt}1(t).$$

The response to a general input is given by the convolution of this impulse response and the input, as follows:

$$y(t) = \int_{-\infty}^{\infty} h(\tau)u(t-\tau)d\tau$$

$$= \int_{-\infty}^{\infty} e^{-k\tau}1(\tau)u(t-\tau)d\tau$$

$$= \int_{0}^{\infty} e^{-kt}u(t-\tau)d\tau. \qquad \blacksquare$$

An immediate consequence of convolution is that an exponential time function input of the form e^{st} results in an output that is also an exponential time function $H(s)e^{st}$ changed only in amplitude by the function $H(s)$. The amplitude of the output exponential when the input is a unit exponential is called the *transfer function* of the system. Notice especially that the constant s may be complex as $s = \sigma + j\omega$, and thus that both the input and the output may be complex. If we let $u(t) = e^{st}$ in Eq. (2.62),

$$y(t) = \int_{-\infty}^{\infty} u(\tau)h(t-\tau)d\tau$$

$$= \int_{-\infty}^{\infty} h(\tau)u(t-\tau)d\tau$$

$$= \int_{-\infty}^{\infty} h(\tau)e^{s(t-\tau)}d\tau$$

$$= \int_{-\infty}^{\infty} h(\tau)e^{st}e^{t-s\tau}d\tau$$

$$= \int_{-\infty}^{\infty} h(\tau)e^{t-s\tau}d\tau e^{st}$$

$$= H(s)e^{st}.$$

EXAMPLE 2.3   We can compute the transfer function for the system of Example 2.2 with input $u = e^{st}$. We take the same system equation

$$\dot{y}(t) + ky(t) = u(t) = e^{st}.$$

We assume that we can express $y(t)$ as $H(s)e^{st}$. With this form, we have $\dot{y} = sH(s)e^{st}$, and the equation reduces to

$$sH(s)e^{st} + kH(s)e^{st} = e^{st}.$$

We can solve for the transfer function $H(s)$ as

$$H(s) = \frac{1}{s + k}.$$

Substituting back, the output is therefore $y = e^{st}/s + k$. ∎

The most common use of the exponential response of linear constant systems is in finding the response to a sinusoid—the frequency response. The calculation goes this way. In order to use superposition, it is necessary first to express the sinusoid as a sum of two exponentials. The relation (due to Euler) is

$$A \cos(\omega t) = \frac{A}{2}(e^{j\omega t} + e^{-j\omega t}).$$

If we let $s = j\omega$ in the basic response formula, we find that the response to $u(t) = e^{j\omega t}$ is $y(t) = H(j\omega)e^{j\omega t}$ and, in the same way, the response to $u(t) = e^{-j\omega}$ is $H(-j\omega)e^{-j\omega t}$. The response to the sum of these two exponentials, which make up the cosine signal, is the sum of the responses, namely

$$y(t) = \frac{A}{2}\left[H(j\omega)e^{j\omega t} + H(-j\omega)e^{-j\omega t}\right].$$

The transfer function $H(j\omega)$ is a complex number that can be represented in polar form or magnitude and phase form as $H(j\omega) = M(\omega)e^{j\phi(\omega)}$ or just $H = Me^{j\phi}$. With this substitution, the expression for $y(t)$ becomes

$$y(t) = \frac{A}{2}M\left[e^{j(\omega t + \phi)} + e^{-j(\omega t + \phi)}\right]$$

$$= AM \cos(\omega t + \phi).$$

EXAMPLE 2.4   For the first-order system we have been using, the transfer function for $s = j\omega$ is

$$H = \frac{1}{j\omega + k},$$

from which $M = 1/\sqrt{\omega^2 + k^2}$ and $\phi = \tan^{-1}(-\omega/k)$. The response of this system to a sinusoid $u = A \cos(\omega t)$ will be

$$y(t) = AM \cos(\omega t + \phi). ∎$$

The conclusion to be drawn from this exercise is that the response of a linear constant system to a sinusoid of frequency ω is a sinusoid of the *same* frequency with amplitude multiplied by the magnitude of the transfer function evaluated at the input frequency and with a phase given by the phase of the transfer function at the input frequency.

The frequency response approach is generalized by the introduction of the

Laplace Transform of a signal $f(t)$ as

$$F(s) = \int_{-\infty}^{\infty} f(t)e^{-st}dt. \tag{2.63}$$

If we apply this transform to the convolution, Eq. (2.62), the result is easily computed to be

$$Y(s) = H(s)U(s). \tag{2.64}$$

With the formula from Eq. (2.64), we have a means to compute the response of linear, constant systems to quite general inputs. Given any input and any system, we compute the transform of the input and the transfer function for the system. The transform of the output is then given by Eq. (2.64) as the product of these two transforms. To get the output time function, we need only compute the inverse transform of $Y(s)$. A number of aids are available to help with each of these steps. For example, in order to help with the computation of the transform of input signals and the computation of the inverse transform to get $y(t)$ from $Y(s)$, there are extensive tables of transforms of elementary functions and also tables of properties that permit us to find complex transforms from simpler ones. Table A.1 in the appendix has a set of useful properties, and Table A.2 has transforms of a few common signals from which, using Table A.1, more complex signal transforms can be developed. Each of the entries in these tables follows from direct application of the definition of the transform. The process of inversion requires the opposite process, whereby a complex transform is reduced to a sum of simpler terms that can be found in the table. The basic tool of this table-based approach to Laplace transform inversion is the partial fraction expansion, which is best illustrated by example.

EXAMPLE 2.5   Suppose we have computed $Y(s)$ and found

$$Y(s) = \frac{(s + 2)(s + 4)}{s(s + 1)(s + 3)}.$$

The elementary entry in the table of transforms is in the form $1/(s + a)$, and we want to expand $Y(s)$ into a sum of such terms. The idea is to find constants C_i so that

$$Y(s) = C_1 \frac{1}{s} + C_2 \frac{1}{s + 1} + C_3 \frac{1}{s + 3}.$$

In order to evaluate the constants, we equate the two forms of $Y(s)$ and solve. The easy way to get the solution is to solve the equations for particular values of s. For example, if we multiply both equations by s and set $s = 0$, we find

$$C_1 = \frac{(s + 2)(s + 4)}{(s + 1)(s + 3)}\bigg|_{s=0} = \frac{8}{3}.$$

In a similar fashion, we evaluate

$$C_2 = \frac{(s+2)(s+4)}{s(s+3)}\Big|_{s=-1} = -\frac{3}{2}$$

and

$$C_3 = \frac{(s+2)(s+4)}{s(s+1)}\Big|_{s=-3} = -\frac{1}{6}.$$

The correctness of the result is readily checked by adding the components again to verify that the original function is recovered. With the partial fractions, the solution can be looked up in the tables at once to be

$$y(t) = \frac{8}{3}1(t) - \frac{3}{2}e^{-t}1(t) - \frac{1}{6}e^{-3t}1(t).$$ ∎

The problem of partial fraction expansion is slightly more complex if the function has repeated roots—in that case, it is necessary to differentiate the equations after multiplying by the appropriate factor.

The Final Value Theorem

A very useful property of the Laplace transform known as the final value theorem follows easily from the development of partial fraction expansion. Suppose we have a rational transform $Y(s)$ of a causal signal $y(t)$ and we wish to know the final value $y(t)$ from $Y(s)$. There are three possibilities for the limit. It can be constant, undefined, or unbounded. If $Y(s)$ has any pole in the right-half plane or a multiple pole on the imaginary axis, then $y(t)$ will grow and the limit is unbounded. If all poles of $Y(s)$ are in the left-half plane except for a simple pair on the imaginary axis, then $y(t)$ will contain a sinusoid that persists forever and the final value is not defined. Only one case provides for a constant final value: if all poles of $sY(s)$ are in the left-half plane, then all terms of $y(t)$ will decay to zero except for the term corresponding to the pole at $s = 0$, and that term is a constant. Thus, in this case the final value is given by the residue of the pole at the origin. We can state this result as a theorem:

If all poles of $sY(s)$ are in the left-half plane, then

$$\lim_{t\to\infty} y(t) = \lim_{s\to 0} sY(s).$$

In addition to having tables of transforms and tables of properties of transforms for handling the computation of $U(s)$ and the inversion of $Y(s)$, it is valuable to have techniques that help in finding the transfer function of the system. This is especially important when the system is a composite of several simpler systems connected together with internal feedback. For this purpose, there is the theory of block diagrams. This topic is sufficiently important to our task in control design that a careful development of the rules and techniques of block diagram reduction is given in a separate section, which follows.

2.7.2 The Block Diagram

To obtain the transfer function, we need to find the Laplace transform of the system equations and solve the resulting algebraic equations for the relation between the input and the output. In many control systems, the system equations can be written so that they are composed of several components that do not interact except that the input of one part is the output of another part. In these cases, it is easy to draw a graph known as a *block diagram* that represents the relationships, and the equations can be solved by a graphical simplification, which is often easier and more informative than algebraic manipulation, even though the methods are in every way equivalent. Drawings of three elementary block diagrams are seen in Fig. 2.30. It is convenient to think of each block as representing an electronic amplifier with the transfer function printed inside. The interconnections of blocks include summing points where any number of signals may be added together. These are represented by a circle with the math symbol of summation, Σ, inside. In Fig. 2.30(a), the block with transfer function $G_1(s)$ is in series with the block with transfer function $G_2(s)$, and the overall transfer function is given by the product $G_2 G_1$. In Fig. 2.30(b), the two component systems are in parallel with their outputs added, and the overall transfer function is given by the sum of G_1 and G_2. These results are very easy consequences of the equations that describe the block diagrams. In Fig. 2.30(c), we have a more complicated case, in that the two blocks are connected in a feedback arrangement so that each feeds into the other. Here it is worthwhile to look at the equations, solve them, and then relate them back to the picture. We have

$$U_1 = R + Y_2$$
$$Y_2 = G_2 G_1 U_1$$
$$Y_1 = G_1 U_1.$$

FIGURE 2.30
Three examples of elementary block diagrams.

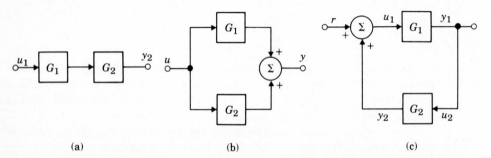

(a)   (b)   (c)

The solution of these equations is

$$Y_1(s) = \frac{G_1}{1 - G_1 G_2} R.$$

(2.65)

We can express the solution by the rule:

The gain of a single-loop feedback system is given by the forward gain divided by 1 minus the loop gain.

The three elementary cases given in Fig. 2.30 can be used in combination to solve for any transfer function defined by a block diagram by repeated reduction. However, the manipulations can become tedious and are subject to error when the topology of the diagram gets complicated. Figure 2.31 shows examples of block diagram algebra that complement those shown in Fig. 2.30.

FIGURE 2.31
Examples of block diagram algebra: (a) moving a pick-off point; (b) moving a summing point; (c) conversion to unity feedback.

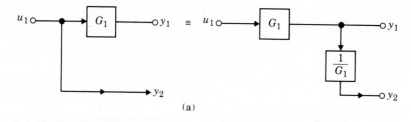

(a)

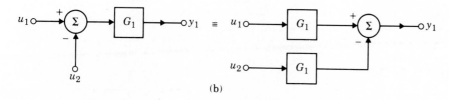

(b)

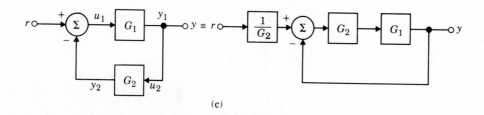

(c)

In all cases, the basic principle is to simplify the topology while maintaining exactly the same relations among the remaining variables of the block diagram. In relation to the algebra of the underlying linear equations, block diagram reduction is a graphical way to solve equations by eliminating variables.

EXAMPLE 2.6   Consider the system shown in Fig. 2.32, and suppose we wish to find the transfer function for the corresponding system.

First, we may simplify the block diagram by replacing the feedback loop involving G_1 and G_3 by its equivalent transfer function (see Fig. 2.30(c)). The next step is to move the pick-off point preceding G_2 to its output (see Fig. 2.31(a)).

FIGURE 2.32
An example of block diagram simplification.

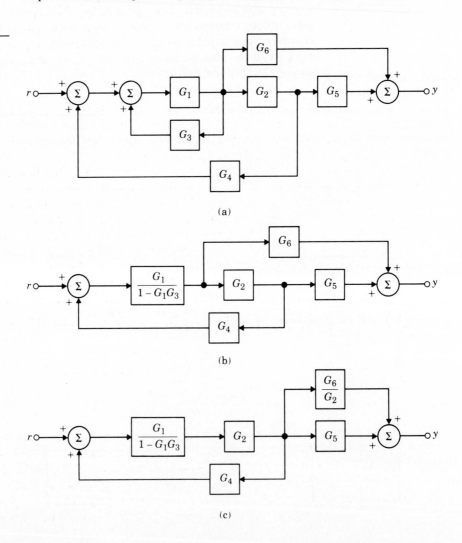

(a)

(b)

(c)

Finally, we reduce the resulting loop and the two subsystems in parallel to obtain:

$$\frac{Y}{U} = \frac{G_1 G_2/(1 - G_1 G_3)}{1 - [G_1 G_2 G_4/(1 - G_1 G_3)]}\left(G_5 + \frac{G_6}{G_2}\right)$$

$$= \frac{G_1 G_2 G_5 + G_1 G_6}{1 - G_1 G_3 - G_1 G_2 G_4}. \qquad\blacksquare$$

Mason's Rule for Block Diagram Reduction

As we have seen, a system of algebraic equations may be represented graphically by a block diagram, which represents individual transfer functions by blocks and has interconnections that correspond to the system equations. A block diagram is a convenient tool to visualize the system as a collection of interrelated subsystems with the relations among the system variables emphasized. An alternative notation to the block diagram is given by the *signal flow graph* introduced by S. J. Mason (1953). As with the block diagram, the signal flow graph offers a visual tool representing the causal relationships among the components of the system. The method consists of characterizing the system by a network of directed branches and associated gains (transfer functions) connected at nodes. Several flow graphs and the corresponding block diagrams are shown in Fig. 2.33.

In this book, we will use the block diagram notation that draws a block around each gain so there is no chance that a given gain will be associated with the wrong branch of the graph. The block diagram has an explicit symbol for the summation function to make the output of the node unmistakable. A reduction rule for any block diagram was given by Mason (1953, 1956), who related the graph to the matrix algebra of the simultaneous equations they represented. We will state the rule and illustrate it for several cases, but for the derivation, based on Cramer's rule for solving linear equations by determinants, the papers of Mason should be consulted. Consider Fig. 2.33(c), where the signal at each node has been given a name and the gains are marked. Then the block diagram (or the signal flow graph) represents the system of equations:

$$x_1 = x_3 + u$$
$$x_2 = g_1 x_1 + g_2 x_2 + g_4 x_3$$
$$x_3 = g_3 x_2$$
$$y = x_3.$$

We refer to the internal signals in the diagram, such as the common input to several blocks or the output of a summing junction, as *nodes*. The system input signal point and the system output point are also nodes. Mason defined a *path* through a block diagram as a sequence of connected blocks, the path passing from one variable to another *in the direction of signal flow of the blocks*, without including any variable more than once. A signal could traverse through the graph

FIGURE 2.33
Example block diagrams and corresponding signal flow graphs.

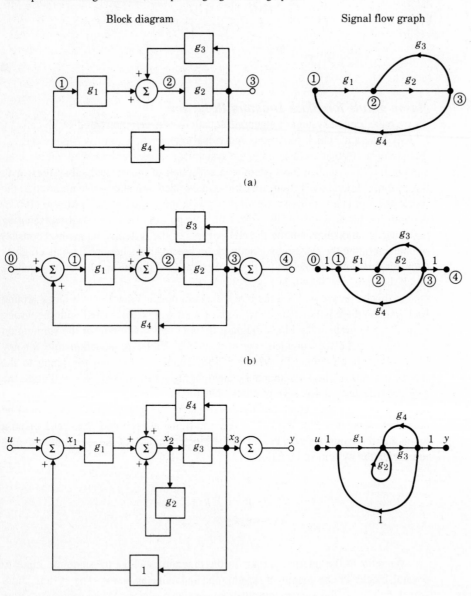

along a path. A *forward path* is a path from the input to the output such that no node is included more than once. If the nodes are numbered in a convenient order, a forward path can be identified by the numbers that are included. Any closed path that returns to its starting node is a loop, and a path that leads from a given variable back to the same variable is defined as a *loop path*. Again, a path is a continuous sequence of nodes, with direction specified by the arrows, with no node repeating. The source, or input, is a node with outgoing branches only and the sink, or output, is a node with incoming branches only. The *path gain* is the product of component gains (transfer functions) making up the path. Similarly the loop gain is the path gain associated with a loop, that is, the product of gains in a loop. If two paths have a common component, they are said to touch. Notice particularly in this connection that the input and the output of a summing junction are not the same and that the summing junction is a one-way device from its inputs to its output.

With these definitions, Mason's rule states that the input-output transfer function associated with a signal flow graph is given by

$$G(s) = \frac{Y(s)}{U(s)} = \frac{1}{\Delta} \sum_i G_i \Delta_i,$$

where G_i = path gain of ith forward path, the system determinant is $\Delta = 1 - \sum$ all individual loop gains $+ \sum$ gain products of all possible two loops which do not touch $- \sum$ of gain products of all possible three loops that do not touch $+ \cdots$, and the ith forward path determinant, Δ_i, is the value of Δ for that part of the block diagram that does not touch the ith forward path.

EXAMPLE 2.7   As an example, for the signal flow graph shown in Fig. 2.34, we have

forward path	*path gain*
1236	$G_1 = 1(1/s)(b_1)(1);$
12346	$G_2 = 1(1/s)(1/s)(b_2)(1)$
123456	$G_3 = 1(1/s)(1/s)(1/s)(b_3)(1)$

loop path

$$232 \quad \ell_1 = -a_1/s$$
$$2342 \quad \ell_2 = -a_2/s^2$$
$$23452 \quad \ell_3 = -a_3/s^3$$

The determinants are

$$\Delta = 1 - (-\frac{a_1}{s} - \frac{a_2}{s^2} - \frac{a_3}{s^3}) + 0;$$

$$\Delta_1 = 1 - (0);$$
$$\Delta_2 = 1 - (0);$$
$$\Delta_3 = 1 - (0),$$

FIGURE 2.34
A block diagram of control canonical form.

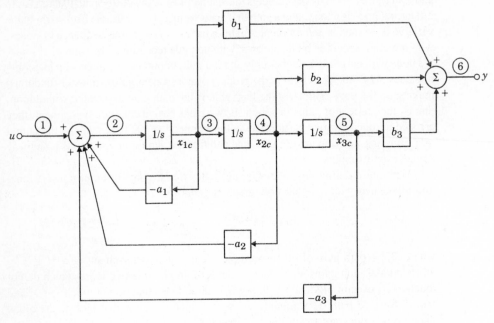

Applying the rule, the transfer function is

$$\frac{Y(s)}{U(s)} = \frac{(b_1/s) + (b_2/s^2) + (b_3/s^3)}{1 + (a_1/s) + (a_2/s^2) + (a_3/s^3)}$$

$$= \frac{b_1 s^2 + b_2 s + b_3}{s^3 + a_1 s^2 + a_2 s + a_3}.$$

■

EXAMPLE 2.8   As another example, consider the block diagram in Fig. 2.35. We have

	forward path	*path gain*
	15	$G_1 = H_1$;
	12345	$G_2 = H_2 H_3 H_4$;

The loop path is

$$232 \qquad \ell_1 = H_3 H_5;$$

The determinants are

$$\Delta = 1 - (H_3 H_5);$$
$$\Delta_1 = 1 - H_3 H_5 \quad \text{(note that path 15 does not touch 232)}$$
$$\Delta_2 = 1 - (0);$$

FIGURE 2.35
An example block
diagram.

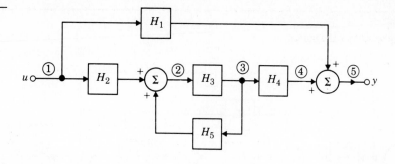

and the overall transfer function:

$$\frac{Y(s)}{U(s)} = \frac{H_1(1 - H_3H_5) + H_2H_3H_4}{1 + H_3H_5}.$$

■

EXAMPLE 2.9 As the third example, consider the block diagram shown in Fig. 2.36.

forward path	*gain*
12456	$G_1 = H_1H_2H_3$;
1236	$G_2 = H_4$;

loop path

242 $\ell_1 = H_1H_5$; (does not touch ℓ_3)
454 $\ell_2 = H_2H_6$
565 $\ell_3 = H_3H_7$ (does not touch ℓ_1)
236542 $\ell_4 = H_4H_7H_6H_5$

determinants

$\Delta = 1 - (H_1H_5 + H_2H_6 + H_3H_7 + H_4H_7H_6H_5) + H_1H_5H_3H_7)$;
$\Delta_1 = 1 - (0)$;
$\Delta_2 = 1 - H_2H_6$;

$$\frac{Y(s)}{U(s)} = \frac{H_1H_2H_3 + H_4 - H_4H_2H_6}{1 - H_1H_5 - H_2H_6 - H_3H_7 - H_4H_7H_6H_5 + H_1H_5H_3H_7}.$$

■

Special Case Mason's rule for the special case where the paths are closely coupled, in that *all* forward paths and loop paths touch, can be stated as follows:

The gain of a closely coupled feedback system is given by the sum of forward
path gains divided by 1 minus the sum of loop gains.

FIGURE 2.36
Another example block diagram.

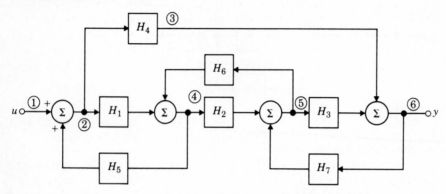

EXAMPLE 2.10   The application of this rule can be illustrated by the block diagram shown in Fig. 2.37. In this case, the forward paths and their gains are given by

forward path	gain
1 2 3 4 5 9	$G_1 G_2 G_5$
1 2 3 6 9	$G_1 G_6$

loop path	gain
2 3 8 2	$G_1 G_3$
2 3 4 7 2	$G_1 G_2 G_4$

and the overall gain is given by the rule as

$$\frac{Y}{U} = \frac{G_1 G_2 G_5 + G_1 G_6}{1 - G_1 G_3 - G_1 G_2 G_4}.$$ ∎

FIGURE 2.37
An example block diagram.

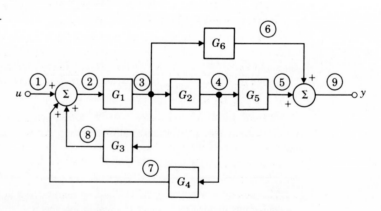

Summary Mason's rule is useful for solving relatively complicated block diagrams by hand. It yields the solution to the graph in the sense that it provides an explicit input-output relationship for the system represented by the diagram. The advantage as compared to path-by-path reduction is that it is systematic and algorithmic rather than ad hoc and problem dependent. However, the availability of computer aids makes determination of transfer functions based on calculations from state-variable models increasingly attractive (see Chapter 6).

2.7.3 Poles and Zeros

Once the transfer function has been worked out by any of the available methods, we are in a position to start the analysis of the response of the system so represented. When the equations are simultaneous ordinary differential equations, the transfer function that results will be a ratio of polynomials, $H(s) = b(s)/a(s)$. For example, suppose

$$H(s) = \frac{2s + 1}{s^2 + 3s + 2},\qquad (2.66)$$

then

$$b(s) = 2s + 1 = 2\left(s + \frac{1}{2}\right),$$

and

$$a(s) = s^2 + 3s + 2$$
$$= (s + 1)(s + 2).$$

If we assume that b and a have no common factors (this is the usual case), values of s such that $a(s) = 0$ will be places where $H(s)$ is infinity, and these values of s are called *poles* of $H(s)$. They are $s = -1$ and $s = -2$ for $H(s)$ of Eq. (2.61). On the other hand, values of s such that $b(s) = 0$ are places where $H(s)$ is zero, and the corresponding s locations are called *zeros*. There is one zero of $H(s)$ at $s = -1/2$. These poles and zeros completely describe $H(s)$ except for a constant multiplier, and one can give a description of a transfer function by plotting the locations of the poles and the zeros in the s plane, as shown in Fig. 2.38 for $H(s)$ as given in Eq. (2.66). Since the impulse response is given by the inverse Laplace transform of the transfer function, we call the impulse response the *natural response* of the system, and we can use the poles and zeros to compute the corresponding time response and thus identify time histories with pole and zero patterns. For example, the poles identify the classes of signals contained in the impulse response, as may be seen by a partial fraction expansion of $H(s)$. For

FIGURE 2.38
Sketch of s plane
showing poles as crosses
and zeros as circles.

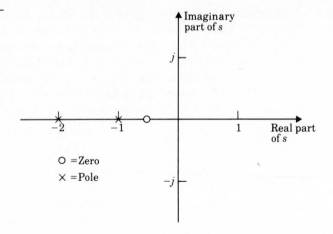

example, we can write

$$H(s) = \frac{2s + 1}{s^2 + 3s + 2}$$

$$= \frac{2s + 1}{(s + 1)(s + 2)}$$

$$= \frac{-1}{s + 1} + \frac{3}{s + 2}.$$

The poles of this $H(s)$ are at $s = -1$ and $s = -2$. From the table of transforms in Appendix A, we can look up the inverse of each term in $H(s)$, which will give us the time function $h(t)$. In this case,

$$h(t) = -e^{-t} + 3e^{-2t} \qquad t \geq 0$$

$$= 0 \qquad\qquad\qquad t < 0. \qquad\qquad (2.67)$$

We see that the shape of the component parts of $h(t)$, which are e^{-t} and e^{-2t}, are determined by the poles that are at $s = -1$ and -2. This is true of more complicated cases as well, of course, and is the basis for the statement that the shape of the natural response is determined by the location of the poles of the transfer function. A sketch of these pole locations and corresponding natural responses is given in Fig. 2.39, along with complex pole locations, which will be discussed shortly. The role of the numerator in the process of partial fraction expansion is to influence the size of the coefficient that multiplies each component.

Because e^{-2t} decays faster than e^{-t}, we say that the pole at -2 is "faster" than the pole at -1. Of course, we mean that the signal corresponding to the pole at -2 decays faster than the signal corresponding to the pole at -1, but it is much easier to speak of "fast poles" and "slow poles." The point to remember is that poles farther to the left in the s plane are associated with natural signals that decay faster than those associated with poles closer to the imaginary axis.

FIGURE 2.39
Time functions
associated with points
in the *s* plane.

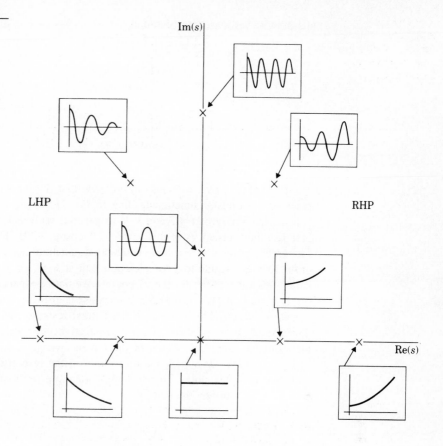

More complicated than the real poles of the previous example are the complex conjugate poles given in the transfer function of Eq. (2.68):

$$H(s) = \frac{2s + 1}{s^2 + 2s + 5}.$$
(2.68)

The poles of this transfer function are located at $s = -1 \pm j2$, and a partial fraction expansion of it can be performed with the result (which the reader should verify) of

$$
\begin{aligned}
H(s) &= \frac{(1 + j4)/j4}{s + 1 + j2} + \frac{(1 - j4)/ - j4}{s + 1 - j2} \\
&= \frac{K_1^*}{s + 1 + j2} + \frac{K_1}{s + 1 - j2} \\
&= \frac{A}{2}\frac{e^{-j\theta}}{s + 1 + j2} + \frac{A}{2}\frac{e^{j\theta}}{s + 1 - j2},
\end{aligned}
$$
(2.69)

where K_1^* is the complex conjugate of K_1. From the table of transforms, as before, the impulse response can be found, which in this case is given by

$$h(t) = \frac{A}{2} e^{-j\theta} e^{-t} e^{-j2t} + \frac{A}{2} e^{j\theta} e^{-t} e^{j2t} \qquad t \geq 0$$

$$= 0 \qquad\qquad\qquad\qquad t < 0$$

$$= \frac{A}{2} e^{-t} \left(e^{-j(2t+\theta)} + e^{j(2t+\theta)} \right) \qquad t \geq 0$$

$$= A e^{-t} \cos(2t + \theta) \qquad\qquad t \geq 0$$

$$= 0 \qquad\qquad\qquad\qquad t < 0.$$

From Eq. (2.69), it is easy to compute that $A = 2.06$ and $\theta = 14.03°$. The poles of the transfer function leading to Eq. (2.68) are complex, with real part -1 and imaginary part $+2$ and -2. In general, with real part $-\sigma$ and imaginary part $j\omega$, the transient is of the form $a e^{-\sigma t} \cos(\omega t + \theta)$. Thus, we see that to any first-order pole at $s = -a$ there corresponds a natural response to initial conditions of the form e^{-at}, and to a complex pole pair at $s = +\sigma \pm j\omega$ there corresponds a natural response of the form $e^{+\sigma t} \cos \omega t$. We notice especially that if σ is positive (in the right-half plane), the natural response is growing with time, and the system is said to be unstable. If $\sigma = 0$, the natural response neither grows nor decays; thus stability is in question, and if σ is negative, the natural response decays, and the system is stable. A sketch of the s plane with this correspondence is shown in Fig. 2.39. We now look more carefully at the response of a transfer function with two complex poles. In the control literature, it is standard to describe the second-order transfer function as

$$H(s) = \frac{\omega_n^2}{s^2 + 2\zeta\omega_n s + \omega_n^2}. \qquad (2.70)$$

The parameter ζ is called the *damping ratio*, and ω_n is called the *undamped natural frequency*.* The poles of this transfer function are located at a radius ω_n in the s plane and at an angle $\theta = \sin^{-1} \zeta$, as shown in Fig. 2.40.

In rectangular coordinates, the poles are at $s = -\sigma - j\omega_d$, where $\sigma = \zeta\omega_n$ and $\omega_d = \omega_n \sqrt{1 - \zeta^2}$. When $\zeta = 0$, we have no damping and ω_d, the damped natural frequency, equals ω_n, the undamped natural frequency. For complex poles, it is informative to present both the step response and the impulse response. Rather than work out the algebra, which is straightforward but tedious and not very informative, we present graphs of the responses of the system described by the transfer function in Eq. (2.70) for various values of the parameter ζ in Fig. 2.41. Notice that for very low damping the response is oscillatory, while for large damping (ζ near 1) the response shows very little oscillation. Also important to notice is the fact that the curves are plotted with normalized time, being plotted

---

*In communications and filter engineering, the standard second-order transfer function is written as $H = 1/[1 + Q(s/\omega_n + \omega_n/s)]$. Here ω_n is called the *band center*, and Q is the *quality factor*. If we compare this with Eq. (2.70), we can see that $Q = 1/2\zeta$.

FIGURE 2.40
s-plane plot for a pair of
complex poles.

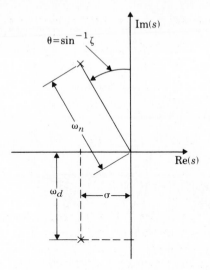

versus $\omega_n t$. Three corresponding pole locations are shown for comparison purposes in Fig. 2.42.

2.7.4 Time-Domain Specifications in Terms of Poles and Zeros

While the time responses of Fig. 2.41 are informative, they are still almost too complex to be remembered. In order to better summarize this important information, we can extract information respecting rise time, settling time, and overshoot from these curves in a simpler form. Definitions of these standard features of a step response are given in Fig. 2.43.

Comparing these definitions with the curves of the standard second-order system of Fig. 2.41, we can relate them qualitatively to the pole-location parameters ζ and ω_n. For example, the curves with moderate overshoot rise in approximately the same time, and, considering the curve for $\zeta = 0.5$, we notice that the rise time from $y = 0.1$ to 0.9 is approximately $\omega_n t_r = 1.8$. Thus, we can say that

$$t_r \approx \frac{1.8}{\omega_n}. \tag{2.71}$$

For the peak magnitude M_p, which expresses the overshoot, we can be more analytical. This value occurs when the derivative is zero, which can be found from calculus. The time history of the curves of Fig. 2.41(a) is found from the inverse Laplace transform of $H(s)/s$ to be

$$y(t) = 1 - e^{-\sigma t}\left(\cos\omega_d t + \frac{\sigma}{\omega_d}\sin\omega_d t\right),$$

where $\omega_d = \omega_n \sqrt{1 - \zeta^2}$ and $\sigma = \zeta\omega_n$.

FIGURE 2.41
(a) Step responses of
second-order systems;
(b) impulse responses of
second-order systems.

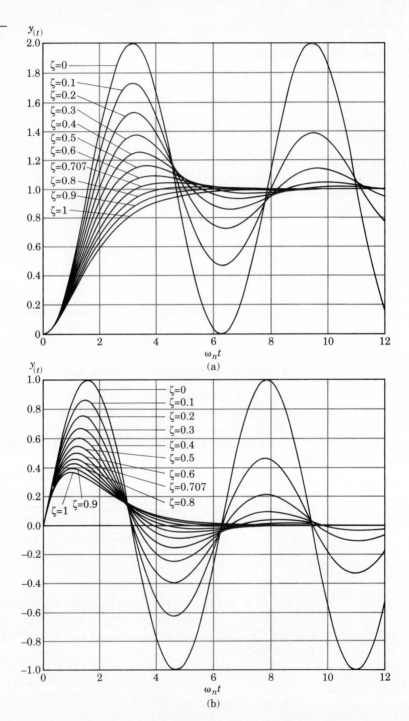

FIGURE 2.42
Pole locations corresponding to three responses of Fig. 2.41.

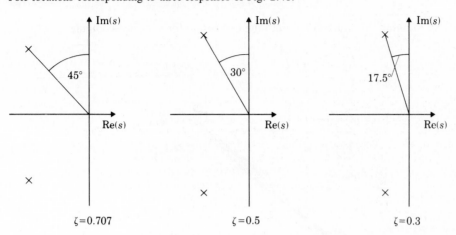

$\zeta = 0.707$ $\zeta = 0.5$ $\zeta = 0.3$

The derivative of this function is set to zero as follows:

$$\dot{y}(t) = \sigma e^{-\sigma t} \left(\cos \omega_d t + \frac{\sigma}{\omega_d} \sin \omega_d t \right) - e^{-\sigma t} (-\omega_d t \sin \omega_d t + \sigma \cos \omega_d t) = 0$$

$$= e^{-\sigma t} \left(\frac{\sigma^2}{\omega_d} \sin \omega_d t + \omega_d \sin \omega_d t \right) = 0,$$

FIGURE 2.43
Definition of rise time
t_r, settling time t_s, and
overshoot M_p.

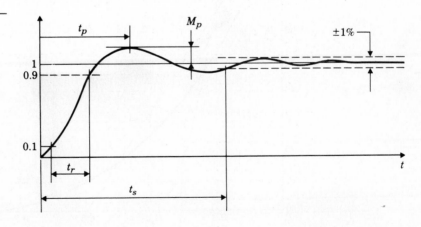

which occurs when $\sin \omega_d t = 0$, or

$$\omega_d t_p = \pi. \tag{2.72}$$

Substituting this value for t_p into the expression for $y(t)$, we compute

$$y(t_p) \triangleq 1 + M_p = 1 - e^{-\sigma\pi/\omega_d}\left(\cos \pi + \frac{\sigma}{\omega_d}\sin \pi\right)$$

$$= 1 + e^{-\sigma\pi/\omega_d}.$$

Thus, we have the formula for the second-order system

$$M_p = e^{-\pi\zeta/\sqrt{1-\zeta^2}} \qquad 0 \leq \zeta < 1 \tag{2.73a}$$

and the approximation

$$\cong 1 - \frac{\zeta}{0.6} \qquad 0 \leq \zeta \leq 0.6. \tag{2.73b}$$

A plot of this function is given in Fig. 2.44. Two frequently used values from this curve are $M_p = 0.16$ for $\zeta = 0.5$ and $M_p = 0.04$ for $\zeta = 0.707$.

FIGURE 2.44
Plot of the peak
overshoot M_p versus
the damping ratio
ζ for the second-order
system.

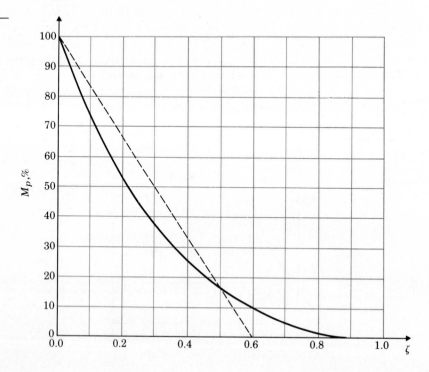

The final parameter of interest from the transient response is the settling time t_s. This is the time required for the transient to decay to a small value so that $y(t)$ is almost in the steady state. Various measures of "smallness" are possible; we will use 1% as a reasonable measure here for illustration. In other cases 2% or 5% are used. For these percentages, the constant in Eq. (2.74) would change accordingly. As an analytic computation, we notice that the deviation of y from 1 is the product of the decaying exponential $e^{-\sigma t}$ and the circular functions sin and cos. The duration of this error is essentially decided by the transient exponential, so we can define (in normalized time) the settling time as that t_s such that

$$e^{-\zeta \omega_n t_s} = 0.01.$$

Therefore,

$$\zeta \omega_n t_s = 4.6 \quad \text{or} \quad t_s = \frac{4.6}{\zeta \omega_n}$$

$$t_s = \frac{4.6}{\sigma}, \tag{2.74}$$

where σ is the real part of the pole, as may be seen from Fig. 2.40.

Equations (2.71), (2.73), and (2.74) characterize the transient response of a system having no finite zeros and two complex poles with undamped natural frequency ω_n, damping ratio ζ, and real part σ. In analysis, we can use them to estimate rise time, overshoot, and settling time for a system that is adequately described as being of this form. In synthesis, we wish to specify t_r, M_p, and t_s and to ask where the poles need to be to satisfy these specifications. In fact, we usually want rise time $\leq t_r$, overshoot $\leq M_p$, and settling time $\leq t_s$. For specified values of t_r, M_p, and t_s, the synthesis form of the equations is then

$$\omega_n \geq \frac{1.8}{t_r} \tag{2.75a}$$

$$\zeta \geq 0.6(1 - M_p) \qquad 0 \leq \zeta \leq 0.6 \tag{2.75b}$$

$$\sigma \geq \frac{4.6}{t_s}. \tag{2.75c}$$

These equations can be graphed in the s plane, as shown in Fig. 2.45(a), (b), and (c), respectively, and will be used in later chapters to guide the selection of pole and zero locations to meet control-system specifications for dynamic response. It is important to keep in mind that the three parts of Eq. (2.75) are qualitative guides and not precise design formulas.

2.7.5 Effects of an Additional Zero and an Additional Pole

Curves such as those shown in Fig. 2.45 are correct for the simple second-order system; for more complicated systems they can only be used as guidelines of the following kind: If a certain design has inadequate rise time (too slow), we must raise the natural frequency. Similarly, if the transient has too much overshoot, then damping needs to be increased, and if the transient persists too long, the poles need to be moved to the left in the s plane.

FIGURE 2.45
Graphs of regions in the s plane delineated by Eqs. (2.75) if certain transient requirements are to be met: (a) rise-time requirements;
(b) overshoot requirements; (c) settling-time requirements; and
(d) composite of all three requirements.

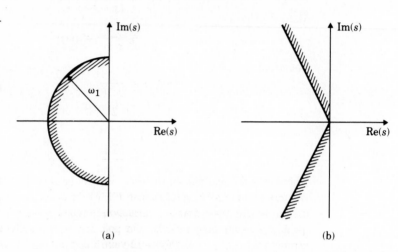

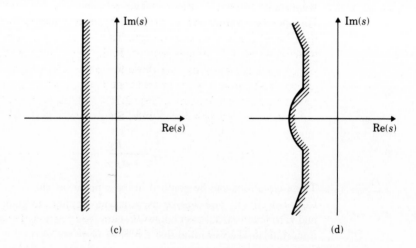

Thus far, only the poles of $H(s)$ have entered into the discussion. Because $H(s) = b(s)/a(s)$, there may be finite values of s for which $b(s) = 0$, and these are called the *zeros* of $H(s)$.* At the level of transient analysis, the zeros exert their influence by modifying the coefficients of the exponential terms whose shape is decided by the poles. To illustrate this, consider the following two transfer functions, which have the same poles but differ in the locations of their zeros. They are normalized to have the same gain at $s = 0$:

$$H_1(s) = \frac{2}{(s + 1)(s + 2)}$$

$$= \frac{2}{s + 1} - \frac{2}{s + 2}; \tag{2.76a}$$

$$H_2(s) = \frac{2(s + 1.1)}{1.1(s + 1)(s + 2)}$$

$$= \frac{2}{1.1}\left(\frac{0.1}{s + 1} + \frac{0.9}{s + 2}\right)$$

$$= \frac{0.18}{s + 1} + \frac{1.64}{s + 2}. \tag{2.76b}$$

Notice that the coefficient of the $s + 1$ term has been modified from 2 in $H_1(s)$ to 0.18 in $H_2(s)$. This dramatic reduction is brought about by the zero at $s = -1.1$ in $H_2(s)$, which almost cancels the pole at $s = -1$. Of course, if we put the zero exactly at $s = -1$, this term will vanish completely.

As an important aside to the issue of dynamic response, this discussion of the effects of the several terms on the natural response raises the possibility that some term or terms may contribute so small a component to the response that they may be omitted and the order of the model reduced, with negligible effect on the resulting control-system design. This issue is especially relevant in the study of the control of flexible structures. In this case, there are an infinity of dynamic modes present, and removing those that have negligible effect is very important. In a general transient term, the net contribution is given by the combination of size (the numerator coefficients in Eq. 2.76b) and the duration of the term, given by the size of the real part of the pole. Suppose we have

$$H(s) = \frac{R_1}{s + P_1} + \frac{R_2}{s + \sigma_2 + j\omega_2} + \frac{\bar{R}_2}{s + \sigma_2 - j\omega_2} + \cdots.$$

---

*We assume that $b(s)$ and $a(s)$ have no common factors. If this is not so, it is possible for $b(s)$ and $a(s)$ to be zero at the same location and $H(s)$ to be not zero there. The implications of this case will be discussed in Chapter 6 when we have a state-space description.

If the contribution of a given term calculated by $|R|/\sigma_j$ is much smaller than the contribution of other terms, then the negligible term can be replaced by its DC gain, $R_2/(\sigma_2 + j\omega_2)$. In Eq. (2.76b), the contributions are

$$\frac{0.18}{1} \qquad \text{and} \qquad \frac{1.64}{2} = 0.82.$$

If the smaller term is neglected, the approximation becomes

$$H_2(s) \approx 0.18 + \frac{1.64}{s + 2}.$$

It is important to notice that in this case it is the *lowest* frequency term that is removed.

To consider the effects of zeros on transient response in order to take them into account in design, we consider transfer functions with two complex poles and one zero. In order to present the responses in a form easy to remember and to plot, we write the transform in a form with normalized time and zero location as follows:

$$H(s) = \frac{(s/\alpha\zeta\omega_n) + 1}{(s/\omega_n)^2 + 2\zeta(s/\omega_n) + 1}. \tag{2.77}$$

The zero is located at $s = -\alpha\zeta\omega_n = -\alpha\sigma$, so, if α is large, the zero is far removed from the poles and will have little effect on the response. If $\alpha = 1$, the zero is at the value of the real part of the poles and could be expected to have a substantial influence on the response. The step response curves for $\zeta = 0.5$ and for several values of α are plotted in Fig. 2.46. We see that the major effect of the zeros is to increase the overshoot M_p with very little influence on the settling time. A plot of M_p versus α is given in Fig. 2.47.

It is informative to explain Fig. 2.46 in terms of Laplace-transform analysis. To do so, we replace s/ω_n with s, which means we present the transfer function in normalized frequency and the corresponding step responses in normalized time, $\tau = \omega_n t$:

$$H(s) = \frac{s/\alpha\zeta + 1}{s^2 + 2\zeta s + 1}.$$

We can write this transfer function as the sum of two terms:

$$H(s) = \frac{1}{s^2 + 2\zeta s + 1} + \frac{1}{\alpha\zeta} \frac{s}{s^2 + 2\zeta s + 1}. \tag{2.78}$$

The first term is the original (no finite zero) term, and the second term, which is introduced by the zero, is a constant $(1/\alpha\zeta)$ times s times the original term. The

FIGURE 2.46
Plots of the step response
of a second-order system
with an extra zero (ζ =
0.5).

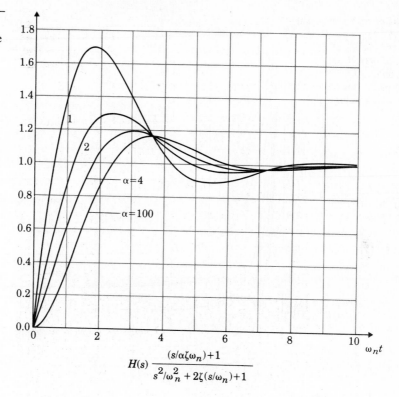

$$H(s) \quad \frac{(s/\alpha\zeta\omega_n)+1}{s^2/\omega_n^2 + 2\zeta(s/\omega_n)+1}$$

Laplace transform of df/dt is $sF(s)$, so this second term corresponds to adding some of the derivative to the original term. The standard term and its derivatives are plotted in Fig. 2.48. Looking at these curves, it is fairly obvious why the zero increased the overshoot—the derivative has a large hump at the early part of the curve, and adding this to $h(t)$ lifts up the response to produce the overshoot. This analysis is also very informative in case the zero is in the right-half plane. (This is the nonminimum phase case discussed in later sections.) In this case, α is negative, and the derivative term is subtracted rather than added. A typical case is sketched in Fig. 2.49.

In addition to studying the effects of an extra zero, it is useful to consider the effects of an extra pole on the standard second-order step response. In this case, we take the transfer function to be

$$H(s) = \frac{1}{(s/\alpha\zeta\omega_n + 1)[(s/\omega_n)^2 + 2\zeta(s/\omega_n) + 1]}. \qquad (2.79)$$

Plots of the step response for this case are shown in Fig. 2.50 for ζ = 0.5 and

FIGURE 2.47
Plot of overshoot M_p as a function of normalized zero location α. At $\alpha = 1$, the real part of the zero equals the real part of the poles.

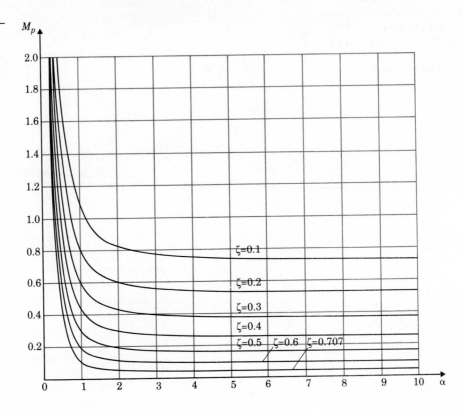

FIGURE 2.48
Plot of the second-order step response and its derivative.

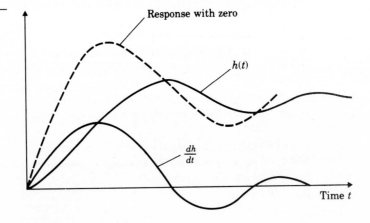

FIGURE 2.49
Plot of the response of a
second-order system with
a right-half plane zero:
a nonminimum phase
system.

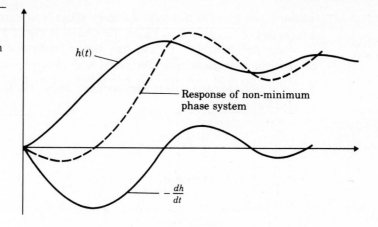

$h(t)$

Response of non-minimum
phase system

$-\dfrac{dh}{dt}$

FIGURE 2.50
Plot of step response for
several third-order
systems with $\zeta = 0.5$.

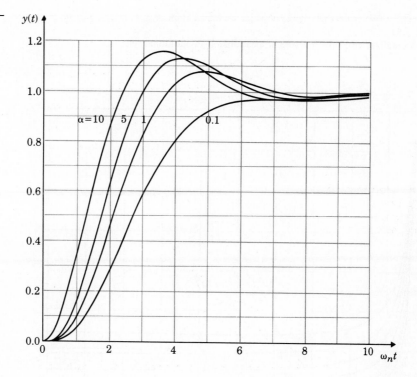

several values of α. In this case, the major effect is to increase the rise time, and a cross-plot of this parameter versus α is shown in Fig. 2.51.

From this discussion, we can draw several qualitative conclusions respecting the dynamic response of a simple system as disclosed by its pole-zero patterns:

1. For a second-order system with no finite zeros, the transient parameters are approximated by

$$\text{Rise time}: \quad t_r \approx \frac{1.8}{\omega_n}$$

$$\text{Overshoot}: \quad M_p \approx \left(1 - \frac{\zeta}{0.6}\right) \qquad 0 \leq \zeta \leq 0.6$$

$$\text{Settling time}: \quad t_s \approx \frac{4.6}{\sigma}$$

FIGURE 2.51
Plot of normalized rise time for several locations of an additional pole.

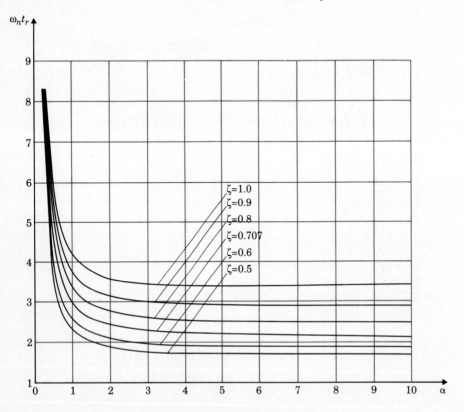

2. An additional zero in the left-half plane will increase the overshoot if the zero is within a factor of 4 of the real part of the complex poles. A plot is given in Fig. 2.47.

3. An additional zero in the right-half plane will depress the overshoot (and may cause the step response to start out in the wrong direction).

4. An additional pole in the left-half plane will increase the rise time significantly if the extra pole is within a factor of 4 of the real part of the complex poles. A plot is given in Fig. 2.51.

□ 2.8 Obtaining Models from Experimental Data

There are several reasons for wishing to obtain a model of a dynamic system to be controlled from experimental data. In the first place, the best of theoretical model building from equations of motion is only an approximation of the reality. Sometimes, as in the case of an essentially rigid spacecraft, the theoretical model is extremely good. And sometimes, as in the case of many chemical processes such as papermaking or metalworking, the theory is very approximate. In every case, before final control design is done, it is important and prudent to verify the theoretical model with experimental data. In cases where the theoretical model is especially complicated or the science of the process is poorly understood, the only reliable information on which to base the control design is the experimental data. Finally, the system is sometimes subject to on-line changes, as occurs when the environment of the system changes (e.g., an aircraft changes altitude; a paper machine is given a different composition of fiber; a nonlinear system moves to a new operating point). On these occasions, we need to change the control parameters and thus "tune" the controller. In order to do this, it is necessary to have a model under the new conditions, and experimental data are often the most effective, if not the only information available to do this.

The experimental data for generating a model are of four kinds: transient, such as comes from an impulse or a step; sinusoidal steady state of various frequencies; stochastic steady state, as might come from a signal derived from random fluctuations of electrons (thermal noise) or from some other natural source in the process itself; and pseudorandom noise, as may be generated in a digital computer. Each of these classes of data has its properties, advantages, and disadvantages.

Transient response data is quick and relatively easy to obtain. It is also often representative of the natural signals to which the system is subjected, and thus a model based on such data can be a reliable basis for the design of a control system. On the other hand, in order for the signal-to-noise level to be sufficiently high, it is necessary for the transient to be highly noticeable. Thus the method is rarely suitable for normal operations, and the data must be collected as part of special tests. Also, the data is not in a form suitable for standard control-system designs, and the model, such as poles and zeros, must be computed from the data.* This computation can be simple in special cases or complex in the general case.

Frequency response data (see Chapter 5) is simple to obtain but substantially more time-consuming than transient response. This is especially so if the time constants of the process are long, as often occurs in the chemical process industries. As with the transient data, it is important to have a good signal-to-noise ratio, so tests to obtain frequency response data can be very expensive. On the other hand, as we will see in Chapter 5, frequency response is exactly what is required for one of the more effective (Bode plot) methods of control-system design, where, once the data is obtained, the control design can immediately proceed.

Normal operating records are an attractive basis for system model making since they are, by definition, nondisruptive and inexpensive to obtain. Unfortunately, the quality of such data is very mixed and is worst just when the control is best, for then the upsets are minimal and the signals are smooth. Unfortunately, at such times some or even most of the system dynamics are hardly excited, and, since they contribute very little to the system output, they will not be found in the resulting model built to explain the signals. The result is a model that represents only part of the system and is sometimes unsuitable for control. In some instances, as occurs when one tries to model the dynamics of the electroencephalogram (or brain waves) of a sleeping or anesthetized person to locate the frequency and intensity of α waves, for example, normal records are the only possibility. Usually they are the last choice for control purposes.

Finally, the pseudorandom signals that can be constructed by digital logic have much appeal. Especially interesting for model making is the pseudorandom binary signal, or PRBS. This is a signal that takes on the value $+A$ or $-A$, according to the output (1 or 0) of a feedback shift register. The feedback to the register is a binary sum of various states of the register so selected that the output period (which must repeat itself in finite time) is as long as possible. With a register of 20 bits, for example, $2^{20} - 1$, or over a million, pulses are produced before the pattern repeats. Analysis beyond our scope has revealed that the resulting signal is almost like a broadband random signal. And yet this signal is entirely under the control of the engineer who can set the level (A) and the length (bits in the register) of the signal. Analysis of the data obtained from a test with a PRBS must be done by machine, and both special-purpose hardware and programs for general-purpose computers have been developed.

2.8.1 Models from Transient Response Data

To obtain a model from transient data, we assume that a step response is available. If the transient is a simple combination of elementary transients, a reasonable low-

---

*Ziegler and Nichols, building on the earlier work of Callender, Hartree, and Porter, use step response directly in design of certain classes of processes. See Chapter 3 for details.

order model can be estimated by hand calculations. For example, consider the step response shown in Fig. 2.52. The response is monotonic and smooth. If we assume that it is given by a sum of exponentials, we can write

$$y(t) = y(\infty) + Ae^{-\alpha t} + Be^{-\beta t} + Ce^{-\gamma t} + \cdots. \tag{2.80}$$

Subtracting off the final value and assuming $-\alpha$ is the smallest (slowest) root, we write

$$y - y(\infty) \cong Ae^{-\alpha t} + \cdots$$
$$\log_{10}[y - y(\infty)] \cong \log_{10} A - \alpha t \log_{10} e$$
$$\cong \log_{10} A - 0.4343\alpha t. \tag{2.81}$$

This is the equation of a line. If we fit a line to the plot of $\log_{10}[y - y(\infty)]$, or $\log[y(\infty) - y]$ if A is negative, then we can estimate A and α. Once these are estimated, we plot $y - y(\infty) - Ae^{-\alpha t}$, computed by subtracting the line $Ae^{-\alpha t}$ from the original data. Now we have a curve that is approximately $Be^{-\beta t}$ and on the log plot is $\log_{10} B - 0.4345\beta t$. We repeat the process until the accuracy of the data is lost. The final model step response should be plotted, and then the error (residual) should be plotted to estimate the quality of the computed model. Notice that we can get a good fit to the step response and yet be far off from the true time constants (poles) of the system. However, the method gives a good approximation for control of processes whose step responses look like Fig. 2.52.

An example of this method may make the procedure more clear. Consider the data of Table 2.1 taken from Sinha and Kuszta (1983). A plot of this data is given in Fig. 2.53. From the line (fitted by eye), the values are

$$\log_{10}|A| = 0.125$$
$$0.4343\alpha = \frac{1.602 - 1.167}{\Delta t} = \frac{0.435}{1} \Rightarrow \alpha \cong 1.$$

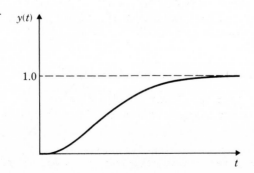

FIGURE 2.52
A step response characteristic of many chemical processes.

TABLE 2.1
A Set of Step Response Data

t	$y(t)$	$\hat{y}(t)$	t	$y(t)$	$\hat{y}(t)$
0.0	0.000	0.00	1.0	0.510	0.507
0.1	0.005	−0.022	1.5	0.700	0.701
0.2	0.034	0.01	2.0	0.817	0.819
0.3	0.085	0.068	2.5	0.890	0.89
0.4	0.140	0.134	3.0	0.932	0.933
0.5	0.215	0.206	4.0	0.975	0.975
				1.000	1.000

Thus

$$A = -1.33$$
$$\alpha = 1.0.$$

We know that A is negative because $y(\infty)$ is greater than $y(t)$. If we now subtract

FIGURE 2.53
Plot of step response data.

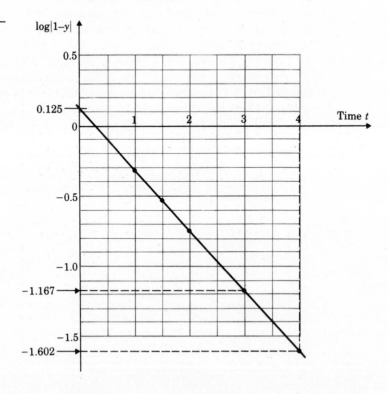

this line from the data and plot the log of the result, we find the plot of Fig. 2.54. Here we estimate

$$\log B = -0.45$$

$$0.4343\beta = \frac{-0.45 - (-1.7)}{0.5} = 2.5$$

$$\beta \cong 5.8$$

$$B = 0.35.$$

Combining these results, we estimate that

$$y(t) \cong 1 - 1.33e^{-t} + 0.35e^{-5.8t}.$$

Since it is clear that $y(0) = 0$, we modify the estimates to reflect this constraint to find

$$\hat{y}(t) \cong 1 - 1.34e^{-t} + 0.34e^{-5.8t}. \tag{2.82}$$

FIGURE 2.54
Plot of $1 - y - Ae^{-\alpha t}$ for step response of Table 2.1.

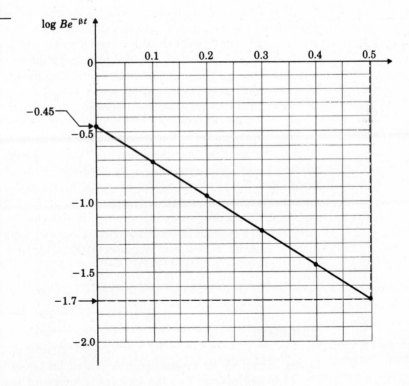

From $\hat{y}(t)$ we compute

$$\begin{aligned}
\hat{Y}(s) &= \frac{1}{s} - \frac{1.34}{s+1} + \frac{0.34}{s+5.8} \\
&= \frac{(s+1)(s+5.8) - 1.34s(s+5.8) + 0.34s(s+1)}{s(s+1)(s+5.8)} \\
&= \frac{-0.63s + 5.8}{s(s+1)(s+5.8)}.
\end{aligned}$$

The resulting transfer function is

$$G(s) = \frac{-0.63s + 5.8}{(s+1)(s+5.8)}. \tag{2.83}$$

Notice that this method has given us a system with a zero in the right-half plane, a fact that may influence the design. This small example has illustrated the sensitivity of pole locations to the quality of the data and emphasizes that to use this data effectively one must have a good signal-to-noise ratio.

If the transient response has oscillatory modes, these can sometimes be estimated by comparison with the standard plots of Fig. 2.41. The period will give the frequency ω_n, and the decay from one period to the next will afford an estimate of the damping ratio. If the response has a mixture of modes not well separated in frequency, more sophisticated methods such as least-squares identification need to be used.

Frequency response data, obtained by exciting the system with a set of sinusoids and plotting $H(j\omega)$, can be used to generate a model. In Chapter 5, we will show how such plots can be used directly for design, or how using straight-line asymptotes, the frequency response can be used to estimate the poles and zeros of a rational transfer function.

The construction of dynamic models from normal stochastic operating records or from the response to a pseudorandom binary sequence can be based either on the idea of cross-correlation or by least-squares fit of a discrete equivalent model. Both these methods require substantial background we have not developed, the presentation of which would take us too far afield at this point. Suffice it to say that the field of systems identification is concerned with this problem, and many effective methods have been developed. An introduction to the topic is to be found in Franklin, Powell, and Workman (1990), Chapter 8, and a comprehensive treatment is given in Ljung (1987).

Summary

In this chapter we have presented a review of background material essential to control-system design. This has included a review of writing the equations of motion for elementary mechanical, electrical, electromechanical, thermal, and fluid systems. Because many of these models are nonlinear and the most effective design methods are based on linear models, the topic of linearization was presented and illustrated with an example of a laboratory experiment in magnetic levitation.

With a model in hand, the control engineer needs to be able to analyze the response; especially the engineer must be able to predict the qualitative character of the response based on the transfer-function poles and zeros. For this purpose we have presented the connection between a pattern of poles and zeros and the time response. This information has also been used to relate time-domain specifications to requirements on pole-zero locations. It is also important for the control engineer to be able to interpret the frequency response of dynamic systems and to understand the state-variable method of describing control systems. These topics of background analysis have been deferred until Chapters 5 and 6, where they will be introduced as needed.

Finally, in recognition of the fact that many dynamic systems needing control are described by complex and poorly understood physical principles, the process and techniques of obtaining a model from experimental data have been briefly described.

Problems Section 2.2

2.1 Write the differential equations for the mechanical system shown in Fig. 2.55.

2.2 Write the equations of motion for a single body of mass M suspended from a fixed point by a spring of force constant k.

2.3 Write the equations of motion of a clock pendulum consisting of a 1-kg bob that may be considered to be a point mass suspended from a fixed point by a thin rod of length ℓ. How long should the rod be for the period to be exactly one second if we take $\sin \theta \sim \theta$?

2.4 Write the equations of motion for a pendulum formed by pinning a uniform stick of length ℓ and mass M at one end to a fixed point. Assume a torque τ can be applied at the pin.

FIGURE 2.55
Example mechanical system.

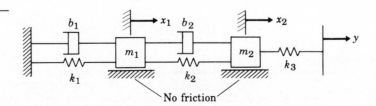

Section 2.3

2.5 Use node analysis to write the equations of motion for the circuits shown in Fig. 2.56.

   a) The lead network.
   b) The lag network.
   c) The notch network.

2.6 Use node analysis to write the equations of motion of the op-amp circuits in Fig. 2.57. Assume ideal operational amplifiers in every case.

   a) A first-order op-amp circuit
   b) A second-order op-amp circuit
   c) The Sallen-Key circuit

FIGURE 2.56
Example electric circuits:
(a) lead, (b) lag, and
(c) notch circuits.

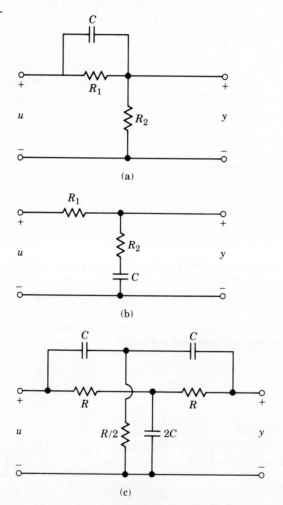

(a)

(b)

(c)

FIGURE 2.57
Op-amp circuits: (a) first-order transfer function; (b) second-order transfer function; (c) Sallen-key circuit.

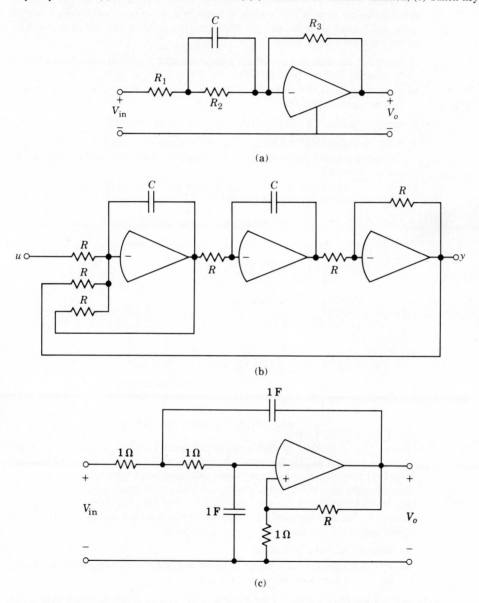

(a)

(b)

(c)

Section 2.4

2.7 The torque constant of a motor is the ratio of torque to current and is often given in ounce-inches per ampere. The electric constant of a motor is the ratio of back emf to speed and is often given in volts per 1000 rpm. In consistent units, the two constants are the same for a given motor.

a) Show that the units ounce-inch per ampere are proportional to volts per 1000 rpm by reducing both to MKS (SI) units.

b) A certain motor has a back emf of 25 V at 1000 rpm. What is its torque constant in ounce-inches per ampere?

2.8 A simplified sketch of a computer tape drive is given in Fig. 2.58.

a) Write the equations of motion in terms of the literal parameters. K and B represent the spring constant and damping of tape stretch, respectively, and ω_1 and ω_2 are angular velocities.

b) Use the values below and write the equations as a set of first-order differential equations (*in state form*). Use the variables $(x_1, \omega_1, x_2, \omega_2)$.

$$J_1 = 4 \times 10^{-5}\,\text{kg} \cdot \text{m}^2,\ \text{motor and capstan}$$

$$B_1 = 1 \times 10^{-2}\,\text{N} \cdot \text{m} \cdot \text{s},\ \text{motor damping}$$

$$r_1 = 2 \times 10^{-2}\,\text{m}$$

$$K_t = 3 \times 10^{-2}\,\text{N} \cdot \text{m/A},\ \text{motor torque constant}$$

$$K = 2 \times 10^4\,\text{N/m}$$

$$B = 20\,\text{N/m/s}$$

$$r_2 = 2 \times 10^{-2}\,\text{m}$$

$$J_2 = 1 \times 10^{-5}\,\text{kg} \cdot \text{m}^2$$

$$B_2 = 1 \times 10^{-2}\,\text{N} \cdot \text{m} \cdot \text{s},\ \text{viscous damping, idler}$$

$$F = 6\,\text{N},\ \text{constant force}$$

$$\dot{x}_1 = \text{tape velocity (variable to be controlled)}.$$

2.9 A very typical problem of electromechanical position control is an electric motor driving a load that has one dominant vibration mode. The problem arises in computer-disk-head control, reel-to-reel tape drives, and many other applications. A schematic diagram is sketched in Fig. 2.59. The motor has electrical constant K_e, torque constant K_t, armature inductance L_a, and resistance R_a. The rotor has inertia J_1 and viscous friction B. The load has inertia J_2. The two inertias are connected by a shaft that has spring constant k and equivalent viscous damping b.

a) Write the equations of motion.

b) Write the equations as a set of simultaneous first-order equations (state-variable form). Use state $\mathbf{x} = [\theta_2 \dot{\theta}_2 \theta_1 \dot{\theta}_1 i_a]^T$.

2.10 Assume the driving force on the hanging crane of Fig. 2.9 is provided by a motor mounted on the cab and with one of the support wheels on the motor armature shaft. The motor constants are K_e, K_t, and R_a. The wheel has radius r. Write the equations of motion for this case.

FIGURE 2.58
Tape drive.

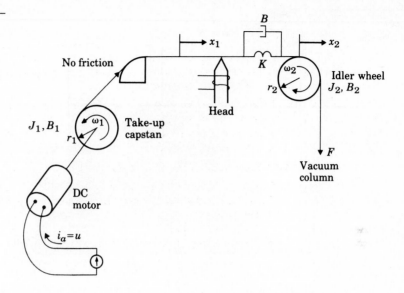

Section 2.5

2.11 A sketch of a laboratory experiment in fluid (water) flow (two-tank fluid flow) is shown in Fig. 2.60. Assume that Eq. (2.36) describes flow through the equal-sized holes at A, B, or C.

a) With holes at A and C and no holes at B, write the equations of motion for this system in terms of h_1 and h_2. Assume that h_3 is 20 cm and h_2 is less than 20 cm. When h_2 is 10 cm, the outflow is 200 g/min.

b) At $h_1 = 30$ cm and $h_2 = 10$ cm, compute a linearized model and the transfer function from pump flow (in cubic centimeters per minute) to h_2.

c) Repeat (**a**) and (**b**) if hole A is closed and B is opened.

FIGURE 2.59
Motor with flexible load.

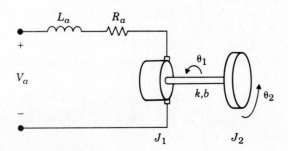

FIGURE 2.60
Fluid flow, two-tank
problem.

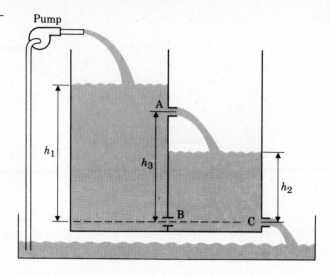

Section 2.6

2.12 Consider the network shown in Fig. 2.61; u_1, u_2 are voltage and current sources, respectively; 1 and 2 are nonlinear resistors with characteristics of:

$$\text{Resistor 1}: i_1 = G(v_1) = v_1^3$$
$$\text{Resistor 2}: v_2 = r(i_2),$$

where the function r is defined in Fig. 2.62.

a) Show that the equations can be written as:

$$\dot{x}_1 = G(u_1 - x_1) + u_2 - x_3$$
$$\dot{x}_2 = x_3$$
$$\dot{x}_3 = x_1 - x_2 - r(x_3).$$

FIGURE 2.61
A nonlinear circuit.

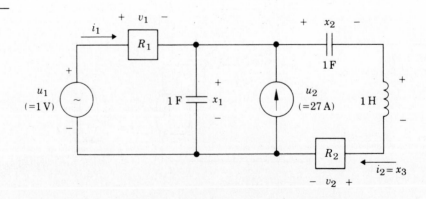

FIGURE 2.62
Nonlinear resistance.

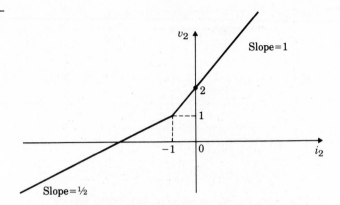

FIGURE 2.62
Nonlinear resistance.

Suppose we have a constant voltage source of 1 V at u_1 and a constant current source of 27 A; i.e., $u_1^0 = 1$, $u_2^0 = 27$. Find the equilibrium state (x_1^0, x_2^0, x_3^0) for the network. Given a particular input $\mathbf{u}^0$, the equilibrium state of $\mathbf{x}^0$ of the system is a constant vector such that the system will stay where it started. In terms of the state equations,

$$\dot{x}_1 = \dot{x}_2 = \dot{x}_3 = 0.$$

b) Due to disturbances, the initial state (capacitance, voltages, and inductor current) is slightly different from the equilibrium and so are the independent sources; that is,

$$u(t) = u^0 + \delta u(t)$$
$$x(t_0) = x^0(t_0) + \delta x(t_0).$$

Do a small-signal analysis of the network about the equilibrium found in (a), displaying the equations in the form

$$\delta \dot{x}_1 = f_{11}\delta x_1 + f_{12}\delta x_2 + f_{13}\delta x_3 + g_1\delta u_1 + g_2\delta u_2.$$

c) Draw the circuit diagram that corresponds to the linearized model. Give the values of the elements.

2.13 Consider the network in Fig. 2.63 with a nonlinear conductance G, where $i_G = g(v_G) = v_G(v_G - 1)(v_G - 4)$. The state differential equations are

$$\frac{di}{dt} = -i + v$$

$$\frac{dv}{dt} = -i + g(u - v),$$

where i and v are the states and u is the input.

a) One equilibrium state occurs when $u^* = 1$ is $i_1^* = v_1^* = 0$. Find the other two pairs of v and i that will produce equilibrium.

b) Find the linearized model of the system about the equilibrium point $u^* = 1$, $i^* = v_1^* = 0$.

c) Find the linearized models about the other two equilibrium points.

FIGURE 2.63
Nonlinear circuit for
Problem 2.13.

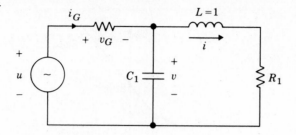

2.14 Show that we can scale the variables so that the equations of motion for the "stick-on-a-cart" problem given in Eq. (2.15) can be written as

$$\ddot{\theta} = \theta + u$$

$$\ddot{y} = \theta - u,$$

where u is a force input.

a) Draw a block diagram of this system with u as the input and y as the output *using only integrators* as the dynamic elements.

b) What is the transfer function from u to y? (Be careful with the positive feedback loop.)

☐ **2.15** Consider the single-spool turbojet described by,

$$\dot{x} = Fx + Gu$$

$$y = Hx,$$

where

$$x = \begin{bmatrix} x_1 \\ x_2 \end{bmatrix} \quad F = \begin{bmatrix} -1 & -0.5 \\ 10 & -5.0 \end{bmatrix} \quad G = \begin{bmatrix} 2 \\ 0 \end{bmatrix} \quad H = \begin{bmatrix} 0.1 & 18 \end{bmatrix}$$

x_1 = Rotor Speed (rpm),
x_2 = Tailpipe Pressure (psi),
u = Fuel Flow (pph),
y = Thrust (lbs).

The scaling matrices are

$$D_x = \begin{bmatrix} 15760 \text{ rpm} & 0 \\ 0 & 27 \text{psi} \end{bmatrix}$$

$$D_u = 1770 \text{ pph}$$

$$D_y = 2062 \text{ lbs}.$$

Compute the normalized system matrices.

Section 2.7

2.16 Compute the poles and zeros of the turbojet engine described in the previous problem.

2.17 Compute the transfer functions for the circuits in

    a) Fig. 2.56(a)
    b) Fig. 2.56(b)
    c) Fig. 2.56(c)
    d) Fig. 2.57(a)
    e) Fig. 2.57(b)
    f) Fig. 2.57(c)

2.18 Compute the transfer function from motor voltage to position θ_2 for the system in Fig. 2.59.

2.19 Compute the transfer function from motor current to tape tension for the tape drive with the values given in Fig. 2.58.

2.20 (Computer problem) Compute the poles and zeros of the tape drive transfer function from current to position.

2.21 Compute the transfer function for the two-tank problem of Fig. 2.60 with holes at A and C.

2.22 Consider the second-order system with the transfer function,

$$G(s) = \frac{3}{(s^2 + 2s - 3)}.$$

    a) Determine the DC gain for this system.
    b) What is the final value of the step response of this system?

2.23 **a)** Compute the transfer function of the block diagram shown in Fig. 2.64 by Mason's rule.
    b) Write the third-order differential equation that relates y and u. (*Hint:* Consider the transfer function.)
    c) Write three simultaneous first-order (state-variable) differential equations using variables x_1, x_2, and x_3, as defined on the block diagram in Fig. 2.64. Notice how the same constant parameters enter the transfer function, the differential equations and the matrices of the state variable form in this case. Because of these simple relationships, the system is said to be in *control canonical form*.

2.24 Find the transfer function for the block diagrams in Fig. 2.65.

2.25 Find the transfer functions of the block diagrams in Fig. 2.66(a), (b), (c), and (d) by the rules of Fig. 2.30 and Fig. 2.31.

2.26 Find the transfer functions of the block diagrams in Fig. 2.66(a), (b), (c) and (d) by Mason's rule.

2.27 Consider the system shown in Fig. 2.67:
    Let

$$G(s) = \frac{1}{(s + 3)s} \qquad G_c(s) = \frac{K(s + z)}{(s + p)}.$$

FIGURE 2.64
Control canonical form
block diagram.

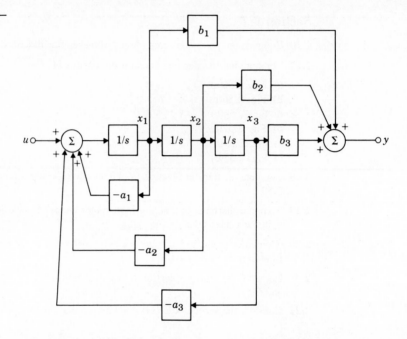

Find K, z, and p such that the closed-loop system has a 5% overshoot to a step input and a settling time of 4/3 s (1% criterion). You may start with $z = 3$, but it is not acceptable as the final answer. (*Note:* Theoretically you can achieve the transient response specifications exactly.)

2.28 Given the following $G(s)$, sketch the step response based on the locations of the poles and zeros.

$$G(s) = \frac{s/2 + 1}{(s/40 + 1)\,[(s/4)^2 + s/4 + 1]}$$

2.29 Draw the region in the s plane that corresponds to meeting specifications on the peak time $t_p < t'_p$, where t'_p has been specified.

2.30 The open-loop transfer function of a unity negative feedback system is

$$\mathbf{G}(s) = \frac{K}{s(s + 2)}.$$

The system response to a step input is specified as follows: peak time $t_p = 1$ and percent overshoot $M_p = 5\%$.

a) Determine whether both specifications can be met simultaneously by selection of K.

b) If the specifications cannot be met simultaneously, determine the value of K such that both specifications are relaxed by the same percentage.

c) Sketch the associated region in the s plane.

2.31 Consider the system described by the single loop shown in Fig. 2.67. It may be imagined as a system with an infinite number of paths. One path passes directly

FIGURE 2.65
Example block diagrams.

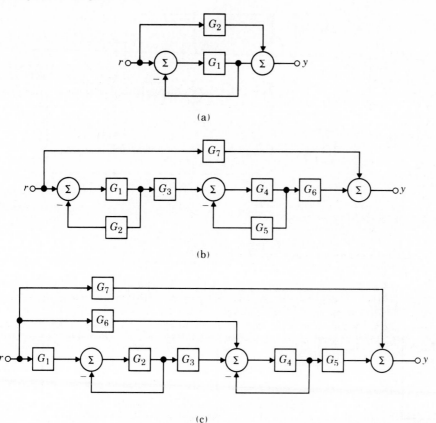

(a)

(b)

(c)

through the node. The other path goes through the loop once before leaving. The third path goes through the loop twice before leaving and so on. Determine the transfer function from this interpretation of the loop.

2.32 Show that for a second-order system, the initial condition response is given by

$$y(t) = y_0 \frac{e^{-\sigma t}}{\sqrt{1 - \zeta^2}} \sin\left(\omega_d t + \cos^{-1} \zeta\right),$$

Prove that for the underdamped case (i.e., $\zeta < 1$), the response oscillations decay at a predictable rate (see Fig. 2.68) called the *logarithmic decrement* (δ), where

$$\delta = \ln\left(\frac{y_0}{y_1}\right) = \sigma \tau_d$$

$$= \ln\frac{\Delta y_1}{y_1} \cong \ln\frac{\Delta y_i}{y_i}$$

FIGURE 2.66
Block diagrams for Problems 2.25 and 2.26.

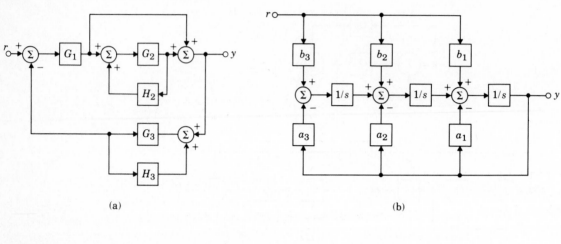

(a)   (b)

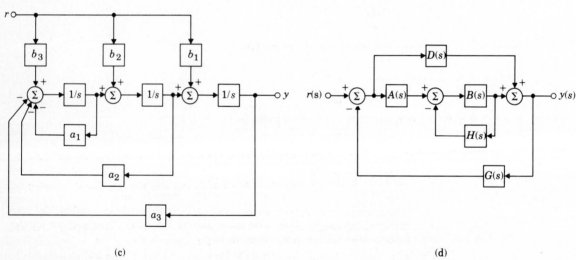

(c)   (d)

FIGURE 2.67
A unity feedback system.

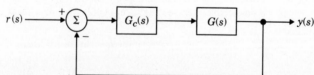

FIGURE 2.68
Definition of logarithmic
decrement.

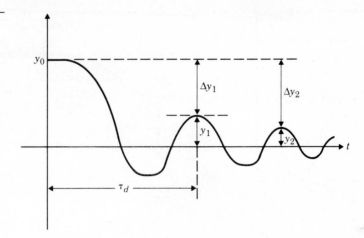

and τ_d is the damped natural period of vibration

$$\tau_d = \frac{2\pi}{\omega_d}.$$

2.33 Consider the two nonminimum phase systems,

(i) $G_1(s) = \dfrac{-2(s-1)}{(s+1)(s+2)}$

(ii) $G_2(s) = \dfrac{3(s-1)(s-2)}{(s+1)(s+2)(s+3)}$

a) Sketch the unit step responses for $G_1(s)$ and $G_2(s)$, paying close attention to the start of the response.
b) Explain the difference in the behavior of the two responses as it relates to the zero locations.
c) Consider a stable strictly proper system (i.e., m zeros, n poles, $m < n$). Let $y(t)$ denote the step response of the system. The step response is said to have an "undershoot" if it initially "starts off in the wrong direction." Prove that a system such as the one described has an undershoot if and only if its transfer function has an *odd* number of *real* right-half plane zeros.

2.34 Consider the second-order system with an extra pole

$$H(s) = \frac{\omega_n^2 p}{(s+p)(s^2 + 2\zeta\omega_n s + \omega_n^2)}.$$

Show that the unit step response is

$$y(t) = 1 + Ae^{-pt} + Be^{-\sigma t}\sin(\omega_d t - \theta),$$

FIGURE 2.69
Time to double.

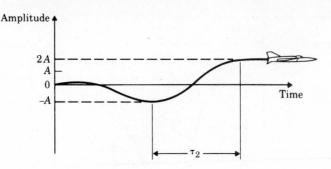

where

$$A = \frac{-\omega_n^2}{\omega_n^2 - 2\zeta\omega_n p + p^2}$$

$$B = \frac{p}{\sqrt{p^2 - \zeta^2(p^2 - 2\zeta\omega_n p + \omega_n^2)}}$$

$$\theta = \tan^{-1} \frac{\sqrt{1 - \zeta^2}}{-\zeta} + \tan^{-1} \frac{\omega_n \sqrt{1 - \zeta^2}}{p - \zeta\omega_n}.$$

a) Which term dominates $y(t)$ as p gets large?
b) Give approximate values for A and B for small p.
c) Which term dominates as p gets small? (Small with respect to what?)

2.35 A measure of the degree of *instability* in an unstable aircraft response is the amount of time it takes for the *amplitude* of the time response to double (see Fig. 2.69). Show that for a first-order system *the time to double* (τ_2) is

$$\tau_2 = \frac{\ln 2}{p},$$

where p is the pole location in the RHP. Similarly, show that the time to double for a second-order system (with two complex poles in RHP) is

$$\tau_2 = \frac{\ln 2}{-\zeta\omega_n}.$$

Section 2.8

2.36 Samples from a step response are given in Table 2.2. Plot this data on a linear [$y(t)$ vs. t] and semilog [$\log(y - y_\infty)$ vs. t] scale and obtain an estimate of the transfer function.

TABLE 2.2

t	y	t	y
0	0		
0.02	0.0001	0.30	0.0771
0.04	0.0005	0.50	0.1979
0.06	0.0014	0.60	0.2624
0.08	0.0031	0.70	0.3253
0.10	0.0057	0.80	0.3851
0.12	0.0091	0.90	0.4409
0.14	0.0135	1.00	0.4924
0.16	0.0187	1.50	0.6904
0.18	0.0248	2.00	0.8121
0.20	0.0138	2.50	0.8860
0.22	0.0395	3.00	0.9309
0.24	0.0480	3.50	0.9581
0.26	0.0571	4.00	0.9746
0.28	0.0668	5.00	0.9907

3

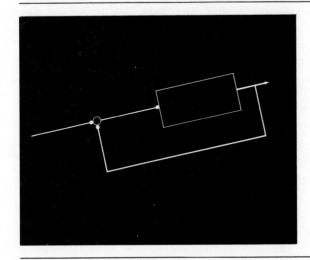

Basic

Principles

of Feedback

The typical control problem begins with a physical dynamic system that has an output that is required to follow some externally-set desired path over time and has a control input that can be set by the designer. The reference path may be a constant, in which case the problem is called one of *regulation,* or the reference path may vary in some unforeseen way, in which case the problem is one of *track following*. In the generic case, the output is prevented from matching the reference exactly because external disturbances that are present influence the controlled output and also because the equations of motion of the system are not known perfectly, so that the designer cannot compute exactly what the effect of a given control action may be. There are two basic system structures for achieving control of dynamic systems: *open-loop control,* as shown in Fig. 3.1, and *closed-loop control,* as shown in Fig. 3.2. Notice that closed-loop control requires introduction of an output sensor, which usually brings noise as well as a measure of the output variable. In this chapter, we will compare these two structures from the point of view of their effectiveness in keeping the system error small in the face of disturbances and model errors. We also will explore the consequences of control on the dynamic response of the system.

We begin with a case study using speed control to illustrate the properties of feedback and open-loop control. We consider first constant gain controllers, which in the process industries are given the name *proportional control*. We next show

FIGURE 3.1
Open-loop control
system.

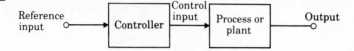

how *integral control* can be introduced to obtain improved steady-state tracking properties, and finally how *derivative control* can correct some of the bad side effects of integral control. Considered together, these are called *PID control* and constitute the elementary heuristic approach to controller design, which has found wide acceptance in the process industries. Before leaving this introduction to the advantages of feedback, we consider the general case of steady-state tracking of polynomial reference inputs. In the final section, we consider the definition of system stability and Routh's test, which can determine if the roots of a characteristic equation are all in the left-half plane.

3.1 A Case Study of Speed-Control

Equations (2.27) and (2.28) describe the dynamics of a DC motor. If we assume that we can neglect the armature inductance and the shaft compliance, the equation in terms of motor speed is

$$J\dot{\Omega} + \frac{K_t K_e}{R_a}\Omega = \frac{K_t}{R_a}v_a + T_\ell, \tag{3.1}$$

where

$$\Omega = \text{rotor speed } (\dot{\theta}_m),$$
$$J = \text{shaft inertia},$$
$$T_\ell = \text{load torque}.$$

By redefining constants, the equation can be simplified to

$$\tau\dot{\Omega} + \Omega = K_0(v_a + K_1 T_\ell), \tag{3.2}$$

where

$$\tau = JR_a/K_t K_e,$$
$$K_0 = 1/K_e,$$
$$K_1 = R_a/K_t.$$

FIGURE 3.2
Feedback control system.

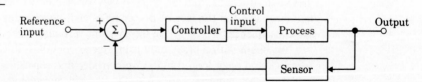

FIGURE 3.3
Feedback speed-control
system.

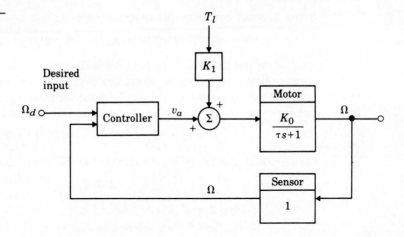

These equations are described by the block diagram seen in Fig. 3.3. The output sensor is a tachometer, which is usually a small, permanent-magnet, DC machine that produces a voltage proportional to the shaft speed, Ω. The controller is an electronic amplifier of gain K, which produces a voltage proportional to the difference between the voltages that represent the reference speed, Ω_r, and the motor speed, Ω. A very similar block diagram would result for many different types of speed-control systems. For example, in a steam engine controlled by a flying-ball governor, the centrifugal weights are the speed sensor, and their motion directly operates a valve (the controller) that changes the steam supplied, which in turn changes the speed of the engine.

3.1.1 Disturbance Rejection

Let's compare feedback control and open-loop control with respect to how well they maintain a constant set reference speed in the face of load or disturbance torques. For the open-loop controller, the amplifier voltage is taken to be

$$v_a = K_c \Omega_r. \tag{3.3}$$

The gain, K_c, is determined so that the output speed equals the reference speed when the load torque is zero. From the equation, this value is easily found to be

$$K_c = \frac{1}{K_0}.$$

In steady state ($\dot{\Omega} = 0$) and with no load torque ($T_\ell = 0$), the output speed is

$$\Omega = K_0 v_a = K_0 \frac{1}{K_0} \Omega_r = \Omega_r.$$

In the feedback controller, the amplifier output voltage is taken to be

$$v_a = K(K_r \Omega_r - K_f \Omega). \tag{3.4}$$

Because the control effort is proportional to the speed signals, this is called proportional control. Combining the motor model given by Eq. (3.2) with the feedback controller added according to Eq. (3.4), the formulas for the (closed-loop) system are

$$\tau \dot{\Omega} + (1 + K_0 K K_f)\Omega = K_0(K K_r \Omega_r + K_1 T_\ell). \tag{3.5}$$

For no load torque ($T_\ell = 0$), the steady-state ($\dot{\Omega} = 0$) value of speed is

$$\Omega = \frac{K_0 K K_r}{1 + K_0 K K_f} \Omega_r. \tag{3.6}$$

The gains K, K_r, and K_f are selected so that $\Omega = \Omega_r$. Usually the sensor and amplifier gains are selected for performance, and the reference gain set by the formula $K_r = (1 + K_0 K K_f)/K_0 K$. At this point, it appears that both approaches provide the desired result—specifically, $\Omega = \Omega_r$ in the steady state. The gains have been set with zero load-torque; let's compute what effect a nonzero torque will have on the steady-state speed in the two cases. Equations (3.2) and (3.3) show that in the open-loop system the steady-state speed is

$$\Omega = K_0(K_c \Omega_r + K_1 T_\ell).$$

Hence, with $K_c = 1/K_0$, the speed with load torque is

$$\Omega = \Omega_r + K_0 K_1 T_\ell,$$

and the variation in speed caused by the load torque is

$$\delta\Omega = K_0 K_1 T_\ell. \tag{3.7}$$

So we see that the speed error is proportional to the disturbing load, and the control designer has no influence over the parameters (K_1 and K_0) that determine the error.

For the feedback case (Eq. 3.5), the steady-state speed for a constant Ω_r and T_ℓ becomes

$$\Omega = \frac{K_0 K K_r}{1 + K_0 K K_f} \Omega_r + \frac{K_0 K_1}{1 + K_0 K K_f} T_\ell. \tag{3.8}$$

If it is possible for the designer to pick K so that $K_0 K K_f \gg 1$ and $K K_f \gg K_1$, no significant error will result with a disturbing load torque. In fact, as long as $K_0 K K_f > 0$, comparison of Eqs. (3.7) and (3.8) reveals that the feedback-controller speed errors due to a load torque will be less than the open-loop speed errors by exactly $1 + K_0 K K_f$. This is our first conclusion about feedback: System errors are less sensitive to disturbances with feedback than they are in open-loop systems.

3.1.2 Effects of Gain Changes

As another comparison of open- and closed-loop control, let's suppose the motor gain K_0 drifts from its original value to a value of $K_0 + \delta K_0$. In the open-loop case, the controller gain remains $K_c = 1/K_0$, and the new steady-state speed would be

$$\Omega = (K_0 + \delta K_0)\frac{1}{K_0}\Omega_r = \left(1 + \frac{\delta K_0}{K_0}\right)\Omega_r.$$

The speed error is

$$\delta\Omega = \frac{\delta K_0}{K_0}\Omega_r.$$

In terms of percent changes, this is

$$\frac{\delta\Omega}{\Omega_r} = 1\frac{\delta K_0}{K_0}. \qquad (3.9)$$

This means that a 10% error in K_0 would yield a 10% error in Ω.

Applying the same change in K_0 to the feedback case (Eq. 3.6) yields the new speed

$$\Omega = \frac{(K_0 + \delta K_0)KK_r}{1 + (K_0 + \delta K_0)KK_f}\Omega_r.$$

We would like to express this speed in terms of δK_0, which we do by expanding the expression in a series, as follows. First, we arrange the denominator to be of the form $1 + x$, where x is small:

$$\Omega = \frac{[K_0KK_r/(1 + K_0KK_f)] + [KK_r\delta K_0/(1 + KK_0K_f)]}{1 + [KK_f\delta K_0/(1 + KK_0K_f)]}\Omega_r.$$

Now

$$\frac{1}{1 + x} \cong 1 - x,$$

so with $x = K_fK\delta K_0/(1 + KK_0K_f)$,

$$\Omega \cong \left(\frac{K_0KK_r}{1 + K_0KK_f} + \frac{KK_r\delta K_0}{1 + KK_0K_f}\right)\left(1 - \frac{KK_f\delta K_0}{1 + KK_0K_f}\right)\Omega_r.$$

Ignoring the second power of δK_0 and defining

$$K_r = \frac{1 + KK_0K_f}{KK_0}$$

as before, we find

$$\Omega \cong \Omega_r + \Omega_r\frac{1}{1 + KK_0K_f}\frac{\delta K_0}{K_0}.$$

Thus

$$\delta\Omega = \Omega_r \frac{1}{1 + KK_0K_f} \frac{\delta K_0}{K_0},$$

or

$$\frac{\delta\Omega}{\Omega_r} = \frac{1}{1 + KK_0K_f} \frac{\delta K_0}{K_0}. \qquad (3.10)$$

Here we see that a 10% change in K_0 will cause only a $1/(1 + KK_0K_f) \times 10\%$ change in the speed Ω, and this is very small if $KK_0K_f \gg 1$.

Bode called the ratio of $\delta\Omega/\Omega$ to $\delta K_0/K_0$ the *sensitivity* of the gain from Ω_r to Ω with respect to K_0. We can compute this result rather more directly if we use differential calculus. Suppose for the moment we set $K_f = K_r = 1$, and the transfer function is

$$T(K_0) = \frac{KK_0}{1 + KK_0}.$$

To the first order, the variation is the derivative, so

$$\delta T = \frac{dT}{dK_0}\delta K_0,$$

and

$$\frac{\delta T}{T} = \left(\frac{K_0}{T} \frac{dT}{dK_0} \right) \frac{\delta K_0}{K_0}$$

$$= (\text{sensitivity}) \frac{\delta K_0}{K_0}.$$

Direct calculation gives, for

$$T = \frac{KK_0}{1 + KK_0}$$

then

$$S_{K_0}^T \triangleq \text{sensitivity of } T \text{ with respect to } K_0$$

$$\triangleq \frac{K_0}{T} \frac{dT}{dK_0},$$

$$S_{K_0}^T = \frac{K_0}{KK_0/(1 + KK_0)} \frac{(1 + KK_0)K - KK_0(K)}{(1 + KK_0)^2}$$

$$= \frac{1*}{1 + KK_0}.$$

---

*The factor $1 + KK_0$ occurs so frequently in feedback design that Bode named it the *return difference* of the feedback path.

This result exhibits another major advantage of feedback: The error in the controlled quantity is substantially less sensitive to variations in the gain of the plant in a feedback loop than when control is done by the open-loop controller. Therefore if the system gain is subject to change, it is still possible to achieve precise control if feedback is used. This is often very important when the system being controlled—for example, an electric motor or steam engine—is a large, high-power device whose gain is not only difficult to compute but is naturally subject to substantial variations. The superiority of feedback control would be equally apparent if we considered errors in the control gain K. This is so because this gain always appears as a product with K_0; therefore, the analysis for a K_0 error applies directly for an error in K.

3.1.3 Dynamic Tracking

The two features we have looked at thus far have been steady-state properties in the presence of constant reference and constant disturbance. The systems we are interested in are dynamic, and the tracking of time-varying inputs is an important role for control. It is also true that the dynamics of the system are changed by feedback. A constant-gain, open-loop controller has no effect on the dynamics of the system for either reference or disturbance inputs. If the open-loop controller includes a filter to change the response to the reference, the filter is in series with the dynamics of the plant, and the unchanged plant dynamics will still determine the response to disturbances. In the case of the speed control, Eq. (3.2) shows that the simple first-order plant dynamics are described by the time constant τ. The dynamics with feedback control are described by Eq. (3.5), and the root of the characteristic equation of this system is at

$$s = -\frac{1 + KK_0K_f}{\tau}.$$

Therefore, the time constant $\tau/(1 + KK_0K_f)$ becomes smaller (faster) as the feedback gain K is increased. It is often true in this way that closed-loop systems become faster as the feedback gain is increased, and if there are no other effects, this is generally desirable. As we will see, higher-order systems typically become less well damped and possibly even unstable as the gain increases. Thus, a limit exists on how high the gain can be made in an effort to reduce the effects of disturbances and plant parameter changes. An early solution to improve steady-state control without extremely high gains was provided by the introduction of integral control.

Integral Control

To see the effects of integral control, we return to Eq. (3.2) and let the voltage be the integral of the error between Ω and Ω_r, as follows:

$$v_a = K \int^t (\Omega_r - \Omega)d\eta.$$

In this case, we set $K_r = K_f$ and include them in the general K. With this choice, the equation of motion is

$$\tau\dot{\Omega} + \Omega = K_0[K\int^t (\Omega_r - \Omega)d\eta + K_1 T_\ell].   \qquad (3.11)$$

To remove the integral in Eq. (3.11), we differentiate once to obtain

$$\tau\ddot{\Omega} + \dot{\Omega} = K_0 K(\Omega_r - \Omega) + K_0 K_1 \dot{T}_\ell$$

$$\tau\ddot{\Omega} + \dot{\Omega} + K_0 K\Omega = K_0 K\Omega_r + K_0 K_1 \dot{T}_\ell.   \qquad (3.12)$$

We can see from this equation that several limitations of proportional control have been removed. If the load torque is a constant, then $\dot{T}_\ell = 0$, and the response to this class of load disturbance is completely eliminated. Furthermore, if the reference speed is a constant, the steady-state speed is given by

$$K_0 K\Omega = K_0 K\Omega_r,$$

and the motor speed equals the reference speed *regardless* of the value of K_0! We need to consider what integral control does to the dynamic response. For this, we need to look at the characteristic equation corresponding to Eq. (3.12), which is

$$\tau s^2 + s + K_0 K = 0.$$

FIGURE 3.4
Locus of roots of Eq. (3.13) vs. K.

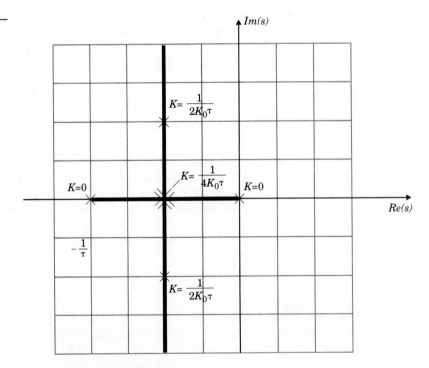

The roots of this equation, which are readily computed, vary as the control gain K is changed. In fact, it is informative to compute these and to plot them as a function of K. By the quadratic formula of algebra, the roots are

$$s = \frac{-1 \pm \sqrt{1 - 4K_0 K\tau}}{2\tau} \tag{3.13}$$

and are plotted in Fig. 3.4 for changing values of the gain K. When the gain is zero, the roots are at $-1/\tau$ and at 0. As the gain is raised, the roots move together until there is a double root at $-1/2\tau$ and then, for higher gains, the roots remain at the same real value while the imaginary part gets larger and larger. Thus, as the integral gain is raised, the damping of the roots becomes less and less.* If the designer wishes to increase the dynamic speed of response with large integral gain, the response becomes very oscillatory. A way to avoid this behavior in some cases is to use both proportional and integral control at the same time.

Proportional Plus Integral (PI) Control

With both proportional and integral control, the control voltage for our example problem becomes[†]

$$v_a = K\left[(\Omega_r - \Omega) + \left(\frac{1}{T_I}\right)\int^t (\Omega_r - \Omega)d\eta\right].$$

In this form, the terminology is to call the parameter K the "proportional gain" and to call the parameter T_I the *integral* or *reset* time. If we substitute this expression for the voltage into Eq. (3.2) and differentiate again to remove the integral, the equation of motion this time is

$$\tau\ddot{\Omega} + \dot{\Omega} = K_0[K\{(\dot{\Omega}_r - \dot{\Omega}) + (\frac{1}{T_I})(\Omega_r - \Omega)\} + K_1\dot{T}_\ell] \tag{3.14}$$

As with integral control, if both the disturbance torque and the reference speed are constant, the steady-state speed is exactly equal to the reference speed, regardless of the plant gain K_0. The characteristic equation corresponding to Eq. (3.14) is

$$\tau s^2 + (1 + K_0 K)s + K_0 K/T_I = 0.$$

It can be seen that by choice of K and T_I the designer can independently set values for the coefficient of s and for the constant term and thus has independent control over the damping ratio and the natural frequency of the roots. PI control is enough for our simple case study example; the motor will keep an exact match to the constant reference speed in spite of the constant load torque and regardless of

---

*This kind of plot, a plot of the roots of the characteristic equation as one parameter is changed, is extremely useful and was developed into a major design tool by W. Evans. He called the plot a *root locus*. We will study Evans' technique in some detail in Chapter 4.

[†] Many designers omit the Ω_r in the proportional term so $\dot{\Omega}_r$ does not appear in Eq. (3.14).

small changes in the motor gain K_0. Furthermore, the designer has complete control over the coefficients of the characteristic equation. For higher order systems, it is necessary to introduce another term in the equation, *derivative control,* to give more control over the dynamics. The result is called the "three-term controller," PID control.

3.2 Types of Feedback

3.2.1 Proportional Feedback

When the feedback control signal is made to be linearly proportional to the error in the measured output, we call the result *proportional feedback:* This is the case for the feedback (Eq. 3.4) in the controller in Section 3.1. The general form of proportional control is

$$u = Ke,$$

and, therefore, the $D(s)$ in Fig. 3.5 is

$$D(s) = K.$$

An arbitrary second-order plant,

$$G(s) = \frac{1}{s^2 + as + b},$$

with proportional feedback yields a closed-loop characteristic equation:

$$s^2 + as + b + K = 0. \tag{3.15}$$

Had the plant been a spring-mass system such as shown in Fig. 3.6, the feedback would provide a control force that would act as if it were an additional spring and would increase the speed at which the system returned to its neutral position. If the system had $b = 0$ (that is, no spring) it would have no tendency to return to the neutral position; thus feedback is essential to its return.

FIGURE 3.5
Simple feedback control block diagram.

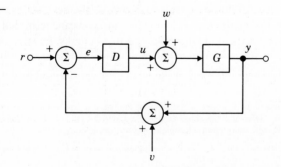

FIGURE 3.6
Spring-mass system.

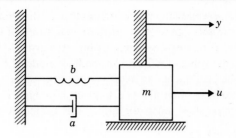

This can be seen by inspecting the roots of Eq. (3.15), with $b = 0$ as a function of the feedback gain K. The solution is given by

$$s = -\frac{a}{2} \pm \sqrt{\left(\frac{a}{2}\right)^2 - K}$$

and is plotted in Fig. 3.7. It shows that the root at $s = 0$ for the open-loop case ($K = 0$) moves increasingly to the left for increasing values of $K (0 < K < a^2/4)$, which implies a decreasing time constant for the system. For values of $K > a^2/4$, the real part of the roots do not change with K; but the imaginary part increases with K, and thus the rise time of the system decreases. However, a decreasing damping ratio goes along with the faster response. For higher-order systems, very large values of the proportional feedback gain can often lead to instability. For most systems, there is typically an upper limit on the proportional feedback gain in

FIGURE 3.7
Locus of roots of
Eq. (3.15) vs. $K (b = 0)$.

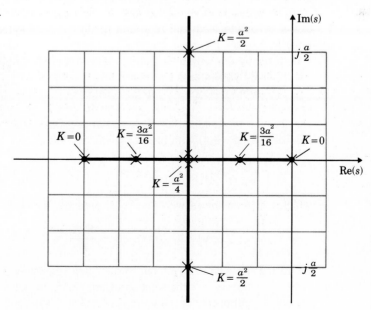

order to achieve a well-damped, stable response. This example is typical of most; proportional feedback, at the least, increases the speed of driving the error to zero and in some cases is essential for eliminating the error.

Increasing the value of K increases the magnitude of DG, which was shown in Section 3.1 to be instrumental in reducing the errors of the system. We see here that the dynamic response often places a limit on how high K can be made; therefore there is a limit on how much the errors can be reduced by using proportional feedback only.

3.2.2 Integral Feedback

As shown in Section 3.1, integral feedback has the form

$$u(t) = \frac{K}{T_I} \int_{t_0}^{t} e \, dt;$$

therefore the $D(s)$ in Fig. 3.5 becomes

$$D(s) = \frac{K}{T_I s},$$

where T_I is called the *integral*, or *reset, time* and $1/T_I$ is referred to as the *reset rate*. This feedback has the primary virtue that it can provide a finite value of u with no error-signal input e. This comes about because u is a function of past values of e rather than of the current value, as in the proportional case. Therefore past errors e will "charge up" the integrator to some value that will remain, even if the error becomes zero. This feature means that disturbances w (see Fig. 3.5) can be accommodated with zero error because it is no longer necessary for e to be finite to produce a control that will negate the (constant) disturbance w. Another way of seeing this is to note that it is now possible for D to have an infinite gain, which would yield a zero y response to a disturbance w.

The primary reason for integral control is to reduce or eliminate steady-state errors, but this benefit typically comes at the cost of reduced stability. For example, consider the speed controller discussed in Section 3.1. Replacing the proportional feedback in Eq. (3.4) with integral feedback yields

$$V_a(s) = \frac{K}{T_I s}(\Omega_r - \Omega),$$

which, when combined with Eq. (3.2), produces the characteristic equation

$$\tau s^2 + s + \frac{K_0 K}{T_I} = 0.$$

Increasing the gain K/T_I of this system will ultimately result in lightly damped roots, as is the case for the solution of Eq. (3.15) for large values of K. However, the speed controller with proportional feedback was shown to yield roots on the negative real axis for any value of the proportional feedback gain K. In general, any system will be made less stable or less damped by the addition of integral control. This characteristic will be discussed in more depth in the following chapters.

3.2.3 Derivative Feedback

Derivative feedback (or rate feedback) has the form

$$u(t) = KT_D \dot{e}.$$

Therefore the $D(s)$ in Fig. 3.5 becomes

$$D(s) = KT_D s,$$

and T_D is called the *derivative time*.

It is typically used in conjunction with proportional and/or integral feedback to increase the damping and generally to improve the stability of a system. Unless the system has a natural proportional term such as a spring or equivalent feature, derivative feedback by itself will not drive the error to zero.

Let's reexamine the system in Fig. 3.6, this time without the spring. The characteristic equation with $D(s) = KT_D s$ becomes

$$s^2 + (a + KT_D)s = 0,$$

which has roots at

$$s = 0, -(a + K_p T_D).$$

The root at $s = 0$ remains there independent of the value of KT_D and signifies that any initial condition on y will not have the tendency to go to the commanded value. However, the feedback does affect the rate at which initial $\dot{y}$ values decay to zero.

3.2.4 PID Controllers

Feedback that has three terms—proportional plus integral plus derivative—is referred to as *PID control*. In the generic unity-sensor-gain topology shown in Fig. 3.5, the controller transfer function is given by

$$D(s) = K\left(1 + \frac{1}{T_I s} + T_D s\right) E(s). \tag{3.16}$$

As suggested at the end of our consideration of the simple first-order case study of speed control, the PID combination is sometimes able to provide an acceptable degree of error reduction simultaneously with acceptable stability and damping. These controllers have been found to be so effective that PID is the standard control action in the process industries, such as petroleum refining, paper making, metal forming, and so on. To design a particular control loop, the engineer merely

FIGURE 3.8
Response of a second-order system $G(s) = 1/[(s + 1)(5s + 1)]$ to a unit disturbance W.

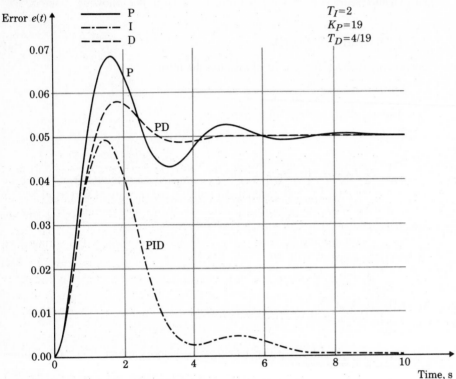

has to adjust the three constants in Eq. (3.16) to arrive at acceptable performance. The adjustment process is so common that it is called "tuning" the controller. Criteria for tuning are based on the ideas presented in the case study: Simply stated, increasing K and $1/T_I$ tend to reduce system errors but may not be capable of also producing adequate stability, and increasing T_D tends to improve stability. Figure 3.8 illustrates the effect of the proportional, PD, and PID feedback on the step response of a second-order plant. The figure shows the reduced oscillatory behavior brought on by addition of the derivative term, and the increased oscillatory behavior but lower error brought on by the integral term.

☐ *Ziegler-Nichols Tuning of PID Regulators*

As we will see in later chapters, sophisticated design methods are available to develop a controller that will meet steady-state and transient specifications for both tracking input commands and rejecting disturbances. These methods require

FIGURE 3.9
Process reaction curve.

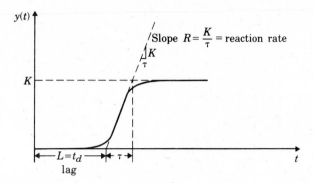

complete models of the dynamics of the process to be controlled in the form of equations of motion or transfer functions. Callender et al. (1936) attempted to select a class of common systems dynamics and widely used PID controllers and to specify satisfactory values for the controller settings, based on estimates of the plant parameters that could be made by an operating engineer from experiments on the process itself. Ziegler and Nichols recognized that the step response of most process control systems has the general S-shaped curve shown in Fig. 3.9, which is called the *process reaction curve* and can be generated experimentally or from dynamic simulation of the plant. The shape of the curve is characteristic of high-order systems, and the plant input-output behavior may be approximated by

$$\frac{Y(s)}{U(s)} = \frac{Ke^{-\tau_d s}}{\tau s + 1},$$

which is simply a first-order system plus a transportation lag.* The constants in the above equation can be determined from the unit step response of the process. If a tangent is drawn at the inflection point of the reaction curve, then the slope of the line is approximately $R = K/\tau$, and the intersection of the tangent line with the time axis identifies the time delay $L = t_d$. If the actual plant output does not fit this sample model very well, other poles may be added. However, the model will most probably be adequate to provide a first attempt at controller design.

Ziegler and Nichols gave two methods for tuning the controller. In the first method, the choice of controller parameters is based on a decay ratio of approximately 0.25, which means that a dominant transient decays to a quarter of its value after one period of oscillation, as shown in Fig. 3.10. A quarter decay corresponds to $\zeta = 0.21$ and is a good compromise between quick response and adequate stability margins. The equations for the system were simulated on an analog computer, and the controller parameters were adjusted until the transients showed a decay of

---

*The transfer function, $e^{-\tau_d s}$, corresponds to a signal delay in time. Physically this usually occurs because some material must be transported from one place to another. Hence the name, transportation lag.

FIGURE 3.10
Quarter decay ratio.

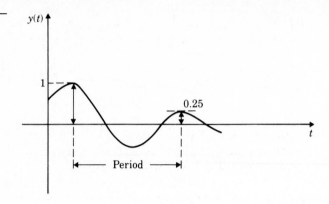

25% in one period. The regulator parameters suggested by Ziegler and Nichols are shown in Table 3.1.

In the second method, the criteria for adjusting the regulator parameters are based on evaluating the system at the limit of stability. Using only proportional control, the gain is increased until continuous oscillations are observed, that is, until the system becomes marginally stable. The corresponding gain K_u (also called the *ultimate gain*) and the period of oscillation P_u (also called the *ultimate period*) are determined as shown in Figs. 3.11 and 3.12. This period should be measured when the amplitude of oscillation is quite small. Then one "backs off" from this gain, as shown in Table 3.2. These parameters can be easily computed from a limit cycle under relay control (see Section 4.6.4).

Experience has shown that the controller settings according to Ziegler-Nichols rules provide a good closed-loop response for many systems. The final tuning of the controller can be done manually by the process operator to yield the "best" control.*

*For a recent revisit, see Hang, Åström, and Ho (1990).

TABLE 3.1
Ziegler-Nichols Tuning for the Regulator,
$D(s) = K\left(1 + \dfrac{1}{T_I s} + T_D s\right)$, for
a Decay Ratio of 0.25

Type of Controller	Optimum Gain
P	$K = \dfrac{1}{RL}$
PI	$K = \dfrac{0.9}{RL}, T_I = \dfrac{L}{0.3}$
PID	$K = \dfrac{1.2}{RL}, T_I = 2L,$ $T_D = 0.5L$

FIGURE 3.11
Determination of the
ultimate gain and period.

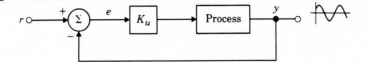

Integrator Antiwindup: A Taste of the Practical World

The dynamic range of practical actuators is usually limited; for example, a valve saturates when it is fully open or closed, and the control surfaces in an aircraft can be deflected only to a certain angle from their nominal positions. If integral control is used with such actuators, a phenomenon known as *integrator windup** may occur. In the event of large commands, the integrator builds up a large value, which results in large overshoots and errors. Consider the feedback system shown in Fig. 3.13. Suppose a large reference step causes the actuator to saturate at u_{max}. The integrator keeps integrating the error e, and the signal u_c keeps growing. However, the input to the plant is still at its maximum value, namely, $u = u_{max}$, and the error remains large. The increase in u_c is not helping anything, as the input to the plant is not changing. The integrator output may become quite large if saturation lasts a long time. It will require considerable error to discharge the integrator to its proper value.

As an example, consider the first-order plant with a proportional-plus-integral (PI) regulator,

$$u = K\left[e + \frac{1}{T_I}\int_0^t e(t)dt\right],\qquad(3.17)$$

and a control that saturates at ± 1. With the system at rest, an input of size 1 will cause the error to equal 1, the control to saturate, and the integral action to be-

*In process control, this is usually referred to as *reset windup,* since integral control is also called *reset control.* The term also arises because, with no integral control, a given setpoint of, say 10, results in a response of less value, say 9.9. The operator must then "reset" to 10.1 to bring the output to the desired value of 10. With integral control, the controller automatically brings the output to 10 with a setpoint of 10; hence the integrator does the reset.

FIGURE 3.12
Marginally stable system.

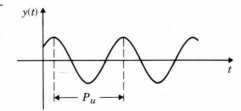

TABLE 3.2
Ziegler-Nichols Tuning for the Regulator,
$D(s) = K(1 + \frac{1}{T_I s} + T_D s)$, **Based on a Stability Boundary**

Type of Controller	Optimum Gain
P	$K = 0.5K_u$
PI	$K = 0.45K_u$
	$T_I = \frac{1}{1.2}P_u$
	$K = 0.6K_u$
PID	$T_I = \frac{1}{2}P_u$
	$T_D = \frac{1}{8}P_u$

gin. The integrator output u_c will grow until the plant output exceeds the command input, and this may take a long time. The result is a response shown by the dotted line in Fig. 3.14.

The solution to this problem is to "turn off" the integral action as soon as the actuator saturates. (This can be done quite easily if the controller is implemented digitally.) Two such "antiwindup" schemes are shown in Fig. 3.15(a) and (b) using a dead-zone nonlinearity with a PI regulator. In this scheme, as soon as the actuator saturates, the feedback loop around the integrator becomes active and moves rapidly to keep the input to the integrator e_1 at zero. During this time the integrator essentially becomes a fast first-order lag, as shown in Fig. 3.15(c). The slope of the dead-zone nonlinearity K should be chosen to be large enough that the antiwindup circuit is capable of following e to keep the output from saturating. The effect of the antiwindup is to reduce both the error overshoot and the control

FIGURE 3.13
Feedback system with actuator saturation.

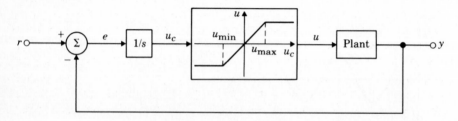

FIGURE 3.14
Effect of antiwindup:
(a) step response;
(b) control effort.

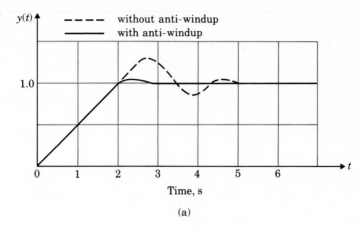

(a)

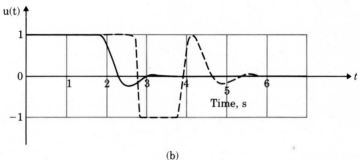

(b)

effort in the feedback system. Figure 3.14 shows a step response and control effort for a saturating system with and without an antiwindup integrator.

3.3 Steady-State Tracking and System Type

In the speed control case-study, we considered constant reference inputs and constant disturbances, and we found that integral control could keep the steady-state system error at zero, even with a motor gain that was different from the one used in the design. In a number of important cases, the reference input is not constant but can be approximated as a linear function of time for a time span long enough for the system to reach steady-state and we would like to know what the steady-state error is in this case. The generalization is to consider the input to be a *polynomial* in time and to consider the steady-state tracking error that results. As we will see, the error is zero for input polynomials up to a certain degree, is a constant for that

FIGURE 3.15
(a) PI controller with antiwindup; (b) alternative PI controller with antiwindup; and (c) first-order lag equivalent of an antiwindup integrator during saturation.

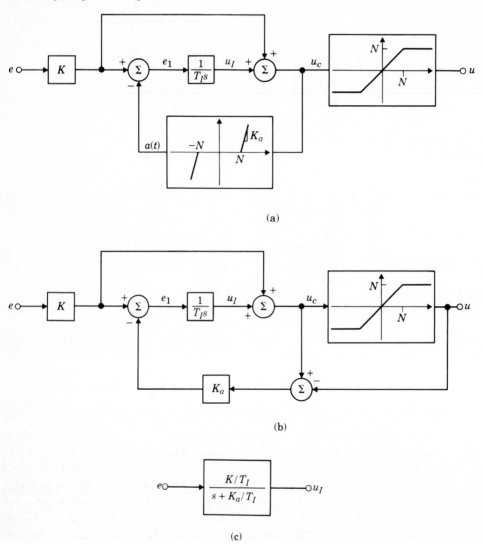

(a)

(b)

(c)

degree polynomial, and is unbounded for higher degrees. A stable system can be classified according to the degree of the polynomial for which the error is constant, and the classification is called the *system type*.

A typical single-loop feedback block diagram is shown in Fig. 3.16, where care has been taken to show the system error, $e(t) = r(t) - y(t)$. The variables

FIGURE 3.16
Typical single-loop
system.

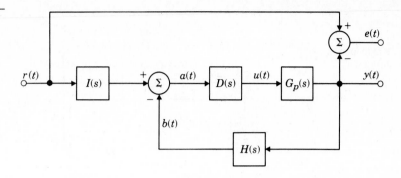

are defined as

$$r(t) = \text{reference input,}$$
$$u(t) = \text{plant input (control signal),}$$
$$y(t) = \text{plant output,}$$
$$e(t) = \text{system error or } r(t) - y(t),$$
$$a(t), b(t) = \text{internal signals; } a(t) \text{ is called the } \textit{actuating signal,}$$
$$G_p(s) = \text{plant transfer function including actuator,}$$
$$H(s) = \text{feedback components including sensor transfer function and dynamic compensation,}$$
$$D(s) = \text{controller transfer function,}$$
$$I(s) = \text{input components, that is, reference signal processing or prefiltering.}$$

The feedback transfer function $H(s)$ typically represents the sensor action to convert the output $y(t)$ to an electrical sensor output signal $b(t)$. This was K_f in the case study. Likewise, we often require a transfer function $I(s)$ to convert the reference input, which may be speed or temperature or whatever, into an electrical signal to be combined with $b(t)$ in the controller to generate the actuating signal $a(t)$. In the case study, we had $I(s) = K_r$. The controller with transfer function $D(s)$ converts the electrical actuating signal into the control signal $u(t)$. Notice that H and I usually have the same dimensions (volts per 1000 rpm or volts per degree Celsius, etc.), and that I can be assumed to be nonzero. Using this, we can simplify Fig. 3.16 to obtain the structure shown in Fig. 3.17.

Let's assume that the reference input is a polynomial of degree k and compute the steady-state error for this general system in terms of the given transfer functions. If $r(t) = (t^k/k!)1(t)$, then the transform of the input is $R(s) = 1/s^{k+1}$. If $k = 0$, the input is a step of unit amplitude; if $k = 1$, the input is a ramp with unit slope; if $k = 2$, the input is a parabola with unit second derivative; and so on. From the common problems of mechanical motion control, these inputs are

FIGURE 3.17
Simplified block diagram
of a typical single-loop
system.

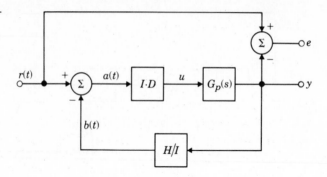

called, respectively, position, velocity, and acceleration inputs. In order to compute the steady-state errors, we need the transfer function from input to system error, then we can apply the final-value theorem. From the block diagram, the transform from r to y is defined as $F(s)$ and is

$$F(s) = \frac{IDG_p}{(1 + HDG_p)} \tag{3.18}$$

and the system error is

$$E(s) = R(s) - Y(s)$$
$$E(s) = R(s) - F(s)R(s).$$

The reference-to-error transfer function is

$$1 - F(s).$$

With the test input, the error transform is

$$E(s) = \frac{1}{s^{k+1}}[1 - F(s)].$$

We assume that the assumptions of the final-value theorem are satisfied, namely that all poles of $sE(s)$ are in the left-half plane. In that case, the steady-state error is given by

$$e_\infty = \lim_{t \to \infty} e(t) = \lim_{s \to 0} sE(s) \tag{3.19}$$

$$e_\infty = \lim_{s \to 0} \frac{[1 - F(s)]}{s^k}. \tag{3.20}$$

By the nature of the limit, we can see that the result of Eq. (3.20) can be zero, can be a constant different from zero, or could not exist, in which case the final-value theorem does not apply, but it is easy to see that $e_\infty = \infty$ in this case anyway because $E(s)$ will have a pole at the origin that is of order higher than one. By convention, the classification is given that if the solution to Eq. (3.20) is a constant different from zero, the system being considered is *type k*.

FIGURE 3.18
Unity feedback system.

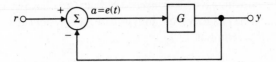

Special Case: Unity Feedback

A further simplification of Fig. 3.17 is possible if $H = I$. In that case, which is quite common, we can model the system as the *unity feedback system* as drawn in Fig. 3.18 and take advantage of the fact that now the actuating signal is the error, $a(t) = e(t)$.* In Fig. 3.18, we have defined $G = IDG_p$ as the overall feed-forward or open-loop transfer function.

If our system is described by the block diagram in Fig. 3.18, then

$$1 - F(s) = \frac{1}{1 + G(s)},$$

and the system error is

$$e_\infty = \lim_{s \to 0} \frac{1}{[1 + G(s)]s^k}. \tag{3.21}$$

If the system is type 0, we set $k = 0$, and the error is

$$e_\infty = \frac{1}{[1 + G(0)]}$$

$$e_\infty = \frac{1}{(1 + K_p)}. \tag{3.22}$$

The constant defined in Eq. (3.22) is called the *position error constant*. This notation derives from reference to the fact that if the signal being controlled is mechanical motion and $k = 0$, then the reference is a position and the error is a position error. The form of the definition leads to the simple computation

$$K_p = \lim_{s \to 0} G(s).$$

In other words, *if the loop is unity feedback* and type 0, the position error constant is the zero-frequency or DC gain.

If we now consider ramp inputs, then, from Eq. (3.21) with $k = 1$,

$$e_\infty = \lim_{s \to 0} \frac{1}{[1 + G(s)]s}$$

$$= \lim_{s \to 0} \frac{1}{sG(s)}.$$

---

*Much of the confusion in the literature (textbooks especially) on this topic arises because some authors have treated the *actuating signal* $a(t)$ as if it were the system error even in the general (nonunity feedback) case.

For this error to exist, it is necessary that $G(s)$ have at least one pole at $s = 0$. Suppose it has exactly one; in that case, the system is type 1 (or type I) and the error is, by definition

$$e_\infty = 1/K_v,$$

where the error constant is the *velocity constant,* and with unity feedback, is given by

$$K_v = \lim_{s \to 0} sG(s).$$

In similar fashion, systems of type 2 (or type II) are found to require that in the unity feedback case, $G(s)$ must have two poles at $s = 0$, the error is defined as

$$e_\infty = \frac{1}{K_a}.$$

The acceleration constant can be computed in the unity-feedback-gain case by

$$K_a = \lim_{s \to 0} s^2 G(s).$$

The situation is summarized in Table 3.3.

In general, the error constants must be computed from Eq. (3.20), but in the special unity feedback gain case, they are given by

$$\lim_{s \to 0} s^k G(s).$$

The definition of system type helps one to identify quickly the ability of a system to track polynomials. In the unity feedback structure, the property is "robust" in the sense that, if the process gain, such as K_0, changes in a type I system because of parameter changes, the velocity error constant changes but the system will still have zero steady-state error to a constant input and will still be type I.

TABLE 3.3
Errors versus System Type for Unity Feedback

	Input		
	Step	**Ramp**	**Parabola**
Type 0	$\dfrac{1}{(1 + K_p)}$	∞	∞
Type 1	0	$\dfrac{1}{K_v}$	∞
Type 2	0	0	$\dfrac{1}{K_a}$

This robustness is the major reason for preferring a unity feedback system over any other.

System type can also be defined with respect to disturbance inputs by considering system error in response to polynomial disturbance inputs. The reader is invited to investigate this property in the problems (see Problem 3.13).

In many cases, the control system block diagram does not fit the standard forms shown in Figs. 3.16, 3.17, or 3.18 and the computation of steady-state errors due to the inputs or disturbances cannot be determined easily from Table 3.3. In these cases, the designer is advised to derive the Laplace transform of the desired error based on the specific input or disturbance to which the error sensitivity is desired and to apply the final value theorem to the result. In other words, the steps taken in arriving at Eqs. (3.18) through (3.21) should be repeated for the specific block diagram being analyzed.

EXAMPLE 3.1   *DC Motor* As an example to illustrate the idea of system type, consider the simplified model of the DC motor employing proportional feedback as shown in Fig. 3.19(a). Let us use the following parameter values,

$$J = 1, \qquad F = 1, \qquad K_m = 1, \qquad K_l = 1.$$

The closed-loop transfer function from r to y (with $\tau_L = 0$) is

$$H(s) = \frac{1}{s(s + 1) + K}.$$

FIGURE 3.19
(a) DC motor;
(b) DC motor with unity feedback.

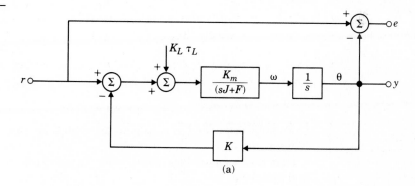

(a)

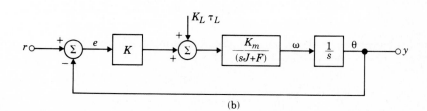

(b)

The system error is

$$E(s) \triangleq R(s) - Y(s)$$

$$= R(s)\left(1 - \frac{Y(s)}{R(s)}\right) = (1 - H(s))R(s) = \frac{s(s+1) + K - 1}{s(s+1) + K}.$$

The system error due to a unit-step input is given by

$$E(s) = \frac{(1 - H(s))}{s}.$$

Using the Final Value theorem,

$$e_\infty = \lim_{s \to 0} \frac{(1 - H(s))}{s} = \frac{K - 1}{K},$$

that is, the feedback system is not capable of following a step with zero steady-state error unless $K = 1$. This is somewhat surprising because the plant itself has a pure integrator. This happens because nonunity feedback renders the integrator ineffective so far as achieving zero steady-state error to a constant input. Furthermore, the error values contained in Table 3.3 no longer apply. However, it is possible to remedy this situation by implementing a unity-feedback configuration as shown in Fig. 3.19(b). Now we have

$$H(s) = \frac{K}{s(s+1) + K}$$

and

$$E(s) = \frac{s(s+1)}{s(s+1) + K}$$

$$e_\infty = 0,$$

hence,

$$y(\infty) = r(\infty).$$

This means that unity feedback provides asymptotic tracking without help from a compensator.

For both setups in Figs. 3.19(a)–(b), the transfer function from the disturbance to the output is

$$\frac{Y(s)}{T_L(s)} = \frac{1}{s(s+1) + K}.$$

For a unit-step disturbance input, the asymptotic output will be

$$y(\infty) = \frac{1}{K},$$

so that the system is incapable of rejecting the disturbance exactly and there is a steady-state offset. The integrator in the plant does not help with this because the disturbance enters the system preceding the integrator.

We may probe a little deeper as to the reason for the tracking properties of the system. Note that the transfer functions from r to e in Fig. 3.19(a) (for $K = 1$) and Fig. 3.19(b) (regardless of the value of K)

$$\frac{E(s)}{R(s)} = \frac{s(s + 1)}{s(s + 1) + K},$$

both have a zero at the origin. This is called a *blocking zero* and is the mechanism behind the tracking properties. Note that no such blocking zeros exist in the transfer function from the disturbance to the output. Hence the system is not capable of rejecting the disturbances exactly.

Let us now investigate the tracking properties of the system with respect to a ramp input signal. For the system in Fig. 3.19(a)

$$e_\infty = \lim_{s \to 0} s \frac{s(s + 1) + K - 1}{s(s + 1) + K} \frac{1}{s^2},$$

which will be unbounded unless $K = 1$, that is, there is unity feedback. In that case,

$$e_\infty = \frac{1}{K}.$$

The system in Fig. 3.19(b) will have a constant error due to ramp input which is the same as computed above.   ∎

EXAMPLE 3.2   *Satellite Attitude Control* The system is as shown in Fig. 3.20(a) with $J =$ moment of inertia, $\tau_d =$ disturbance torque, $K =$ sensor gain, and $D(s)$ is the compensator. If the compensator is proportional plus derivative (PD) feedback, $D(s) = K(1 + T_D s)$, then we may redraw the system as shown in Fig. 3.20(b). The system error for a ramp input is

$$e_\infty = \lim_{s \to 0} s \frac{Js^2}{Js^2 + K(1 + T_D s)} \cdot \frac{1}{s^2},$$

which is zero. For a parabolic input, where $R(s) = 1/s^3$, there will be a constant error

$$e_\infty = \frac{J}{K}.$$

Note the presence of the two blocking zeros at the origin. The system has a finite

FIGURE 3.20
(a) Satellite attitude
control; (b) proportional
plus derivative (PD)
control; (c) PID control.

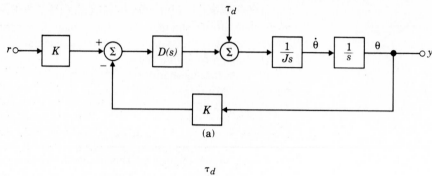

(a)

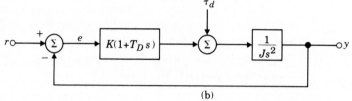

(b)

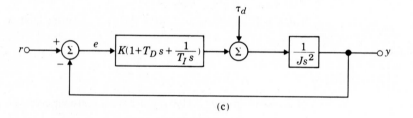

(c)

error for a constant disturbance torque,

$$y(\infty) = \lim_{s \to 0} s \frac{1}{Js^2 + K(1 + T_D s)} \frac{1}{s} = \frac{1}{K}.$$

Suppose we use a PID type controller,

$$D(s) = K\left(1 + T_D s + \frac{1}{T_I s}\right),$$

as shown in Fig. 3.20(c). It may be easily shown that the system now has zero error for steps, ramps, and parabolic-inputs. Furthermore, the system now has no error to a constant disturbance because,

$$y(\infty) = \lim_{s \to 0} s \frac{T_I s}{JT_I s^3 + K(T_D s^2 + T_I s + 1)} \frac{1}{s} = 0,$$

and the integrator from the compensator creates a blocking zero from the disturbance to the output.   ∎

Truxal's Formula

We have derived formulas for the error constants in terms of the system transfer function. Truxal (1955) derived a formula for the velocity constant in terms of the *closed-loop* poles and zeros, a formula that connects the steady-state error to the dynamic response. Since much of control design requires a trade-off between these characteristics, the formula can be very useful and is well worth knowing. The derivation is quite direct. Suppose the closed-loop transfer function of a type I system is $F(s)$ given by

$$F(s) = K \frac{(s + z_1)(s + z_2)\ldots(s + z_m)}{(s + p_1)(s + p_2)\ldots(s + p_n)}. \tag{3.23}$$

Furthermore, since the steady-state error to a step of a type I system is zero, the DC gain is unity, that is,

$$F(0) = 1. \tag{3.24}$$

The system error is given by

$$E(s) \triangleq R(s) - Y(s)$$

$$= R(s)\left(1 - \frac{Y(s)}{R(s)}\right) = R(s)[1 - F(s)]. \tag{3.25}$$

The system error due to a unit ramp input is given by

$$E(s) = \frac{[1 - F(s)]}{s^2}. \tag{3.26}$$

Using the final-value theorem,

$$e_\infty = \lim_{s \to 0} \frac{[1 - F(s)]}{s}. \tag{3.27}$$

Using L'Hôpital's rule, we may rewrite Eq. (3.27) as

$$e_\infty = -\lim_{s \to 0}\left(\frac{dF}{ds}\right) \tag{3.28}$$

or

$$e_\infty = -\lim_{s \to 0} \frac{dF}{ds} = \frac{1}{K_v}. \tag{3.29}$$

Note that Eq. (3.29) implies that the reciprocal of K_v is given by the slope of the transfer function at the origin (this fact will be used again in Chapter 5). Using

Eq. (3.24), we can rewrite Eq.(3.29) as

$$e_\infty = -\lim_{s \to 0} \frac{dF}{ds} \frac{1}{F} \tag{3.30}$$

or

$$e_\infty = -\lim_{s \to 0} \frac{d}{ds} \ln F(s). \tag{3.31}$$

Substituting Eq.(3.23) into Eq.(3.31),

$$e_\infty = -\lim_{s \to 0} \frac{d}{ds} \ln K \frac{\prod_{i=1}^{m}(s + z_i)}{\prod_{i=1}^{n}(s + p_i)} \tag{3.32}$$

$$e_\infty = -\lim_{s \to 0} \frac{d}{ds} \left[\sum_{i=1}^{m} \ln(s + z_i) - \sum_{i=1}^{n} \ln(s + p_i) \right] \tag{3.33}$$

or

$$\frac{1}{K_v} = \sum_{i=1}^{n} \frac{1}{p_i} - \sum_{i=1}^{m} \frac{1}{z_i}. \tag{3.34}$$

We observe from Eq. (3.34) that the farther the closed-loop poles are from the origin, the larger is K_v, but that it may also be increased by having closed-loop zeros close to the origin. Truxal's formula will be used in Chapter 6 to aid with dynamic compensator design.

3.4   Stability

A system that always gives responses appropriate to the stimulus is considered to be stable. In testing the stability of a dynamic system, we must carefully define the terms "response," "stimulus," and "appropriate." For example, it is possible for a system to have a well-behaved output while an internal variable is growing without bound. If we consider only the output as the "response," we might call such a system stable; however, if we consider the internal variable to be the response, we might just as reasonably call this system unstable. The stimulus might be a signal persisting for all time, or it might be only a set of initial conditions. Finally, we might require that the response not grow if the stimulus is removed, or we might require that the response go to zero if the stimulus is removed. For systems that are nonlinear or time-varying, the number of definitions of stability is large, and some of the definitions can be quite complicated in order to account for the varieties of responses the systems can display. We will describe a few elementary results, particularly those results useful in the study of linear constant systems. In addition to the results discussed in this section, Nyquist's frequency-response stability test will be presented in Chapter 5, and Lyapunov stability will be covered in Chapter 6.

3.4.1 Bounded-Input–Bounded-Output Stability

A system is said to have bounded-input–bounded-output (BIBO) stability if every bounded input results in a bounded output (regardless of what goes on inside the system). A test for this property is readily found using convolution. If the system has input $u(t)$, output $y(t)$, and impulse response $h(t)$, then

$$y(t) = \int_{-\infty}^{\infty} h(\tau)u(t - \tau)d\tau. \tag{3.35}$$

If $u(t)$ is bounded, there is a constant M such that $|u| \leq M < \infty$, and the output is bounded by

$$
\begin{aligned}
|y| &= \left| \int hu \, d\tau \right| \\
&\leq \int |h||u|d\tau \\
&\leq M \int_{-\infty}^{\infty} |h(\tau)|d\tau.
\end{aligned}
$$

Thus, the output will be bounded if $\int_{-\infty}^{\infty} |h|d\tau$ is bounded. On the other hand, suppose the integral is *not* bounded, and we consider the bounded input $u(t - \tau) = +1$ if $h(\tau) > 0$, and $u(t - \tau) = -1$ if $h(\tau) < 0$. In this case,

$$y(t) = \int_{-\infty}^{\infty} |h(\tau)|d\tau, \tag{3.36}$$

and the output is not bounded. We conclude that

The system with impulse response $h(t)$ is BIBO stable if and only if $\int_{-\infty}^{\infty} |h(\tau)|d\tau < \infty$.

Consider, as an example, the capacitor driven by a current source as sketched in Fig. 3.21. The capacitor voltage is the output, and the current is the input. One would expect that a simple ideal capacitor circuit would be stable. Wrong! The

FIGURE 3.21
Capacitor driven by
current source.

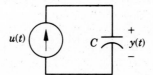

impulse response of this setup is $h(t) = 1(t)$, the unit step. For this response,

$$\int_{-\infty}^{\infty} |h(\tau)|d\tau = \int_{0}^{\infty} d\tau \qquad (3.37)$$

is not bounded. The capacitor is not BIBO stable. Notice that the transfer function of the system is $1/s$ and has a pole on the imaginary axis. Physically, we can see that constant input current will cause the voltage to grow, and thus the system response is not bounded, and the system is not stable. As a matter of fact, if a constant system has any pole on the imaginary axis or in the right-half plane, the response will not be BIBO stable; if every pole is inside the left-hand plane, the response will be BIBO stable. Thus for these systems, pole locations can be used to check for stability. An alternative to computing the integral of the impulse response or even to locating the roots of the characteristic equation is given by Routh's stability criterion.

3.4.2 Routh's Stability Criterion

Consider the characteristic equation of an nth-order system:*

$$a(s) = s^n + a_1 s^{n-1} + a_2 s^{n-2} + \cdots + a_{n-1}s + a_n. \qquad (3.38)$$

It is possible to make certain statements about the stability of the system without actually solving for the roots of the polynomial. This is a classical problem, and several methods exist for the solution. One such method was proposed by Routh in 1874. An equivalent technique was independently derived by Hurwitz in 1895 and similar results obtained by Lyapunov in 1893. We will discuss the version developed by Routh. Before one considers the full test, one should examine the equation to see that it satisfies a simple necessary condition. A *necessary* condition for stability of the system, that is, that all of the roots of Eq. (3.38) have negative real parts—is that all the $\{a_i\}$'s be positive.[†] If any of the coefficients are missing (i.e., are zero) or negative, the system will have at least one pole located outside the left-half plane (LHP). This condition can be checked by inspection. Once the elementary necessary conditions have been satisfied, we need a more powerful test. Routh showed that a *necessary and sufficient condition* for stability is that all of the elements in the first column of the so-called Routh array be positive.

To determine the Routh array, the coefficients of the characteristic polynomial are arranged in two rows, each beginning with the first and second coefficients and followed by the even-numbered and odd-numbered coefficients as follows:

$$
\begin{array}{cccc}
s^n : & 1 & a_2 & a_4 & \cdots \\
s^{n-1} : & a_1 & a_3 & a_5 & \cdots .
\end{array}
$$

---

*The polynomial is assumed to be monic (the coefficient of the highest power of s is 1) without loss of generality.

[†] This is easy to see if one imagines the polynomial as a product of first- and second-order factors.

The following rows are subsequently added to complete the Routh array:

$$
\begin{array}{lll}
\text{row } n & s^n: & 1 \quad a_2 \quad a_4 \quad \cdots \\
\text{row } n-1 & s^{n-1}: & a_1 \quad a_3 \quad a_5 \quad \cdots \\
\text{row } n-2 & s^{n-2}: & b_1 \quad b_2 \quad b_3 \quad \cdots \\
\text{row } n-3 & s^{n-3}: & c_1 \quad c_2 \quad c_3 \quad \cdots \\
& \vdots & \vdots \\
\text{row } 2 & s^2: & * \quad * \\
\text{row } 1 & s: & * \\
\text{row } 0 & s^0: & *
\end{array}
$$

where the elements from the $(n-2)$th row on are computed as follows:

$$
b_1 = \frac{-\det\begin{bmatrix} 1 & a_2 \\ a_1 & a_3 \end{bmatrix}}{a_1} = \frac{a_1 a_2 - a_3}{a_1},
$$

$$
b_2 = \frac{-\det\begin{bmatrix} 1 & a_4 \\ a_1 & a_5 \end{bmatrix}}{a_1} = \frac{a_1 a_4 - a_5}{a_1},
$$

$$
b_3 = \frac{-\det\begin{bmatrix} 1 & a_6 \\ a_1 & a_7 \end{bmatrix}}{a_1} = \frac{a_1 a_6 - a_7}{a_1};
$$

and

$$
c_1 = \frac{-\det\begin{bmatrix} a_1 & a_3 \\ b_1 & b_2 \end{bmatrix}}{b_1} = \frac{b_1 a_3 - a_1 b_2}{b_1},
$$

$$
c_2 = \frac{-\det\begin{bmatrix} a_1 & a_5 \\ b_1 & b_3 \end{bmatrix}}{b_1} = \frac{b_1 a_5 - a_1 b_3}{b_1},
$$

$$
c_3 = \frac{-\det\begin{bmatrix} a_1 & a_7 \\ b_1 & b_4 \end{bmatrix}}{b_1} = \frac{b_1 a_7 - a_1 b_4}{b_1}.
$$

Note that the elements of the $(n-2)$th row and of rows thereafter are formed from the two previous rows using determinants with the two elements in the first column and other elements for successive columns. Normally, there will be $n+1$ elements in the first column when the array terminates. These must all be positive if all the roots of the characteristic polynomial are in the LHP. However, if the elements of the first column are not all positive, the number of roots in the RHP equals the number of sign changes in the first column. A pattern of $+, -, +$ is counted as *two* sign changes, one going from $+$ to $-$ and the other from $-$ to $+$.

As an example, consider the polynomial

$$a(s) = s^6 + 4s^5 + 3s^4 + 2s^3 + s^2 + 4s + 4,$$

which satisfies the necessary condition for stability since all the $\{a_i\}$'s are positive and nonzero. The Routh array for this polynomial is

$$
\begin{array}{llll}
s^6: & 1 & 3 & 1 & 4 \\
s^5: & 4 & 2 & 4 & 0 \\
s^4: & \dfrac{5}{2} = \dfrac{(4)(3)-(1)(2)}{4} & 0 = \dfrac{(4)(1)-(4)(1)}{4} & 4 = \dfrac{(4)(4)-(1)(0)}{4} & \\
s^3: & 2 = \dfrac{(5/2)(2)-(4)(0)}{5/2} & -\dfrac{12}{5} = \dfrac{(5/2)(4)-(4)(4)}{5/2} & 0 & \\
s^2: & 3 = \dfrac{(2)(0)-(5/2)(-12/5)}{2} & 4 = \dfrac{(2)(4)-(5/2)(0)}{2} & & \\
s: & \dfrac{-76}{15} = \dfrac{(3)(-12/5)-(8)}{3} & 0 & & \\
s^0: & 4 = \dfrac{(-76/15)(4)-0}{-76/15}. & & &
\end{array}
$$

It is seen that the polynomial has RHP roots, since the elements of the first column are not all positive. In fact, there are two poles in the RHP since there are two sign changes.* Note that in computation of Routh's array, a row may be multiplied or divided by a positive constant if this simplifies the rest of the calculations. Also notice that the last nonzero element in each column of the Routh array is simply a_n and that the last two rows each have one term. If the first term in one of the rows is zero or if an entire row is zero, the standard Routh array cannot be formed, and one has to use the special techniques described below.

Special Cases

If only the first element in one of the rows is zero, we can replace the zero with a small positive constant $\varepsilon > 0$ and proceed as before. The stability criterion is then applied by taking the limit as $\varepsilon \to 0$. Consider the polynomial

$$a(s) = s^5 + 3s^4 + 2s^3 + 6s^2 + 6s + 9,$$

---

*The actual roots of the polynomial are $-3.2644, 0.6797 \pm j0.7488, -0.6046 \pm j0.9935$, and -0.8858, which agree with our conclusion.

for which the Routh's array is

$$
\begin{array}{llll}
s^5: & 1 & 2 & 6 \\
s^4: & 3 & 6 & 9 \\
\text{new } s^4: & 1 & 2 & 3 \leftarrow \text{ divide previous row by 3} \\
s^3: & 0 & 3 & 0 \\
\text{new } s^3: & \varepsilon & 3 & 0 \leftarrow \text{ replace zero by } \varepsilon \\
s^2: & \dfrac{2\varepsilon - 3}{\varepsilon} & 3 & 0 \\
s: & 3 - \dfrac{3\varepsilon^2}{2\varepsilon - 3} & 0 & 0 \\
s^0: & 3 & 0. &
\end{array}
$$

For small ε, there are two sign changes in the first column of the array, which implies that there are two poles not in the LHP.* An alternative procedure is to define the auxiliary variable,

$$
z \triangleq \frac{1}{s}, \tag{3.39}
$$

and convert the characteristic equation so that it is in terms of z. This often produces a Routh's array whose elements of the first column are nonzero. The stability properties of the system are then deduced from this new array.

Another special case occurs if an entire row of the Routh array is zero. This indicates that there are complex conjugate pairs of roots that are mirror images of each other with respect to the imaginary axis. If the ith row is zero, an auxiliary equation is formed from the previous (nonzero) row as follows:

$$
a_1(s) = \beta_1 s^{i+1} + \beta_2 s^{i-1} + \beta_3 s^{i-3} + \cdots, \tag{3.40}
$$

where $\{\beta_i\}$'s are the coefficients of the $(i + 1)$th row in the array. The ith row is then replaced by the coefficients of the *derivative* of the auxiliary polynomial, and the array is completed; however, the roots of the auxiliary polynomial given above are also roots of the characteristic equation, and these must be tested separately.

As an example, consider

$$
a(s) = s^5 + 5s^4 + 11s^3 + 23s^2 + 28s + 12,
$$

---

*The actual roots are at -2.9043, $.6567 \pm j1.2881$, $-.7046 \pm j.9929$.

for which the Routh's array is

$$
\begin{array}{llll}
s^5: & 1 & 11 & 28 \\
s^4: & 5 & 23 & 12 \\
s^3: & 6.4 & 25.6 & 0 \\
s^2: & 3 & 12 & \\
s: & 0 & 0 & \leftarrow\ a_1(s) = 3s^2 + 12 \\
\text{new } s: & 2 & 0 & \leftarrow\ \dfrac{d\,a_1(s)}{ds} = 2s \\
s^0: & 12. & &
\end{array}
$$

Note that the auxiliary polynomial was divided by 3 for simplification. There are no sign changes in the first column, hence all the roots have negative real parts except for a pair on the imaginary axis. This may be deduced as follows. If the zero in the first column is replaced by $\varepsilon > 0$ there are no sign changes. If $\varepsilon < 0$ there would be two sign changes. Thus if $\varepsilon = 0$, there would be two poles on the imaginary axis. These are the roots of

$$
a_1(s) = s^2 + 4 = 0
$$

or

$$
s = \pm j2.^*
$$

Routh's method is also useful in determining the range of parameters for which a feedback system remains stable. Consider the system shown in Fig. 3.22. The stability properties of the system are a function of the feedback gain K. The characteristic equation for the system is given by

$$
1 + K\,\frac{s + 1}{s(s - 1)(s + 6)} = 0
$$

* The actual roots are at $-3, \pm j2, -1, -1$.

FIGURE 3.22
A feedback system for
stability test.

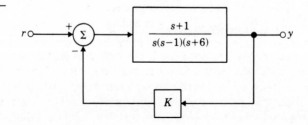

or

$$s^3 + 5s^2 + (K - 6)s + K = 0.$$

The corresponding Routh's array is

$$
\begin{array}{ccc}
s^3 : & 1 & K - 6 \\
s^2 : & 5 & K \\
s : & \dfrac{4K - 30}{5} & \\
s^0 : & K. &
\end{array}
$$

If the system is to remain stable we must have

$$\frac{4K - 30}{5} > 0$$

$$K > 0$$

or

$$K > 7.5.$$

Summary

In this chapter, some essential properties of feedback have been discussed. Feedback can be used to stabilize systems, speed up transient response, improve steady-state characteristics, provide disturbance rejection, and decrease sensitivity to parameter variations. Some simple types of feedback control have been considered. Proportional feedback reduces errors, but high gains may destabilize the system. Integral control improves the steady-state error and provides robustness with respect to parameter variations, but it also reduces stability. Derivative control generally increases damping and improves stability. The combination of all three, the PID controller, is ubiquitous in the process-control industry, and the concepts of increasing gain to reduce error and adding dynamics to improve response is the basic ingredient in virtually all control designs. A set of useful guidelines for tuning PID controllers has been presented. We have also discussed the definition of system type, which provides a figure of merit indicative of the ability of a system to follow polynomial time functions with zero steady-state error. Finally, BIBO stability and the Routh test for determining stability for linear constant systems have been presented.

Problems   Section 3.1

3.1 Consider a system with the configuration of Fig. 3.5. Let the gain of the controller be D and of the process be G. The nominal values of these gains are $D = 5$ and $G = 7$. Suppose a constant disturbance, w, is added to the control input before the signal goes to the process. The designer is able to increase the loop gain of this system by a factor of six before the dynamic response goes out of specifications.

a) Compute the gain from w to y.

b) Where should the designer place the extra gain if the objective is to minimize the system error, $r - y$, due to the disturbance? That is, the gain D can be raised to 30 or the gain G can be raised to 42, and so on; which is best?

3.2 Bode defined the sensitivity function of a transfer function G to one of its parameters k as the ratio of percent change in k to percent change in G. We define the reciprocal

FIGURE 3.23

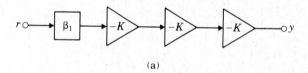

(a)

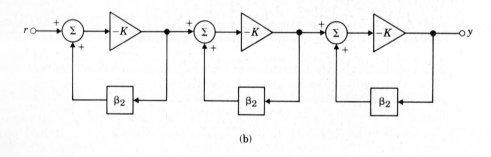

(b)

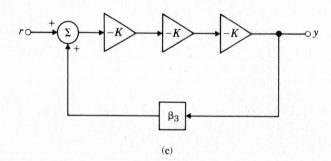

(c)

of Bode's function as

$$S_k^G = \frac{dG/G}{dk/k} = \frac{d\ln G}{d\ln k} = \frac{k}{G}\frac{dG}{dk}.$$

Thus, if the parameter k changes by 1%, S tells us what percent change to expect in G. For control systems, we are almost always interested in zero frequency, or $s = 0$ in sensitivity calculations. The purpose of this exercise is to examine the effect of feedback on sensitivity. In particular, we would like to compare the topologies shown in Fig. 3.23 for connecting three amplifier stages of gain $-K$ each into a single amplifier of gain -10.

a) For each case, compute β_i so that, if $K = 10$, $y = -10r$.
b) For each case, compute S_K^G when $G = y/r$. [Use the respective $\{\beta_i\}$'s found in **(a)**]. Which case is the *least* sensitive?
c) Compute the sensitivities of Fig. 3.23(b) and (c) to β_2 and β_3. Comment on the relative requirements for precision in sensors and actuators from these cases.

3.3 Compare the two structures shown in Fig. 3.24 with respect to sensitivity to changes in the overall gain due to changes in the amplifier gain. Use the definition

$$S_K^F = \frac{d\ln F}{d\ln K} = \frac{K}{F}\frac{dF}{dK}$$

as the measure. Select H_1 and H_2 so that the nominal values are $F_1 = F_2$.

3.4 A DC-motor speed control is described by the equation:

$$\dot{\Omega} + 60\Omega = 600V_a - 1500T_1,$$

FIGURE 3.24

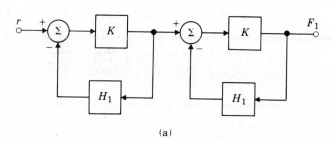

(a)

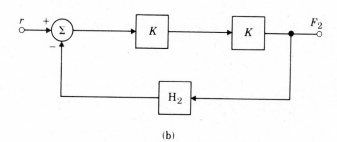

(b)

FIGURE 3.25

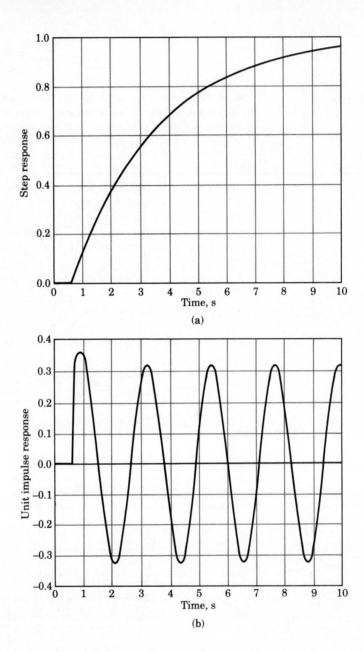

(a)

(b)

where Ω is the motor speed, V_a is the armature voltage, and T_1 is the load torque. Let

$$V_a = -K\left(\Omega + \frac{1}{T_I}\int^t \Omega \, d\tau\right).$$

a) Compute the transfer function from T_1 to Ω as a function of K and T_I.
b) Compute the values of K and T_I so that the characteristic equation of the closed-loop system will have roots at $-60 \pm j60$.

Section 3.2

3.5 The unit-step response of a paper machine is shown in Fig. 3.25(a). The time delay and the steepest slope may be determined from this transient response.

a) Find the P-, PI-, and PID-controller parameters using the Zeigler-Nichols transient response method.
b) The system with proportional feedback has the unit-impulse response shown in Fig. 3.25(b). The gain $K_u = 8.556$ corresponds to the system being on the verge of instability. Determine the P-, PI-, and PID-controller parameters according to the Zeigler-Nichols ultimate sensitivity method.

3.6 A heat exchanger has the transfer function

$$G(s) = \frac{e^{-5s}}{(10s + 1)(60s + 1)}.$$

a) Find the PID-controller parameters according to the Zeigler-Nichols tuning rules.
b) The system becomes marginally stable for a proportional gain of $K_u = 15.25$ as shown by the unit-impulse response in Fig. 3.26. Find the optimal PID-controller parameters according to the Zeigler-Nichols tuning rules.

FIGURE 3.26

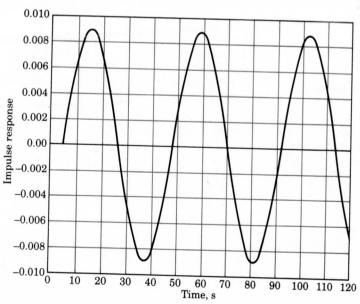

Section 3.3

3.7 A certain control system has the following specifications: rise time $t_r \leq 0.010s$ and percent overshoot $M_p \leq 17$; and steady-state error to unit ramp $e_{ss} \leq 0.005$.

a) Sketch the allowable region in the s plane for the dominant second-order poles of an acceptable system.

b) If y/r for this system is $G/(1 + G)$, what is the behavior of $G(s)$ near $s = 0$ for an acceptable system?

3.8 a) Compute the steady-state error to a unit-step, reference input for the system of Problem 3.4(**b**).

b) Compute the steady-state error to a unit-ramp, reference input for the system of Problem 3.4(**b**).

c) Compute the steady-state error to a unit-step, disturbance input for the system of Problem 3.4(**b**).

d) Compute the steady-state error to a unit-ramp, disturbance input for the system of Problem 3.4(**b**).

3.9 *System type for disturbance inputs:* In most control systems, disturbances of one type or another exist. In practice, these disturbances are represented by polynomial time functions such as steps or ramps. This would suggest definitions of system type with respect to disturbance rejection analogous to those suggested in this chapter for tracking. Consider the unity feedback system shown in Fig. 3.27 with disturbance inputs W_1, W_2, and W_3. The closed-loop system is asymptotically stable and

$$G_1(s) = \frac{K_1 \prod_{i=1}^{m_1}(s + z_{1i})}{s^{l_1} \prod_{i=1}^{m_1}(s + p_{1i})},$$

$$G_2(s) = \frac{K_2 \prod_{i=1}^{m_1}(s + z_{2i})}{s^{l_2} \prod_{i=1}^{m_1}(s + p_{2i})}.$$

a) Show that the system is of type 0, type l_1, and type $(l_1 + l_2)$ with respect to disturbance inputs W_1, W_2, and W_3, respectively.

b) Consider the multivariable system shown in Fig. 3.28. Find y_1 and y_2 for constant disturbances. What is the system type with respect to disturbances at W_1? At W_2?

FIGURE 3.27 ·
Single input-single ouput
unity feedback system
with disturbance units.

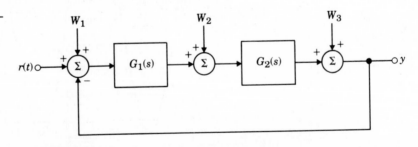

FIGURE 3.28

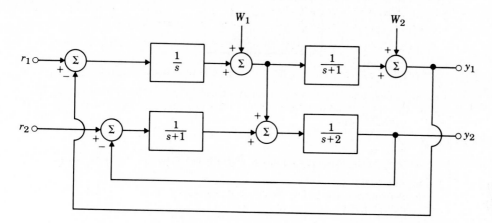

3.10 One possible representation of an automobile speed-control system, with integral control, is shown in Fig. 3.29.

a) With zero command input ($v_c = 0$), find the transfer function relating the output v to the wind disturbance w.

b) What is the steady-state response of v if w is a unit ramp function?

c) What type is this system in relation to command inputs? What is the value of the static error constants K_p and K_v?

d) What type is this system in relation to disturbance w? What is the error constant due to the disturbance?

3.11 For the feedback system shown in Fig. 3.30, find α to make the system type I for $K = 5$. Show that the system is not robust by using this value of α and computing the tracking error $e = r - y$ to a step command for $K = 4$ and $K = 6$.

FIGURE 3.29

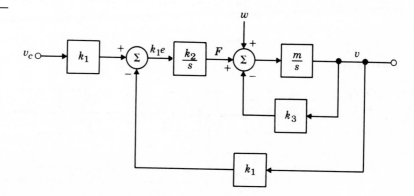

FIGURE 3.30

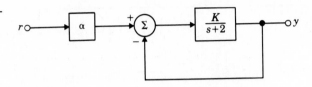

3.12 A position-control system has the overall transfer function (meter/meter) given by

$$\frac{y}{r} = \frac{b_0 s + b_1}{s^2 + a_1 s + a_2}.$$

Suppose we are able to select all the parameters. Choose them so that

a) Rise time is $t_r \leq 0.1$ s.
b) Percent overshoot $M_p \leq 20$.
c) Setting time $t_s \leq 0.5$ s.
d) Steady-state error to a constant command is zero.
e) Steady-state to a ramp of 0.1 m/s is not more than 1 mm.

3.13 Two feedback systems are shown in Fig. 3.31.

a) Determine values of K_1, K_2, and K_3 so that both systems exhibit zero steady-state error to step inputs (i.e., Type I) and such that $K_v = 1$ in both cases if $K_0 = 1.0$.
b) Suppose K_0 changes to $K_0 = K_0 + \delta K_0$ (where δK_0 is a small change in the nominal value of gain K_0). Prove that (a) is still a Type I system in spite of the perturbation in K_0.
c) If $K_0 = K_0 + \delta K_0$, show that the system in Fig. 3.31(b) is no longer Type I.
d) Explain why control engineers prefer the system in Fig. 3.31(a) over that in Fig. 3.31(b).

FIGURE 3.31

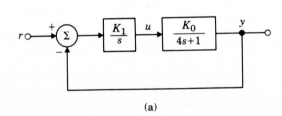

(a)

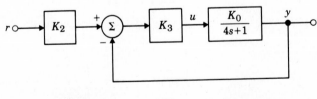

(b)

FIGURE 3.32

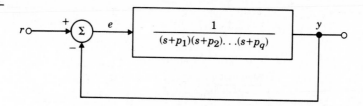

e) Estimate the transient response of both systems to a step reference input and compute t_s, t_r, and M_p. In your opinion, which system has a better transient response at the nominal parameter values? Consider rise time and overshoot.

3.14 Consider the system shown in Fig. 3.32.

a) Find the transfer function from the input to the tracking error.

b) It is desired that this system respond to inputs of the form $r(t) = t^n (n \ll q)$ with zero steady-state error. If this requirement is to be met, what must be true about the open-loop poles $p_1, p_2, \ldots, p_q$?

3.15 The feedback control system shown in Fig. 3.33 is to be designed to satisfy the following specifications: (1) steady-state error to a ramp input less than 10% of the input magnitude, (2) maximum percent overshoot for a unit step input less than 5%, and (3) settling time (1%) less than 3 s.

a) Compute the closed-loop transfer function.

b) Find the error due to a unit ramp input.

c) What does (1) imply about the possible values of K_1?

d) What does (3) imply about the closed-loop poles?

e) Sketch the region in the complex plane where the closed-loop poles may lie.

f) Suppose $K_1 = 32$. Find the values of K_2 such that the poles are on the right-hand boundary of the feasible region.

g) Estimate the settling time of the system.

3.16 The transfer functions of a magnetic-tape-drive speed-control system are shown in Fig. 3.34. The speed sensor is fast enough that its dynamics can be neglected.

a) What is the steady-state error due to a step disturbance torque of 1 N·m? (Here assume $\omega_r = 0$.) What must the amplifier gain K be in order to make the steady-state error $e_{ss} \leq 0.01$ rad/s?

FIGURE 3.33

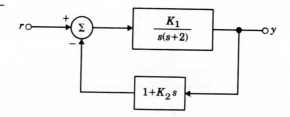

FIGURE 3.34

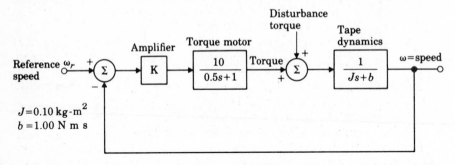

$J = 0.10$ kg-m^2
$b = 1.00$ N m s

b) Plot in the complex plane the roots of the closed-loop system, and accurately sketch the time response $\omega(t)$ for a step input ω_r for the gain K in (**a**). Are these roots desirable? Why or why not?

c) Plot the region in the complex plane corresponding to the specifications

$$t_s = \text{settling time} \leq 0.1s,$$

$$M_p = \text{percent overshoot} \leq 5.$$

d) Suggest a simple control scheme that can be added to the proportional control system above to meet the specifications in (**c**), and show that the new scheme does meet the specifications.

e) How would the disturbance-induced steady-state error change with the new control scheme in (**d**)? How could the steady-state error be eliminated entirely?

3.17 A linear ordinary differential equation model of the DC motor with negligible armature inductance ($L_a = 0$) and disturbance torque Q_L is given by

$$\frac{JR_a}{K_m}\ddot{\theta} + K_m\dot{\theta} = V_a + \frac{R_a}{K_m}Q_L.$$

Dividing by the coefficient of $\ddot{\theta}$, we obtain

$$\ddot{\theta} + a_1\dot{\theta} = b_0V_a + c_0Q_L,$$

where

$$a_1 = \frac{K_m^2}{JR_a}, \quad b_0 = \frac{K_m}{JR_a}, \quad c_0 = \frac{1}{J}.$$

With rotating potentiometers, it is possible to measure the positioning error between θ and the reference angle θ_r. Call this error $e = \theta_r - \theta$. With a tachometer, we can measure motor speed $\dot{\theta}$. We would like to consider feedback of error e and output velocity $\dot{\theta}$ in the form

$$V_a = K(e - T_D\dot{\theta}).$$

a) Draw a block diagram of the resulting feedback system.

b) Suppose the numbers work out so that $a_1 = 65$, $b_0 = 200$, and $c_0 = 10$. If there is *no* load torque ($Q_L = 0$), what speed (rpm) results from $V_a = 100V$? (In the equations above, θ is measured in radians.)

c) Using the parameter values given in (b), find K and T_D so that a step change in θ_r with zero load torque results in a transient, having approximately 17% overshoot, that settles to no more than 5% of steady-state in 0.05 s.

d) Compute an expression for the steady-state error to an arbitrary reference angle, and give its value for $\theta_r = 1$ rad for your design from (c).

e) Compute an expression for the steady-state error to a disturbance torque when $\theta_r = 0$ and give its value for $Q_L = 1.0$ for your design from (c).

3.18 We wish to design an automatic speed control for an automobile. Assume the pertinent characteristics are (1) the car has mass m of 1000 kg, (2) the accelerator is the control U and supplies a force on the automobile of 10 N per degree of accelerator motion, and (3) air drag provides a friction force (proportional to velocity) of 10 N $\cdot$s^2/m.

a) Obtain the transfer function from U to velocity of the automobile.

b) Regardless of the answer you get for (a), assume the velocity changes are given by

$$V(s) = \frac{1}{s + 0.02}U(s) + \frac{0.05}{s + 0.02}G(s),$$

where V is in m/s and G is the grade in %. Design a proportional control law that will maintain a velocity error of less than 1 m/s in the presence of a constant 2% grade.

c) Discuss what advantage (if any) integral control would have on this problem.

d) Assuming pure integral control (i.e., no proportional term) is advantageous, select the feedback gain so that the roots have critical damping ($\zeta = 1$).

Section 3.4

3.19 Use the Routh criterion to determine if the closed-loop systems under unity feedback corresponding to the following open-loop transfer functions are stable.

a) $KG(s) = \dfrac{4(s + 2)}{s(s^3 + 2s^2 + 3s + 4)}$.

b) $KG(s) = \dfrac{2(s + 4)}{s^2(s + 1)}$.

c) $KG(s) = \dfrac{4(s^3 + 2s^2 + s + 1)}{s^2(s^3 + 2s^2 - s - 1)}$.

3.20 Use Routh's stability criterion to determine how many roots with positive real parts each of the following equations has.

a) $s^4 + 8s^3 + 32s^2 + 80s + 100 = 0$.

b) $s^5 + 10s^4 + 30s^3 + 80s^2 + 344s + 480 = 0$.

c) $s^4 + 2s^3 + 7s^2 - 2s + 8 = 0$.

d) $s^3 + s^2 + 20s + 78 = 0.$

e) $s^4 + 6s^2 + 25 = 0.$

3.21 For what range of K would all the roots of the following polynomial be in the left-half plane?

$$s^5 + 5s^4 + 10s^3 + 10s^2 + 5s + K = 0.$$

3.22 A typical transfer function for a tape-drive system would be (with time in milliseconds)

$$G(s) = \frac{K(s + 4)}{s[(s + 0.5)(s + 1)(s^2 + 0.4s + 4)]}.$$

From Routh's criterion, what is the range of K for which this system is stable if the characteristic equation is $1 + G(s) = 0$?

3.23 Automatic ship steering is particularly useful in heavy seas when it is important to maintain the ship along an accurate path. Such a control system for a large tanker is shown in Fig. 3.35, with the plant transfer function relating heading changes to rudder deflection.

a) Write the differential equation that relates the heading angle to rudder angle for the ship *without* feedback.

b) Is the control system stable as shown (simple proportional control with a gain of 1)? *Hint:* Use Routh's criterion.

c) Is it possible to stabilize the system by changing the proportional gain from 1 to some lower value?

d) Suggest another possible control scheme that might improve matters. Investigate its stability.

FIGURE 3.35

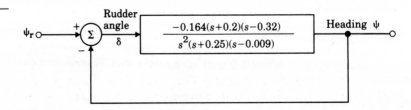

4

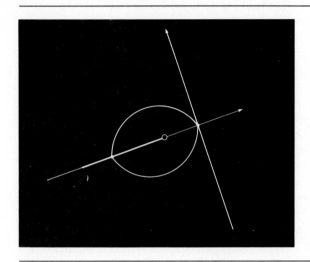

The Root-Locus

Design

Method

4.1 Introduction

In Chapter 2, we discussed models of dynamic systems and the relationships between the time-domain response and the transform-domain parameters of poles and zeros. We related the features of a step response such as rise time, overshoot, and settling time to pole locations in the s plane characterized by natural frequency ω_n, damping ζ, and real part σ. We also examined the changes in these transient-response features when a pole or a zero is added to a given simple transfer function. In Chapter 3, we saw how feedback can influence stability and dynamic response by changing the system's pole locations. In this chapter, we present a specific technique developed to help in the design of feedback systems, a technique based on studying the effects of a single parameter on closed-loop pole locations. While the method is a general one, permitting the effects of any one parameter to be studied, the most common case involves study of open-loop gain, and the method will be presented and first used in this context. The reader should keep in mind that any single real parameter can be used; extensions of the method based on this fact are described in Section 4.6. The method was developed by W. R. Evans (1948) and is called the root-locus method.

We begin with the basic feedback loop shown in Fig. 4.1. For this system, the transfer function is

$$\frac{Y(s)}{R(s)} = \frac{K_A K_P G(s)}{1 + K_A K_P G(s)},\tag{4.1}$$

and the characteristic equation, whose roots are the poles of this transfer function, is

$$1 + K_A K_P G(s) = 0.\tag{4.2}$$

Clearly, the closed-loop poles depend on the amplifier gain K_A, and we will have some influence over the closed-loop dynamic response through our choice of K_A. Evans suggested that the control-system designer construct the *locus* of all possible roots of Eq. (4.2) as K_A varies from zero to infinity; a study of the resulting plot then aids in the selection of the best value of K_A. Furthermore, by studying the effects of additional poles and zeros on a given locus, we can examine the consequences of additional dynamics in the loop; we thus have a tool not only for gain selection but for dynamic-compensation design as well. Similarly, we can examine the effect of plant-parameter changes to aid in achieving best overall control design. The graph of all possible roots of Eq. (4.2) versus some parameter is called the *root-locus*, and the design technique based on it is called the *root-locus method of Evans*. We begin with the mechanics of construction of a root locus, using amplifier gain as the parameter.

To set the notation for our study, we assume here that $K_P G(s)$, the process transfer function, is a rational function with numerator $K_P b(s)$, where $b(s)$ is a monic* polynomial of degree m, and the denominator polynomial is $a(s)$ of degree

*Monic means the coefficient of the highest power of s is 1.

FIGURE 4.1
Basic closed-loop block diagram used to develop the root-locus method.

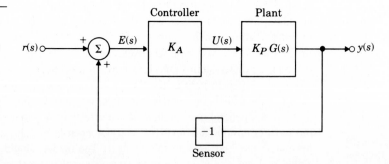

n, where $n \geq m$, and

$$b(s) = s^m + b_1 s^{m-1} + \cdots + b_m$$
$$= (s - z_1)(s - z_2) \cdots (s - z_m) \qquad (4.3a)$$

$$= \prod_{i=1}^{m}(s - z_i); \qquad (4.3b)$$

$$a(s) = s^n + a_1 s^{n-1} + \cdots + a_n \qquad (4.3c)$$

$$= \prod_{i=1}^{n}(s - p_i). \qquad (4.3d)$$

For our initial study, we assume that K_P is positive and take the product of K_A and K_P to be the root-locus parameter K so that

$$K = K_A K_P. \qquad (4.4)$$

The roots of $b(s) = 0$ are labeled z_i, and they are the zeros of $G(s)$; the roots of $a(s) = 0$ are labeled p_i, and they are the poles of $G(s)$. The root locus problem expressed in Eq. (4.2) may now be stated in several equivalent but useful ways. Each of the following equations has the same roots:

$$1 + KG(s) = 0, \qquad (4.5a)$$

$$1 + K\frac{b(s)}{a(s)} = 0, \qquad (4.5b)$$

$$a(s) + Kb(s) = 0, \qquad (4.5c)$$

$$G(s) = -\frac{1}{K}. \qquad (4.5d)$$

The root locus is the locus of values of s for which Eq. (4.5) holds for some positive real value of K (and K_A).

Before looking further into the general problem, let's solve a specific case by direct calculation. Suppose we have normalized the transfer function of the DC motor position control so that

$$G(s) = \frac{1}{s(s + 1)}.$$

In terms of the notation introduced thus far, we have $m = 0$, $n = 2$, and

$$K_P = 1,$$
$$b(s) = 1,$$
$$a(s) = s^2 + s,$$
$$p_i = 0, -1.$$

From Eq. (4.5c), the root locus is a graph of the roots of the quadratic equation

$$s^2 + s + K = 0. \tag{4.6}$$

By formula, we can immediately express the roots of Eq. (4.6) as

$$r_1, r_2 = -\frac{1}{2} \pm \frac{\sqrt{1 - 4K}}{2}. \tag{4.7}$$

A plot of the corresponding root locus is shown in Fig. 4.2. For $0 \leq K \leq 1/4$, the roots are real between -1 and 0. At $K = 1/4$, there are two roots at $-1/2$, and for $K > 1/4$ the roots become complex, with a real part $-1/2$ and an imaginary part that increases essentially in proportion to the square root of K. The dashed lines in Fig. 4.2 correspond to roots with damping ratio $\zeta = 0.5$. The poles of $G(s)$ at $s = 0$ and $s = -1$ are marked $\times$, and the points where the locus crosses the 0.5 damping-ratio lines are marked with a dot (•). These points correspond to where the roots of Eq. (4.6) will be for the specific value of $K = 1$. We can compute K in this instance because we know that, if $\zeta = 0.5$, the magnitude of the imaginary part of the root is $\sqrt{3}$ times the magnitude of the real part. Since the size of the real part is $1/2$, then from Eq. (4.7),

$$\frac{\sqrt{4K - 1}}{2} = \frac{\sqrt{3}}{2},$$

which is true if $K = 1$.

FIGURE 4.2
Root locus for $G(s) = 1/[s(s + 1)]$ as a function of open-loop gain K.

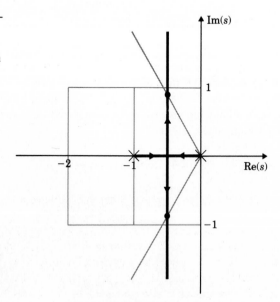

We can observe several features of this simple locus by looking at Eqs. (4.6) and (4.7) and at Fig. 4.2. First, there are two roots and thus two branches to the root locus. At $K = 0$, these branches begin at the poles of $G(s)$ (which are at 0 and -1) as they should, since for $K = 0$ the system is open loop. As K is increased, the roots move toward each other, coming together at $s = -1/2$. At that point, they break away from the real axis. After the breakaway point, the roots move off to infinity with equal real parts, so the sum of the two roots is always -1. From the viewpoint of design, we see that, by choice of gain K, we can cause the closed-loop poles to be at any point along the cross that makes up the root locus in Fig. 4.2. If some points along this locus correspond to satisfactory transient response, we can complete the design by choice of the corresponding value of K; otherwise we will be forced to consider more complex compensation.

As we pointed out earlier, the root-locus technique is not limited to finding root loci versus the system gain (K in the preceding example). All the same ideas are applicable for finding loci versus any parameter in a system characteristic equation. For example, let's examine the same $G(s)$ as before, but with K fixed ($= 1$) and with the motor time constant as the parameter of interest. That is, we select the open-loop transfer function

$$G(s) = \frac{1}{s(s + c)}. \tag{4.8}$$

The corresponding closed-loop characteristic equation is

$$1 + G(s) = 0$$

or

$$s^2 + cs + 1 = 0. \tag{4.9}$$

The same forms of Eq. (4.5), with the associated definitions of poles and zeros, will apply if we assign

$$a(s) = s^2 + 1,$$
$$b(s) = s,$$
$$K = c.$$

The solutions to Eq. (4.9) are easily computed as

$$r_1, r_2 = -\frac{c}{2} \pm \frac{\sqrt{c^2 - 4}}{2}, \tag{4.10}$$

and the locus of solutions versus the parameter c is shown in Fig. 4.3, with the poles again indicated by $\times$'s and the zero by $\bigcirc$. Note that when $c = 0$, the roots are at the $\times$'s and are oscillatory, while the damping ratio ζ grows as c grows.

Of course, computing the root locus for a quadratic equation is easy to do since we can solve the characteristic equation Eq. (4.6 or 4.9) for the roots, as is done

FIGURE 4.3
Root locus vs. damp-
ing factor c for the plant
$G(s) = 1/[(s)(s + c)]$.

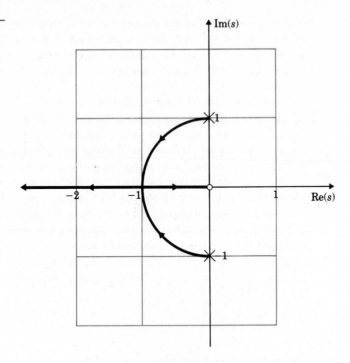

FIGURE 4.3
Root locus vs. damp-
ing factor c for the plant
$G(s) = 1/[(s)(s + c)]$.

in Eq. (4.7) and (4.10), and directly plot these as a function of the parameter K or c. To be useful, the method must be suitable for much higher-order systems, where explicit solutions are impossible. General properties of the root locus and rules for the construction of complex graphs of roots were developed by Evans, and together they constitute the mechanics of drawing a root locus, the core of the method. The general method will be presented in guidelines that can be followed in a step-by-step manner to sketch any root locus. The techniques will be developed assuming the loop gain K as the parameter of interest. In Section 4.6, the modifications for other parameters will be developed.

4.2 Guidelines for Sketching a Root Locus

Our goal in this section is to study the equations defined by Eq. (4.5) and to extract as much information as possible from them to aid us in sketching a reasonably accurate picture of the root locus for the given $G(s)$. It is important to keep in mind the reasons for making the plot as the guidelines are presented. The purpose of the root locus is to show in a graphical form the general trend of the roots of the closed-loop system as the parameter is varied. There are two reasons for learning how to generate a root locus by hand—first, so that we can use the manual method for design of fairly simple systems, and second, so that we can verify and understand computer-generated loci. The latter reason will become increasingly important as the cost of computers continues to decrease and control-design software becomes more available.

In this case, we are interested more in the general shape of the locus than in specific values, and some of the following rules may be omitted. Important features from a computer-generated locus can be judged to be correct or not by a good understanding of how loci must behave. For a manual control design to be feasible, we need to be able to sketch a locus quickly so that several attempts at the trial-and-error process are possible. Reasonable accuracy is possible with experience, and good estimates can be made as to how an additional pole or zero will change a locus and permit a better system design.

As a starting point for our guidelines, we begin with an alternative definition of the root locus. Earlier, we defined it this way:

Definition I: The root locus is the locus of values of s for which $1 + KG(s) = 0$ is satisfied as the real parameter K varies from zero to infinity. Usually $1 + KG(s)$ is the denominator of a transfer function of interest, so roots on the locus are *closed-loop poles* of the system.

Let's look at Eq. (4.5d). If K is to be real and positive,* $G(s)$ must be real and negative. In other words, if we arrange $G(s)$ in polar form as magnitude and phase, the phase of $G(s)$ must be the opposite of that of K to complete the requirement that Eq. (4.5) be satisfied. We can thus define the root locus as follows.

Definition II: The root locus of $G(s)$ is the locus of points in the s plane, where the phase of $G(s)$ is $180°$.†

Since phase is unchanged if an integral multiple of $360°$ is added, we can express definition II as $\angle G = 180° + l360°$, where l is any integer. The immense merit of definition II is that, while it is very difficult to solve a high-order polynomial, computation of phase is relatively easy. From definition II we can, in principle, sketch a root locus for a complex transfer function by measuring phase and marking down those places where we find $180°$. This direct approach can be illustrated by considering the example

$$G(s) = \frac{s + 1}{s\{[(s + 2)^2 + 4](s + 5)\}}. \tag{4.11}$$

---

*This is the usual case, but K could be real and negative, in which case $G(s)$ would be real and positive.

† The phase of $G(s)$ is $0°$ if K is negative. See Section 4.6.

In Fig. 4.4, the poles of this $G(s)$ are marked $\times$ and the zeros are marked $\bigcirc$. Suppose we select the test point $s_0 = -1 + j2$. The vector $s_0 + 1$ is found by drawing a line from the zero location -1 to the test point s_0. In this case, the line is vertical and has a phase angle, marked ψ_1 on Fig. 4.4. In similar fashion, the vector s_0 is shown with angle ϕ_1, and the angles of the two vectors from the complex poles at $-2 \pm j2$ to s_0 are shown with angles ϕ_2 and ϕ_3. The phase of the vector $(s_0 + 5)$ is shown with angle ϕ_4. The total phase of $G(s)$ at $s = s_0$ is found as the sum of the phases of the zero terms minus the phases of the denominator terms corresponding to the poles:

$$\angle G = \psi_1 - \phi_1 - \phi_2 - \phi_3 - \phi_4$$
$$= 90° - 116.6° - 0° - 76° - 26.6°$$
$$= -129.2°.$$

Since the phase of G is not $180°$, we conclude that s_0 is *not* on the root locus, and we must select another point and try again. Clearly, although measuring phase is easy, measuring phase at every point in the s plane is hardly the practical guide to the sketch we need. We must back off a bit more and seek general guidelines as to where the root locus is. For simpler reference later, we will present the results as a list of steps that, if followed, will go far toward indicating where a root locus is. To illustrate the guidelines, we will use first the transfer function

$$G(s) = \frac{1}{s[(s + 4)^2 + 16]}. \tag{4.12}$$

FIGURE 4.4
Measuring the phase of
$G(s) = (s + 1)/\{s[(s + 2)^2 + 4](s + 5)\}$.

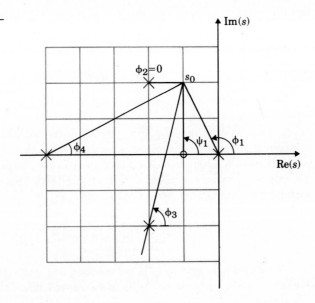

STEP 1: *Draw the axes to the s plane to a suitable scale and enter an* × *on this plane for each pole and a* ○ *for each zero. See Fig. 4.5(a).*

STEP 2: *Find the real axis portions of the loci.* If we take a test point on the real axis, such as s_0 in Fig. 4.5(b), we find that the angles of the two factors in $a(s)$ due to the complex poles are ϕ_1 and ϕ_2 such that $\phi_1 = -\phi_2$, and they cancel each other. The same would be true, of course, for the angles from complex zeros. The

FIGURE 4.5
Step-by-step development of a root locus.

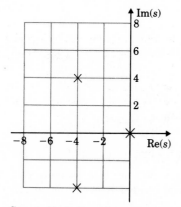

Step 1: Mark the poles and zeros.
(a)

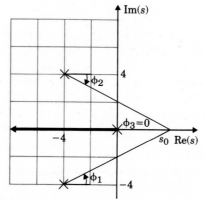

Step 2: Find the real axis part of the locus.
(b)

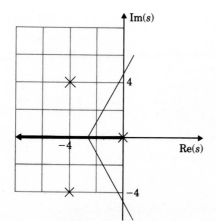

Step 3: Draw the asymptotes.
(c)

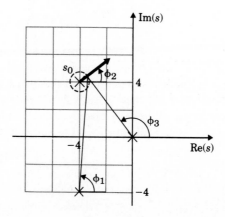

Step 4: Compute the departure and arrival angles.
(d)

result is that the angle of $G(s_0)$ for s_0 on the real axis is given by the angles from poles and zeros *on the real axis*. But these angles are $0°$ if the test point is to the right and $180°$ if the test point is to the left of the pole or zero. For the angle to add to $180° + l360°$, the test point must be to the left of an *odd* number of real axis poles plus zeros. In our test case, there is one real axis pole at the origin so the locus is on the real axis for all s_0 to the left of the origin. Accordingly, we mark this segment with a heavy line in Fig. 4.5(b).

STEP 3: *Draw the asymptotes for large K.* As K goes to infinity, the equation $1 + KG(s) = 0$ can be satisfied only if $G(s) = 0$. This can occur in two apparently different ways. In the first instance, since $G = b(s)/a(s)$ we have $G = 0$ if $b(s) = 0$. Therefore, for large K, a root of $1 + KG = 0$ will be found near the zeros of G, which is where $b(s) = 0$. To see the second manner in which $G(s)$ may go to zero, let's repeat Eq. (4.5b):

$$1 + K\frac{b(s)}{a(s)} = 0. \tag{4.5b}$$

Using Eq. (4.3), this may be written as

$$1 + K\frac{s^m + b_1 s^{m-1} + \cdots + b_m}{s^n + a_1 s^{n-1} + \cdots + a_n} = 0. \tag{4.13}$$

We are interested in the case of large K. If $n > m$,* then it is clear that $G(s)$ goes to zero as s goes to infinity. In fact, for very large s, the poles and zeros would appear to be in a cluster, m zeros would essentially cancel the effects of m on the poles, and the rest ($n - m$ poles) would appear to be at the same place. We conclude that our equation may be approximated by

$$1 + K\frac{1}{(s - \alpha)^{n-m}} \tag{4.14}$$

for large s and K.† We say that the locus of Eq. (4.13) is asymptotic to the locus of Eq. (4.14) for large K and s. We need to compute α and to find the locus for the asymptotic system. To find the locus, we let our search point s_0 be such that $s_0 = Re^{j\phi}$ for some large fixed R and variable ϕ. Since all poles of this simple system are in the same place, the angle of its transfer function is $180°$ if all $n - m$ angles, each equal to ϕ_l, add to $180°$. ϕ_l is given by

$$(n - m)\phi_l = 180° + l360°$$

---

*This must be the case if $G(s)$ represents a physical system and K is the open-loop gain.

†You can also obtain this approximation by dividing $a(s)$ by $b(s)$ and matching the dominant two terms (highest powers in s) to the expansion $(s - \alpha)^{n-m}$.

for some integer l and for any R. Thus, the asymptotic root locus consists of radial lines at the $n - m$ distinct angles given by

$$\phi_l = \frac{180° + l360°}{n - m}, \qquad l = 0, 1, 2, \ldots, n - m - 1. \qquad (4.15)$$

For our example system, $n - m = 3$ and $\phi_{1,2,3} = 60°, 180°$, and $300°$. The lines of the asymptotic locus come from $s_0 = \alpha$ on the real axis. To determine α, we make use of a simple property of polynomials. Suppose we have a monic polynomial with coefficients a_i and roots p_i, as in Eq. (4.3c and d), and we equate the polynomial form with the factored form

$$s^n + a_1 s^{n-1} + a_2 s^{n-2} + \cdots + a_n = (s - p_1)(s - p_2)(\cdots)(s - p_n).$$

If we multiply out the factors on the right-hand side of this equation, it is easy to see that the coefficient of s^{n-1} is $-p_1 - p_2 - \cdots - p_n$. From the left side, we see that this term is a_1. Thus, $a_1 = -\sum p_i$; in other words, the coefficient of the *second*-highest term in a monic polynomial is the negative sum of its roots—in this case, the poles of $G(s)$. Applying this result to the polynomial $b(s)$, we find the negative sum of the zeros to be b_1. These results can be written as

$$-b_1 = \sum z_i,$$

$$-a_1 = \sum p_i. \qquad (4.16)$$

Finally, we apply this result to the closed-loop characteristic polynomial obtained from Eq. (4.13):

$$s^n + a_1 s^{n-1} + \cdots + a_n + K(s^m + b_1 s^{m-1} + \cdots + b_m) = 0.$$

Here the sum of the roots is the negative of the coefficient of s^{n-1}, which is a_1, *independent* of K if $m < n - 1$. But since this is the closed-loop characteristic equation, this coefficient is the negative of the sum of the roots of the *closed-loop system*, $\sum r_i$. We have thus shown that the center point of the roots does not change with K if $m < n - 1$, and the common open-loop and closed-loop sum is $-a_1$, which can be expressed as

$$-\sum r_i = -\sum p_i. \qquad (4.17)$$

For large K, m of the roots r_i coincide with the z_i and $n - m$ are from the asymptotic $1/(s - \alpha)^{n-m}$ system whose poles add up to $(n - m)\alpha$. Combining these results, we have that the sum of all the roots equals the sum of those that go to infinity plus the sum of those that go to the zeros of $G(s)$. We have shown that

$$+\sum r_i = +(n - m)\alpha + \sum z_i = +\sum p_i.$$

From this we solve for α as

$$\alpha = \frac{\sum p_i - \sum z_i}{n - m}. \tag{4.18}$$

Notice that the imaginary parts *always* add to zero, so Eq. (4.18) gives new information about the real parts only. For our example system,

$$\alpha = \frac{-4 - 4 + 0}{3 - 0}$$

$$= -\frac{8}{3} = -2.67.$$

These asymptotes are sketched in Fig. 4.5(c). Notice that the asymptotes cross the imaginary axis at $\pm j(1.67)\sqrt{3} = \pm j4.62$.

STEP 4: *Compute the departure and arrival angles.* At this point we know that the loci begin at the points marked $\times$ —the poles of $G(s)$—and that they go either to the points marked $\bigcirc$ —the zeros of $G(s)$—or they go to infinity approaching the radial asymptote lines. We next compute the angle by which a branch of the locus departs from one of the poles. For this purpose, we take a test point very near pole 2 at $-4 + j4$ and compute the angle of $G(s_0)$. The situation is sketched in Fig. 4.5(d). We select the test point to be so close to the pole that the other angles (ϕ_1 and ϕ_3 in this case) do not change significantly as the test point moves around. We have the situation where all angles but one are fixed, and that one takes on whatever value will be necessary to meet the root-locus angle condition. For our illustrative example, the situation is as shown in Fig. 4.5(d), and the values are

$$\phi_1 = 90° \qquad \text{and} \qquad \phi_3 = 135°.$$

Thus, the angle condition is

$$-90° - \phi_2 - 135° = +180° + l360°, \tag{4.19}$$

where the integer l is selected so that the result for ϕ_2 is between $\pm 180°$. For a single angle, l will take on one value. For a multiple pole or zero of order n, l takes on n distinct values corresponding to the n branches of the root locus. Continuing with the example, we take $l = -1$ and find that

$$-\phi_2 = 90° + 135° + 180° - 360°,$$
$$\phi_2 = -45°.$$

If we had let $l = 0$, we would compute $\phi_2 = -405°$, which is the same as $-45°$. By the complex conjugate symmetry of the plots, the angle of departure of the locus near pole 1 at $-4 - j4$ will be $+45°$. We, of course, found that pole 3, at the origin on the real axis, departs at $180°$ along the real axis in step 2.

The nature of the calculation of a departure angle for small K, as shown in Fig. 4.5(d), is also valid to compute the angle by which a root locus "arrives" at a zero of $G(s)$ for large K. These angles should also be computed at this step and the process will be illustrated in our next example.

STEP 5: *Estimate (or compute) the points where loci cross the $s = j\omega$ axis.** A root of the characteristic equation in the right-half plane implies that the closed-loop system is unstable, a fact that can be tested by Routh's criterion (Chapter 3). If we compute the Routh array with K as a parameter, we can locate those values of K such that an incremental change will cause the number of roots in the right-half plane to change. Such values must correspond to a root locus crossing the imaginary axis. Given the facts that K is known and that a root exists for $s_0 = j\omega_0$, it is often not hard to solve for the frequency ω_0.

For the third-order example being used, the characteristic equation is

$$1 + \frac{K}{s[(s + 4)^2 + 16]} = 0,$$

which is equivalent to

$$s^3 + 8s^2 + 32s + K = 0. \tag{4.20}$$

The Routh array for this polynomial is

$$\begin{array}{cc} 1 & 32 \\ 8 & K \\ \dfrac{(8)(32) - K}{8} & 0 \\ K. & \end{array}$$

In this case, we see that the equation has no roots in the right-half plane for $0 < K < (8)(32) = 256$. Since we are interested in the locus for positive K, the upper bound is the point of interest here. For $K < 256$, there are no roots in the right-half plane, and if $K > 256$, the Routh test indicates that there are two roots in the right-half plane. Thus $K = 256$ must correspond to a solution at $s = j\omega_0$ for some ω_0. If we substitute this data into Eq. (4.20), we find

$$(j\omega_0)^3 + 8(j\omega_0)^2 + 32(j\omega_0) + 256 = 0. \tag{4.21}$$

If Eq. (4.21) is to be true, both real and imaginary parts must equal zero.[†] This gives us the equations

$$-8\omega_0^2 + 256 = 0,$$
$$-\omega_0^3 + 32\omega_0 = 0. \tag{4.22}$$

---

*For systems of order above 4, this step can be tedious and is often omitted in manual plotting. Computer-generated loci, of course, easily calculate the points of crossing.

[†] Did you notice that we could have substituted an unknown K in Eq. (4.21) and computed that $K = 256$ without using the Routh array in this case?

The solutions to Eq. (4.22) are $\omega_0 = \pm \sqrt{32} = \pm 5.66$. We note that the asymptotic locus crosses the imaginary axis at $s = j4.62$, which is below the actual locus crossing.

STEP 6: *Estimate locations of multiple roots, especially on the real axis, and determine the arrival and departure angles at these locations.** As with any polynomial, it is possible for a characteristic polynomial of degree greater than 1 to have multiple roots. In the second-order locus we plotted in Fig. 4.2, there were two roots at $s = -1/2$ when $K = 1/4$. At this point, we found that the loci come together and break away from the real axis, becoming complex for $K > 1/4$. The loci arrive at $0°$ and $180°$ and depart at $+90°$ and $-90°$. For simple systems, it is possible to use a formula for points of multiple roots, but the formula is only slightly easier to solve than the original root locus, so the best approach is to estimate these points based on a study of several prototype cases to be given soon. However, we will first demonstrate the formula that is useful for simple cases and which, if a means of solving higher-degree polynomials is available, can be used in complicated cases.

If we look again at the second-order root locus of Fig. 4.2 and consider what happens to the gain K between -1 and 0 along the real axis, we realize that $K = 0$ at $s = -1$, K increases steadily to $K = 1/4$ at $s = -1/2$, and, *staying on the real axis*, K decreases back to zero at $s = 0$. The gain has a maximum at the point of multiple roots, $s = -1/2$. Since $G(s)$ is a smooth function except at a pole and from the characteristic equation $K = -1/G$, the change in K must be smooth. Therefore, if K is to have a maximum at this point, the derivative of K with respect to s must be zero at the locus "breakaway," or multiple-root, point. Using the relation $K = -1/G$, we thus obtain the formula

$$\frac{d}{ds}\left(-\frac{1}{G}\right)_{s=s_0} = 0. \tag{4.23}$$

Since $G = b/a$, Eq. (4.23) requires that[†]

$$\frac{d}{ds}\left(-\frac{a}{b}\right) = -\frac{1}{b^2}\left(b\frac{da}{ds} - a\frac{db}{ds}\right) = 0. \tag{4.24}$$

---

*Like step 5, this step is also not often done by hand.

[†]Notice that $dG/ds = 0$ and $d/ds \ln G = 0$ as well. This latter form gives

$$\frac{d}{ds}[\ln(s - z_1) + \ln(s - z_2) + \cdots - \ln(s - p_1) - \cdots - \ln(s - p_n)]$$

$$= \frac{1}{s_0 - z_1} + \frac{1}{s_0 - z_2} + \cdots - \frac{1}{s_0 - p_1} - \cdots - \frac{1}{s - p_n} = 0.$$

Evans used this form to estimate s_0 on the real axis. One can also derive Eq. (4.23) from the observation that, at a point of multiple roots, the root-locus equation has the form

$$a(s) + Kb(s) = (s - r_1)^P(s - r_2)\cdots.$$

For the second-order case, we have

$$b = 1, \qquad \frac{db}{ds} = 0;$$

$$a = s^2 + s, \qquad \frac{da}{ds} = 2s + 1.$$

Substituting into Eq. (4.24), we find

$$2s_0 + 1 = 0$$

$$s_0 = -\frac{1}{2}, \qquad (4.25)$$

which confirms our previous observation of the breakaway point. For the third-order case given by the plots in Fig. 4.5, the polynomials are

$$b = 1, \qquad \frac{db}{ds} = 0;$$

$$a = s^3 + 8s^2 + 32s, \qquad \frac{da}{ds} = 3s^2 + 16s + 32.$$

The points of possible multiple roots are at

$$3s^2 + 16s + 32 = 0$$

$$s_0 = -2.67 \pm j1.89. \qquad (4.26)$$

However, our plot of Fig. 4.5(e) shows that these points are *not* on the root locus and are extraneous. This emphasizes that Eq. (4.24) is derived *if* s_0 is a multiple root on the locus. The derivative condition is necessary but not sufficient to indicate a breakaway, or multiple-root, situation.

In order to compute (estimate) the angles of arrival and departure from a point of multiple roots, it is useful to introduce two additional features of the root locus. The first concept we need is that of the continuation locus. One can imagine plotting a root locus for an initial range of K, perhaps for $0 \le K \le K_0$. Then, if we let $K = K_0 + K_1$, a new locus can be plotted with parameter K_1, a locus whose starting poles are the roots of the original system at $K = K_0$. To see how this goes, we return to the second-order root locus of Eq. (4.6) and let K_0 be the

---

Clearly, if we differentiate this equation with respect to s and set $s = r_1$, we'll get zero, so we have

$$\frac{da}{ds} + K\frac{db}{ds} = 0 \qquad \text{at } s = r_1.$$

We let $K = -a/b$, and Eq. (4.24) results.

FIGURE 4.5 *(cont.)*
Step-by-step development
of a root locus.

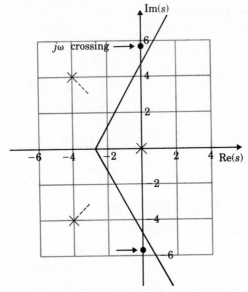

Step 5: Find the Im(s) axis crossing points.

(e)

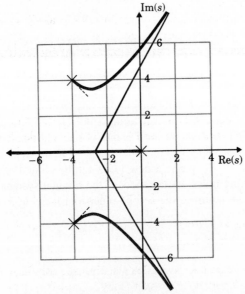

Step 7 : Complete the sketch.

(f)

value corresponding to the breakaway point $K_0 = 1/4$. If we let $K = 1/4 + K_1$, we have the locus equation

$$s^2 + s + \frac{1}{4} + K_1 = 0$$

$$\left(s + \frac{1}{2}\right)^2 + K_1 = 0. \tag{4.27}$$

The steps for plotting this locus are, of course, the same as for any other, except now the initial departure of the K_1 locus of Eq. (4.27) corresponds to the breakaway point of the K locus of Eq. (4.6)! Applying the rules of departure angles to the double pole at $s = -1/2$, we find

$$-2\phi_l = -180° + l360°,$$

$$\phi_l = 90° + l180°,$$

$$\phi_{1,2} = +90°, -90° \text{ (departure angles at breakaway)}. \tag{4.28}$$

In this case, the arrival angles at $s = -1/2$ are, from the original root locus, along the real axis and are clearly 0° and 180°. We can derive this result from the idea of the continuation root locus if we allow one additional change. We consider the possibility that K_1 is negative. In this case, we have the locus of Eq. (4.27), but, if K_1 is negative and $K = 1/4 + K_1$, as K_1 grows in magnitude we will trace the original locus back until, at $K_1 = -1/4$, we have a point corresponding to $K = 0$. Our purpose here is to find the angles along which the roots move as they arrive at the multiple-root location $s = -1/2$. Again, we apply the technique for departure angles, but we need the rule for negative gains. If we return to the basic equation

$$1 + KG(s) = 0,$$

it is evident that, if K is negative and real, $G(s)$ must be positive—that is, the angle of $G(s)$ must be 0°, rather than 180°. Considering this, the angles of arrival are

$$2\phi_l = 0° + l360°,$$

$$\phi_{1,2} = 0°, 180° \text{ (arrival angles at breakaway)}. \tag{4.29}$$

We will compute arrival and departure angles for more complicated cases in some of our later examples.

STEP 7: *Complete the sketch*. The final locus for the third-order example is drawn in Fig. 4.5(f). It combines all the results found so far—that is, the real axis segment, the angles of departure from the poles, the number of asymptotes and their angles, and the imaginary-axis crossing point.

Summary of Guidelines for Plotting a Root Locus

STEP 1: Mark poles $\times$ and zeros $\bigcirc$.

STEP 2: Draw the locus on the real axis to the left of an odd number of real poles plus zeros.

STEP 3: Draw $n - m$ asymptotes, centered at α and leaving at angles ϕ_l, where

$$\alpha = \frac{\sum p_i - \sum z_i}{n - m} = \frac{-a_1 + b_1}{n - m};$$

$$\phi_l = \frac{180° + l360°}{n - m}, \qquad l = 0, 1, 2 \cdots n - m - 1.$$

STEP 4: Compute loci departure angles from the poles and arrival angles at the zeros.

STEP 5:* Assume $s_0 = j\omega_0$ and compute the point(s) where the locus crosses the imaginary axis for positive K.

STEP 6:* The equation has multiple roots at points on the locus where

$$b\frac{da}{ds} - a\frac{db}{ds} = 0.$$

If s_0 is on the real axis, these are points of breakaway or break-in. Compute the angles of arrival and the angles of departure for any points of multiple roots.

STEP 7: Complete the locus, using the facts developed in the previous steps and making reference to the illustrative loci for guidance.

4.3   Selected Illustrative Root Loci

In order to gain facility with the root-locus method, it is helpful to sketch a number of locus plots and visualize the various alternative shapes a locus of a given complexity can take. As a start on this experience, we will go through the seven-step procedure for several cases selected for the way they illustrate important features

*These are optional steps.

of root loci. As a first case, we consider the locus $1 + KG(s)$, where

$$G(s) = \frac{s + 1}{s^2(s + p)} \tag{4.30}$$

for various values of p.

EXAMPLE 4.1 First we will sketch the root locus for p very large—namely, the locus for $G(s) = (s + 1)/s^2$. We will go through each of the seven steps of the guidelines.

STEP 1: Mark the poles and zeros on the s plane.

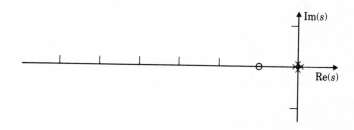

STEP 2: Draw the locus on the real axis to the left of an odd number of poles plus zeros.

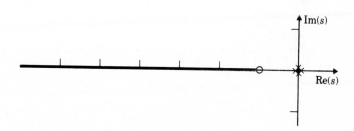

STEP 3: Draw $n - m$ asymptotes:

$$\phi_l = \frac{180° - l360°}{2 - 1}$$
$$= 180°.$$

α does not matter, since there is only one asymptote.

STEP 4: Compute the departure and arrival angles. For the two poles at $s = 0$, we draw a small circle around them. The angles from the two poles are equal, and the angle from the zero is essentially zero, so the angle condition reduces to

$$-2\phi_l + 0° = 180° + l360°,$$
$$\phi_{1,2} = -90°, +90°.$$

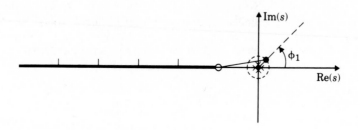

The locus leaves vertically, one up and one down, as we indicate by arrows:

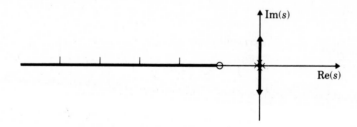

STEP 5: Compute the points where the locus crosses the imaginary axis. The characteristic equation is

$$s^2 + Ks + K = 0,$$

and the Routh array is

$$
\begin{array}{lll}
s^2: & 1 & K \\
s^1: & K & \\
s^0: & K. &
\end{array}
$$

Since the entries in the first column are all positive if K is positive, the equation's roots are all in the left-half plane on the root locus and do *not* cross the imaginary axis.

FIGURE 4.6
Root locus for $G(s) = (s + 1)/s^2$.

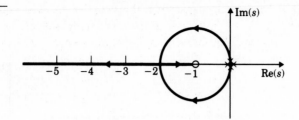

STEP 6: Locate the possible multiple roots according to Eq. (4.24). Here we have

$$b(s) = s + 1,$$
$$a(s) = s^2,$$
$$\frac{db}{ds} = 1;$$
$$\frac{da}{ds} = 2s.$$

And the condition is

$$-s^2(1) + (s + 1)(2s) = 0,$$
$$s^2 + 2s = 0,$$
$$s = 0, -2.$$

The root at $s = 0$ corresponds to the two poles of $G(s)$ that are on the locus at $K = 0$. The other case at $s = -2$ corresponds to a multiple root, since in step 2 we found that $s = -2$ is on the locus.

STEP 7: Sketch the locus as shown in Fig. 4.6. In this case, we can prove that the locus is, in fact, a circle, the proof being left as a problem. ∎

EXAMPLE 4.2   Now we will plot the root locus for $1 + KG(s)$, where $G(s)$ is given by Eq. (4.30), with $p = 4$.

STEP 1: Mark the poles and zeros on the s plane.

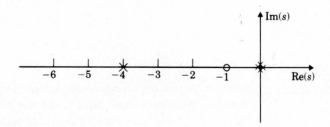

STEP 2: Draw the locus on the real axis to the left of an odd number of poles plus zeros.

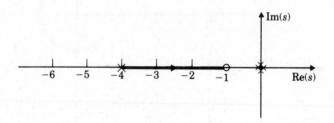

STEP 3: Draw $n - m$ asymptotes:

$$\phi_l = \frac{180° + l360°}{n - m}$$

$$= \frac{180° + l360°}{3 - 1}$$

$$= 90° + l180°;$$

$$\phi_{1,2} = +90°, -90°;$$

$$\alpha = \frac{\sum p_i - \sum z_i}{n - m}$$

$$= \frac{-4 - (-1)}{3 - 1}$$

$$= \frac{-3}{2}.$$

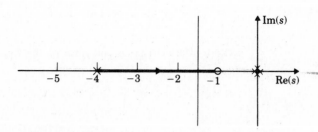

STEP 4: Compute the departure angles from the poles and the arrival angles toward the zeros. We draw a small circle around the two poles at $s = 0$. The angles from the zero at -1 and from the pole at -4 are both zero, and the angles from the two

poles at the origin are the same. Therefore the root-locus condition is

$$-2\phi_l - 0° + 0° = 180° + l360°,$$
$$\phi_l = -90° - l180°$$
$$= -90°, +90°.$$

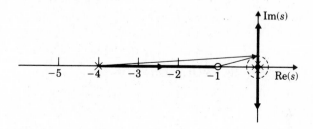

STEP 5: Assume $s = j\omega_0$ and compute the point(s) where the locus crosses the imaginary axis:

$$1 + K\frac{s + 1}{s^2(s + 4)} = 0,$$
$$s^3 + 4s^2 + Ks + K = 0.$$

Let $s = j\omega$:

$$-j\omega^3 - 4\omega^2 + Kj\omega + K = 0.$$

Set real and imaginary parts separately to zero:

$$-4\omega^2 + K = 0,$$
$$K = 4\omega^2;$$
$$-\omega^3 + K\omega = 0,$$
$$K = \omega^2.$$

The two equations have no solution except $K = \omega = 0$, so we conclude that the loci do not cross the imaginary axis for $K > 0$. We can confirm this finding by the Routh array test. In the present case, the array is

$$
\begin{array}{ll}
s^3 : & 1 \qquad K \\
s^2 : & 4 \qquad K \\
s^1 : & \dfrac{4K - K}{4} \\
s^0 : & K.
\end{array}
$$

The entries in the first column are all positive if K is positive, so the equation has no roots in the right-half plane for positive K.

STEP 6: Locate the points of multiple roots, which will include breakaway and break-in points:

$$b(s) = s + 1,$$
$$a(s) = s^3 + 4s^2,$$
$$\frac{db}{ds} = 1;$$
$$\frac{da}{ds} = 3s^2 + 8s.$$

Substituting into Eq. (4.24),

$$-(s^3 + 4s^2)(1) + (s + 1)(3s^2 + 8s) = 0,$$
$$-s^3 - 4s^2 + 3s^3 + 11s^2 = 8s = 0,$$
$$2s^3 + 7s^2 + 8s = 0,$$
$$s(2s^2 + 7s + 8) = 0.$$

The possible locations are

$$s = 0 \quad \text{and} \quad s = -1.75 \pm j0.968.$$

The root at $s = 0$ is on the locus and corresponds to the two poles of $G(s)$ at $s = 0$, which are on the locus for $K = 0$. The two other roots are not on the

FIGURE 4.7
Root locus for $G(s) = (s + 1)/[s^2(s + 4)]$.

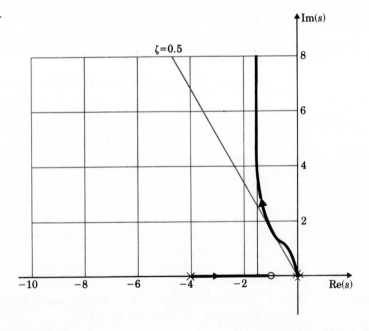

root locus and therefore do not correspond to multiple roots. To confirm this, it would be necessary to compute the phase of $G(-1.75 \pm j0.97)$, which is not worth the trouble.

STEP 7: The complete root locus drawn is Fig. 4.7. ■

EXAMPLE 4.3   As a third example, we will sketch the root locus for $1 + KG(s)$, where

$$G(s) = \frac{s + 1}{s^2(s + 12)}.$$

STEP 1: Mark the poles and zeros.

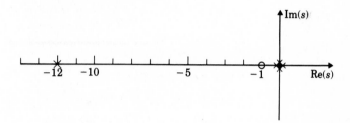

STEP 2: Draw the real axis loci.

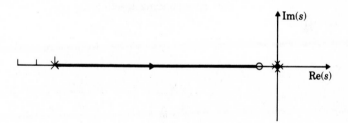

STEP 3: Draw the asymptotes:

$$\phi_l = \frac{180° + l360°}{3 - 1}$$

$$= +90°, -90°;$$

$$\alpha = \frac{-12 - 0 - (-1)}{3 - 1}$$

$$= \frac{-11}{2}.$$

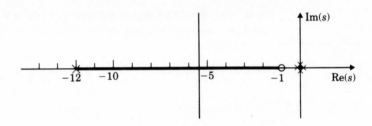

STEP 4: Compute the departure and arrival angles at poles at $s = 0$, $\phi_{1,2} = \pm 90°$.

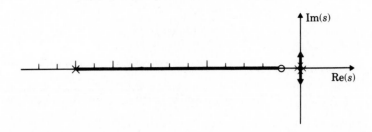

STEP 5: Compute the imaginary axis crossing. The characteristic polynomial is

$$s^3 + 12s^2 + Ks + K = 0,$$

and the corresponding Routh array is

$$
\begin{array}{ccc}
s^3 : & 1 & K \\
s^2 : & 12 & K \\
s^1 : & \dfrac{12K - K}{4} & \\
s^0 : & K. &
\end{array}
$$

Again, we find that the signs of the terms in the first column are all positive if K is positive, so there are *no* crossings of the imaginary axis.

STEP 6: Locate the multiple roots:

$$
\begin{aligned}
b(s) &= s + 1, \\
a(s) &= s^3 + 12s^2, \\
\frac{db}{ds} &= 1; \\
\frac{da}{ds} &= 3s^2 + 24s.
\end{aligned}
$$

Using Eq. (4.24) again, we find the possible points of multiple roots at the solution of

$$-(s^3 + 12s^2)(1) + (s + 1)(3s^2 + 24s) = 0,$$
$$-s^3 - 12s^2 + 3s^3 + 27s^2 + 24s = 0,$$
$$2s^3 + 15s^2 + 24s = 0.$$

The roots of this equation are

$$s = 0, \qquad -5.18, \qquad -2.31.$$

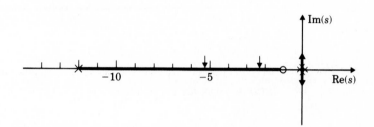

In this case, all three roots are on the locus and are thus points of multiple roots. The complete locus is drawn in Fig. 4.8. Note that the multiple roots at -2.31 are at a break-in point, and at -5.18 the roots break away from the real axis, approaching the vertical asymptotes as K gets large. Notice that the locus both breaks in and breaks away at right angles. ∎

FIGURE 4.8
Root locus for $G(s) = (s + 1)/[s^2(s + 12)]$.

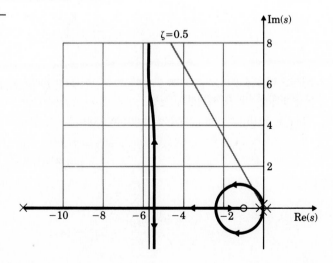

EXAMPLE 4.4   As the fourth example in this series based on the transfer function of Eq. (4.30), we consider

$$G(s) = \frac{s + 1}{s^2(s + 9)}.$$

STEPS 1, 2: Mark the axes and the real axis segments.

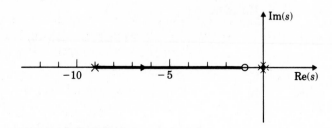

STEP 3: Draw the asymptotes:

$$\phi_l = \frac{180° + l360°}{3 - 1} = \pm 90°;$$

$$\alpha = \frac{-9 - 0 - (-1)}{3 - 1} = -4.$$

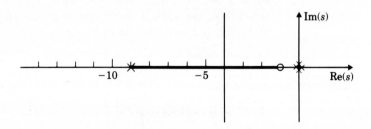

STEPS 4, 5: Again, we have departure of the two poles at $s = 0$ at $\pm 90°$ and no crossing of the imaginary axis.

STEP 6: Estimate the possible multiple roots:

$$b(s) = s + 1$$
$$a(s) = s^3 + 9s^2,$$
$$\frac{db}{ds} = 1;$$
$$\frac{da}{ds} = 3s^2 + 18s.$$

The possible multiple roots are at

$$-(s^3 + 9s^2)(1) + (3s^2 + 18s)(s + 1) = 0,$$
$$-s^3 - 9s^2 + 3s^3 + 21s^2 + 18s = 0,$$
$$2s^3 + 12s^2 + 18s = 0,$$
$$s = 0, -3, -3.$$

Again, the points of multiple roots are on the locus, but in this case we have *repeated* roots in the derivative, which indicates not only $dG/ds = 0$ but $d^2G/ds^2 = 0$, so we have *three* roots at the same place. Applying the rule of departure angles $(K_1 > 0)$ to the triple root at $s = -3$, we find

$$-3\phi_l + 180° = +180° + l360°,$$
$$\phi = 0°, \pm 120°.$$

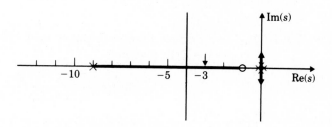

The arrival angle rule $(K_1 < 0)$ to the triple root yields

$$-3\phi_l + 180° = 0° + l360°,$$
$$\phi = \pm 60°, 180°.$$

STEP 7: The complete locus is sketched in Fig. 4.9.   ∎

If we now compare the three root loci in Figs. 4.7, 4.8, and 4.9, it is evident that when the third pole is near the zero, we have a modest distortion of the second-order locus for $1/s^2$. As p is increased, the distortion becomes more extreme, until, at $p = 9$, the locus breaks in at -3 in a triple multiple root. As the pole p is moved beyond -9, the locus exhibits distinct break-in and breakaway points, approaching, as p gets very large, the circle locus of one zero and two poles. The locus of Fig. 4.9 when $p = -9$ is thus a transition locus between the two second-order extremes, which occur at $p = 1$ (where the zero is canceled) and $p = \infty$ (where the extra pole has no effect).

FIGURE 4.9
Root locus for $G(s) =$
$(s + 1)/[s^2(s + 9)]$.

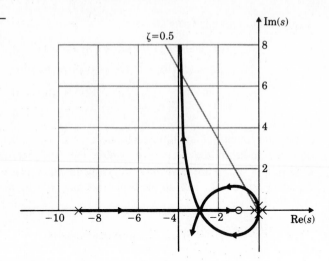

EXAMPLE 4.5 This next example illustrates multiple roots off the real axis. Consider the root locus of $1 + KG(s)$, where

$$G(s) = \frac{1}{s\left\{(s + 2)\left[(s + 1)^2 + 4\right]\right\}}.$$

STEPS 1, 2: Mark the s plane and draw in the real axis segments.

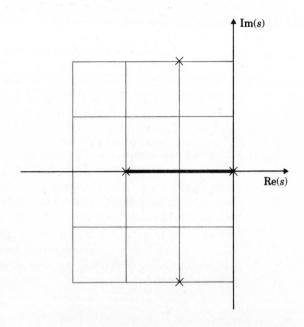

STEP 3: Draw the asymptotes:

$$\phi_l = \frac{180° + l360°}{4 - 0}$$

$$= 45° + l90°$$

$$= 45°, 135°, -45°, -135°.$$

$$\alpha = \frac{-2 - 1 - 1 - 0 + 0}{4 - 0}$$

$$= -1.$$

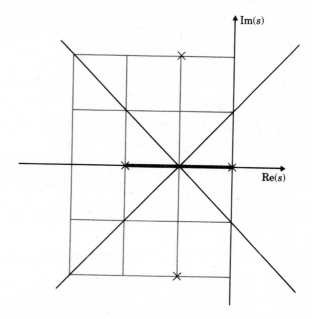

STEP 4: We consider a test point near the complex pole at $-1 + j2$. (See the figure at the top of page 186.)

The angle condition on $G(s)$ is

$$-\phi_1 - \phi_2 - \phi_3 - \phi_4 = +180° + l360°,$$

where

$$\phi_1 = \tan^{-1} \frac{2}{-1} = 116.6°,$$

$$\phi_2 = \tan^{-1} \frac{2}{1} = 63.4°,$$

$$\phi_3 = \text{unknown},$$

$$\phi_4 = 90°;$$

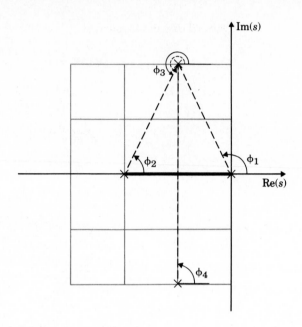

$$-116.6° - 63.4° - \phi_3 - 90° = 180° + l360°,$$
$$-\phi_3 = +450° + l360°,$$
$$= 90°,$$
$$\phi_3 = -90°.$$

In fact, we can observe at once that along the line $s = -1 + j\omega$, ϕ_2, and ϕ_1 are angles on an isosceles triangle and always add to 180°, and ϕ_4 is 90°, so the entire line from one complex pole to the other is on the locus *in this special case.*

STEP 5: Compute the crossing of the imaginary axis. The characteristic equation is

$$s^4 + 4s^3 + 9s^2 + 10s + K = 0.$$

If we try a solution for $s = j\omega_0$, we find ω_0 and K must satisfy

$$\omega_0^4 - 4j\omega_0^3 - 9\omega_0^2 + j10\omega_0 + K = 0.$$

Equating the real and imaginary parts to zero, we get the equations

$$\omega_0^4 - 9\omega_0^2 + K = 0, \qquad\qquad (i)$$
$$-4\omega_0^3 + 10\omega_0 = 0. \qquad\qquad (ii)$$

From (*ii*), $\omega_0^2 = 5/2$ and thus $\omega_0 = 1.58$. From (*i*) and (*ii*),

$$K = 9\left(\frac{5}{2}\right) - \frac{25}{4}$$

$$= \frac{90 - 25}{4} = 16.25.$$

This value of K can also be verified by the Routh array.

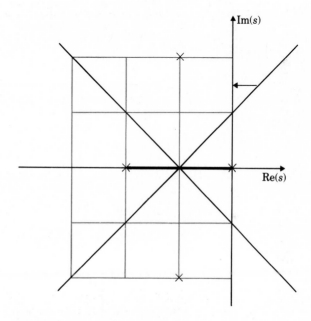

STEP 6: Locate possible multiple roots:

$$b = 1, \qquad \frac{db}{ds} = 0;$$

$$a = s^4 + 4s^3 + 9s^2 + 10s, \qquad \frac{da}{ds} = 4s^3 + 12s^2 + 18s + 10.$$

The condition reduces to $da/ds = 0$. We could find solutions to this cubic, but from step 4 we notice that the line at $s = -1 + j\omega$ is on the locus, so there must be a breakaway at $s = -1$, which can be divided out. That is, we can easily show that

$$4s^3 + 12s^2 + 18s + 10 = (s + 1)(4s^2 + 8s + 10).$$

The quadratic has roots $-1 \pm j\sqrt{3/2} = -1 \pm j1.22$. Since these points are on the line between the complex poles, they are true points of multiple roots on the locus.

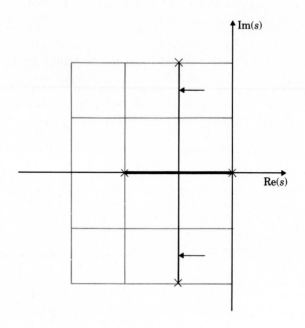

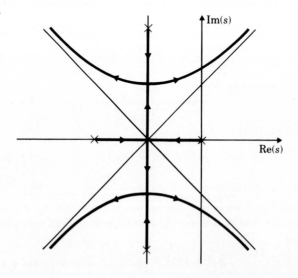

FIGURE 4.10
Root locus for $G(s) = 1/\{s(s+2)[(s+1)^2 + 4]\}$.

STEP 7: Using this data, we sketch the locus in Fig. 4.10. Notice that here we have complex multiple roots. Branches of the locus come together at $-1 \pm j1.22$ and break away at $0°$ and $180°$. It is interesting to sketch the loci as the pole at -2 is moved in to -1 and removed to -10 to see why this locus also is a transition between alternatives. ∎

EXAMPLE 4.6   In the next two examples, we will examine the importance of the departure and arrival angles. The first case is a root locus typical of servomechanisms with a flexible load when the position sensor is on the motor shaft. The locus is $1 + KG(s)$, where

$$G(s) = \frac{(s + 0.1)^2 + 16}{s\,[(s + 0.1)^2 + 25]}.$$

Again, we follow the steps.

STEPS 1, 2:

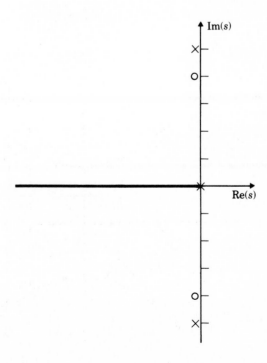

STEP 3: Draw the asymptotes:

$$\phi_l = \frac{180° + l360°}{3 - 2} = 180°,$$

$$\alpha = \text{not relevant.}$$

STEP 4: Compute the departure and arrival angles. For the arrival angle of the zero, the angle condition is

$$-\phi_1 - \phi_2 - \phi_3 + \psi_1 + \psi_2 = 180° + l360°,$$

where

$$\phi_1 \cong 90°; \phi_2 \cong -90°; \phi_3 \cong 90°; \psi_2 \cong 90°.$$

Thus $\psi_1 = 180°$, and the root arrives from the left. At the pole, the departure angle is found from the same formula, except now ϕ_2 is unknown and $\phi_1 \cong \phi_3 \cong \psi_1 \cong \psi_2 = 90°$. Thus,

$$-90° - \phi_2 - 90° + 90° + 90° = 180° + l360°.$$

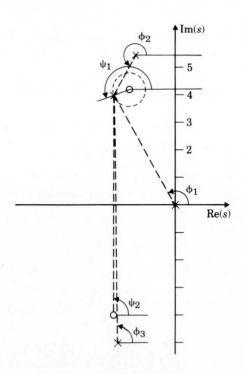

Thus, $\phi_2 \cong 180°$ (with $l = 0$) and departure is also toward the left. We can pause here and sketch in our best guess at the locus, as follows:

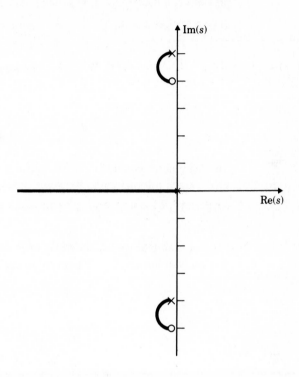

At this point, we would expect to find no root crossing of the imaginary axis and no multiple root points.

STEP 5: Compute the point where the locus crosses the imaginary axis. The characteristic equation is

$$s^3 + (K + 0.2)s^2 + (0.2K + 25.01)s + 16.01K = 0,$$

and the Routh array is

$$
\begin{array}{ll}
s^3 : & 1 \qquad 0.2K + 25.01 \\
s^2 : & K + 0.2 \qquad 16.01K \\
s^1 : & \dfrac{(K + 0.2)(0.2K + 25.01) - 16.01K}{K + 0.2} \\
s^0 : & 16.01K.
\end{array}
$$

The leading (and only) term in the s^1 line is a quadratic in K equal to $0.2K^2 + 9.04K + 5.0$, which is clearly not zero for any positive K, so we confirm our previous conjecture that there is no crossing of the imaginary axis.

STEP 6: Determine the possible multiple roots:

$$b(s) = s^2 + 0.2s + 16.01,$$
$$a(s) = s^3 + 0.2s^2 + 25.01s,$$
$$\frac{db}{ds} = 2s + 0.2;$$
$$\frac{da}{ds} = 3s^2 + 0.4s + 25.01.$$

The formula for possible multiple roots is a fourth-degree polynomial and does not give enough information to be worth solving.

STEP 7: The complete locus is sketched in Fig. 4.11.   ■

EXAMPLE 4.7   For this example, we use the same basic configuration as in Example 4.6, except that we interchange the complex poles and zeros. We take $G(s)$ to be

$$G(s) = \frac{(s + 0.1)^2 + 25}{s\,[(s + 0.1)^2 + 16]}.$$

FIGURE 4.11
Root locus for $G(s) = [(s + 0.1)^2 + 16]/[s(s + 0.1)^2 + 25]$.

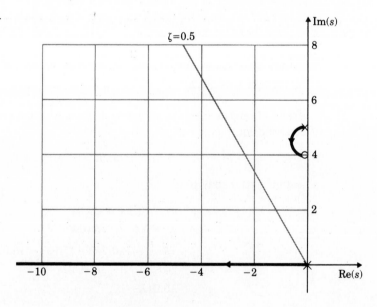

Since this example is very much like Example 4.6, we will only repeat the departure- and arrival-angle calculations and the imaginary-axis crossing calculation.

STEP 4: Compute the angles of arrival and departure. In this case, we have the angle condition for departure at pole $2(-0.1 + j4)$ as $-\phi_1 - \phi_2 - \phi_3 + \psi_1 + \psi_2 = 180° + l360°$, where $\phi_1 \cong \phi_3 \cong \psi_3 = 90°; \psi_1 = -90°$. Thus (with $l = -1$)

$$-90° - \phi_2 - 90° + 90° - 90° = 180° + l360°,$$

$$\phi_2 = 0.$$

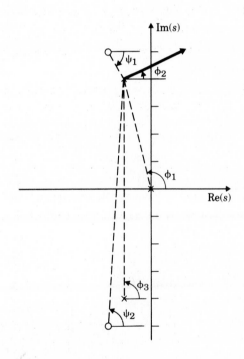

For arrival at the zero $(-0.1 + j5)$, the angles are $\phi_1 \cong \phi_2 \cong \phi_3 \cong \psi_2 = 90°$, and the angle condition for the root locus is

$$-90° - 90° - 90° + 90° + \psi_1 = 180° + l360°,$$

$$\psi_1 = 0°.$$

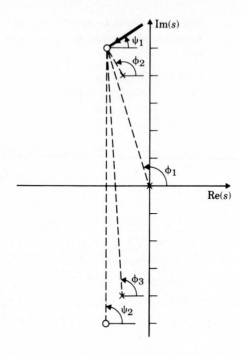

STEP 5: From step 4, we see that both departure and arrival are *toward* the right-half plane so we would expect to find a crossing of the imaginary axis. The characteristic equation is

$$s^3 + (K + 0.2)s^2 + (0.2K + 16.01)s + 25.01K = 0.$$

The Routh array is

$$
\begin{array}{ll}
s^3: & 1 \qquad 0.2K + 16.01 \\
s^2: & K + 0.2 \qquad 25.01K \\
s^1: & \dfrac{(0.2K + 16.01)(K + 0.2) - 25.01K}{K + 0.2} \\
s^0: & 25.01K.
\end{array}
$$

In this case, stability is decided (for $K > 0$) by the term in the s^1 row. Multiplying this out, the imaginary axis will be crossed for real positive solutions to

$$0.2K^2 - 8.96K + 3.20 = 0.$$

These are $K = 44.4$ and 0.362. With these values, the frequencies may be found from the real part of the characteristic equation, namely,

$$(K + 0.2)\omega^2 = 25.01K,$$

$$\omega = \sqrt{\frac{25.01K}{K + 0.2}}$$

$$= 4.99 \qquad \text{and} \qquad 4.01.$$

Notice that the crossings are very close to the (open-loop) poles and zeros.

STEPS 6, 7: Again, we omit the calculation of possible multiple roots and plot the root-locus in Fig. 4.12. ■

The major point of Examples 4.6 and 4.7 is that if the system has poles or zeros near the unstable boundary, the departure angles will quickly show if, in fact, these will cause closed-loop stability problems. Furthermore, if one compares Figs. 4.11 and 4.12, it is obvious that shifting the relative position of poles and zeros can have a dramatic influence on stability. In Fig. 4.11, the zero is at a lower frequency than the pole and the locus stays in the left-half plane, whereas in Fig. 4.12 their positions are reversed, and that portion of the locus represents an unstable system for almost any value of K; thus the entire system is unstable for almost any value of K.

FIGURE 4.12
Root locus for $G(s) = [(s + 0.1)^2 + 25]/\{s[(s + 0.1) + 16]\}$.

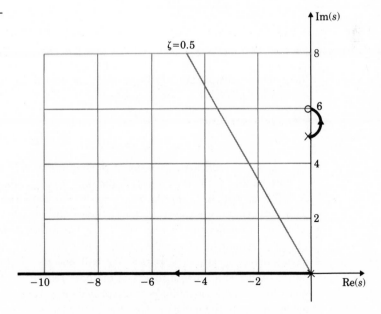

4.4   Selecting Gain from the Root Locus

The root locus is a plot of *all possible locations* for roots to the equation $1 + KG(s) = 0$ for some real positive value of the parameter K. The purpose of design is to select a particular value of K that is suitable according to the static- and dynamic-response specifications. We now turn to the question of selection of K and identification of the value of the parameter that will cause the roots to be at specific places. Using the phase definition (II) of the locus, we developed seven steps to sketch a root locus from the *phase* of $G(s)$ alone. If the equation is actually to have a root at a particular place when the phase of $G(s)$ is 180°, a *magnitude* condition must also be satisfied. If we repeat Eq. (4.5d) we can readily see that this condition is

$$G(s) = -\frac{1}{K},$$

which, of course, is equivalent to

$$K = -\frac{1}{G(s)}.$$

For values of s on the root locus the phase of G is 180°, and we can write the magnitude condition as

$$K = \frac{1}{|G|}. \tag{4.31}$$

Equation (4.31) has both an algebraic and a graphical interpretation. To see the graphical interpretation of Eq. (4.31), consider the locus of $1 + KG(s)$ where, as seen in Eq. (4.12),

$$G(s) = \frac{1}{s\,[(s + 4)^2 + 16]}. \tag{4.32}$$

For this transfer function, the locus is plotted in Fig. 4.5, and is repeated showing all parts in Fig. 4.13. On Fig. 4.13, the lines corresponding to a damping ratio of $\zeta = 0.5$ are sketched, and the points where the loci cross these lines are marked with dots (•). As we know, the locus shows all possible root locations. Suppose we wish to set the gain so that the actual locations of the poles are at the dots. This corresponds to selecting the gain so that two of the closed-loop system poles have a damping ratio of $\zeta = 0.5$. We will find the third pole shortly. The question is, What is K when a root is at the dot? From Eq. (4.31), the value of K is given by 1 over the magnitude of $G(s_0)$, where s_0 is the coordinate of the dot. On the figure, we have plotted three vectors marked $s_0 - s_1$, $s_0 - s_2$, and $s_0 - s_3$, which

FIGURE 4.13
Root locus for $G(s) = 1/\{s[(s + 4)^2 + 16]\}$ showing calculation of gain K.

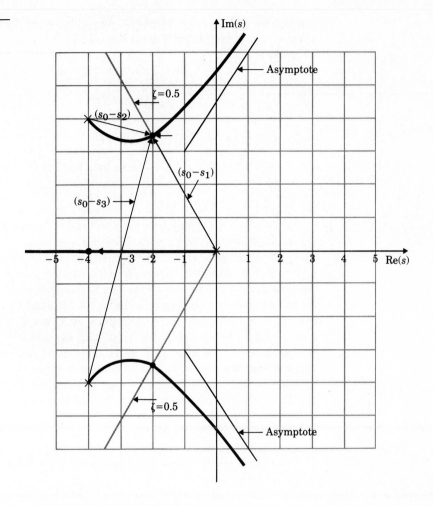

are the vectors from the poles of $G(s)$ to the point s_0. (Since $s_1 = 0$, the first vector equals s_0.) Algebraically, we have

$$G(s_0) = \frac{1}{s_0\,[(s_0 - s_2)(s_0 - s_3)]},\qquad(4.33)$$

and, using Eq. (4.31),

$$K = \frac{1}{|G(s_0)|} = |s_0||s_0 - s_2||s_0 - s_3|.\qquad(4.34)$$

The graphical interpretation of Eq. (4.34) shows that its three magnitudes are the lengths of the corresponding vectors drawn on Fig. 4.13. Hence, we can compute the gain to place the roots at the dot ($s = s_0$) by measuring the lengths of these vectors and multiplying the lengths together. Using the scale of the figure, we estimate that

$$|s_0| = 3.95,$$
$$|s_0 - s_2| = 2.1,$$
$$|s_0 - s_3| = 7.7.$$

Thus, the gain is estimated to be

$$K = (4.0)(2.1)(7.7)$$
$$= 65.$$

We conclude that if K is set to the value 65, then a root of $1 + KG$ will be at s_0, which has a damping ratio of 0.5.* Another root is at the conjugate of s_0, of course. Where is the third root? From the root locus, the third branch is along the negative real axis. Ordinarily we would need to guess a test point, compute a trial gain, and correct the guess until we found the point where $K = 65$. In this case, we can use the property of the root locus discovered when we were drawing the locus asymptotes and expressed in Eq. (4.17)—that is, the sum of the roots is *constant* (does not change as K changes) if $m < n - 1$. Thus the unknown root must be moved far enough to the *left* to keep the sum fixed. From the figure, we estimate that $s_0 = -2.0 + j3.4$. Since the starting point was at $s = -4 + j4$, this root has moved $4 - 2.0$, or two units to the right. The conjugate has moved an equal distance, so the third root must have moved $(2)(2)$, or four units to the *left*. Since the third root began at $s = 0$, it is now located at -4, where we have marked the spot with the third dot.

A process with the transfer function given by Eq. (4.32) has one integrator and, in a unity feedback configuration, will be a Type-I control system. In this case, the steady-state error in tracking a ramp input is given by the velocity constant K_v, which is in this case,

$$K_v = \lim_{s \to 0} sKG(s)$$

$$= \lim_{s \to 0} s \frac{K}{s\,[(s + 4)^2 + 16]}$$

$$= \frac{K}{32}. \tag{4.35}$$

With the gain set for complex roots at damping $\zeta = 0.5$, the root locus gain K is 65; by Eq. (4.35) the velocity constant is $K_v = 65/32 \cong 2$. If the closed-loop dynamic response as determined by the root locations is satisfactory and the steady-state accuracy as measured by K_v is good enough, then the design can

---

*The value (65) is, of course, an approximation but is sufficiently close that a finer tuning would have to be done experimentally on the actual system, since the $G(s)$ we are using is also subject to error.

be completed by gain selection alone. If, as is typically the case, no value of K satisfies all the constraints, then additional dynamics are necessary to meet the system specifications.

4.5 Dynamic Compensation: The Lead and Lag Networks

If the process dynamics are of such a nature that a satisfactory design *cannot* be obtained by a gain adjustment alone, some modification of compensation for the process dynamics is indicated. While the variety of such compensation is great, two categories have been found to be particularly simple and effective. These are the lead and the lag networks.* The lead network acts mainly to modify the dynamic response to raise bandwidth and lower rise time and also to decrease the transient overshoot; in a crude way, a lead network approximates derivative control. The lag network is usually used to raise the low-frequency gain and thus to improve the steady-state accuracy of the system; the lag network is an approximate integral control. We will consider these schemes to improve control-system performance separately, starting with the lead network.

To explain the effect of a lead network on the root locus, we consider again the simplest case of a second-order system with transfer function:

$$KG(s) = \frac{K}{s(s + 1)},$$

which has the root locus shown in Fig. 4.2, repeated now as the solid-line locus in Fig. 4.14. Also shown in Fig. 4.14 is the locus produced by $D(s)G(s)$, where $D(s)$ represents a zero added at $s = -2$. The modified locus is the circle sketched with dotted lines. Notice that the effect of the zero is to move the locus to the left, toward the more stable part of the plane. Clearly, if our speed-of-response specification calls for $\omega_n \cong 2$, then gain alone with $G(s) = 1/[s(s + 1)]$ can only produce a very low value of damping ratio ζ, and hence at the required gain the transient overshoot will be substantial. However, by addition of the zero we can move the locus to a position having closed-loop roots at $\omega_n = 2$ and $\zeta \geq 0.5$. We have "compensated" the given dynamics by use of $D(s) = s + 2$.

The trouble with compensation with a zero is that the physical realization of a transfer function $s + 2$ can only be approximate, and the inevitable high-frequency noise present at the input to the $D(s)$ will be greatly amplified by such a transfer function. However, since the effect of the zero is most important near $\omega_n = 2$, the action of the compensation will not be greatly reduced if a higher-frequency pole is added, perhaps at $s = -20$ to give $D(s) = (s + 2)/(s + 20)$. The resulting transfer function is that of a lead network.

---

*The names of these devices are derived from their frequency (sinusoidal) responses, where the output leads the input in one case (positive phase shift) and lags the input in the other (negative phase shift).

FIGURE 4.14
Root locus for $KG(s) = K/[s(s + 1)]$, with an additional zero $K_D G(s) = [K(s + 2)]/[s(s + 1)]$.

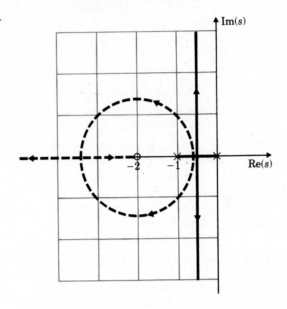

To see the effect of the pole on the compensation, the root loci for two cases are shown in Fig. 4.15(a and b). In Fig. 4.15(a), the root locus is sketched for $D(s) = (s + 2)/(s + 20)$, and in Fig. 4.15(b), the locus is sketched for $D(s) = (s + 2)/(s + 10)$. The important fact about these loci is that for small gains, before the real part of the root reaches -2, they are almost identical with the locus of Fig. 4.14, where $D(s) = s + 2$. To emphasize this point, in Fig. 4.15(c) all three loci are superimposed on the same plot. As can be easily seen from that figure, the effect of the pole is to press the locus toward the right, but for the early part of the locus, the effect of the pole is not great.

The method of selection of the pole and zero of the lead network can be made more analytical if we select the desired closed-loop pole location first. Then we select the zero of the lead-network on the basis of overshoot response and use the root-locus angle condition to select the pole. The method can be illustrated by the same plant. Suppose we wish to force the root locus to pass through the point $-1 + j\sqrt{3}$ corresponding to $\omega_n = 2$, $\zeta = 0.5$. The corresponding s plane is sketched in Fig. 4.15(d). The root-locus angle condition is satisfied if the angle ϕ (the net contribution of the lead pole and zero) is 30°. (The reader should verify this claim.) Thus, if we place the zero at -2 (we expect an overshoot of about 30%), the closed-loop pole will be at $-1 + j\sqrt{3}$ if the lead network pole is at -4. Other combinations of lead zero and pole can also be selected that yield the required 30°; the angle criterion for a specific closed-loop pole location only specifies the relationship between the pole and zero.

FIGURE 4.15
(a) and (b): Root loci
for two specific cases of
$G(s) = 1/[s(s + 1)]$.

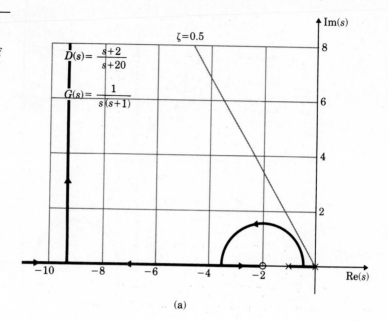

$D(s) = \dfrac{s+2}{s+20}$

$G(s) = \dfrac{1}{s(s+1)}$

$\zeta = 0.5$

(a)

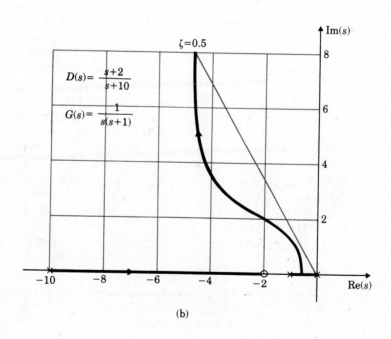

$D(s) = \dfrac{s+2}{s+10}$

$G(s) = \dfrac{1}{s(s+1)}$

$\zeta = 0.5$

(b)

FIGURE 4.15 *(cont.)*
(c) Root loci for three compensators for $G(s) = 1/[s(s + 1)]$; and (d) construction for a specific point on the root locus.

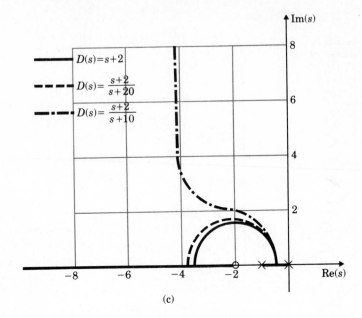

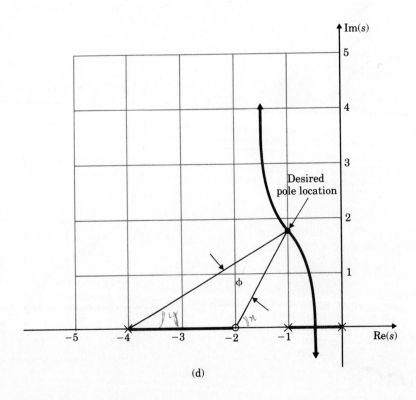

As stated earlier, the name "lead network" is a reflection of the fact that to sinusoidal signals these transfer functions impart phase lead. The details of this will be treated in Chapter 5, but we can quickly point out here that the phase of

$$D(s) = \frac{s + z}{s + p}$$

at $s = j\omega$ is given by

$$\phi = \tan^{-1}\frac{\omega}{z} - \tan^{-1}\frac{\omega}{p}. \tag{4.36}$$

If $z < p$, ϕ is positive, which is, by definition, phase lead.

The selection of exact values of z and p in particular cases is done by experience and by trial and error. In general, the zero is placed in the neighborhood of the closed loop ω_n, as determined by rise-time or settling-time requirements, and the pole is located at a distance from 3 to 10 times the value of the zero location. The choice of pole location is a compromise between the conflicting effects of noise suppression and compensation effectiveness. In general, if the pole is too close to the zero, then, as seen in Fig. 4.15(c), the root locus moves back too far toward its uncompensated shape, and the zero is not successful in doing its job. On the other hand, if the pole is too far to the left, the magnification of noise at the output of $D(s)$ is too great and the motor or other actuator of the process will be overheated by noise energy.

The physical realization of a lead network can be accomplished in many ways. A most common method is by means of an operational amplifier, of which an example is shown in Fig. 4.16. The transfer function of the circuit of Fig. 4.16 is readily obtained as

$$D(s) = -K_D\frac{T_1s + 1}{\alpha T_1s + 1}, \tag{4.37}$$

FIGURE 4.16
Circuit of a lead network.

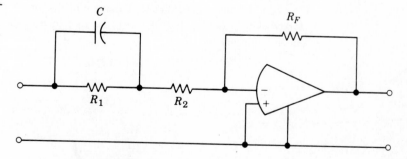

where

$$K_D = \frac{R_F}{R_1 + R_2} = 1 \quad \text{if} \quad R_F = R_1 + R_2,$$

$$T_1 = R_1 C,$$

$$\alpha = \frac{R_2}{R_1 + R_2}.$$

Notice that the zero is located at $z = -1/T_1$ and that the pole is located at $p = -1/\alpha T_1$, so the parameter α sets the separation distance between pole and zero, typically a factor of 3 to 10.

Once satisfactory dynamic response has been obtained, perhaps by the use of one or more lead networks, the designer may discover that the low-frequency gain—the value of the relevant steady-state error constant such as K_v—may be still too low. In order to increase this constant, the equivalent of another integration at near-zero frequency is indicated. The improvement is thus made by a pole near $s = 0$, but usually we include a zero nearby so that the pole-zero pair (called a *dipole*) does not significantly interfere with the overall system dynamic response as determined by the lead network(s). Thus, we want a $D(s)$ that has significant gain at $s = 0$ to raise K_v (or another steady-state error coefficient) and that is nearly unity (no effect) at the higher frequencies where dynamic response is determined. The result is the transfer function

$$D(s) = \frac{s + z}{s + p}, \tag{4.38}$$

where z and p are small (perhaps $z = 0.1$ and $p = 0.01$), yet $D(0) = z/p = 10-$ or generally in the range of 3 to 10, depending on the extent to which the steady-state gain requires boosting. The transfer function of Eq. (4.38) with $z > p$ has phase given by Eq. (4.36), but now, since $z > p$, ϕ is negative, corresponding to phase lag. Hence, the device with this transfer function is called a *lag network*.

The effects of a lag network on dynamic response can also be studied by looking at the corresponding root locus. Again, we take $G(s) = 1/[s(s + 1)]$, include the lead network $D_1(s) = (s + 2)/(s + 20)$, and raise the gain until the closed-loop roots correspond to a damping ratio of $\zeta = 0.707$. At this point, marked with a dot on Fig. 4.17(a), the root locus gain is about 31. Thus, the velocity constant is

$$K_v = \lim_{s \to 0} sKDG$$

$$= \lim_{s \to 0} s(31) \frac{s + 2}{s + 20} \frac{1}{s(s + 1)}$$

$$= \frac{31}{10}$$

$$= 3.1.$$

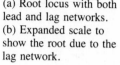

FIGURE 4.17
(a) Root locus with both lead and lag networks.
(b) Expanded scale to show the root due to the lag network.

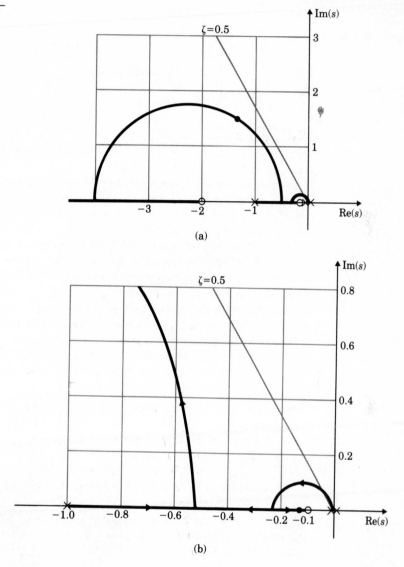

(a)

(b)

Suppose we now add a lag network of $D(s) = (s + 0.1)/(s + 0.01)$ to give an increase in velocity constant of about 10. The root locus is plotted in Fig. 4.17.

In Fig. 4.17(a), the locus is plotted showing the dominant roots at $-1.35 \pm j1.35$. Notice the very small circle near the origin, however. This region is expanded in Fig. 4.17(b) and is seen to be a result of the lag network. Notice that a closed-loop root remains very near to the lag-network zero at $-0.1 + j0$. This root

FIGURE 4.18
Circuit diagram of a lag
network.

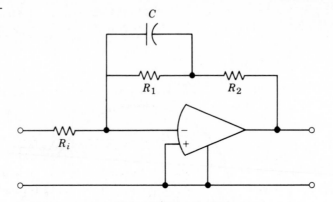

will correspond to a very slowly decaying transient, which has a small magnitude because the zero will almost cancel the pole in the transfer function. Still the decay is so *slow* that this term may seriously influence the settling time. Because of this effect, the lag pole-zero combination is placed at as high a frequency as possible without causing major shifts in the dominant root locations. Also important to notice is the fact that the transfer function from a plant disturbance to system error will *not* have the zero, and thus disturbance transients can be very long in duration in a system with lag compensation. A circuit diagram of a lag network is given in Fig. 4.18. The transfer function of this circuit is easily shown to be

$$D(s) = -\frac{R_2}{R_i} \frac{1}{\alpha} \frac{\alpha T s + 1}{T s + 1}, \tag{4.39}$$

where

$$T = R_1 C,$$

$$\alpha = \frac{R_2}{R_1 + R_2}.$$

Usually $R_i = R_2$, so the high-frequency gain is unity, but some gain adjustment can be made by other selections for R_i.

□ 4.6 Extensions of the Root Locus

The root-locus technique is a graphical scheme to show locations of possible roots of an equation as a real parameter varies. Thus far, we have considered only polynomial equations, control-system loop gain as the parameter, and positive values of the gain as possibilities. Each of these assumptions can be altered to guide us to a far wider range of applications of this useful technique.

4.6.1 Time Delay

Consider first the problem of design of the control system for a heat exchanger described by the transfer function

$$G(s) = \frac{e^{-5s}}{(10s + 1)(60s + 1)}. \tag{4.40}$$

The delay term is caused by the temperature sensor being physically somewhat downstream from the exchanger so that its reading is delayed a few seconds. The corresponding root-locus equations would be

$$1 + KG(s) = 0,$$

$$1 + K \frac{e^{-5s}}{(10s + 1)(60s + 1)} = 0, \tag{4.1}$$

$$600s^2 + 70s + 1 + Ke^{-5s} = 0. \tag{4.41}$$

Equation (4.41) is not a polynomial and thus does not fall in the same class as our other examples so far. How would one plot the root locus corresponding to Eq. (4.41)? There are two basic approaches, which we will describe: approximation and direct application of the phase condition.

In the first approach, we reduce the given problem to one we have previously solved by approximating the nonrational function e^{-5s} with a rational function. Since we are concerned with control systems and typically with low frequencies, we want an approximation that will be good for $s = 0$ and nearby.* The most common means used to find such an approximation is attributed to Padé and is based on matching the series expansion of the transcendental function e^{-5s} with that of a rational function, a ratio of numerator polynomial of degree p and denominator polynomial of degree q. The result is called a (p, q) *Padé approximant to* e^{-5s}. For our purposes, we will consider only the case $p = q$. Also, for generality, we will compute the approximants to e^{-s}, and in the result Ts may be substituted for s to allow for any desired delay.

To illustrate the process, we begin with the $(1,1)$ approximant. In this case, we wish to select b_0, b_1, and a_0 so that the error

$$e^{-s} - \frac{b_0 s + b_1}{a_0 s + 1} = \varepsilon \tag{4.42}$$

is small. For the Padé approximant, we expand both e^{-s} and the rational func-

---

*The nonrational function e^{-5s} is analytic for finite s and so may be approximated by a rational function. If nonanalytic functions such as $\sqrt{s}$ are involved, great caution is needed.

tion into a McLauren series and match as many *initial* terms as possible. The series are

$$e^{-s} = 1 - s + \frac{s^2}{2} - \frac{s^3}{3!} + \frac{s^4}{4!} - \cdots,$$

$$\frac{b_0 s + b_1}{a_0 s + 1} = b_1 + (b_0 - a_0 b_1)s - a_0(b_0 - a_0 b_1)s^2$$
$$+ a_0^2(b_0 - a_0 b_1)s^3 + \cdots.$$

Matching coefficients, we must solve the equations

$$b_1 = 1,$$
$$b_0 - a_0 b_1 = -1,$$
$$-a_0(b_0 - a_0 b_1) = \frac{1}{2},$$
$$a_0^2(b_0 - a_0 b_1) = -\frac{1}{6}.$$

Now we notice that we have an infinite number of equations but only three parameters. The Padé approximant results when we match the *first* three coefficients, from which we obtain

$$e^{-s} \cong \frac{1 - (s/2)}{1 + (s/2)}. \tag{4.43}$$

If we assume $p = q = 2$, we have five parameters, and a better match is possible. The first few approximants are given in Table 4.1.

The comparison of these three approximants is best seen from their pole-zero configuration and use in a root locus (Fig. 4.19). The locations of the poles are shown in the left-half plane; the zeros are in the right-half plane at the reflections of the poles. A root locus for the heat exchanger is drawn in Fig. 4.20 using the

TABLE 4.1 Three Padé Approximants to e^{-s}	$p = q$	$H_q(s)$
	1	$\dfrac{1 - (s/2)}{1 + (s/2)}$
	2	$\dfrac{1 - (s/2) + (s^2/12)}{1 + (s/2) + (s^2/12)}$
	3	$\dfrac{1 - (s/2) + (s^2/10) - (s^3/120)}{1 + (s/2) + (s^2/10) + (s^3/120)}$

FIGURE 4.19
Poles and zeros of the
Padé approximants to
e^{-s}. The order is marked
as a superscript—that is,
$\mathbf{x}^2$ for the (2,2)
approximant.

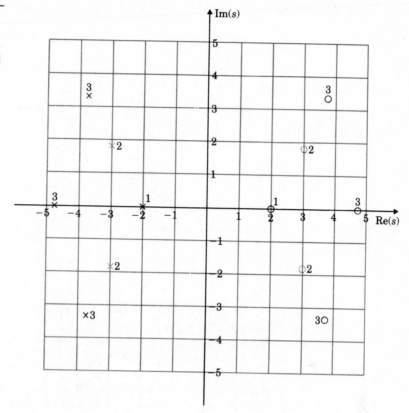

(2,2) approximant to e^{-5s}. For comparison, the exact locus is also drawn. Notice that for low gains and up to a damping ratio of about 0.5, the two curves are very close. However, the Padé curve crosses the imaginary axis at $\omega = 1$, while the true curve crosses the imaginary axis at $\omega = 1.4$. If greater accuracy is required, the (3,3) approximant or even a higher order can be used.

While the Padé approximation leads to a rational transfer function and is especially useful for analog-computer simulation, it is not necessary for plotting a root locus. The phase-angle condition is not changed if the process is nonrational, so one still must search for values of s for which the phase is $180° + l360°$. If we write the transfer function as

$$G(s) = e^{-\lambda s}\overline{G}(s),$$

the phase of $G(s)$ is the phase of $\overline{G}(s)$ minus $\lambda\omega$ for $s = \sigma + j\omega$. Thus, one can formulate a root-locus problem as searching for locations where the phase of $\overline{G}(s)$ is $180° + \lambda\omega + l360°$. To sketch such a locus, one would fix ω and search

FIGURE 4.20
Root locus for heat
exchanger.

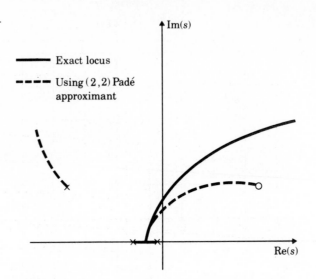

along a horizontal line in the s plane until a point is found, then raise ω, change the target angle, and repeat. Similarly, the departure angles are modified by $\lambda\omega$, where ω is the imaginary part of the pole where the departure is being computed. A more convenient method, and the means by which the exact plot of Fig. 4.20 is obtained, is to have a computer program that performs the search.

Before leaving the topic of systems with delay, it is interesting to consider the matter of where the loci of such systems cross the imaginary axis. Consider the loci of

$$1 + K\frac{e^{-s}}{s} = 0. \tag{4.44}$$

For $s = j\omega$, the phase condition may be written as

$$-90° - \omega = -180° - l360°,$$
$$\omega = 90° + l360°$$
$$= \frac{\pi}{2} + l2\pi. \tag{4.45}$$

Equation (4.45) follows from the fact that the phase of $e^{-j\omega}$ is $-\omega$, and for $\sigma = 0$, the phase of $1/s$ is $90°$, or $\pi/2$. At $\sigma = -\infty$ (far, far into the left-half plane), the phase of $1/s$ is $180°$, so the root-locus formula for phase is

$$-180° - \omega = -180° - l360°,$$
$$\omega = l360° \tag{4.2}$$
$$= l2\pi. \tag{4.46}$$

FIGURE 4.21
Sketch of root locus
for system with delay
$G(s) = e^{-s}/s$.

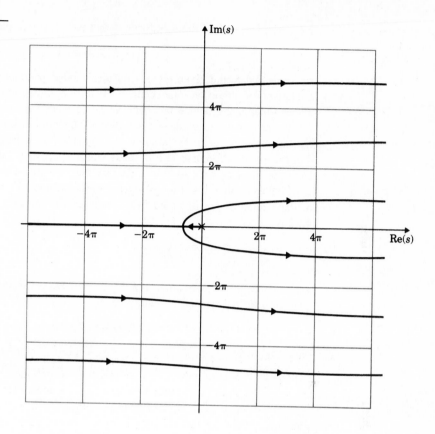

Finally, for $\sigma = +\infty$ (far to the right), the phase of $1/s$ is $0°$, so the root-locus angle condition is

$$-\omega = -180° - l360°,$$
$$\omega = 180° + l360° \tag{4.3}$$
$$= \pi + l2\pi. \tag{4.47}$$

If we combine Eqs. (4.45), (4.46), and (4.47), we obtain the locus sketched in Fig. 4.21. Notice that there is an infinity of branches of this locus, from left (where $e^{-\lambda s} = \infty$) to right (where e^{-s} is zero). Of course, the principal branch of this locus starts at $s = 0$, has a breakaway at $s = -1$, and crosses the imaginary axis at $j(\pi/2)$, going to the horizontal asymptote at $s = j\pi$.

4.6.2 Loci versus Other Parameters

A second extension of the root locus, discussed in the chapter introduction but beyond the developments detailed so far, is to consider parameters other than loop gain. The key idea of the root locus is that the locus is a plot of solutions to

$$1 + KG(s) = 0 \qquad (4.48)$$

for some real K and given $G(s)$. Consider another parameter. A block diagram of a relatively common servomechanism structure is shown in Fig. 4.22. In this case, a speed-measuring device has been added (a tachometer), and the problem is to use the root locus to guide the selection of K_T, the tachometer gain, as well as K_A, the amplifier gain. We assume that the amplifier gain has already been set to 4 at this point. Of course, a root locus with respect to K_A is the standard case we have been dealing with all along. We take the view that a principal function of K_T is to influence the closed-loop system poles, and thus we will consider the locus of these roots as K_T is varied. The characteristic equation of the system of Fig. 4.22 is

$$1 + \frac{4}{s(s+1)} + \frac{K_T}{s+1} = 0,$$

which is not in the standard $1 + KG(s)$ form. However, if we clear fractions, the characteristic polynomial is seen to be

$$s^2 + s + 4 + K_T s = 0. \qquad (4.49)$$

Now Eq. (4.49) is in suitable form for a root-locus study. We need to identify $G(s)$, which we do by writing the equivalent to Eq. (4.49) as

$$1 + K_T \frac{s}{s^2 + s + 4} = 0. \qquad (4.50)$$

Thus, for root-locus purposes, the "zeros" are at $s = 0$, and the "poles" are at $-1/2 \pm j1.94$. A sketch of the locus is shown in Fig. 4.23. The designer is now free to select K_T for a specific damping ratio or whatever else might guide the selection of K_T. Having selected a trial value of K_T, one could return to K_A and consider the effects of a change from 4 in that parameter. With respect to K_A the locus is governed by Eq. (4.49). If $K_T = 1$ is selected, and we wish to consider changes from the nominal value of $K_A = 4$, the equation can be written as

$$s^2 + (1 + K_T)s + 4 + K = 0,$$

FIGURE 4.22
Block diagram of a servomechanism structure including velocity, or tachometer feedback.

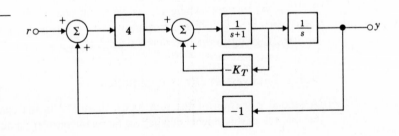

FIGURE 4.23
Root locus of closed-loop
poles of the system in
Fig. 4.22 vs. K_T.

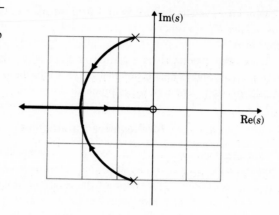

where we have taken $K + 4 = K_A$. With respect to K, the locus (for $K_T = 1$) is
governed by

$$s^2 + 2s + 4 + K = 0. \tag{4.51}$$

Now we see that the "poles" of the locus corresponding to Eq. (4.51) are the roots
of the previous locus, which was drawn versus K_T! The situation is sketched in
Fig. 4.24, with the previous locus versus K_T left dotted. It should be clear that one

FIGURE 4.24
Root locus vs. K after
K_T is selected to be 1.

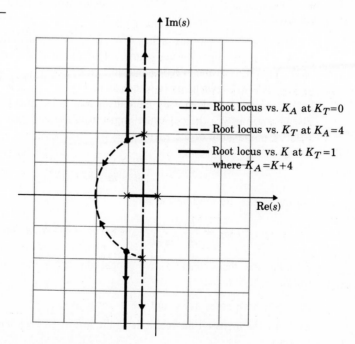

Root locus vs. K_A at $K_T=0$

Root locus vs. K_T at $K_A=4$

Root locus vs. K at $K_T=1$
where $K_A=K+4$

could draw a locus versus K for a while, stop, resolve the equation, and continue the locus versus K_T; in other words, one could go back and forth between these parameters at will! To summarize, the technique for handling any single parameter is to write out the characteristic equation, collect terms that do *not* multiply the parameter as $a(s)$, and collect terms that *do* multiply the parameter as $b(s)$. After this formulation stage, the root locus proceeds as before.

4.6.3 Zero-Degree Loci for Negative Parameters

We now consider the extension of the root locus to consider negative values of the parameter. We have already briefly seen this case when we worked out the angles near a multiple root point. In the case of the example servomechanism considered in Fig. 4.22, we sketched a root locus versus K, where $K_A = 4 + K$. It would seem reasonable to consider values of K_A less than 4 as well as values of K_A greater than 4. In order to make K_A less than 4 we need to make K negative. What are the rules for a root locus with a negative parameter? Consider again the basic equation

$$1 + KG(s) = 0. \tag{4.52}$$

If this time we ask that Eq. (4.52) be satisfied for negative values of K, it must be that $G(s)$ is real and positive. In other words, for the negative-parameter root locus (or *negative locus*), the condition is

---

The angle of $G(s)$ is $0° + l360°$ for s on the negative locus.

The steps for plotting a negative locus are essentially the same as for the positive locus except that we search for $0°$ modulo $360°$ instead of $180°$ modulo $360°$. To return to the servomechanism case, consider again the problem of Eq. (4.51). The negative locus is sketched in Fig. 4.25. This time, we find that the locus is to the left of an *even* number of real poles plus zeros (zero being even). For an angle of $0°$, the departure angle from the complex root at $-1 + j1.73$ is $-90°$. Computation of the asymptotes for large s is

$$\alpha_0 = \frac{\sum p_i - \sum z_i}{n - m} \qquad \text{(as before),} \tag{4.53}$$

but the angles are modified to be

$$\phi_l = \frac{l360°}{n - m} \qquad \text{shifted by } 180° \text{ from } 180° \text{ locus.}$$

To complete the extension, we repeat the guidelines for plotting a negative-parameter root locus.

1. Mark the n poles as $\times$ and the m zeros as $\bigcirc$.

2. Draw the locus on the real axis to the left of an *even* number of real poles plus zeros.

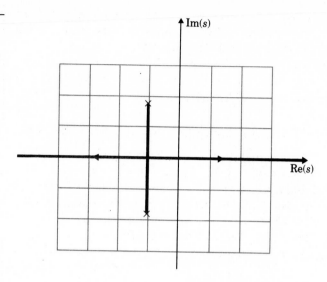

FIGURE 4.25
Negative root locus
corresponding to $G(s) = 1/(s^2 + 2s + 4)$.

3. Draw $n - m$ radial asymptotes centered at α and with angles ϕ_l where

$$\alpha = \frac{\sum p_i - \sum z_i}{n - m} = \frac{-a_1 + b_1}{n - m},$$

$$\phi_l = \frac{l360°}{n - m}, \qquad l = 0, 1, 2, 3, \ldots, n - m - 1.$$

4. Compute departure angles from poles and arrival angles to zeros by searching for points *around* the pole or zero, where the phase of $G(s)$ is $0°$.

5. Assume $s = j\omega_0$ and compute the points where the locus crosses the imaginary axis for negative K.

6. The equation has multiple roots at points *on the locus* [where the angle of $G(s)$ is $0°$ modulo $360°$] where, in addition, $dG/ds = 0$ or

$$b\frac{da}{ds} - a\frac{db}{ds} = 0.$$

7. Fill in the locus using these calculation guides, and the locus is complete.

The result of the extension of the guidelines for construction of root loci to include negative parameters is that we can visualize the root locus as a set of continuous curves showing the location of possible solutions to the equation $1 + KG(s)$ for *all real* K, both positive and negative. Furthermore, the continuity of these curves encourages us to imagine a root starting at a pole of $G(s)$ and moving along a curve (a branch of the root locus) until the parameter has some value K_0. If we now let $K = K_0 + \overline{K}$, a new locus with respect to $\overline{K}$ is just the continuation of the old locus with the same zeros, but with the roots of $1 + K_0G(s) = 0$ as the

FIGURE 4.26
Nonlinear elements
with no memory:
(a) saturation, (b) relay,
(c) relay with dead
zone, (d) gain with dead
zone, (e) preloaded
spring or coulomb plus
viscous friction, and
(f) quantization.

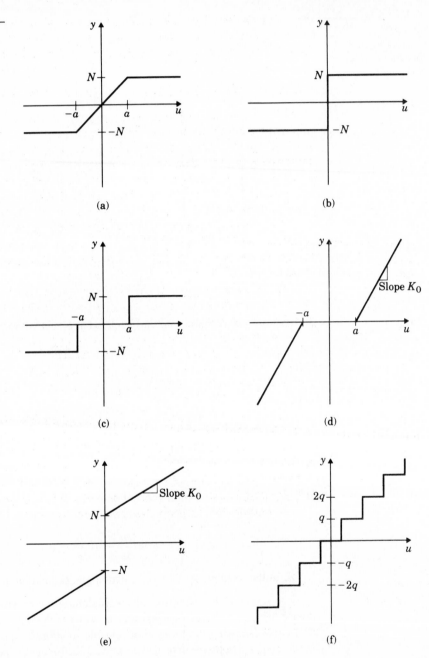

new set of poles. Thus, we visualize the roots moving along the locus branches; the *sensitivity* of the closed-loop poles to the parameter K at $K = K_0$ is given by how the roots move for both positive and negative $\overline{K}$ for $K = K_0 + K$.

4.6.4 Use of the Root Locus in the Analysis of Nonlinear Systems

As we have tried to make clear, every real control system is nonlinear, and the linear analysis and design methods we have described are used with linear approximations to the real models. There is, however, one important category of nonlinear systems for which some significant analysis (and design) can be done. These are the systems in which the nonlinearity has no dynamics and is well approximated as a gain that varies as the size of its input signal varies. Sketches of a few such nonlinear characteristics and their common names are shown in Fig. 4.26.

The sketches suggest symmetry of these functions that is not really necessary to use the tools to be described. The behavior of systems containing any one of these zero-memory nonlinearities can be qualitatively described by considering the nonlinear element as a signal-dependent varying gain.

For example, with the saturation element, it is clear that for input signals of magnitudes less than a, the nonlinearity is linear with gain N/a. However, for signals larger than a, the output size is bounded by N, while the input size can get much larger than a. It is clear that the ratio of output to input goes down, and thus saturation has the gain characteristics shown in Fig. 4.27. We will see presently that one can make an analytical definition of equivalent gain that results in an exact plot of K_{eq} versus $|u|$, but for the qualitative results desired here the shape shown is adequate.

To illustrate the method with an example, consider the system shown in Fig. 4.28. The root locus of this system versus a gain at the location of the saturation is given in Fig. 4.29. At a gain of 1 (the saturation gain for inputs less than 0.4 in magnitude), the damping ratio is $\zeta = 0.5$. As the gain is reduced, the locus shows that the roots move toward the origin of the s plane with less and less damping. Plots of the step responses of this system with input steps of size 2, 4, 6, 8, 10, and 12 are given in Fig. 4.30. Notice that as the input gets larger, the response

FIGURE 4.27
General shape of equivalent gains of saturation.

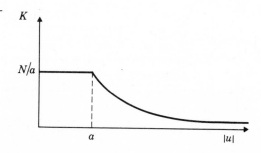

FIGURE 4.28
Dynamic system with a
gain in saturation.

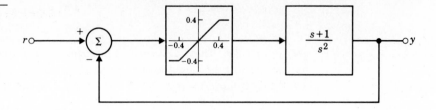

FIGURE 4.29
Root locus of system
in Fig. 4.28, with
equivalent gain in
location of saturation.

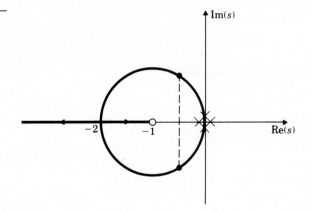

FIGURE 4.30
Step responses of system
in Fig. 4.28 for various
input step sizes.

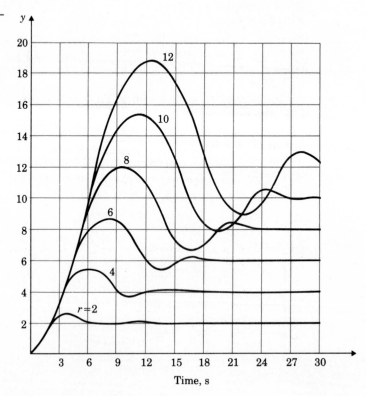

FIGURE 4.31
Block diagram of a
conditionally stable
system.

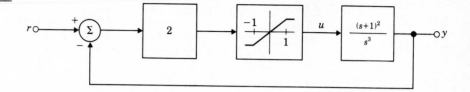

has more and more overshoot and slower and slower recovery. This is exactly as is explained by movement on the root locus with less and less gain, as if the larger signals resulted in a smaller equivalent gain of the saturation element.

As a second example of a nonlinear response described by signal-dependent gain, consider the system whose block diagram is drawn in Fig. 4.31 and whose characteristic-equation root locus is plotted in Fig. 4.32. From this locus, it is readily calculated that the imaginary axis crossing occurs at $\omega_0 = 1$ and $K = 1/2$. Systems such as this, which are stable for (relatively) large gains but unstable for smaller gains, are called *conditionally stable systems*. If our analysis is right, the system should show response corresponding to a damping action of about 0.5 for small signals, get worse as signal strength is increased, and go unstable with oscillations near 1 rad/s for larger signals. Plots of step responses for input steps of size 1, 2, 3, and 3.4 are drawn in Fig. 4.33. These responses confirm the qualitative description above.

As this system goes unstable, the period is very near 2π, as predicted from the consideration of the saturation as a variable gain. However, we have no real

FIGURE 4.32
Root locus for system in
Fig. 4.31.

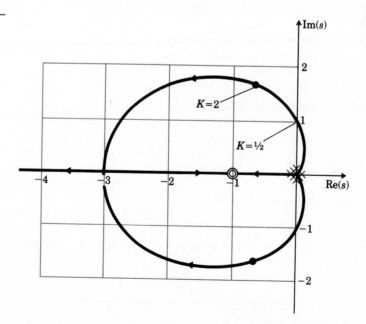

FIGURE 4.33
Step responses of system
in Fig. 4.31.

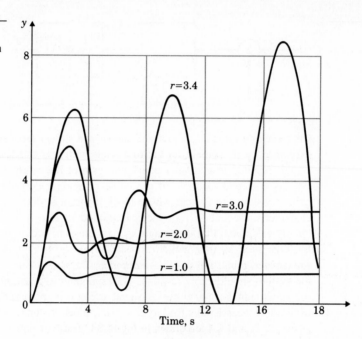

FIGURE 4.33
Step responses of system
in Fig. 4.31.

basis for predicting that the system will go unstable for a step input of size 3.4 but
not for a step of size 3. We will return to this point later.

The final illustration of the use of the root locus to give a qualitative description of the response of a nonlinear system is based on the block diagram seen
in Fig. 4.34. This system is typical of electromechanical control problems where
the designer perhaps at first is not aware of the oscillatory mode, shown here at
$\omega = 1$ and $\zeta = 0.1$. The root locus for this system is sketched in Fig. 4.35.
The imaginary-axis crossing can be verified to be at $\omega_0 = 1, K_0 = 0.2$; thus the
gain of 0.5 is enough to force the roots of the oscillatory mode into the right-half
plane, as shown by the dots. Here our analysis predicts an unstable system but a

FIGURE 4.34
Block diagram of a system with oscillatory mode.

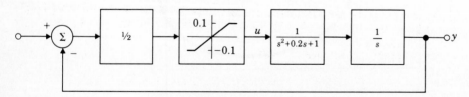

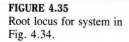

FIGURE 4.35
Root locus for system in
Fig. 4.34.

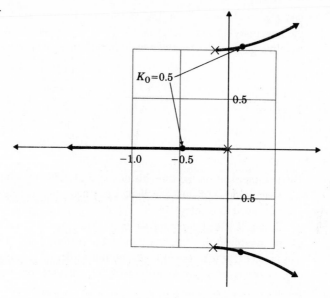

system that becomes stable as gain decreases. Thus, we would expect the response to build up until the equivalent gain is 0.2 *and then stop growing!*

Plots of the step responses for three steps of size 1, 4, and 8 are shown in Fig. 4.36, and again our heuristic analysis is exactly correct: The error builds up to a fixed amplitude oscillation at 1 rad/s and holds this amplitude at any DC equilibrium value (1, 4, or 8).

In this case, the response always approaches a periodic solution of fixed amplitude known as a *limit cycle*. (The response is cyclic and is approached in the limit as time grows large.) In 1950, Dr. Kochenburger proposed that, for this situation, the Fourier series be used to define the equivalent gain in the following way (Truxal, 1955, p. 566). For all the nonlinear characteristics of Fig. 4.26, if the input $u(t)$ is sinusoidal of amplitude A, the output $y(t)$ will be periodic with a fundamental period equal to that of the input and consequently with a Fourier series $y(t) = \sum_1^\infty b_n \cos(n\omega_0 t + \phi_n)$.

Kochenburger suggested that the nonlinear system could be described by the first fundamental component of this series as if it were a linear system with gain b_1/A and phase ϕ_1. He called this approximation the *describing function*. Computation of the describing function for the nonlinear characteristics of Fig. 4.26 is generally straightforward but tedious. The simple case is the relay where the output is a square wave of amplitude N for *every* size input; thus $b_1 = 4N/\pi$*

---

*From the theory of Fourier series,

$$b_1 = \int_0^{1/2} (N \sin 2\pi t)dt - N \int_{1/2}^1 (\sin 2\pi t)dt = \frac{4N}{\pi}.$$

and $K_{eq} = 4N/\pi A$. From the root locus of our example (Fig. 4.35), the gain at the imaginary-axis crossover is 0.2; if we approximate the saturation as a relay, we have the gain as $K_{eq} = 0.2$. Also, in this case, $N = 0.1$, so, substituting values,

$$K_{eq} = 0.2 = \frac{(4)(0.1)}{\pi A}.$$

Thus

$$A = \frac{(0.1)(4)}{(0.2)\pi} = \frac{2}{\pi} = 0.636.$$

By measurement on the time history of Fig. 4.36, the amplitude of the oscillation has magnitude 0.65, an excellent agreement considering that the saturation was approximated as a relay!

Using Kochenburger's describing function, we can return to Fig. 4.33 and be easily convinced that the first transient to a step of size 3 is nearly a sinusoid. We can predict that the system is just on the border of stability for the equivalent gain corresponding to a root-locus gain of 1/2 when the locus crosses into the right-half plane.

FIGURE 4.36
Step response of system
in Fig. 4.34.

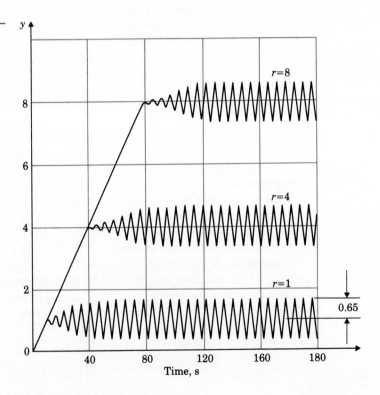

FIGURE 4.37
Block diagram of a system including a "notch" network.

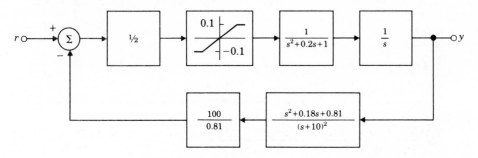

As an example of the use of this method in design, our analysis of the system shown in Fig. 4.34 would suggest that the limit cycle will not occur if the root locus can be kept in the left-half plane. This can be done by addition of a "notch" network, as shown in Fig. 4.37, for which the root locus is given in Fig. 4.38 and the step responses in Fig. 4.39. These figures show plainly that the analysis is borne out by the simulation.

FIGURE 4.38
Root locus of system in
Fig. 4.37.

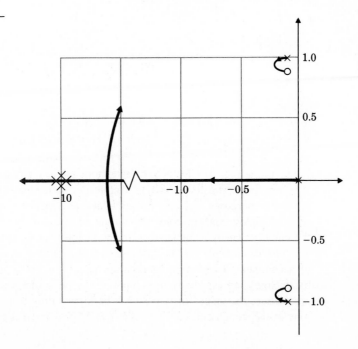

FIGURE 4.39
Step responses of system
in Fig. 4.37.

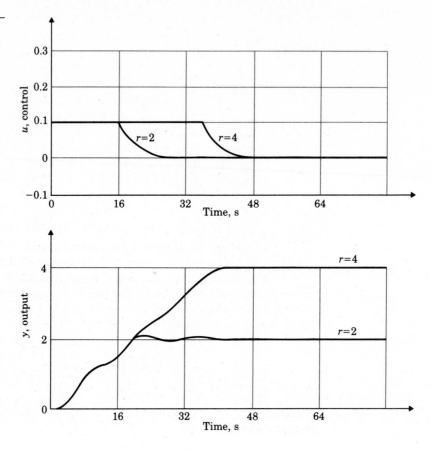

□ 4.7 Computer-Aided Graphing of the Root Locus

In order to use the root locus as a design tool and to verify computer-generated loci, it is very important to be able to sketch root loci. One can then quickly predict, for design purposes, the effect of an added zero or pole, or even several of them, or can quickly confirm computer output in a qualitative sense. For this reason, the engineer needs to understand the guidelines for sketching loci and should be able to plot example loci by hand. Given this skill, it is valuable to have the aid of a computer in determining accurate loci, in establishing exact values for the parameters, and especially in computing the closed-loop pole sensitivity to those parameters, the values of which may be known only to a certain tolerance at the time of the design and may be subject to drift over the life of the system.

There are two basic approaches to machine computation of the root locus. In the first approach, the root-locus problem is formulated as a polynomial in the form $a(s) + Kb(s) = 0$, and for a sequence of values of K varying from near zero to a large value, the polynomial is solved for its n roots by any of many available numerical methods. This method has the advantage that all roots are computed for each value of the parameter, and one can be sure to guarantee that a specific value is included—for example, that value of the loop gain that gives the design velocity constant. One of the disadvantages of the method is that the resulting root locations are very unevenly distributed in the s plane. For example, near a point of multiple roots, the sensitivity of the root locations to the parameter value is very great, and the roots just fly through such points. On the other hand, near a zero the root moves very slowly, since it takes an infinite value of the parameter to push the root all the way into the zero. A second disadvantage of this method is the requirement that the problem be a polynomial. As we saw, with pure time delay a transcendental equation is involved, and an approximation such as the Padé method must be used to reduce the given problem to an equivalent polynomial. Such approximations limit the range of values of the parameter for which the results are accurate, and checking the accuracy is difficult unless a means is available to solve the true equation at critical points. A final disadvantage is that many algorithms are not easily applied in solving polynomials at points of multiple roots. This problem is related to the great sensitivity of the roots to the parameter at these points, as mentioned earlier. A method related to polynomial solving is possible when a state-variable formulation of the equations of motion is available. In terms of state variables, an idea to be treated in Chapter 6, the closed-loop poles are eigenvalues of the state matrix, and computer scientists have developed an algorithm called QR that can be used with great accuracy to solve problems of substantial complexity. For systems described by high-order ordinary differential equations, this is the most powerful method available.

The alternative to polynomial solving is a method based on curve tracing. The basic idea is that a point on the positive root locus is a point where the phase of $G(s)$ is 180°. Thus, given a point on the locus at s_0 with parameter values K_0, one can draw a circle of radius δ around s_0 and search for a new place where the angle condition is met and the new gain is larger than K_0. This method can be easily arranged to include a delay term as $e^{-\lambda s}$, and the resulting points will be spaced δ radians apart, a value that the designer can specify. One disadvantage of this method is that only one branch of the locus is generated at a time (although computer logic can be easily set up to step through each of the open-loop poles to produce a complete graph). A second disadvantage of the method is that the designer needs to monitor the selection of δ in order to ensure convergence of the search for 180° at some points and to avoid wasting too much time with a small δ at less critical points. With either method, it is very important to have a means to graph the results, since the essential information in the locus is graphical.

Many of the root loci plotted in this chapter were plotted by a computer program developed at Stanford and, in this case, implemented on the HP-85. Other programs that include root-locus capability are CTRL-C of System Control Technology; MATRIX$_X$ of Integrated Systems, Inc.; CC of Systems Technology, Inc.; and Matlab by the Mathworks, Inc.

Summary

In this chapter, we have discussed a very useful graphical technique for studying the effect of the changes of a real parameter on the poles of the closed-loop system. Most often the parameter is the open-loop gain, and hence the root locus is a plot of the closed-loop poles parameterized as a function of the open-loop gain. The root locus can be viewed as a general technique for the graphical solution of any polynomial. Two types of root loci have been discussed: the $0°$ and the $180°$ root locus. Although the $180°$ locus is encountered most often in feedback control systems, it is useful to understand the concept of the $0°$ locus, and it will be discussed again in Chapter 6. A seven-step procedure has been outlined for plotting either locus; it allows a designer to sketch quickly a root locus without any computer aids. This in turn facilitates the search for many different candidate solutions in the synthesis process: different gains, different pole-zero placements for a lead or lag network, different feedback variables, and different characteristics of the sensor, actuator, and plant.

The root loci of systems with time-delay and nonlinearities have been discussed, and we have demonstrated the utility of the method in the design of much larger systems than the linear, rational transfer function systems discussed through most of the chapter. Finally, some of the considerations involved in implementing the root locus in a computer-aided design procedure have been addressed.

Problems   Section 4.2

4.1 Set up the following characteristic equations in the form suited to the Evans root-locus method:

a) $s + (1/\tau) = 0$ versus the parameter τ;
b) $s^2 + bs + b + 1 = 0$ versus the parameter b;
c) $(s + b)^3 + A(Ts + 1) = 0$:

    (1) versus the parameter A.
    (2) versus the parameter T.
    (3) Can you set it up versus the parameter b? Why not? Can a locus be drawn versus b for given constant values of A and T? Explain.

4.2 Roughly sketch the root loci for the pole-zero maps shown in Fig. 4.40. Show asymptotes, centroids, a rough evaluation of arrival and departure angles, and the loci for positive values of the parameter K. Each map is from a characteristic equation of the form

$$1 + K \frac{b(s)}{a(s)} = 0,$$

where the roots of the numerator $b(s)$ are shown as small circles, and the roots of the denominator $a(s)$ are shown as $\times$'s on the s plane.

4.3 Consider the root locus for the equation

$$1 + \frac{K}{s[(s + 1)(s + 5)]} = 0.$$

a) Show the real-axis segments clearly.
b) Sketch the asymptotes for $K \to \infty$.
c) For what value of K are the roots on the imaginary axis?

FIGURE 4.40

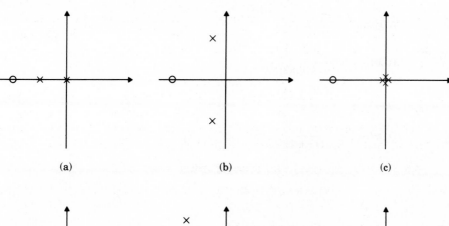

(a)   (b)   (c)

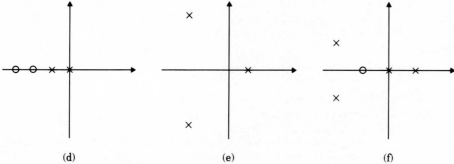

(d)   (e)   (f)

4.4 Sketch the root locus for the following systems versus K. Be sure to give the asymptotes, arrival and departure angles, and imaginary-axis crossings, if any.

a) $KG(s) = \dfrac{K}{s(s^2 + 2s + 10)}$.

b) $KG(s) = \dfrac{K(s^2 + 2s + 8)}{s(s^2 + 2s + 10)}$.

c) $KG(s) = \dfrac{K(s^2 + 2s + 12)}{s(s^2 + 2s + 10)}$.

d) $KG(s) = \dfrac{K(s + 3)}{s[(s + 1)(s^2 + 4s + 5)]}$.

e) $KG(s) = \dfrac{K(s + 2)}{s^4}$.

4.5 Sketch the root locus for the following systems versus K.

a) $KG(s) = \dfrac{10K}{s^2 + 3s + 7}$.

b) $KG(s) = \dfrac{10K}{s(s^2 + 3s + 7)^2}$.

c) $KG(s) = \dfrac{K(s + 1)}{s(s + 2)}$.

d) $KG(s) = \dfrac{K(s^2 + 1)}{s(s^2 + 100)}$.

e) $KG(s) = \dfrac{K(s^2 + 1)}{s(s^2 + 4)}$.

f) $KG(s) = \dfrac{K(s^2 + 4)}{s(s^2 + 1)}$.

4.6 Sketch the root locus for the following systems versus K.

a) $KG(s) = \dfrac{K[(s + 1.5)(s + 4)]}{s[(s + 1)(s + 2.5)]}$.

b) $KG(s) = \dfrac{K[(s + 1.5)(s + 5)]}{(s + 0.5)(s + 2)(s + 3)}$.

c) $KG(s) = \dfrac{K(s + 1)}{s[(s - 1)(s^2 + 2s + 5)]}$.

4.7 Sketch the complete root locus for the following systems versus K.

a) $KG(s) = \dfrac{K(s + 2)}{s[(s + 0.1)(s^2 + 2s + 2)]}$.

b) $KG(s) = \dfrac{2K}{s(s^2 + 6s + 13)}$.

FIGURE 4.41

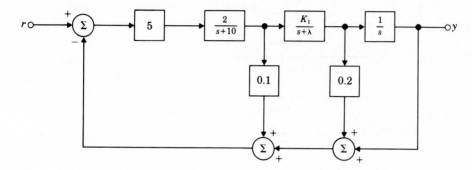

c) $KG(s) = \dfrac{2K(s^2 + 2s + 2)}{s\,[(s^2 + 6s + 13)(s^2 + s + 2)]}$.

d) $KG(s) = \dfrac{K[(s + 0.5)(s + 1.5)]}{s\,[(s^2 + 2s + 2)(s + 5)(s + 15)]}$.

e) $KG(s) = \dfrac{K[(s + 1)(s - 0.2)]}{s\,[(s + 1)(s + 3)(s^2 + 5)]}$.

4.8 a) For the system given in Fig. 4.41, plot the root locus as the parameter K_1 is varied from 0 to ∞ with $\lambda = 2$.

b) Repeat **(a)** with $\lambda = 5$. What is special about this value?

c) Repeat **(a)** for $K_1 = 2$ and λ varying from 0 to ∞.

4.9 Determine Evans' root-locus form of the characteristic equation for the control system shown in Fig. 4.42, with b as the parameter of interest. Sketch the root locus to scale, showing the direction of increasing b on the locus.

4.10 Prove that a combination of two poles and one zero to the left of both of them on the real axis results in a root locus that is a circle centered at the zero with radius given by $\sqrt{|(p_1 - z)||(p_2 - z)|}$.

4.11 The loop transmission of a system has two poles at $s = -1$ and a zero at $s = -2$. There is a third real-axis pole located somewhere to the left of the zero. Several different root loci are possible, depending upon the exact location of the third pole. The extreme cases occur when the pole is located at infinity or when it is located at $s = -2$. Sketch several different possible loci.

FIGURE 4.42

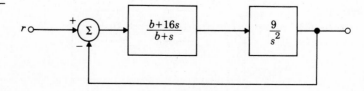

FIGURE 4.43

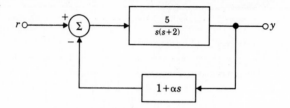

4.12 Sketch the root locus with respect to α for the system shown in Fig. 4.43. Sketch the step responses for $\alpha = 0, 0.5, 2$.

Section 4.3

4.13 Consider the system with

$$KG(s) = \frac{K[(s+1)^2 + 1]}{s^2[(s+2)(s+3)]}.$$

We wish to investigate the root locus versus K.

a) Plot the location of poles and zeros of $G(s)$ showing the segment of the root locus on the real axis.

b) What are the arrival angles at the complex zeros?

c) What is the breakaway point?

d) What are the asymptotes?

e) Sketch the root locus.

4.14 Draw the root locus for the system with

$$G(s) = \frac{K(s+2)}{s[(s-1)(s+6)^2]}$$

and verify the imaginary-axis crossings using Routh's array.

4.15 Using root-locus techniques, determine the values of $\alpha < 0$ for which the polynomial $s^3 + 2s^2 + \alpha = 0$ has repeated roots.

4.16 Sketch the root locus for the system $G(s) = K/[(s+1)(s+2)(s^2 + 4s + 8)]$.

4.17 Sketch the root locus for the system

$$KG(s) = \frac{-8K}{(s-1)[(s+2)^2 + 3]}.$$

FIGURE 4.44

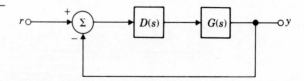

FIGURE 4.45

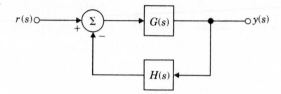

4.18 A simplified model of the longitudinal motion of a helicopter near hover has the transfer function

$$G(s) = \frac{9.8(s - 0.25 \pm j2.4975)}{(s + 0.6565)(s - 0.1183 \pm j0.3678)}.$$

Sketch the root locus for this system.

4.19 Let

$$G(s) = \frac{1}{(s + 2)(s + 3)} \quad \text{and} \quad D(s) = K\frac{s + a}{s + b}.$$

Using root-locus techniques, determine zero and pole locations of the compensator (a and b) that will produce closed-loop poles at $s = -1 \pm 1j$ for the system shown in Fig. 4.44.

4.20 Use asymptotes, center of asymptotes, angles of departure and arrival, and Routh's array to sketch root loci for the following feedback control systems (Fig. 4.45) for $K > 0$.

a) $G(s) = \dfrac{K}{s[(s + 1 + 3j)(s + 1 - 3j)]}; \quad H(s) = \dfrac{s + 2}{s + 8}.$

b) $G(s) = \dfrac{K(s + 1)}{s^2(s + 3)}; \quad H(s) = 1.$

c) $G(s) = \dfrac{K[(s + 5)(s + 7)]}{(s + 1)(s + 3)}; \quad H(s) = 1.$

d) $G(s) = \dfrac{K[(s + 3 + 4j)(s + 3 - 4j)]}{s[(s + 1 + 2j)(s + 1 - 2j)]}; \quad H(s) = 1.$

4.21 Consider the system in Fig. 4.46.

a) Using the Routh criterion, determine all values of K for which the system is stable.
b) Sketch the root locus (include angles of departure and arrival and K and s values at breakpoints and crossings of the imaginary axis).

FIGURE 4.46

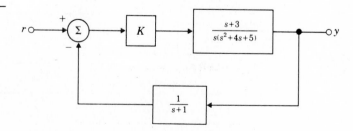

FIGURE 4.47

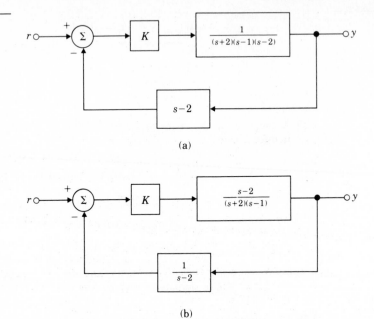

(a)

(b)

4.22 Consider the two systems given in Fig. 4.47. Note that each system has an unstable plant and the same loop transfer function. Can either system be made stable for some positive value of K? Show your reasoning, using a root-locus diagram.

4.23 Using the root locus, find the range of gain K for which the systems in Fig. 4.48 are stable.

FIGURE 4.48

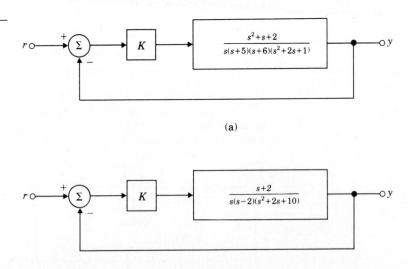

(a)

(b)

FIGURE 4.49

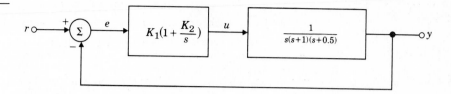

4.24 Consider the root-locus problem where

$$G(s) = \frac{(s^2 + s + 9.25)(s + 2)}{s\,[(s + 1)(s + 20)(s^2 + s + 16.25)]}.$$

a) Sketch the portions of the real axis corresponding to the positive root locus.
b) Sketch the asymptotes for $K \to \infty$ and give the angles of the asymptotes and crossing with the real axis.
c) Estimate the departure angle at the pole at $s = -(1/2) + j4$.
d) Estimate the arrival angle at the zero at $s = -(1/2) + j3$.
e) Give the polynomial whose roots include all points of multiple roots such as break-in and breakaway points.
f) Sketch the root locus.

4.25 Consider the system in Fig. 4.49.

a) Use Routh's criterion to determine regions in the (K_1, K_2) plane such that the system remains stable.
b) Use root locus to verify the results in (a).

4.26 An elementary magnetic suspension scheme is sketched in Fig. 4.50. For small motions near the reference position, the voltage e is proportional to ball displacement x (in meters) such that $e = 100x$. The upward force (in newtons) on the ball caused by the current i (in amperes) may be approximated by $f = 0.5i + 20x$. The mass

FIGURE 4.50

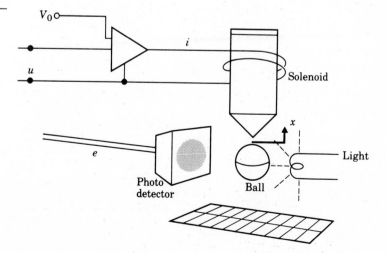

of the ball is 20 g, and the gravity force is 9.8 N/kg. The power amplifier is a voltage-to-current device so that $i = u + V_0$ in amperes.

a) Write the equations of motion for this setup.
b) Give the value of bias V_0 so that the ball is in equilibrium at $x = 0$.
c) What is the transfer function from u to e?
d) Sketch the root locus of the closed loop as a function of K if $u = -Ke$.
e) Suggest a different control law (transfer function from e to u) to give improved control over that shown in **(d)**.

Section 4.4

4.27 Sketch the root locus for the system

$$G(s) = \frac{K(s+1)}{s[(s+1)(s+2)]},$$

and determine the value of the root-locus gain such that the complex conjugate poles have a damping ratio of 0.5.

4.28 For the system in Fig. 4.51:

a) Find the locus of closed-loop roots versus K.
b) Is there a value of K that will cause all roots to have a damping ratio greater than 0.5?
c) Assuming the locus goes through a point where the damping ratio $\zeta = 0.707$, find the values of K that result in these roots.

4.29 Consider the rocket-positioning system shown in Fig. 4.52. First show that lead compensation

$$H(s) = -K \frac{s+2}{s+4}$$

stabilizes the system with a perfect measurement of x. Then assume that there is a sensor that measures x and that it is modeled as a single pole with a 1/10-s time constant (and unity DC gain). Using the root-locus procedure, find K that will provide maximum damping ratio.

FIGURE 4.51

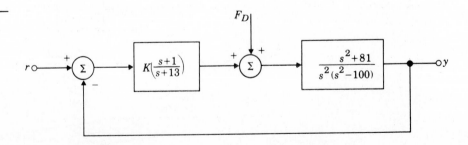

FIGURE 4.52

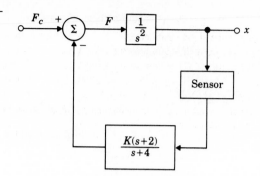

4.30 A unity feedback system has the plant transfer function

$$KG(s) = \frac{K}{s(s^2 + 6s + 12)}.$$

We wish to investigate the root locus versus K.

a) Plot the location of poles and zeros of $G(s)$ showing segments of the root locus on the real axis.

b) What are the departure angles from the complex poles?

c) Where are the breakaway and break-in points?

d) Sketch the root locus.

e) What is the value of K at the point where the closed-loop complex roots have damping ratio $\zeta = 0.5$?

4.31 For the system in Fig. 4.53:

a) Sketch the locus of closed-loop roots versus K.

b) Find the maximum value of K at which the system is stable. (Assume that the locus crosses the imaginary axis at $s = \pm j5.5$.)

FIGURE 4.53

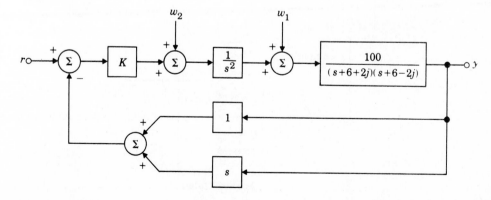

Assume $K = 2$ for the remaining parts of this problem.

c) What is the steady-state error $(e = r - y)$ for a step in r?

d) What is the steady-state error in y for a constant disturbance w_1?

e) What is the steady-state error in y for a constant disturbance w_2?

f) If you wished to have more damping, what changes would you make in the system?

4.32 Consider the positioning servomechanism system shown Fig. 4.54, where

$$e_i = K_{pot}\theta_i, \quad e_0 = K_{pot}\theta_0, \quad \text{and } K_{pot} = 10 \text{ V/rad};$$

$$\text{Motor torque } T = k_m i_a;$$

$$k_m = 0.1 \text{ N} \cdot \text{m/A};$$

$$R_a = 10\Omega (armature\ resistance);$$

$$\text{Gear ratio} = 1 : 1;$$

$$J_L + J_m = 10^{-3}\text{kg} \cdot \text{m}^2 (\text{total inertia});$$

$$C = 200\mu\text{f};$$

$$v_a = K_A(e_i - e_f).$$

a) What is the range of the gain K_A for which the system is stable? Calculate the upper limit graphically from your root locus.

b) Choose a gain K_A that gives roots at $\zeta = 0.7$. Where are all three closed-loop root locations?

FIGURE 4.54
(Reprinted from Clark, 1962, with permission.)

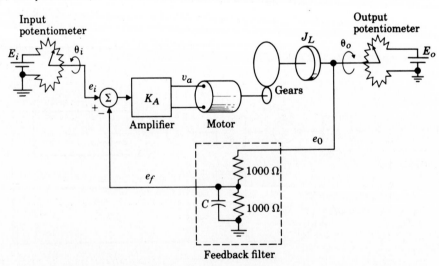

Section 4.5

4.33 We wish to design a velocity control for a tape-drive servo. The transfer function from current to velocity (in millimeters per millisecond per ampere) is

$$\frac{V(s)}{I(s)} = \frac{15(s^2 + 0.9s + 0.8)}{(s + 1)(s^2 + 1.1s + 1)}.$$

We wish to design a feedback system so that the result is Type I and the transient satisfies

$$t_r \leq 4 \text{ ms}, \qquad t_s \leq 15 \text{ ms}, \qquad \text{and} \qquad M_p \leq 0.05.$$

a) Assume a compensation K/s to achieve Type I behavior, and sketch the root locus versus K. Show on the same plot the region of acceptable pole locations corresponding to the specifications.

b) Assume compensation of the form $K(s + \alpha)/s$ and select the best possible values of K and α you can find. Sketch the root-locus plot of your design, giving values for K and α and the velocity constant K_v achieved. Indicate the closed-loop poles with Δ and include the boundary of the region of acceptable root locations.

4.34 Consider the Type I system in Fig. 4.55. We would like to design the compensation $D(s)$ to meet the following requirements: (1) Steady-state error in y due to a constant unit disturbance w should be less than 4/5, and (2) damping ratio ζ equals 0.7. Using root-locus techniques:

a) Show that proportional control alone is not adequate.
b) Show that proportional-plus-derivative control will work.
c) For $D(s) = K + K_d s$, find gains K and K_d to meet the design specifications.

4.35 Consider the 270 ft. United States Coast Guard WMEC *Tampa* (WMEC 902) shown in Fig. 4.56. Parameter identification based on sea trials data (Trankle, 1987) was used to estimate the hydrodynamic coefficients in the equations of motion. The result is that the system may be described by the second-order transfer functions:

$$\frac{v}{\delta} = \frac{0.2196(s + .0019)}{(s + 0.2647)(s + 0.0063)}$$

$$\frac{r}{\delta} = \frac{-0.0184(s + 0.0068)}{(s + 0.2647)(s + 0.0063)}$$

$$\frac{r}{w} = \frac{0.0000064}{(s + 0.2647)(s + 0.0063)},$$

FIGURE 4.55

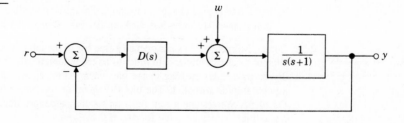

FIGURE 4.56
USCGC Tampa (WMEC 902)

where

v = lateral velocity (m/sec),

r = yaw rate (rad/sec),

δ = rudder angle (rad),

w = wind (m/sec)

a) Determine the open loop settling time of the system for both v and r.

b) In order to regulate the heading angle ψ, where

$$\dot{\psi} = r,$$

design a compensator using the measurement provided by a yaw rate gyro (i.e., r). The settling time is specified to be less than fifty seconds. Design a compensator to get a steady-state heading angle error of less than 0.5 degree for wind gusts of 10m/sec.

c) For a five degree change in heading, the maximum allowable rudder angle deflection is specified to be less than ten degrees. Repeat (**b**) with this requirement.

4.36 Golden Nugget Airlines has opened a free bar in the tail of each of their airplanes in an attempt to lure customers. In order to adjust automatically for the sudden weight shift due to passengers rushing to the bar when it first opens, the airline is mechanizing a pitch-attitude autopilot. The block diagram of the proposed arrangement is seen in Fig. 4.57. We assume a step function for the passenger moment $M_p(s) = M_0/s$ and a maximum expected value for M_0 of 0.6.

FIGURE 4.57

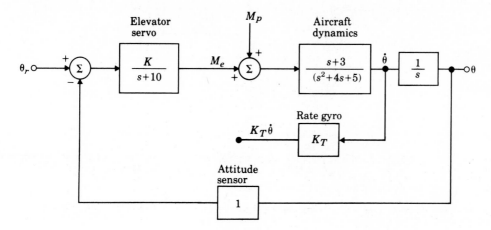

a) What value of K is required to keep the steady-state error in θ less than 0.02 rad($\approx$ 1°)? (Assume the system is stable.)

b) Draw a root locus versus K.

c) Based on your root locus, what is the value of K when the system goes unstable?

d) Suppose the value of K required for acceptable steady-state errors is 600. Show that this value yields an unstable system with roots at (approximately)

$$s = +1 \pm j6.6, \ -3, \ -14.$$

e) You are given a black box with "Rate gyro" written on the side and told that when installed it provides a perfect measure of $\dot{\theta}$, with output $K_T\dot{\theta}$.

    (1) Draw a block diagram indicating how you would incorporate this into the autopilot (include transfer functions in boxes).

    (2) Sketch a root locus versus K_T (the gain associated with the $\dot{\theta}$ term).

    (3) What is the maximum damping ratio of the complex roots obtainable with this configuration?

    (4) What is the value of K_T for (3)?

f) Suppose you aren't satisfied with the steady-state errors and damping ratio of the system with a rate gyro. Discuss the advantages and disadvantages of adding an integral term and extra lead networks in the control law. Support your comments with rough root-locus sketches.

4.37 Consider the instrument servomechanism with parameters as given in Fig. 4.58. For each case draw a root locus with respect to the parameter K and indicate the location of the roots corresponding to your final design.

a) *Lead Network:* Let

$$G_s(s) = 1, \qquad G_a(s) = K\frac{s+z}{s+p}, \qquad \text{and} \qquad \frac{p}{z} = 6.$$

FIGURE 4.58

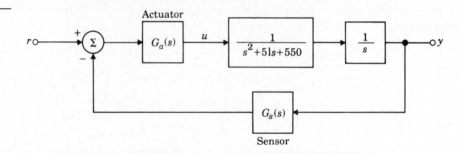

Select z and K so that the roots nearest the origin (the dominant roots) have $\zeta \geq 0.4$, $Re(\lambda_i) \leq -7$, $K_v \geq 16\frac{2}{3}$. $Re(\lambda_i)$ denotes the real part of the pole λ_i.

b) *Output Velocity (Tachometer) Feedback:* Let

$$G_s(s) = 1 + K_T s \qquad \text{and} \qquad G_a(s) = K.$$

Select K_T and K so that the dominant roots are in the same location as those of case I. Compute K_v. Give, if you can, the physical reason explaining the reduction in K_v when output derivative feedback is used.

c) *Lag Network:* Let

$$G_s(s) = 1 \qquad \text{and} \qquad G_a(s) = K\frac{s+1}{s+p}.$$

When we use proportional control, we obtain a $K_v = 12$ at $\zeta = 0.4$. Select K and p so that the dominant roots correspond to the proportional control case but with a K_v of 100 rather than 12.

Section 4.6

4.38 For the third-order system shown in Fig. 4.59,

a) Sketch the locus of the closed-loop system pole locations versus the parameter K using Evans' methods. Show your calculations for sketching (angle of asymptotes, departure angles, etc.).

b) Using graphical techniques, locate carefully the point at which the locus crosses the imaginary axis. What is the value of K at that point?

FIGURE 4.59

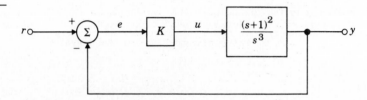

FIGURE 4.60

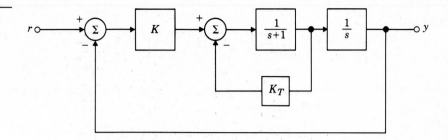

c) Assume that (due to some unknown mechanism) the amplifier output is a function of the magnitude of e. Specifically, let the control be

$$u = e \qquad \text{for } |e| \leq 1;$$
$$u = 1 \qquad \text{for } e > 1;$$
$$u = -1 \qquad \text{for } e < -1.$$

Qualitatively describe how you would expect the system to respond to a unit step input.

4.39 A block diagram of a position servomechanism is shown in Fig. 4.60.

a) Sketch the root locus versus K for K_T (tachometer feedback) $= 0$.

b) Indicate on the locus of **(a)** the root locations corresponding to $K = 4$. For these locations, estimate the transient-response parameters t_r, M_p, and t_s.

c) For $K = 4$, draw the root locus versus K_T.

d) For $K = 4$ and K_T set so that $M_p = 0.05$ ($\zeta = 0.707$), estimate t_r and t_s.

e) For the parameters of **(d)**, what is the velocity constant K_v of this system?

4.40 Consider the mechanical system shown in Fig. 4.61 where g and a_0 are gains. The feedback path containing gs controls the amount of rate feedback. For a fixed value of a_0, adjusting g corresponds to varying the location of a zero in the s plane.

a) For $g = 0$, find a value for a_0 such that the poles are complex. *Hint:* Let $a_0 = 1$, $\tau = 1$.

b) Fix a_0 at this value and construct root loci that demonstrate the effect of varying g.

FIGURE 4.61

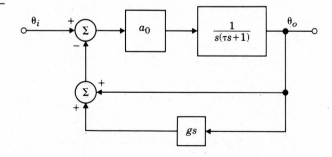

FIGURE 4.62

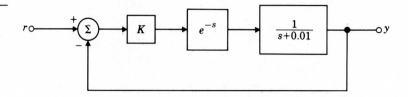

4.41 Sketch the $0°$ (negative) root locus for the systems in Problem 4.2 versus K.

4.42 Sketch the $0°$ root locus for systems in Problem 4.4 versus K.

4.43 State and prove an expression for the angles of the asymptotes of the negative ($0°$) root locus.

4.44 Sketch the root locus for the system in Fig. 4.62 versus K. What is the range of K for which the system is unstable?

4.45 Prove that the plant $G(s) = 1/s^3$ *cannot* be made unconditionally stable if pole cancellation is forbidden.

5

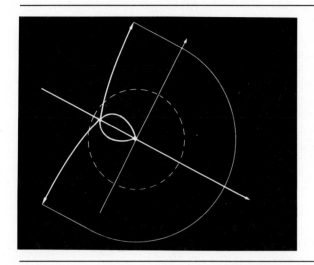

The Frequency-

Response Design

Method

5.1 Introduction

The design of feedback control systems in industry is probably accomplished using frequency-response methods more often than any other method. The popularity of these methods with practicing engineers has remained high over the years despite the development of alternate design methods, such as root locus, state-space, and optimal control. The primary reason for the popularity of frequency-response design is that it provides good designs in the face of uncertainty in the plant model. For example, systems with poorly known or changing high-frequency resonances can have their feedback compensation tempered to alleviate the effects of those uncertainties. Currently this tempering is carried out more easily using frequency-response design rather than any other method.

For complex systems with more than one input or output, frequency-response methods have been extended (called H^∞ design) to aid in design and especially to establish measures of sensitivity to plant uncertainties. The study of frequency-response design for the single-input, single-output systems that is the subject of this book is important for later understanding of more advanced topics.

Another advantage of using frequency response is the ease with which experimental information can be used for design purposes. Raw measurements of the output amplitude and phase of a plant undergoing a sinusoidal input are sufficient to design a suitable feedback control. No intermediate processing of the data to arrive at the system model (poles and zeros or system matrices) is required. The wide availability of computers has rendered this advantage less important now than it was years ago; however, for relatively simple systems, frequency response is often still the most cost-effective design method.

A disadvantage of the method is that the underlying theory is rather complicated and requires a rather broad knowledge of complex variables. However, the methodology of design is easy, and the insights gained by learning the theory are well worth the struggle.

5.2   Frequency Response

In this section we will describe how a system responds to sinusoidal input—called the system's *frequency response*—and how this response can be obtained from a knowledge of its pole and zero locations.

To introduce the ideas, consider a system described by

$$\frac{Y(s)}{U(s)} = G(s)$$

where the input $u(t)$ is a sine wave with an amplitude of U_o, that is,

$$u(t) = U_o \sin \omega t,$$

which has a Laplace transform

$$U(s) = \frac{U_o \omega}{s^2 + \omega^2}.$$

The Laplace transform of the output is

$$Y(s) = G(s)\frac{U_o \omega}{s^2 + \omega^2}.$$

A partial fraction expansion of the preceding will result in an equation with the form

$$Y(s) = \frac{\alpha_1}{s + a_1} + \frac{\alpha_2}{s + a_2} + \cdots + \frac{\alpha_n}{s + a_n} + \frac{\alpha_o}{s + j\omega} + \frac{\alpha_o^*}{s - j\omega} \qquad (5.1)$$

where $a_1, a_2, \ldots, a_n$ are the poles of $G(s)$, where α_o would be found by performing the partial fraction expansion as outlined in Problem 5.1, and α_o^* is the complex conjugate of α_o. The time response that corresponds to $Y(s)$ is

$$y(t) = \alpha_1 e^{-a_1 t} + \alpha_2 e^{-a_2 t} + \ldots + \alpha_n e^{-a_n t} + 2|\alpha_o| \sin(\omega t + \phi) \qquad (5.2)$$

where $\phi = \tan^{-1} \text{Re}(\alpha_o)/\text{Im}(\alpha_o)$. If all the poles of the system represent stable behavior ($a_1, a_2, \ldots, a_n \geq 0$), the natural, unforced response will die out eventually and therefore the steady-state response of the system is due solely to the sinusoidal term in Eq. (5.2) caused by the sinusoidal excitation. For example, Fig. 5.1 shows the response of the system, $G(s) = 1/(s + 1)$, to the input, $u = \sin(10t)$. It shows how the natural part of the response associated with the $G(s)$, that is, the e^{-t} portion, disappears after several time constants and so the pure sinusoidal response is all that remains. Problem 5.1 shows that the remaining sinusoidal term in Eq. (5.2) can be expressed as

$$y(t) = U_o A \sin(\omega t + \phi) \qquad (5.3)$$

where

$$A = |G(j\omega)| = |G(s)|\Big|_{s=j\omega} \qquad \text{and} \qquad \phi = \tan^{-1} \frac{\text{Im}(G(j\omega))}{\text{Re}(G(j\omega))} = \angle G(j\omega).$$

$$(5.4)$$

Equation (5.3) shows that a stable system with transfer function $G(s)$ excited by a sinusoid with unit amplitude and frequency ω will, in steady-state, exhibit a

FIGURE 5.1
Response of $G(s) = 1/(s + 1)$ to $\sin(10t)$.

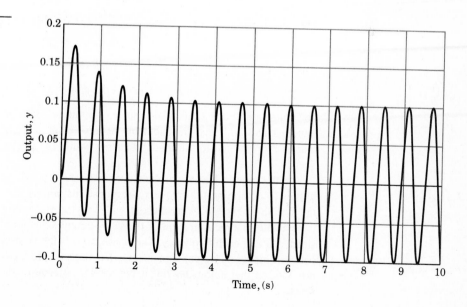

Output, y

Time, (s)

sinusoidal output with magnitude $A(\omega)$ and phase $\phi(\omega)$ at frequency ω. The fact that the output y is a sinusoid with the same frequency as the input u and that the magnitude A and phase ϕ of the output are independent of the amplitude U_o of the input are consequences of $G(s)$ being a linear continuous system. If the system being excited were a nonlinear or discrete system, the output might contain frequencies other than the input frequency, and the output magnitude might be dependent on the input magnitude.

The *magnitude* A is given by $|G(j\omega)|$ and the *phase* ϕ, is given by $\angle G(j\omega)$; that is, the magnitude and angle of the complex quantity $G(s)$ evaluated with s taking on values along the imaginary axis. We are not only interested in analyzing the frequency response of a system in order to understand how it responds to a sinusoidal input, but also because evaluating $G(s)$ with s taking on values along the $j\omega$ axis will prove to be very useful in the determination of stability of a closed-loop system. The $j\omega$ axis is the boundary between stability and instability and we will see in Section 5.5 that the evaluation of $G(j\omega)$ provides information that allows a closed-loop stability determination from the open loop $G(s)$.

As an example of frequency response, consider the capacitor described by the equation

$$i = C\frac{dv}{dt},$$

where the input is v and the output is i. The transfer function of this circuit is

$$G(s) = Cs,$$

and

$$G(j\omega) = Cj\omega. \tag{5.5}$$

Computing the magnitude and phase, we find that

$$A = |Cj\omega| = C\omega \quad \text{and} \quad \phi = \angle(Cj\omega) = 90°. \tag{5.6}$$

For a unit amplitude sinusoidal input v, the output i will be a sinusoid with magnitude A and will lead the input by $90°$. Note for this example that the magnitude is proportional to the input frequency while the phase is independent of frequency.

In the cases where a designer does not have a good model of the system and wishes to determine the frequency-response magnitude and phase experimentally, the system is excited by a sinusoid with varying frequency. The magnitude $A(\omega)$ is obtained by a measurement of the ratio, output sinusoid to input sinusoid, in the steady state at each frequency. The phase $\phi(\omega)$ is the measured difference in phase between input and output signals. Hewlett-Packard produces an instrument (Model no. 3563A) that automates this experimental procedure and greatly speeds up the process.

A great deal more can be learned about the dynamic response of a system from knowledge of the magnitude $A(\omega)$ and the phase $\phi(\omega)$ of its transfer function. In the obvious case, if the signal is a sinusoid, then A and ϕ completely describe the response. Furthermore, if the input is periodic, then a Fourier series can be constructed to decompose the input into a sum of sinusoids, and again $A(\omega)$ and $\phi(\omega)$ can be used with each component to construct the total response. For transient inputs, our best path to understanding the meaning of A and ϕ is to relate the frequency response $G(j\omega)$ to the transient responses calculated by the Laplace transform. For example, in Fig. 2.35 we have plotted the step response of a system having the transfer function

$$G(s) = \frac{1}{(s/\omega_n)^2 + 2\zeta(s/\omega_n) + 1} \tag{5.7}$$

for various values of ζ. To compare, we should plot $A(\omega)$ and $\phi(\omega)$ for these same values of ζ and see which features of frequency response correspond to the transient characteristics. In order to make the comparisons effective, we need to consider an efficient and meaningful form in which to make frequency-response plots.

5.2.1 Bode Plot Techniques

Display of frequency response is a problem that has been studied for a long time, and the most useful technique for our purposes was developed by H. W. Bode at Bell Laboratories between 1932 and 1942. This technique allows a plotting by hand that is quick and yet sufficiently accurate for control-system design. Most control-system designers will have access to computer programs that will diminish the need for hand plotting; however, it is still important to develop good intuition so that erroneous computer results are quickly identified, and for this one needs the ability to check results by hand.

The first idea is to plot magnitude curves using a logarithm scale and phase curves using a linear scale. This allows the plotting of a high-order $G(j\omega)$ by simple graphical addition of the separate terms, as discussed in Appendix B. This follows because a complex expression with zero and pole factors can be written as

$$G(j\omega) = \frac{\vec{s_1}\vec{s_2}}{\vec{s_3}} = \frac{r_1 e^{j\theta_1} r_2 e^{j\theta_2}}{r_3 e^{j\theta_3}} = \left(\frac{r_1 r_2}{r_3}\right) e^{j(\theta_1 + \theta_2 - \theta_3)}. \tag{5.8}$$

Equation (5.8) shows that the phases of the individual terms are added directly to obtain the phase of the composite expression. Furthermore, since

$$|G(j\omega)| = \frac{r_1 r_2}{r_3},$$

then

$$\log_{10}|G(j\omega)| = \log_{10} r_1 + \log_{10} r_2 - \log_{10} r_3, \qquad (5.9)$$

and we see that addition of the logs of the individual terms provides the log of the magnitude of the composite expression. The frequency response is typically present as two curves, a plot of logarithm of magnitude versus log ω and a plot of the phase versus log ω. This is called a *Bode plot* of the system. Since

$$\log_{10} A e^{j\phi} = \log_{10} A + j\phi \log_{10} e, \qquad (5.10)$$

the Bode plot is seen to be essentially the real and imaginary parts of the logarithm of $G(j\omega)$. In communications it is standard to measure power gain in decibels.*

$$|G|_{dB} = 10 \log_{10} \frac{P_2}{P_1}.$$

Since power is the square of the voltage, the voltage gain is

$$|G|_{dB} = 20 \log_{10} \frac{V_2}{V_1}.$$

Hence one can present a Bode plot giving the magnitude in decibels versus log ω and the phase in degrees versus log ω.[†] In this book, we will present a Bode plot as log $|G|$ versus log ω and also mark an axis in decibels on the right-hand side of the magnitude plot in order to give the reader a choice in working with whichever representation is preferred. If the magnitude plot is carried out using log $|G|$, it is common to plot on a log scale and mark the scale in terms of $|G|$ magnitude directly. If the plot is carried out using decibels, the vertical scale will be a linear one with each decade of $|G|$ magnitude representing 20 decibels. An example of Bode plots of magnitude and phase will be seen later.

Advantages of working with frequency response in terms of Bode plots include:

1. Bode plots of systems in series simply add.
2. Bode's phase-gain relationship is in terms of logarithms of phase and gain.
3. A much wider range of the system's behavior can be displayed; that is, both low- and high-frequency behavior can be displayed in one plot.
4. Bode plots can be determined experimentally.
5. Compensator design can be based entirely on Bode plots.

In Chapter 4, we always wrote the open-loop transfer function in the form

$$KG(s) = K\frac{(s + z_1)(s + z_2)\dots}{(s + p_1)(s + p_2)\dots} \qquad (5.11)$$

---

*Researchers for the telephone company first defined the unit of power gain as a "bel." However, this unit proved to be too large, and hence a decibel (i.e., one-tenth of a bel) was selected as the unit, which is, of course, named after Alexander Graham Bell.

[†] Henceforth we will drop the base of the logarithm; it is understood to be 10.

because this was the most convenient way to determine the degree of stability from the root locus versus the gain K. In working with frequency response, it is most convenient to replace s with $j\omega$ and to write the transfer functions in the "Bode" form

$$KG(j\omega) = K\frac{(j\omega\tau_1 + 1)(j\omega\tau_2 + 1)\ldots}{(j\omega\tau_a + 1)(j\omega\tau_b + 1)\ldots} \qquad (5.12)$$

because the gain K in this form is directly related to the transfer function magnitude at very low frequencies. In fact, for type I systems, the K in Eq. (5.12) is equal to A at $\omega = 0$, and is also equal to the DC gain of the system. Note that if a transfer function is given in the form of Eq. (5.11), it is a straightforward calculation to arrive at an equivalent transfer function in the form of Eq. (5.12); however, note that the gain K will not have the same value in the two expressions.

As an example of rewriting a transfer function according to Eqs. (5.8) and (5.9), suppose

$$KG(j\omega) = K\frac{j\omega\tau_1 + 1}{(j\omega)^2(j\omega\tau_a + 1)}. \qquad (5.13)$$

Then

$$\angle KG(j\omega) = \angle K + \angle(j\omega\tau_1 + 1) - \angle(j\omega)^2 - \angle(j\omega\tau_a + 1), \qquad (5.14)$$

and

$$\log|KG(j\omega)| = \log|K| + \log|j\omega\tau_1 + 1| - \log|(j\omega)^2| - \log|j\omega\tau_a + 1|,$$

or in decibels,

$$|KG(j\omega)|_{dB} = 20\log|K| + 20\log|j\omega\tau_1 + 1| - 20\log|(j\omega)^2|$$
$$- 20\log|j\omega\tau_a + 1|.$$

All transfer functions for the kinds of systems we've talked about so far are composed of three *classes* of terms:

1. $K(j\omega)^n$,
2. $(j\omega\tau + 1)^{\pm 1}$,
3. $\left[\left(\frac{j\omega}{\omega_n}\right)^2 + 2\zeta\frac{j\omega}{\omega_n} + 1\right]^{\pm 1}$.

First we'll discuss the plotting of each individual term and how the terms affect the composite plot including all the terms; then we'll discuss how the composite curve can be drawn.

1. $K(j\omega)^n$ Since

$$\log K|(j\omega)^n| = \log K + n\log|j\omega|,$$

FIGURE 5.2
Magnitude of $(j\omega)^n$.

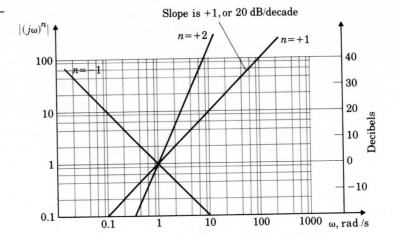

the magnitude plot of this term is a straight line with a slope of n. Some examples are shown in Fig. 5.2. This term is the only one that affects the slope at the lowest frequencies because all other terms are constant in that region. The easiest way to draw the curve is to locate $\omega = 1$ and plot log K at that frequency. Then draw the line with slope n through that point.* The phase of $(j\omega)$ is $n \times 90°$; it is independent of frequency and thus a horizontal line: $-90°$ for $n = -1$; $-180°$ for $n = -2$; $+90°$ for $n = +1$; and so on.

2. $j\omega\tau + 1$ The magnitude of this term approaches one asymptote at very low frequencies and another asymptote at very high frequencies:

a) For $\omega\tau \ll 1$, $j\omega\tau + 1 \cong 1$.

b) For $\omega\tau \gg 1$, $j\omega\tau + 1 \cong j\omega\tau$.

If we call $\omega = 1/\tau$ the *breakpoint*, then we see that at frequencies below the breakpoint, the magnitude curve is approximately constant $(= 1)$, while above the breakpoint, the magnitude curve behaves similarly to $K(j\omega)^n$, the term just described as class 1. The example plotted in Fig. 5.3, $G(s) = 10s + 1$, shows how the two asymptotes cross at the breakpoint and how the actual magnitude curve is a factor of 1.4, or 3 dB above that crossing point (or a factor of 0.707, or 3 dB below if the term is in the denominator). Note that this term will have only a small effect on the composite magnitude curve below the breakpoint because its value is equal to 1 $(= 0$ dB$)$ in this region.

---

*In decibel units, the slopes are $n \times 20$ dB per decade or $n \times 6$ dB per octave. (An octave is a change in frequency by a factor of 2.)

FIGURE 5.3
Magnitude plot for $j\omega\tau + 1$.

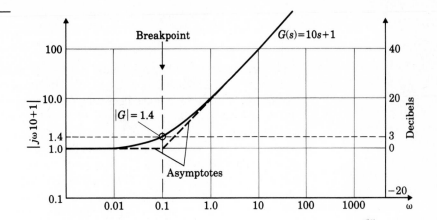

The phase curve can also be drawn easily by using the low- and high-frequency asymptotes:

a) For $\omega\tau \ll 1$, $\angle 1 = 0°$.
b) For $\omega\tau \gg 1$, $\angle j\omega\tau = 90°$.
c) For $\omega\tau \cong 1$, $\angle(j\omega\tau + 1) \cong 45°$.

For $\omega\tau \cong 1$, the $\angle(j\omega + 1)$ curve is tangent to an asymptote going from 0° at $\omega\tau = 0.2$ to 90° at $\omega\tau = 5$, as shown in Fig. 5.4. The figure also illustrates the three asymptotes (shown dashed) used for the phase plot and how the actual curve deviates from the asymptotes by 11° at their intersections. Both the composite phase and magnitude curves are unaffected by this class of term at frequencies a factor of 10 or so below the breakpoint because its magnitude is 0 dB and phase is 0°.

FIGURE 5.4
Phase plot for $j\omega\tau + 1$.

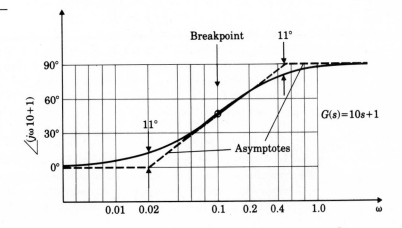

3.

$$\left[\left(\frac{j\omega}{\omega_n} \right)^2 + 2\zeta \frac{j\omega}{\omega_n} + 1 \right]^{\pm 1}$$

This term behaves in a manner similar to the term above. The differences are that the magnitude changes slope by +2 or +40 dB per decade (−2 or −40 dB per decade if the term is in the denominator) at the breakpoint, the breakpoint being at $\omega = \omega_n$; the phase changes by ±180°; and the transition through the breakpoint region varies with the damping ratio ζ. Figure 5.5 shows the magnitude and phase for several different damping ratios, assuming the term is in the denominator. Note that the magnitude asymptote for frequencies above the breakpoint has a slope of −2 or −40 dB per decade and that the transition through the breakpoint region has a large dependence on the damping ratio. A rough sketch of this transition can be made by noting that

$$|G(j\omega)| = \frac{1}{2\zeta} \qquad \text{at} \quad \omega = \omega_n$$

for this class of second-order term in the denominator. If the term is in the numerator, the magnitude would be the reciprocal of the curve plotted in Fig. 5.5(a).

No such handy rule exists for sketching the transition for the phase curve; therefore one would have to resort to Fig. 5.5(b) for an accurate plot of the phase. However, a very rough idea of the transition can be gained by noting that the transition is a step for $\zeta = 0$, while it obeys the rule for two first-order terms, (i.e., the class 2 terms just described) when $\zeta = 1$ with simultaneous break frequencies. All intermediate values of ζ fall between these two extremes.

We can now compare the frequency-response curves of Fig. 5.5 with the transient-response curves of Fig. 2.41. The transient curves are normalized with respect to time as $\omega_n t$. From the frequency response, we see that ω_n is the break frequency and is approximately the bandwidth—the place where the gain starts to fall off from its low-frequency value. Therefore the rise time can be estimated from the bandwidth. We also see that the peak overshoot in frequency is approximately $1/2\zeta$ for $\zeta < 0.5$, so the peak overshoot in step response can be estimated from the peak overshoot in frequency response. Thus we see that essentially the same information is contained in the frequency-response curve as is found in the transient responses.

When the system has several poles and several zeros, plotting the frequency response requires that the components be combined into a composite curve. To plot the composite magnitude curve, it is useful to note that the slope of the asymptotes is equal to the sum of the slopes of the individual curves. Therefore the composite asymptote curve has integer slope changes at each break frequency: +1 for

FIGURE 5.5
(a) Magnitude and
(b) phase of
$1/[(j\omega/\omega_n)^2 + 2\zeta(j\omega/\omega_n) + 1]$.

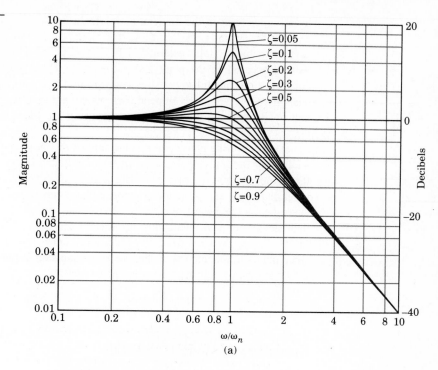

(a)

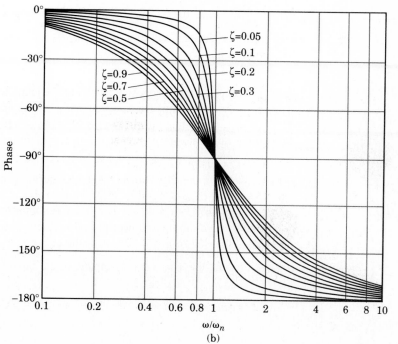

(b)

a first-order term in the numerator, -1 for a first-order term in the denominator, and ± 2 for second-order terms. Furthermore, the lowest-frequency portion of the asymptote has a slope determined by the value of n in the $(j\omega)^n$ term and is located by plotting the point $K\omega^n$ at $\omega = 1$. Therefore the complete procedure consists of plotting the lowest-frequency portion of the asymptote, then sequentially changing the asymptote's slope at each breakpoint in order of ascending frequency, and finally drawing the actual curve by using the transition rules discussed with classes 2 and 3.

The composite phase curve is the sum of the individual curves. The graphical addition of the individual phase curves is facilitated by locating the curves so that the composite phase approaches the individual curve as closely as possible. A quick but crude sketch of the composite phase can be found by starting the phase curve below the lowest breakpoint and setting it equal to $n \times 90°$. The phase is then stepped at each breakpoint in order of ascending frequency. The amount of the phase step is $\pm 90°$ for a first-order term and $\pm 180°$ for a second-order term. Breakpoints in the numerator indicate a positive step in phase while breakpoints in the denominator indicate a negative phase step.*

Summary of Bode Plot Rules

STEP 1: Manipulate the transfer function into the form given by Eq. (5.12).

STEP 2: Determine the value of n for the $K(j\omega)^n$ class of term. Plot the low-frequency magnitude asymptote through the point K at $\omega = 1$ with a slope of n (or $n \times 20$ dB per decade).

STEP 3: Complete the composite magnitude asymptotes by extending the low-frequency asymptote until the first frequency breakpoint, then stepping the slope by ± 1 or ± 2, depending on whether the breakpoint is from a first- or second-order term in the numerator or denominator, and continuing through all breakpoints in ascending order.

STEP 4: Sketch in the approximate magnitude curve by increasing from the asymptote by a factor of 1.4 ($+3$ dB) at first-order numerator breaks and decreasing it by a factor of 0.707 (-3 dB) at first-order denominator breaks. At second-order breakpoints, sketch in the resonant peak (or valley) according to Fig. 5.5(a) using the relation that $|G(j\omega)| = 1/(2\zeta)$ at the break.

STEP 5: Plot the low-frequency asymptote of the phase curve, $\phi = n \times 90°$.

*This approximate method was pointed out to us by our Parisian friends.

STEP 6: As a guide, sketch in the approximate phase curve by changing the phase by $\pm 90°$ or $\pm 180°$ at each breakpoint in ascending order. For first-order terms in the numerator, the change of phase is $+90°$; in the denominator, the change is $-90°$. For second-order terms, the change is $\pm 180°$.

STEP 7: Locate the asymptotes for each individual phase curve so that their phase change corresponds to the steps in the phase from the approximate curve indicated by step 6. Sketch in each individual phase curve as indicated by Fig. 5.4 or Fig. 5.5(b).

STEP 8: Graphically add each phase curve. Use dividers if an accuracy of about $\pm 5°$ is desired. If lesser accuracy is acceptable, the composite curve can be done by eye, keeping in mind that the curve will start at the lowest-frequency asymptote and end on the highest-frequency asymptote, and will approach the intermediate asymptotes to an extent that is determined by the proximity of the breakpoints to each other.

EXAMPLE 5.1 As an example of Bode plotting, suppose we wish to plot the magnitude and phase of

$$G(s) = \left. \frac{2000(s + 0.5)}{s[(s + 10)(s + 50)]} \right|_{s = j\omega}. \tag{5.15}$$

STEP 1: Convert to the form in Eq. (5.12). This yields

$$G(j\omega) = \frac{2[(j\omega/0.5) + 1]}{j\omega\{[(j\omega/10) + 1][(j\omega/50) + 1]\}}. \tag{5.16}$$

STEP 2: The term in $j\omega$ is first order and in the denominator; hence $n = -1$. Therefore, the low-frequency asymptote (valid for $\omega < 0.5$ because the lowest breakpoint is at $\omega = 0.5$) is defined by the first term:

$$G(j\omega) = \frac{2}{j\omega}. \tag{5.17}$$

The magnitude plot of this term has the slope of -1 or -20 dB per decade and is located by passing the line through the value 2 at $\omega = 1$ even though the composite curve will not go through the point because of the breakpoint at $\omega = 0.5$. This is shown in Fig. 5.6(a).

STEP 3: The remainder of the asymptotes, also shown in Fig. 5.6(a), were obtained by next drawing a line with 0 slope that intersects the original -1 slope at $\omega = 0.5$; then drawing a -1 slope line that intersects the previous one at $\omega = 10$; and finally drawing a -2 slope line that intersects the previous -1 slope at $\omega = 50$.

STEP 4: The actual curve is then sketched in to be approximately tangent to the asymptotes when far away from the breakpoints, a factor of 1.4 above ($+3$ dB)

FIGURE 5.6
Composite plots:
(a) magnitude, (b) phase,
and (c) approximate
phase.

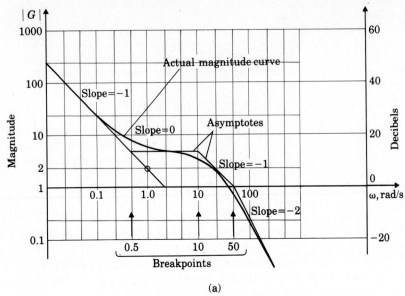

(a)

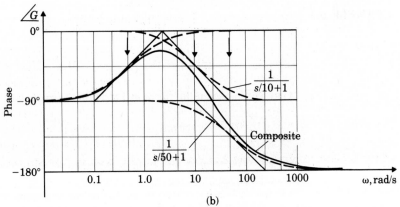

(b)

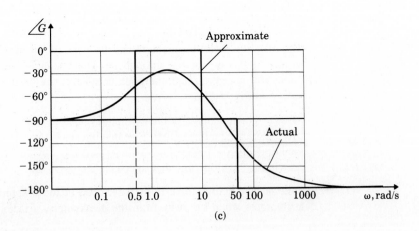

(c)

256

the asymptote at the $\omega = 0.5$ breakpoint, and a factor of 0.7 down (-3 dB) from the asymptote at the $\omega = 10$ and $\omega = 50$ breakpoints.

STEP 5: Since the phase of Eq. (5.17) is $-90°$, the phase curve in Fig. 5.6(b) starts at $-90°$ at the lowest frequencies.

STEP 6: The result is shown in Fig. 5.6(c).

STEP 7: The individual phase curves, shown dashed in Fig 5.6(b), have the correct phase change for each term and are located vertically so that their phase change corresponds to the steps in the phase from the approximate curve in Fig. 5.6(c). Note that the composite curve approaches each individual term.

STEP 8: The graphical addition of each dotted curve results in the solid composite curve in Fig. 5.6(b). As can be seen from the figure, the vertical placement of each individual phase curve makes the required graphical addition particularly easy because the composite curve approaches each individual phase curve in turn. ■

EXAMPLE 5.2   As a second example, we show the Bode plots for a system with second-order terms. An example like this is more difficult to plot than the previous example because the transition between asymptotes is dependent on the damping ratio; however, the same basic ideas previously illustrated apply. The transfer function represents a mechanical system with two equal masses coupled with a lightly damped spring. The applied force and position measurements are collocated on the same mass. For the transfer function, the time scale has been chosen so that the resonant frequency of the complex zeros is equal to 1. The transfer function is

$$G(s) = \frac{0.01(s^2 + 0.01s + 1)}{s^2[(s^2/4) + 0.02(s/2) + 1]}.$$

Proceeding through the steps, we start with the low-frequency asymptote, $0.01/\omega^2$. It has a slope of -2 (-40 dB per decade) and passes through magnitude $= 0.01$ at $\omega = 1$ as shown in Fig. 5.7(a). At the break frequency of the zero, $\omega = 1$, the slope shifts to zero (break up by 2) until the break of the pole is located at $\omega = 2$, when the slope again breaks down to a slope of -2. To interpolate the true curve, we plot the point at the zero breakpoint, $\omega = 1$, with a magnitude ratio below the asymptote of $2\zeta = 0.01$. At the pole breakpoint, the magnitude ratio above the asymptote is $1/(2\zeta) = 1/0.02 = 50$. The magnitude curve is a "doublet" of a negative pulse followed by a positive pulse. The phase curve for this case starts at $-180°$ corresponding to the $1/s^2$ term, jumps $+180°$ to $\phi = 0$ at $\omega = 1$ due to the zeros and falls $180°$ back to $\phi = -180°$ at $\omega = 2$ due to the poles as shown in Fig. 5.7(b). With such small damping ratios, the square-

FIGURE 5.7
Bode plot for transfer
function with complex
poles and zeros:
(a) magnitude and
(b) phase.

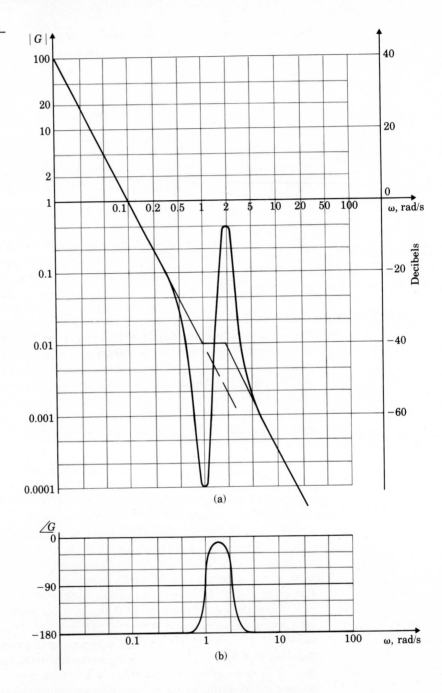

wave approximation is quite good as suggested by Fig. 5.5(b). The true composite phase curve is nearly a square pulse between $\omega = 1$ and $\omega = 2$, as shown in Fig. 5.7(b).   ∎

5.2.2 Steady-State Errors

We saw in Section 3.3 that the steady-state error of a feedback system decreases as the gain of the open-loop transfer function increases. In plotting a composite magnitude curve, we saw in the previous section that the low-frequency asymptote is given by

$$KG(j\omega) = K(j\omega)^n. \tag{5.18}$$

Therefore, one can conclude that the larger the value of the magnitude on the low-frequency asymptote, the lower the steady-state errors will be for the closed-loop system. This idea is very useful in the design of compensation where we often want to evaluate several alternate ways to improve stability. It is also useful in being able to quickly see the effects of changes in the compensation on the steady-state errors.

For a system of the form given by Eq. (5.12), that is, where $n = 0$ in Eq. (5.18) and often referred to as a type 0 system, the low-frequency asymptote is a constant and the gain K of the open-loop system is equal to the position error constant, K_p. For a unity feedback system with a unit step input, Table 3.3 shows that the steady-state error is

$$e_\infty = \frac{1}{1 + K_p}.$$

For a system where $n = -1$ in Eq. (5.18), often referred to as a type I system, the low-frequency asymptote has a slope of -1. The magnitude of the low-frequency asymptote is related to the gain according to Eq. (5.18) and, therefore, we can again read the gain K directly from the Bode magnitude plot. Section 3.3 shows that the velocity error coefficient K_v is equal to the gain K in Eq. (5.18) for this case. For a unity feedback system with a unit ramp input, the steady-state error is then found from

$$e_\infty = \frac{1}{K_v}.$$

The easiest way of determining the value of K_v in a type I system is to read the magnitude of the low-frequency asymptote at $\omega = 1$ rad/s because $K_v = K$ at this frequency. In some cases, the lowest-frequency breakpoint will be below $\omega = 1$ rad/s, therefore the asymptote can be extended to $\omega = 1$ rad/s in order to read K_v directly. Alternately, the magnitude can be read at any frequency on the low-frequency asymptote and K_v computed from Eq. (5.18).

FIGURE 5.8
Determination of K_v
from the Bode plot.

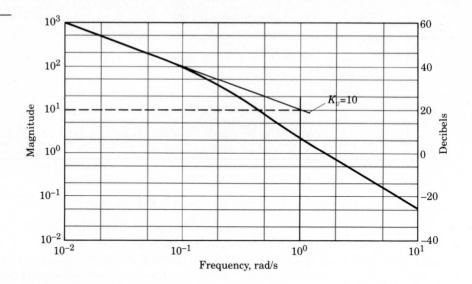

EXAMPLE 5.3   As an example of the determination of steady-state errors, a Bode magnitude plot
of an open-loop plant plus compensation is shown in Fig. 5.8. Because the slope
at the low frequencies is -1, we know that the system is type I. The extension of
the low-frequency asymptote crosses $\omega = 1$ rad/s at a magnitude of 10. Therefore,
$K_v = 10$ and the steady-state error to a unit ramp for a unity feedback system
would be 0.1.   ∎

5.3 Specifications

A natural specification for system performance in terms of frequency response is the
bandwidth, defined to be the maximum frequency at which the output of a system
will track an input sinusoid in a satisfactory manner. By convention, for the system
shown in Fig. 5.9 with a sinusoidal input r, the bandwidth is the frequency of r
at which the output y is attenuated to a factor of 0.707 times the input (or down 3
dB) relative to the zero-frequency gain. Figure 5.10 shows the idea graphically.

FIGURE 5.9
System description.

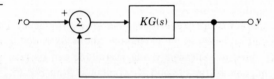

FIGURE 5.10
Bandwidth and resonant peak definition.

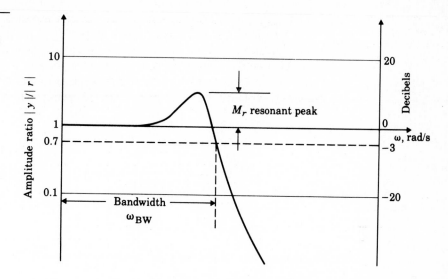

Bandwidth is a measure of speed of response and is therefore similar to time-domain measures like rise time and peak time or the s-plane measure of dominant-root(s) natural frequency. In fact, if the $KG(s)$ in Fig. 5.9 is such that the closed-loop response is given by Fig. 5.5, it can be seen that the bandwidth will be equal to the closed-loop root, natural frequency (that is ω_n) for a closed-loop damping ratio of $\zeta = 0.7$. For other damping ratios, the bandwidth is approximately equal to the natural frequency of the closed-loop roots, with an error typically less than a factor of 2.

Another specification related to the frequency response is the resonant peak magnitude M_r. It is defined in Fig. 5.10 and is shown in Fig. 5.5 to be generally related to the damping of a system. In practice, M_r is rarely used; instead, the gain and/or phase margin of a system is the preferred specification to indicate the degree of damping or stability of a system. The definitions of these margins are given in Section 5.6; they depend on stability ideas to be described in the next two sections.

5.4 Stability

In the early days of electronic communications, most applications were judged in terms of their frequency response. It is therefore natural that when the feedback amplifier was introduced, techniques to determine stability in the presence of feedback were based on this response.

Suppose the closed-loop transfer function of a system is known. We can determine the stability of a system by simple inspection of the denominator in factored
form (the factors give the system roots)—observing whether the real parts are positive or negative. But the closed-loop transfer function is not usually known; in
fact, the whole idea of the root-locus technique is to find the denominator factors
of the closed-loop transfer function, given the open-loop transfer function. It is also

FIGURE 5.11
Stability example:
(a) system definition and
(b) root locus.

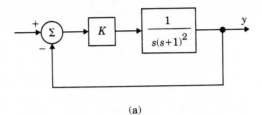

(a)

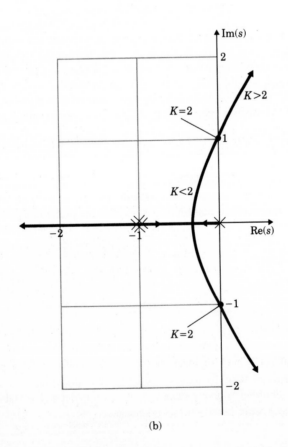

(b)

possible to determine closed-loop stability by evaluating the frequency response of the *open-loop* transfer function $KG(j\omega)$ (assuming unity feedback*) and then performing a simple test on that response. Note that factoring the denominator of the closed-loop transfer function is not required, just evaluation of $KG(j\omega)$. Now let's explain how this comes about.

Suppose we have a system whose root locus behaves as shown in Fig. 5.11 — that is, instability results if K has a sufficiently high value. The neutrally stable points lie on the imaginary axis, that is, where $s = j\omega$. Furthermore, we saw in Chapter 4 that all points on the locus have the property that

$$|KG(s)| = 1$$

and

$$\angle G(s) = 180°.$$

Therefore at the point of neutral stability we see that these root-locus conditions hold for $s = j\omega$, which means that

$$|KG(j\omega)| = 1 \quad \text{and} \quad \angle G(j\omega) = 180°. \qquad (5.19)$$

Thus a plot of the frequency-response magnitude and phase of a system that is neutrally stable (that is, with K so that the closed-loop roots fall on the imaginary axis) will satisfy the conditions of Eq. (5.19). Figure 5.12 shows the frequency response for the system whose root locus is plotted in Fig. 5.11. The magnitude response corresponding to $K = 2$ passes through 1 at the same frequency ($\omega = 1$ rad/s) at which the phase passes through 180°, as indicated by the conditions in Eq. (5.19).

For the example shown in Fig. 5.11, it can be seen from the root locus in Fig. 5.11 that any K less than the value at the neutrally stable point will result in a stable system. The frequency response in Fig. 5.12 shows plots of $|KG(j\omega)|$ with various values of K. At the frequency ω where the phase $\angle G(j\omega) = -180°$, ($\omega = 1$ rad/s) the magnitude $|KG(j\omega)|$ will be < 1 for stable values of K and > 1 for unstable values of K. Therefore, we have a stability condition based on the character of the open-loop frequency response,

$$|KG(j\omega)| < 1 \quad \text{at} \quad \angle G(j\omega) = -180°. \qquad (5.20)$$

This stability criterion holds for all systems where increasing gain leads to instability, the most common situation. However, there are systems where an increasing gain can lead from instability to stability (conditionally stable systems). The con-

---

*The unity feedback assumption does not limit the application of this concept because, for purposes of stability analysis, one can transform any system to the unity feedback configuration by simply including the transfer function of the feedback path as part of the open-loop transfer function $KG(j\omega)$.

FIGURE 5.12
Frequency-response
magnitude and phase.

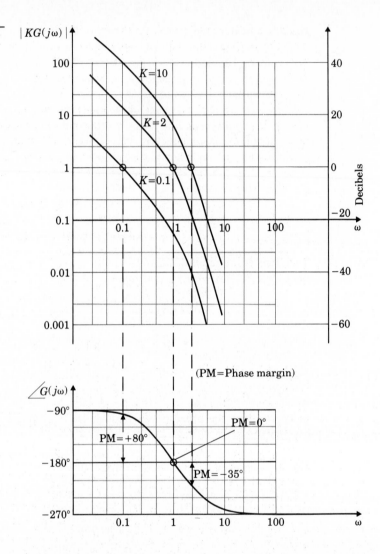

ditions for neutral stability given by Eq. (5.19) *always* hold; therefore, the only question in the stability criterion of Eq. (5.20) is whether $|KG(j\omega)|$ should be < 1 or > 1 at the frequency where $\angle G(j\omega) = -180°$. One way to resolve the ambiguity is to perform a rough sketch of the root locus to resolve the key question: does increasing gain lead to stability or instability? Another more rigorous way to resolve the ambiguity is to use the Nyquist stability criterion, the subject of the next section.

Quantities that measure the *degree of stability* have been defined that are directly related to the stability criterion of Eq. (5.20). The factor by which the gain is less than the neutral stability value is referred to as the *gain margin*. It can be read directly from the frequency response in Fig. 5.12 by measuring the distance between the $|KG(j\omega)|$ curve and the magnitude = 1 line at the frequency where $\angle G(j\omega) = -180°$. We see from the figure that the $K = 0.1$ case is stable and the gain margin (GM) is = 20 (26 dB). The $K = 2$ case is neutrally stable and the GM = 0 while the $K = 10$ case is unstable and has a GM = 0.2 $(-14$ dB). Note that the GM is the factor by which the gain K can be raised before instability results. A GM with a magnitude < 1 (< 0 if in dB) indicates an unstable system. The GM can also be determined from a root locus versus K by noting K at the point the locus crosses the $j\omega$ axis and K at the design point. The GM is the ratio of these two values of K.

Another measure that is used to indicate the degree of stability is the *phase margin* or PM. It is the amount by which the phase of $G(j\omega)$ exceeds $-180°$ when $|KG(j\omega)| = 1$, which is an alternate way of measuring the degree to which the stability conditions of Eq. (5.20) are met. From Fig. 5.12 we see that the phase margin (PM) is approximately 80° for the $K = 0.1$ case and $-35°$ for $K = 10$. A positive PM is required for stability. Note that the two stability measures, PM and GM, together determine how far that the complex quantity $KG(j\omega)$ passes from the -1 point, which is another way of stating the neutral stability point specified by Eq. (5.19).

In summary, the stability of a closed-loop system can be determined by examining the magnitude and phase of the open-loop transfer function. The stability criterion is given by Eq. (5.20) for systems where increasing gain eventually causes instability, the most common case. For systems that are unstable for low values of gain and become stable as the gain is raised past some critical point, the inequality in Eq. (5.20) needs to be reversed to establish a stability criterion. Resolution of this ambiguity can be accomplished by making a rough sketch of the system's root locus versus the gain K. Alternately, one can study a system's stability using the Nyquist criterion described in the next section, a criterion with no ambiguities. However, the Nyquist criterion is fairly complex and one can easily lose sight during its study that, for most systems, there is a simple relationship that exists between closed-loop stability and the open-loop frequency response. Study of the Nyquist criterion will allow the student to determine stability from the frequency response of complex systems, perhaps with one or more resonances, where the magnitude curve crosses 1 several times and/or the phase crosses $-180°$ several times.

5.5   The Nyquist Stability Criterion

For most systems an increasing gain eventually causes instability. Therefore, in the very early days of feedback control design, this relationship between gain and stability margins was assumed to be the correct one. However, designers would

FIGURE 5.13
Contour evaluations: (a) s-plane plot of poles and zeros of $H_1(s)$ and the contour C_1, (b) $H_1(s)$ for s on C_1, (c) s-plane plot of poles and zeros of $H_2(s)$ and the contour C_1, (d) $H_2(s)$ for s on C_1.

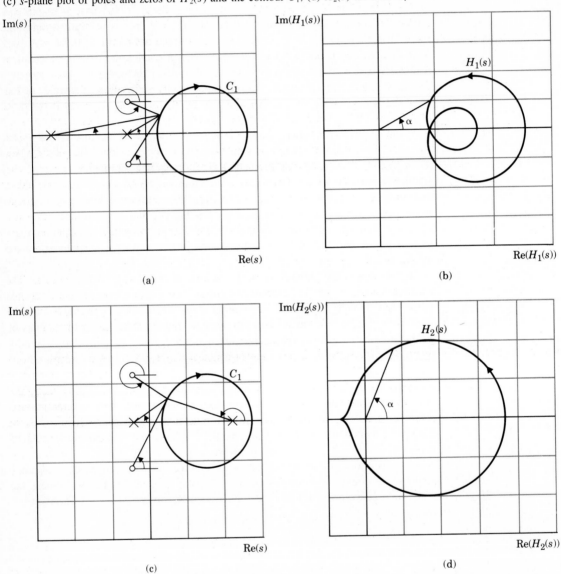

occasionally find in the laboratory that this relationship would reverse itself and an amplifier would go unstable if gain was decreased. This situation had caused some confusion and motivated Harry Nyquist of the Bell Telephone Laboratories to study the problem in 1932. His study explained the occasional reversals and resulted in a more sophisticated analysis with no loopholes. Not surprisingly, his test has come to be called the *Nyquist stability criterion*. The basic concept used in it is based on the argument principle, as explained below and in more detail in Appendix B.

Suppose the $H_1(s)$ whose poles and zeros are indicated in Fig. 5.13(a) is evaluated for s taking on the values of the contour C_1. The angle α of $H_1(s)$ in Fig. 5.13(b) will change some as s traverses C_1 but will not undergo a net change of $360°$ as long as there are no poles or zeros within C_1. This means that the plot of $H_1(s)$ [Fig. 5.13(b)] will not encircle the origin. This follows from the fact that α is the sum of the angles indicated in Fig. 5.13(a), and the only way that α can be changed by $360°$ after s executes one full traverse of C_1 is for C_1 to contain a pole or zero. $H_2(s)$ has a singularity within C_1 as shown in Fig. 5.13(c). Here, the angle from the pole within C_1 undergoes a net change of $360°$ after one full traverse of C_1, and therefore the argument of $H_2(s)$ undergoes the same change, causing H_2 to encircle the origin, as shown in Fig. 5.13(d). Thus we have the essence of the argument principle: *A contour evaluation of a complex function will only encircle the origin if the contour contains a singularity of the function.* The principle can be extended by allowing multiple singularities (poles or zeros) within the contour. The number and direction of origin encirclements then change. For example, if the number of poles and zeros within C_1 were the same, then there would be no encirclement of the origin.

To apply the principle to control design we let the C_1 contour in the s plane encircle the entire right-half plane (RHP). (See Fig. 5.14.) The resulting evaluation of an $H(s)$ will only encircle the origin if $H(s)$ has a right-hand plane pole or zero.

As stated already, a key idea that makes all this useful is that a contour evaluation of an open-loop $KG(s)$ can be used to determine stability of the closed-loop system. Specifically, for the system in Fig. 5.9, the closed-loop transfer function is

$$\frac{Y(s)}{R(s)} = F(s) = \frac{KG(s)}{1 + KG(s)}.$$

Therefore the closed-loop roots are the solutions of

$$1 + KG(s) = 0.$$

If the evaluation contour of s enclosing the RHP contains a zero or pole of $1 + KG(s)$, then the evaluated contour of $1 + KG(s)$ will encircle the origin. But $1 + KG(s)$ is simply $KG(s)$ shifted to the right by 1. (See Fig. 5.15.) Therefore, if $1 + KG(s)$ encircles the origin, $KG(s)$ will encircle -1 and we then know that

FIGURE 5.14
A s-plane plot of a C_1
contour that encircles the
entire right-hand plane.

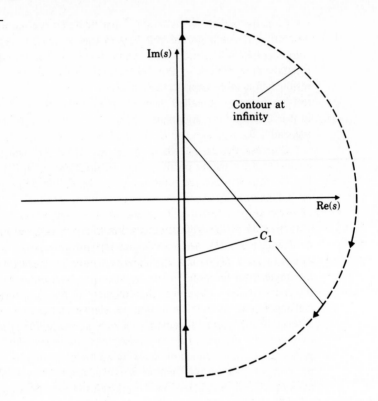

FIGURE 5.15
Evaluations of $KG(s)$
and $1 + KG(s)$: Nyquist
plots.

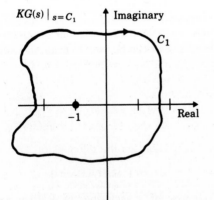

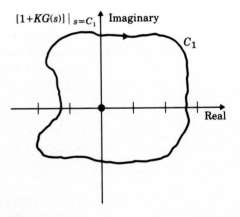

$1 + KG(s)$ contains a pole or zero. Presentation of the evaluation of $KG(s)$ in this manner is often referred to as a *Nyquist plot*.

To unscramble whether an encirclement is due to a pole or zero, we write $1 + KG(s)$ in terms of poles and zeros of $KG(s)$; that is,

$$1 + KG(s) = 1 + K\frac{b(s)}{a(s)} = \frac{a(s) + Kb(s)}{a(s)}, \tag{5.21}$$

and we see that the poles of $1 + KG(s)$ are also the poles of $G(s)$. Since the poles of $G(s)$ [or factors of $a(s)$] are known, the rare existence of any of these in the RHP can be accounted for. Assuming for now that there are no poles of $G(s)$ in the RHP, an encirclement of -1 by $KG(s)$ indicates an unstable root of the closed-loop system.

We can embellish this basic idea by noting that a clockwise C_1 contour enclosing a zero of $1 + KG(s)$—that is, closed-loop-system root—will result in a clockwise $KG(s)$ encirclement of -1. Likewise, if C_1 encloses a pole of $1 + KG(s)$—that is, if there is an unstable open-loop pole—there will be a counterclockwise $KG(s)$ encirclement of -1. Furthermore, if two poles or two zeros of $1 + KG(s)$ are in the RHP, $KG(s)$ will encircle -1 twice, and so on. The net number of clockwise encirclements, N, equals the number of zeros (closed-loop-system roots) in the RHP, Z, minus the number of poles in the RHP, P, that is,

$$N = Z - P.$$

A simplification results from the fact that any $KG(s)$ that represents a physical system will have zero response at infinite frequency. This means that the big arc of C_1 at infinity (Fig. 5.14) results in $KG(s)$ being an infinitesimal point at the origin for that portion of C_1. Therefore, a complete evaluation of a physical $KG(s)$ is accomplished by letting s traverse the imaginary axis from $-j\infty$ to $+j\infty$. The evaluation of $KG(s)$ from $s = 0$ to $+j\infty$ has already been discussed in Section 5.2 under the context of finding the frequency response of $KG(s)$. Because $KG(s)$ is a real function, the remainder of $KG(s)$, $s = -j\infty$ to 0, can be obtained by reflecting the $s = 0$ to $+j\infty$ portion about the real axis, and the entire $KG(s)$ plot is easily obtained. Hence we see that closed-loop stability can be determined in all cases by examination of the frequency response of the open-loop transfer function.

In practice, many systems behave like those discussed in the previous section, and designers need not carry out a complete evaluation of $KG(s)$ with subsequent inspection of the -1 encirclements. A simple look at the frequency response suffices to determine stability based on the gain and phase margins as discussed in the previous section. Alternatively, the stability can be defined in terms of the -1 encirclements, that is, the Nyquist stability criterion. In the case of a complex system where the rules become ambiguous, a designer will want to perform the complete analysis, the process for which can be summarized as follows:

1. Plot $KG(s)$ for $s = -j\infty$ to $+j\infty$. Do this by first evaluating points from $s = 0$ to $+j\infty$ (the frequency response), then reflecting the image about the real axis and adding it to the preceding image.

FIGURE 5.16
Definition of Example
5.4.

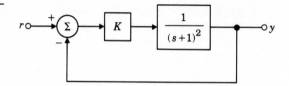

2. Evaluate the number of clockwise encirclements of -1 and call that number N. Do this by drawing a straight line in *any* direction from -1 to ∞. Then count the net number of clockwise crossings of the straight line by $G(s)$. If encirclements are in the counterclockwise direction, N is negative.

3. Determine the number of unstable (RHP) poles of $G(s)$ and call that number P.

4. The number of unstable closed-loop roots, called Z, is then

$$Z = N + P. \qquad (5.22)$$

Let's now follow a rigorous application of this procedure through some examples.

EXAMPLE 5.4   The first example is defined in Fig. 5.16. As can be seen from the root locus in Fig. 5.17, it is stable for all values of K. The magnitude of the frequency response of $KG(s)$ is plotted in Fig. 5.18(a) for $K = 1$, and the phase is plotted in Fig. 5.18(b); this is the typical method of presenting frequency response and represents

FIGURE 5.17
Root locus of Example
5.4 versus K.

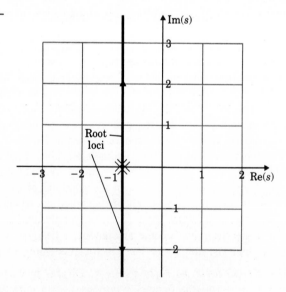

FIGURE 5.18
Open-loop frequency
response of Example 5.4:
(a) magnitude and
(b) phase.

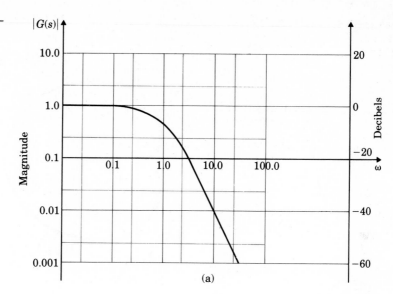

(a)

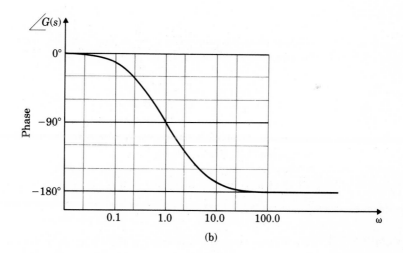

(b)

the evaluation of $G(s)$ from $s = 0$ to $+j\infty$. The same information is replotted in Fig. 5.19 in the Nyquist plot form. The information from Figs. 5.18(a) and (b) becomes the arc from $G(s) = +1$ ($\omega = 0$) to $G(s) = 0$ ($\omega = \infty$) that lies below the real axis. The portion of the C_1 arc at infinity shown in Fig. 5.14 transforms into $G(s) = 0$; therefore a continuous evaluation of $G(s)$ with s traversing C_1 is completed by simply reflecting the lower arc about the real

FIGURE 5.19
Nyquist plot of the
evaluation of $KG(s)$
for $s = C_1$ and
$K = 1$ in Example
5.4.

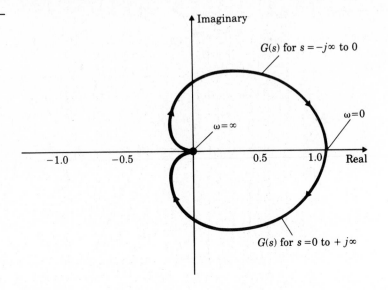

FIGURE 5.19
Nyquist plot of the
evaluation of $KG(s)$
for $s = C_1$ and
$K = 1$ in Example
5.4.

axis. This creates the portion of the contour above the real axis and completes the Nyquist plot. We note from the plot that it does not encircle the -1 point and, therefore, $N = 0$. Also, there are no poles of $G(s)$ in the RHP, thus, $P = 0$; therefore, $Z = 0$, indicating there are no unstable roots of the closed-loop system for $K = 1$.

Although we could replot $KG(s)$ for other values of K, it is more convenient to rescale the Nyquist plot and continue to examine $G(s)$ to determine stability for all values of K. This is possible because an encirclement of -1 by $KG(s)$ is equivalent to an encirclement of $-1/K$ by $G(s)$. For this example, $G(s)$ never crosses the negative real axis except at $G(s) = 0$; therefore it will never encircle $-1/K$ as long as $K > 0$. This result is identical to that shown by the root locus, as it should be! ∎

EXAMPLE 5.5   The second example is defined in Fig. 5.20. It is the same system discussed in Section 5.4, and its root locus in Fig. 5.11 shows that this system is stable for

FIGURE 5.20
Definition of Example
5.5.

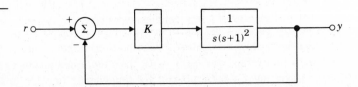

FIGURE 5.21
Frequency response of
Example 5.5:
(a) magnitude and
(b) phase.

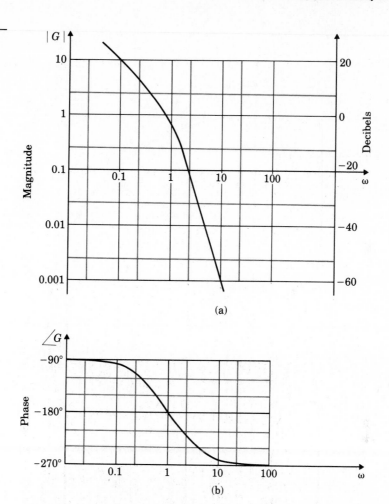

(a)

(b)

small values of K but unstable for large values of K. As explained in Example 5.4, $G(s)$ alone will be evaluated instead of $KG(s)$ to eliminate the need for repeated evaluations with different values of K. The magnitude and phase of $G(s)$ in Fig. 5.21 are transformed to the Nyquist plot shown in Fig. 5.22. It is interesting to note the large arc at infinity that arises from the open-loop pole at $s = 0$. This pole creates an infinite magnitude of $G(s)$ at $\omega = 0$, and, in fact, any pole or zero on the imaginary axis will create an arc at infinity. To correctly determine the number of -1 encirclements, it is necessary to draw the arc at infinity on the proper side. That is, should it cross the positive real axis, as shown in Fig. 5.22, or should it cross the negative real axis? It is also necessary to assess whether the arc should sweep out 180° (as seen in Fig. 5.22), 360°, or 540°.

FIGURE 5.22
Nyquist plot for Example 5.5.

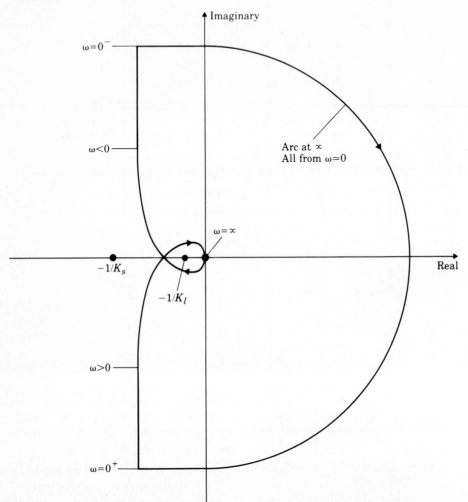

A simple artifice suffices to answer all these questions. Modify the C_1 contour to take a small detour around the pole. It makes no difference to the final stability question which way, but it is more convenient to go to the right, as shown in Fig. 5.23, because then no open-loop poles are introduced within the C_1 contour (which means the value of P remains equal to 0). Since the phase of $G(s)$ is the negative of the sum of the angles from all poles, it can be seen that the evaluation results in a Nyquist plot going from $+90°$ for s just below the pole at $s = 0$, across the

FIGURE 5.23
Description of C_1
contour enclosing RHP
for Example 5.5.

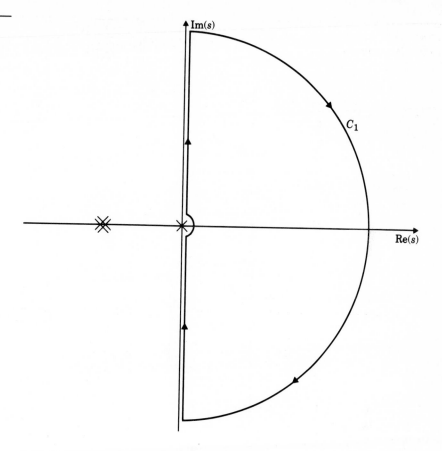

positive real axis to $-90°$ for s just above the pole. Had there been two poles at $s = 0$, the Nyquist plot at infinity would have executed a full $360°$ arc, and so on for three or more poles. Furthermore, for a pole elsewhere on the imaginary axis, a $180°$ clockwise arc will also result but will be oriented differently than the example in Fig. 5.22.

There are two possibilities for the location of $-1/K$—either inside the two loops of the Nyquist plot or outside the Nyquist contour completely. For large values of K (K_l in Fig. 5.22), $-1/K$ will lie inside the two loops; hence $N = 2$ and therefore $Z = 2$, which indicates two unstable roots. For small values of K (K_s in Fig. 5.22), $-1/K$ lies outside the loops, thus $N = 0$, $Z = 0$, and all roots are stable. Again, all this is in agreement with the root locus.

For the prior example system and many similar systems, it can be seen that the encirclement criterion reduces to a very simple test for stability based on the open-loop frequency response: The system is stable if $|KG(j\omega)| < 1$ when the phase of $G(j\omega)$ is $-180°$. This is identical to the stability criterion given in Eq. (5.20);

FIGURE 5.24
Definition of Example
5.6.

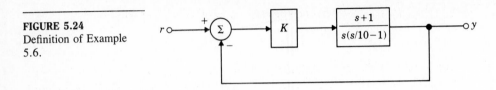

FIGURE 5.25
Root locus for Example 5.6.

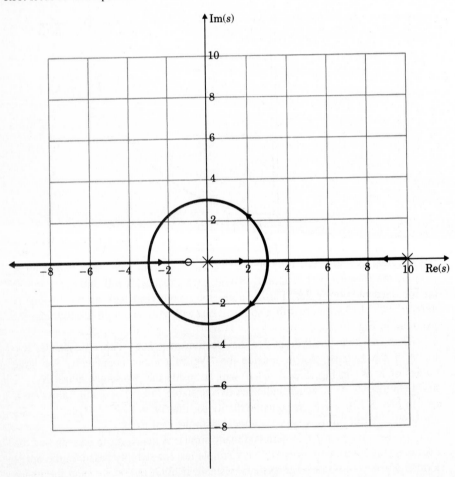

however, in this case we don't require the root locus to determine if $|KG(j\omega)|$ should be < 1 or > 1.   ■

EXAMPLE 5.6   The third example is defined in Fig. 5.24, and its root locus is sketched in Fig. 5.25. The open-loop magnitude and phase are shown in Fig. 5.26. By examining the $|KG(j\omega)|$ and $\angle G(j\omega)$ graphically, one can see that the pole being in the RHP causes the magnitude to behave just as if the pole were in the LHP, while it causes the phase to increase by 90° instead of the usual decrease at a pole. Any system with a pole in the RHP is unstable; hence it would be impossible to determine its frequency response experimentally because the system would never reach a steady

FIGURE 5.26
Frequency response
of Example 5.6:
(a) magnitude and
(b) phase.

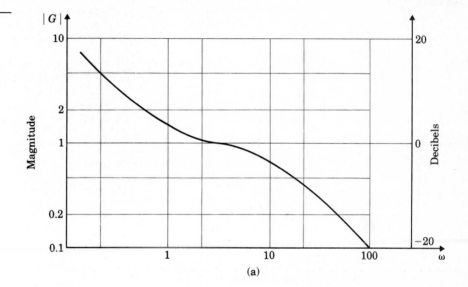

(a)

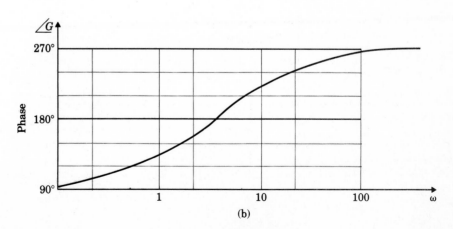

(b)

FIGURE 5.27
Nyquist plot for Example 5.6.

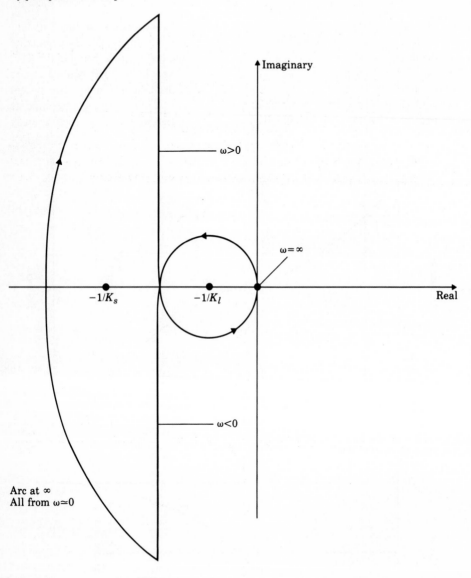

sinusoidal response to the sinusoidal input. It is, however, possible to compute the magnitude and phase of the transfer function according to the rules in Section 5.2. The pole in the RHP affects the Nyquist encirclement criterion because the value of P in Eq. (5.22) is $+1$.

Translation of the frequency-response information to the Nyquist plot (Fig. 5.27) is accomplished as in the previous examples; as before, the C_1 detour around the pole at $s = 0$ (Fig. 5.28) creates a large arc at infinity. This arc crosses the *negative* real axis because of the 180° phase contribution of the pole in the RHP.

For large values of K (K_l in Fig. 5.27), there is one counterclockwise encircle-ment; hence $N = -1$. But since $P = 1$ from the RHP pole, then $Z = N + P = 0$, and there are no unstable system roots. For small values of K (K_s in Fig. 5.27), $N = +1$ because of the clockwise encirclement, and $Z = 2$, indicating two unstable roots. These results agree with the root locus, as they should. The stability bound-ary occurs at $|KG(j\omega)| = 1$ for the phase of $\angle G(j\omega) = -180°$, as will always occur;

FIGURE 5.28
C_1 contour for Example 5.6.

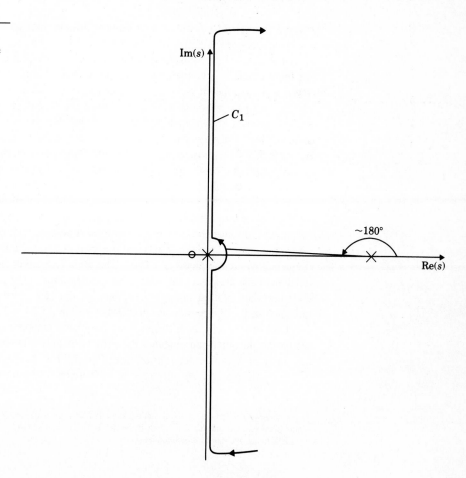

however, in this case, $|KG(j\omega)|$ must be > 1 for the correct number of -1 encirclements to achieve stability.   ∎

The existence of the RHP pole in the last example affected the Bode plotting rules of the phase curve and affected the relationship between encirclements and unstable closed-loop roots because $P = 1$ in Eq. (5.22). For systems with an RHP zero, referred to as a *nonminimum-phase* system,* the Bode phase plot *decreases* at the zero breakpoint instead of the usual phase increase that occurs for an LHP zero. With the exception of this difference in the phase plot, a nonminimum-phase zero has no effect on the Nyquist stability criterion.

5.6   Stability Margins

A large fraction of control-system designs behave in a pattern roughly similar to that of the example system in Section 5.4 and Example 5.5 in Section 5.5; similar in that the systems go unstable if the gain increases past a certain critical point. For these cases, it is convenient to define the gain margin (GM) and phase margin (PM) as was done in Section 5.4. Based on the ideas in Section 5.5, we see that the margins are measures of how close the Nyquist plot comes to encircling the -1 point as shown in Fig. 5.29. It shows that the *gain margin* is a measure of how much the gain can be raised before instability results for a system like Example 5.5. The *phase margin* is the difference between the phase of $G(j\omega)$ and $-180°$ when $|KG(j\omega)|$ crosses the magnitude $= 1$ circle, with the positive value assigned to the stable case (i.e., no Nyquist encirclements for a system like Example 5.5).

It is easiest to determine these margins directly from the magnitude and phase frequency-response plots. The phase margin is defined with respect to the frequency where the gain is unity, or 0 dB, and the gain margin is defined with respect to the frequency where the phase is $-180°$. The term "crossover frequency" is often used to refer to the frequency at which the gain is unity, or 0 dB. Figure 5.30 shows the same data plotted in Fig. 5.21, but for the case with $K = 1$. The margins indicated in the figure indicate that PM $= 22°$ and GM $= 2$.

One of the very useful aspects of frequency-response design is the ease with which the designer can evaluate the effects of gain changes. In fact, the PM for any value of K can be determined from Fig. 5.30 without redrawing the magnitude or phase information. We only need to indicate on the figure where $|KG(j\omega)|$ would

---

*A system with a zero in the RHP undergoes a net change in phase, when evaluated for frequency inputs between zero and infinity which is greater for a given magnitude plot than if all zeros were in the LHP; hence the name, *nonminimum phase*.

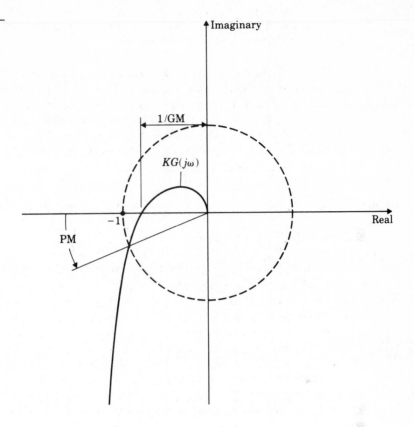

FIGURE 5.29
Nyquist plot for GM and
PM definition.

equal 1 for selected trial values of K. This is done by the dashed lines in Fig. 5.31, and we see that $K = 5$ yields an unstable PM of $-22°$ while a gain of $K = 0.5$ yields a PM of $+45°$. Furthermore, if you wish a certain PM, say $70°$, read the value of $|G(j\omega)|$ that is at the frequency that would create the desired PM ($\omega = 0.2$ rad/s yields $70°$, where $|G(j\omega)| = 5$), and note that the magnitude at this frequency is $1/K$. Therefore, Fig. 5.31 shows that a PM of $70°$ will be achieved with $K = 0.2$.

Both margins offer a measure of the degree of stability and are sometimes used directly to specify control-system performance. It is therefore of interest to relate the margins to other measures of the degree of stability. For simplicity, let's take an open-loop second-order system,

$$G(s) = \frac{\omega_n^2}{s(s + 2\zeta\omega_n)}.$$

FIGURE 5.30
GM and PM from the
magnitude and phase
plots.

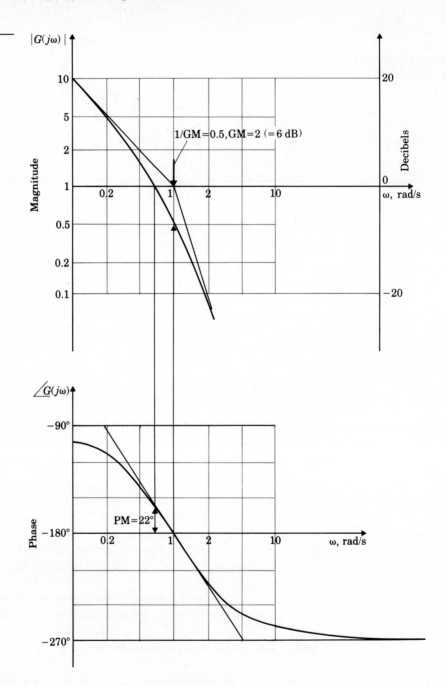

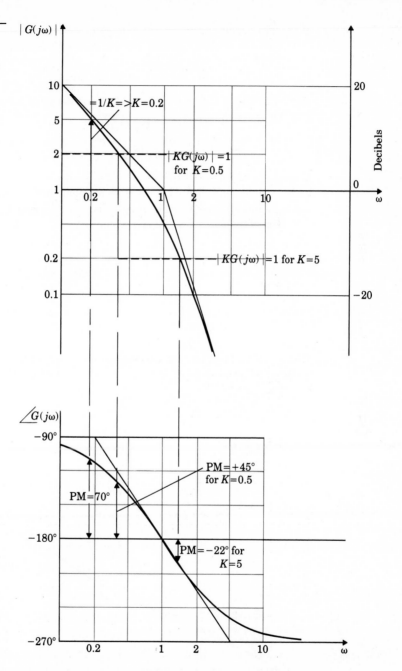

FIGURE 5.31
PM versus K from the frequency-response data.

With unity feedback, the closed-loop system

$$F(s) = \frac{\omega_n^2}{s^2 + 2\zeta\omega_n s + \omega_n^2} \tag{5.23}$$

results. It can be shown (Problem 5.21) that the relationship between the PM and ζ in the system above is

$$PM = \tan^{-1}\left(2\zeta \bigg/ \sqrt{\sqrt{1 + 4\zeta^4} - 2\zeta^2}\right) \tag{5.24}$$

and it is plotted in Fig. 5.32. Note from the figure that the function is approximately a straight line up to a PM of about 60°. The dashed line in the figure shows a straight line approximation to the function where

$$\zeta \cong \frac{PM}{100}. \tag{5.25}$$

It is clear that the approximation only holds for phase margins below about 70°. Furthermore, Eq. (5.24) is only accurate for the second-order system of Eq. (5.23). In spite of these limitations, the relationship in Eq. (5.25) is often used as a rule of thumb in relating the degree of stability in terms of PM. It is useful as a starting point; however, the designer should always check the actual damping of a design as well as other aspects of the performance before calling the design complete.

FIGURE 5.32
Damping ratio versus
phase margin (PM).

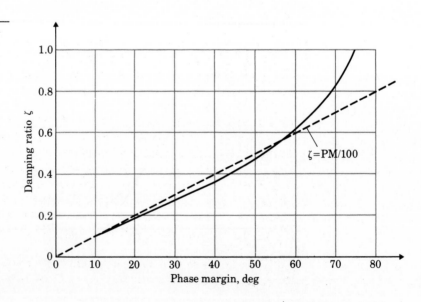

FIGURE 5.33
Transient response over-
shoot and frequency
response: resonant peak
versus phase margin
(PM) for second-order
system.

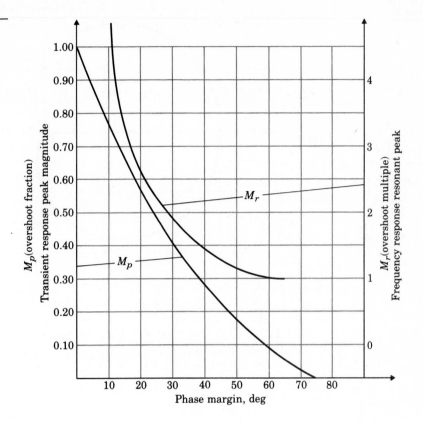

By using the relation between M_r and ζ seen in Fig. 5.5, which was derived for the same system (Eq. 5.23), we can transform the information in Fig. 5.32 to relate M_r to the PM. This is depicted in Fig. 5.33, along with the step response overshoot M_p, and provides additional rule-of-thumb data to aid in the evaluation of a control system based on its PM.

Many engineers think directly in terms of PM in judging whether a control system is adequately stabilized. In using PM this way, a phase margin of 30° is often judged to be the lowest acceptable PM, with values above 30° desirable. In addition to being assured that a system design is sufficiently stable via the PM, a designer typically would also be concerned with meeting some kind of "speed of response" type of specification such as bandwidth as discussed in Section 5.3. In terms of the frequency-response parameters discussed so far, the *crossover frequency* would best describe a system's speed of response. This idea will be discussed further in Sections 5.7 and 5.8.

In some cases, the PM and GM notions break down. For first- and second-order systems, the phase never crosses the −180° line; hence the GM is always ∞

FIGURE 5.34
Example of system where increasing gain leads from instability to stability: (a) root locus and (b) Nyquist plot.

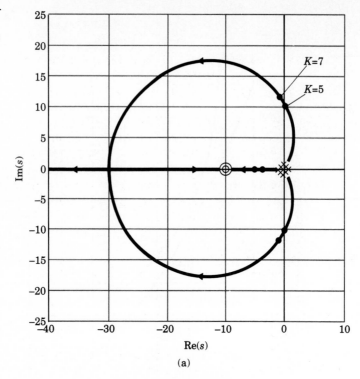

(a)

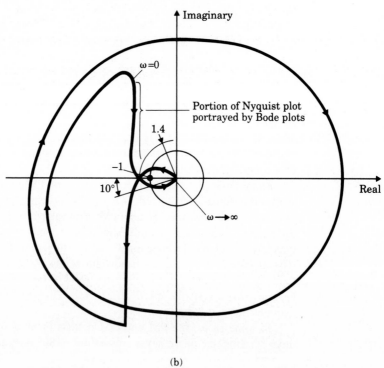

(b)

and not a useful design parameter. For higher-order systems, it is possible to have more than one frequency where $|KG(j\omega)| = 1$ or where $\angle KG(j\omega) = -180°$, and the margins as previously defined would need clarification. An example of this can be seen in Fig. 7.13 where the magnitude crosses 1 three times. The figure indicates that the PM is defined by the first crossing, a decision that was made because the PM at this crossing was the minimum of the values and thus the most conservative assessment of stability. A Nyquist plot based on the data in Fig. 7.13 would show that the portion of the Nyquist curve closest to the -1 point is the critical indicator of stability and, therefore, use of the crossover frequency yielding the minimum value of PM is the logical choice.

EXAMPLE 5.7   An example of a system where increasing gain causes a transition from instability to stability has an open-loop transfer function

$$G(s) = \frac{K(s + 10)^2}{s^3}.$$

The root locus in Fig. 5.34(a) shows that the system is unstable for $K < 5$ and stable for $K > 5$. The Nyquist plot in Fig. 5.34(b) was drawn with $K = 7$, which is a stable value. Determination of the margins according to Fig. 5.29 yields a PM of $+10°$ and a GM of 0.7. According to the rules above, these two margins yield conflicting signals on the system's stability. To resolve the issue, a count of the Nyquist encirclements show that the $K = 7$ case drawn has one clockwise and one counterclockwise encirclement of the -1 point, hence no net encirclements, and the system is confirmed to be stable. A system such as this, where for stability the gain must be higher than some value, is referred to as a *conditionally* stable system.   ∎

EXAMPLE 5.8   A Nyquist plot for

$$G(s) = \frac{85(s + 1)(s + 1 \pm 6.5j)}{s^2(s + 1 \pm 9j)(s + 1 \pm 10j)}$$

is shown in Fig. 5.35. It shows a situation where there are two crossover frequencies with approximately equal values of PM, but the key indicator of stability is the proximity of the Nyquist plot as it approaches the -1 point while crossing the real axis. In this case, the most relevant margin to use for design as a measure of the degree of stability would be the GM because it alone indicates the marginal stability of this system.   ∎

FIGURE 5.35
Nyquist plot of a
complex system.

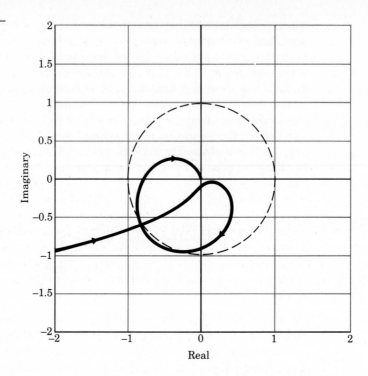

FIGURE 5.36
Definition of the vector
margin on the Nyquist
plot.

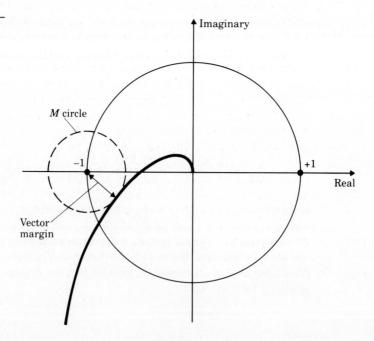

Unstable open-loop systems (see the example in Fig. 5.27) exhibit stability criteria that are different than those defined by Fig. 5.29 because there needs to be one -1 encirclement for each unstable pole in order for the closed-loop system to be stable. The GM and PM as previously defined can be misleading and should therefore be modified by reverting back to the Nyquist stability criterion.

It should be clear that a designer needs to be judicious in the application of the margin definitions as described in Fig. 5.29. The stability margin of a system can only be rigorously assessed by examining the Nyquist plot encirclements and by determining the plot's closest approach to the -1 point. To aid in this analysis, a *vector* margin was introduced by O. J. M. Smith (1958) that was defined to be the distance to the -1 point from the closest approach of the Nyquist plot. Figure 5.36 shows the idea graphically. Although use of this margin definition would remove all ambiguities in the assessment of stability, it is more difficult to compute than phase and gain margins and has not been used extensively in design over the years. With the widespread use of computer-aided calculations, the idea of having a single margin parameter to describe the degree of stability is quite feasible.

□ 5.6.1 Multivariable Systems

Multivariable systems, where there are more than one input and/or more than one output (MIMO systems), have posed difficulties for designers because there are so many aspects of the design that all apparently need attention. This text has dealt with closing one loop and has considered aspects of stability, tracking errors, and response characteristics of that one loop. For MIMO systems, the same considerations need to be addressed for *each* loop; however, it is not always straightforward to identify a "loop" because there are transfer functions from each input to all outputs. For example, for a 3-input, 3-output system, there are 9 transfer functions between inputs and outputs.

The *singular value* decomposition of the matrix of transfer functions evaluated for $s = j\omega$ has proven useful in resolving the complexities of MIMO design (Doyle and Stein, 1981; Safonov, Laub, and Hartmann, 1981). The essential idea is to reduce the matrix of transfer functions to one critical gain versus frequency, that is, the maximum singular value of the matrix, $\bar{\sigma}$, and to examine that quantity in a manner similar to the one used to examine the frequency-response magnitude of a single-input, single-output (SISO) transfer function G. Stability margins in terms of $\bar{\sigma}$ have been defined for the MIMO case following the stability margin ideas presented in this Section 5.6 for the SISO case. Section 5.11 discusses the

FIGURE 5.37
An approximate gain-
phase relationship
demonstration.

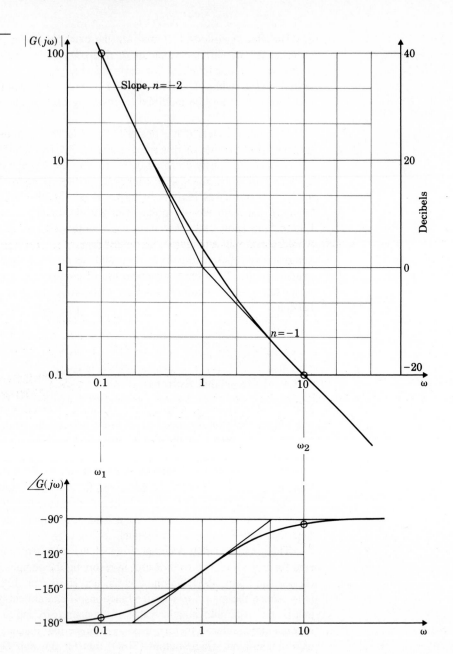

sensitivity of control systems and refines the MIMO design ideas introduced here. A discussion and definition of the singular value decomposition of a matrix is contained in Appendix C.

5.7 Bode's Gain-Phase Relationship

One of Bode's important contributions is his theorem that states: "For any stable minimum phase system (meaning one with no RHP zeros *or poles*), the phase of $G(j\omega)$ is uniquely related to the magnitude of $G(j\omega)$."

When the slope of $|G(j\omega)|$ versus ω on a log-log scale persists at a constant value for something close to a decade of frequency, the relationship is particularly simple:

$$\angle G(j\omega) \cong n \times 90°, \tag{5.26}$$

where n is the slope of $|G(j\omega)|$ in units of decade of amplitude per decade of frequency. For example, in considering the magnitude curve alone in Fig. 5.37, it is seen that Eq. (5.26) can be applied to the two points shown as ω_1 and ω_2, which are a decade removed from the change in slope, to yield the approximate values of phase, $-180°$ and $-90°$. The exact phase curve shown in the figure verifies that, indeed, the approximation is quite good. It also shows that the approximation will degrade if the evaluation is performed at frequencies closer to the changes in slope.

An exact statement of the Bode gain-phase theorem is

$$\angle G(j\omega_0) = \frac{1}{\pi} \int_{-\infty}^{+\infty} \left(\frac{dM}{du} \right) W(u)\, du \text{ radians}, \tag{5.27}$$

where

$$M = \text{log magnitude}, \ln|G(j\omega)|,$$
$$u = \text{log normalized frequency} = \ln(\omega/\omega_0),$$
$$dM/du \cong \text{slope } n, \text{ as defined in Eq.(5.26)},$$
$$W(u) = \text{weighting function} = \ln(\coth|u|/2).$$

Figure 5.38 is a plot of the weighting function $W(u)$ and shows how the phase is most dependent on the slope at ω_0: It is also dependent, though to a lesser degree, on slopes at neighboring frequencies. The figure also suggests that the weighting could be approximated by an impulse function centered at ω_0, which is precisely the approximation made that arrives at Eq. (5.26).

FIGURE 5.38
Weighting function in
Bode's gain-phase
theorem. (From Clark,
1962.)

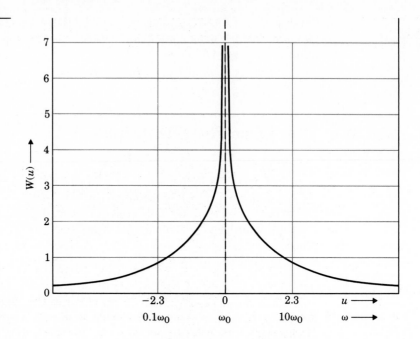

FIGURE 5.38
Weighting function in
Bode's gain-phase
theorem. (From Clark,
1962.)

In practice, Eq. (5.27) is never used. But the approximate relation in Eq. (5.26) is used as a guide to infer stability from $|G(j\omega)|$ alone. When $|KG(j\omega)| = 1$,

$$\angle G(j\omega) \cong -90° \qquad \text{if } n = -1,$$
$$\angle G(j\omega) \cong -180° \qquad \text{if } n = -2.$$

For stability we want $\angle G(j\omega) > -180°$ for a PM > 0. Therefore we adjust the $|KG(j\omega)|$ curve so that it has a slope of -1 at the $|KG(j\omega)| = 1$ crossover frequency. If the slope is -1 for a decade above and below the crossover frequency, the PM would be approximately 90°; however, to ensure a reasonable PM, it is usually only necessary to insist on a -1 slope (-20 dB per decade) persisting for a decade in frequency that is centered at the crossover frequency.

EXAMPLE 5.9   To illustrate the power of this idea, let's apply it to the design of compensation for the spacecraft attitude-control problem. We wish to find a suitable $D(s)$ in Fig. 5.39 that will provide good damping and a bandwidth of approximately 0.2 rad/s. The magnitude of the frequency response of the spacecraft, $|G(j\omega)|$, is plotted in

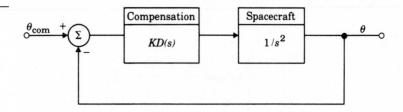

Fig. 5.40 and clearly requires some reshaping since it has a slope of -2 or -40 dB per decade everywhere. The simplest compensation to do the job consists of using proportional-plus-derivative terms, which produces

$$KD(s) = K(T_D s + 1).$$

The gain K will be adjusted to produce the desired bandwidth, while the break-point $\omega_1 = 1/T_D$ will be adjusted to provide the -1 or -20 dB per decade slope at crossover. For design purposes, it is convenient to assume that the crossover frequency and the closed-loop system bandwidth are the same, although we will need to verify this when we are through. The actual design process to achieve the desired specifications is now very simple: Pick K to provide a crossover at 0.2 rad/s and pick ω_1 about a factor of 4 lower than the crossover frequency so that

FIGURE 5.40
Spacecraft frequency-response magnitude.

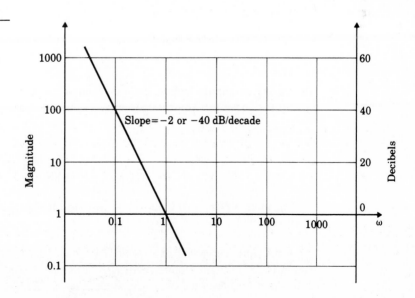

FIGURE 5.41
Compensated open-loop
transfer function.

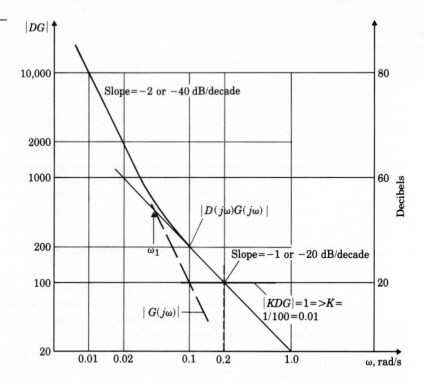

the slope is -1 in the vicinity of crossover. Figure 5.41 shows the steps taken to arrive at the final compensation:

1. Plot $|G(j\omega)|$.
2. Modify the plot to include $|D(j\omega)|$, with $\omega_1 = 0.05$ rad/s ($T_D = 20$).
3. Find that $|DG| = 100$ where it crosses $\omega = 0.2$ rad/s.
4. Compute

$$K = \frac{1}{|DG|_{\omega=0.2}} = \frac{1}{100} = 0.01.$$

Therefore $KD(s) = 0.01(20s + 1)$ will meet the specifications and the design is complete.

If the phase curve is drawn for the DG above, it will be found that the phase margin is $75°$, certainly quite adequate. Furthermore, Fig. 5.42 is a plot of the closed-loop frequency-response magnitude, that is, $|KDG/(1 + KDG)|$. It shows

FIGURE 5.42
Closed-loop frequency
response.

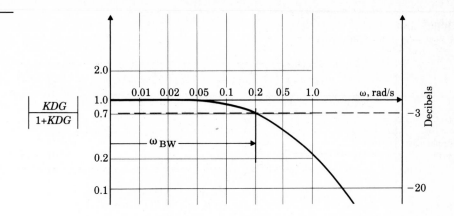

FIGURE 5.42
Closed-loop frequency
response.

that, indeed, the crossover frequency and the bandwidth are almost identical in this case. ∎

5.8 Closed-Loop Frequency Response

Generally the match between the crossover frequency and the bandwidth is not as good as in Example 5.9. We can help establish a more exact correspondence by making a few observations. For the simplified system shown in Fig. 5.43, where $|G(j\omega)|$ has the typical property that

$$|G(j\omega)| \gg 1 \qquad \text{for } \omega \ll \omega_c,$$
$$|G(j\omega)| \ll 1 \qquad \text{for } \omega \gg \omega_c,$$

where ω_c = crossover frequency, the closed-loop frequency-response magnitude is approximated by

$$|F| = \left| \frac{G(j\omega)}{1 + G(j\omega)} \right| \begin{cases} \cong 1 & \text{for } \omega \ll \omega_c \\ \cong |G| & \text{for } \omega \gg \omega_c. \end{cases}$$

In the vicinity of crossover where $|G(j\omega)| \cong 1$, $|F|$ depends heavily on the PM. One of 90° means that $\angle G(j\omega_c) = -90°$, and therefore $|F(j\omega_c)| = 0.707$. A

FIGURE 5.43
Simplified system
definition.

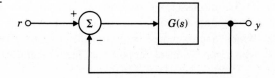

FIGURE 5.44
Closed-loop bandwidth versus PM.

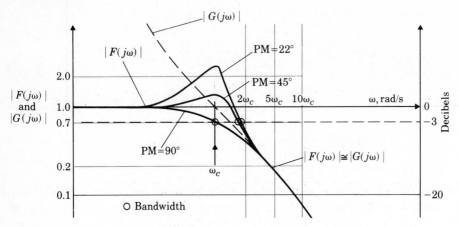

PM of 45°, on the other hand, yields $|F(j\omega_c)| = 1.31$. In fact, the values of M_r plotted in Fig. 5.33 will usually be quite close to $|F(j\omega_c)|$.* These approximations are used to generate the curves of $|F(j\omega)|$ in Fig. 5.44. It shows that the bandwidth for smaller PMs is typically somewhat greater than ω_c though usually less than $2\omega_c$.

5.9 Compensation

As already discussed in Chapters 3 and 4, dynamic elements (or compensation) are often added to feedback control systems to improve their stability or error characteristics. Here, we wish to discuss compensation again in terms of its frequency-response characteristics.

It is desirable to have the ability to analyze compensation in terms of its effect on the frequency response of a system as well as its effect on a root locus as we did in Chapter 4. This ability greatly aids the designer in reaching the best compromise in selecting the parameters of the compensation and in gaining insight into the ramifications of alternate compensation designs.

Section 3.2 discusses the basic types of feedback: proportional, derivative, and integral. Section 4.5 discusses the ideas of dynamic compensation: the lead network, which approximates proportional-plus-derivative feedback and the lag network, which approximates proportional-plus-integral control.

*From the geometry of the Nyquist plot, it is straightforward to show that $|F(j\omega_c)| = 1/(2\sin(\text{PM}/2))$ for any phase margin (see Problem 5.33).

5.9.1 PD Compensation

We will start the discussion of compensation design via the frequency response with proportional-plus-derivative control. Its transfer function is given by

$$D(s) = K(T_D s + 1) \tag{5.28}$$

and it was shown in Fig. 4.14 to have a stabilizing effect on the root locus of a second-order system. The frequency-response characteristics of Eq. (5.28) are shown in Fig. 5.45. A stabilizing influence is apparent by the increase in phase at frequencies above the breakpoint $1/T_D$. This compensation is used by locating $1/T_D$

FIGURE 5.45
Frequency response of proportional-plus-derivative control.

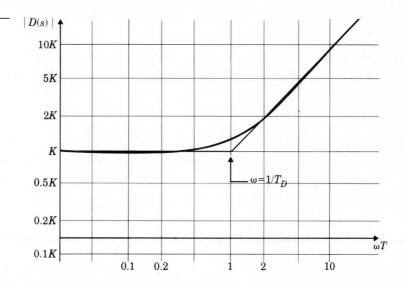

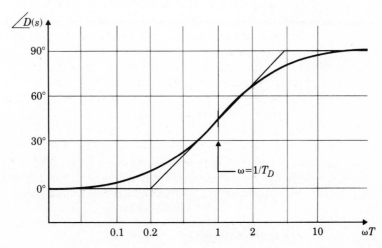

FIGURE 5.46
Lead-compensation
frequency response
with $1/\alpha = 10$.

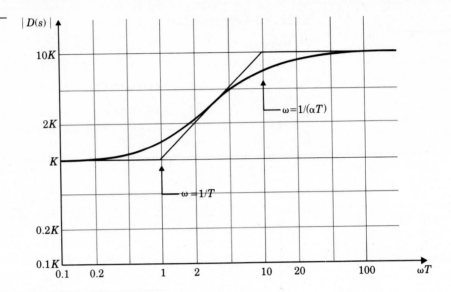

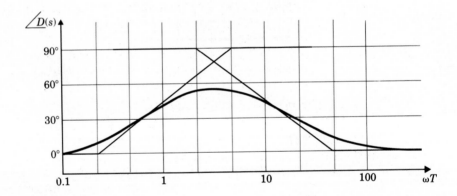

FIGURE 5.46
Lead-compensation
frequency response
with $1/\alpha = 10$.

so that the increased phase occurs in the vicinity of crossover $[|D(s)G(s)| = 1]$, thus increasing the phase margin. Also note that the magnitude of the compensation continues to grow with increasing frequency. This feature is undesirable since it amplifies the high-frequency noise that is typically present in any real system. It is also the reason for the statement in Section 4.5 that this pure derivative form of compensation gives trouble.

5.9.2 Lead Compensation

In order to alleviate the high-frequency amplification and the associated noise sensitivity of PD compensation, a first-order pole is introduced in the denominator at frequencies much higher than the numerator breakpoint. Thus the phase increase (or "lead") still occurs, but the amplification at high frequencies is limited. The resulting *lead compensation* has a transfer function of

$$D(s) = K \frac{Ts + 1}{\alpha Ts + 1},$$ (5.29)

where $\alpha < 1$. Figure 5.46 shows the frequency response of this lead compensation. Note that a significant amount of phase lead is still provided, but with much less amplification at high frequencies. The compensation shown in Fig. 5.46 produces the maximum amount of phase lead midway between the breakpoints, $1/T$ and $1/\alpha T$, on the log scale. This will be the case for any lead compensation of the form seen in Eq. (5.29). If two points on a log scale lie at ω_1 and $k\omega_1$, midway between them is at $\sqrt{k}\omega_1$. Thus we see that the maximum phase lead occurs at

$$\omega = \frac{1}{\sqrt{\alpha}T}.$$ (5.30)

For example, a lead compensator with a zero at $s = 2$ ($T = 0.5$) and a pole at $s = 10$ ($\alpha T = 0.1$) would yield the maximum phase lead at $s = 4.47$. The amount of phase lead at the midpoint depends only on α and is plotted in Fig. 5.47. It shows that a designer can achieve increasing phase lead with higher values of $1/\alpha$, sometimes referred to as the *lead ratio;* however, Fig. 5.46 shows that increasing values of $1/\alpha$ also produce greater amplifications at higher frequencies. Thus the designer's task is to select a value of $1/\alpha$ that is a good compromise between acceptable phase margin and noise sensitivity at high frequencies.

Even if a system had negligible amounts of noise present and the pure derivative compensation of Eq. (5.28) were acceptable, the actual compensation would look more like Eq. (5.29) than Eq. (5.28) because of the impossibility of building a *pure* differentiator. No physical system, mechanical or electrical, responds with anything like infinite amplitude at infinite frequencies and there will be a limit in the frequency range (or bandwidth) for which derivative information (or phase lead) can be provided.

FIGURE 5.47
Maximum phase increase
for lead compensation.

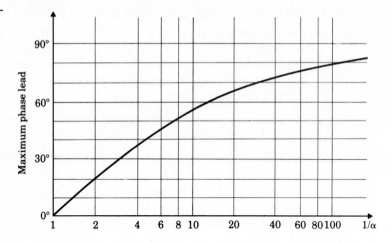

EXAMPLE 5.10   As an example of design using frequency response, let's repeat the design of
compensation for

$$G(s) = \frac{1}{s(s + 1)}$$

that was carried out in Section 4.5. This time we wish to obtain a steady-state error
of less than 0.1 for a unit ramp input. Furthermore, we desire an overshoot (M_p)
of less than 25%.

The steady-state error is given by

$$e(\infty) = \lim_{s \to 0} s \left[\frac{1}{1 + D(s)G(s)} \right] R(s),$$

where $R(s) = 1/s^2$ for a unit ramp, which reduces to

$$e(\infty) = \lim_{s \to 0} \left\{ \frac{1}{s + D(s)[1/(s + 1)]} \right\} = \frac{1}{D(0)}.$$

Therefore we find that $D(0)$, the steady-state gain of the compensation, must not
be less than 10 to meet the error criterion. So let's pick $K = 10$ in Eq. (5.29). To
relate the overshoot requirement to phase margin, Fig. 5.33 shows that a PM of 45°
should suffice. The frequency response of $KG(s)$ in Fig. 5.48(a) shows that a PM of
20° results if no phase lead is added by compensation. If it were possible to simply

FIGURE 5.48
Example of lead-compensation design: (a) frequency response.

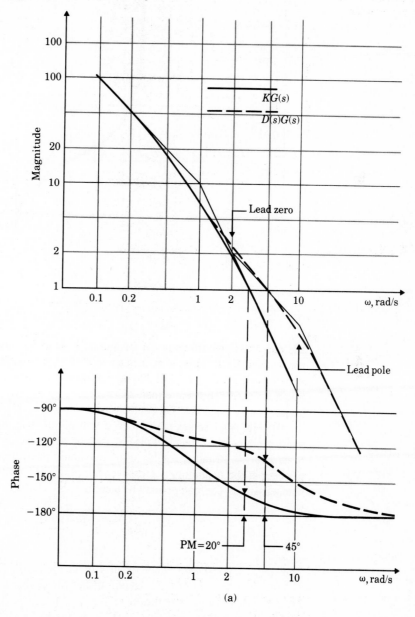

(a)

FIGURE 5.48 *(cont.)*
Example of lead-
compensation design:
(b) root locus.

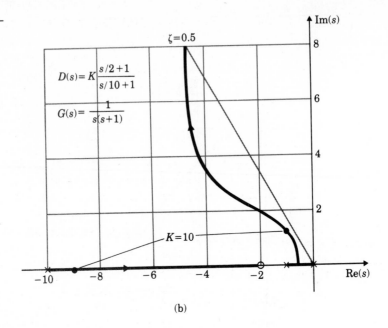

(b)

add phase without affecting the magnitude, we would only need an additional 25°
phase at the $KG(s)$ crossover frequency of $\omega = 3$ rad/s. However, maintaining the
same low-frequency gain and adding a compensator zero will increase the crossover
frequency, and hence more than 25° will be required from the lead network. So to
be safe, let's design the lead compensator so that it supplies a maximum phase lead
of 40°. Fig. 5.47 shows that $1/\alpha = 5$ will accomplish that goal. With some trial
and error, it can be found that placing the zero at $\omega = 2$ rad/s, thus causing the pole
to be at $\omega = 10$ rad/s, results in the maximum phase lead from the compensator
to occur at the magnitude 1 crossover frequency and maximizes the benefit of the
compensation. The compensation, therefore, is

$$D(s) = 10\frac{(s/2) + 1}{(s/10) + 1}.$$

The frequency-response characteristics of $D(s)G(s)$ in Fig. 5.48 can be seen to
yield a phase margin of 45°, which satisfies the design goals. Recall that the root
locus for this design was drawn in Fig. 4.15(b). The figure is repeated as Fig.
5.48(b) with the root locations for $K = 10$ marked and it verifies that this choice
of compensation parameters yields the desired damping ratio.   ∎

A summary of the design procedure for Example 5.10 is: (1) determine the low-frequency gain so that the steady-state errors are within specification, and (2) select the combination of lead ratio $1/\alpha$ and zero location that achieves an acceptable PM at crossover. This design procedure will apply to many cases; however, the designer should keep in mind that the specific procedure followed in any particular design may need to be tailored to its particular set of specifications.

In the design example, there were two specifications: peak overshoot and steady-state error. We transformed the overshoot specification to PM in order to use it directly in the frequency-response design. The steady-state error specification was usable directly. No speed-of-response type of specification was given; however, it would have impacted the design in the same way that the steady-state error specification did. The speed of response or bandwidth of a system is directly related to the crossover frequency as was pointed out in Section 5.8. Fig. 5.48(a) shows that the crossover frequency is 3 rad/s and could have been increased by raising the gain K and increasing the frequency of the lead compensator pole and zero in order to keep the slope $= -1$ at the crossover frequency. Raising the gain would also have had the effect of decreasing the steady-state error so that it was less than the specified limit. The gain margin was never introduced into the problem because the stability was adequately specified by the phase margin alone. Furthermore, the gain margin is not useful for this system because the phase never crosses the $-180°$ line and the GM is always infinite.

5.9.3 PI Compensation

Let's now turn to a discussion of the design of proportional-plus-integral control via its frequency response. It has a transfer function of

$$D(s) = \frac{K}{s}\left(s + \frac{1}{T_I}\right) \qquad (5.31)$$

which results in the frequency-response characteristics shown in Fig. 5.49. The desirable aspect of this compensation is the infinite gain at zero frequency, which causes the steady-state error in many cases to vanish. This is accomplished, however, at the cost of a phase decrease below the breakpoint at $\omega = 1/T_I$. Therefore $1/T_I$ is usually located at a frequency substantially less than the crossover frequency so that the system phase margin is not affected very much.

5.9.4 Lag Compensation

Another similar compensation, an approximation to proportional-plus-integral, is called *lag compensation*. Its transfer function is

$$D(s) = K\frac{Ts + 1}{\alpha Ts + 1}. \qquad (5.32)$$

FIGURE 5.49
Frequency response of
proportional-plus-integral
control.

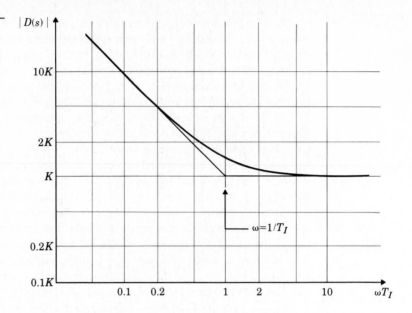

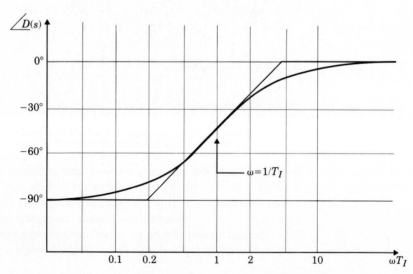

where $\alpha > 1$. Although this looks identical to the lead compensation in Eq. (5.29), the fact that $\alpha > 1$ causes the pole to have a lower breakpoint frequency than the zero, thus producing the low-frequency increase in amplitude and phase decrease (lag) apparent in the frequency-response plot in Fig. 5.50 and giving the compensation the essential feature of integral control: a high low-frequency gain.

FIGURE 5.50
Frequency response of
lag compensation with
$\alpha = 10$.

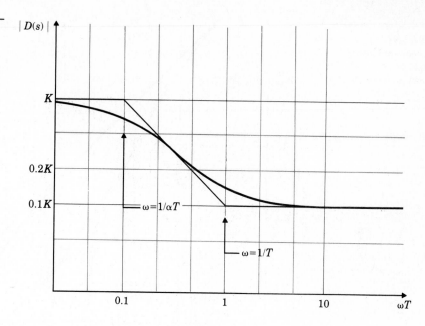

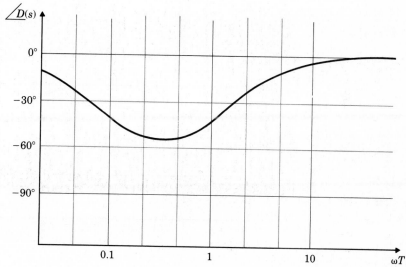

EXAMPLE 5.11   Now let's repeat the same design example once more using lag compensation. The frequency response of the system $KG(s)$, with the required steady-state gain of $K = 10$, is shown in Fig. 5.51. The designer's task is to select the lag compensation breakpoints so that the crossover frequency is lowered and a more favorable phase margin results. To prevent detrimental effects from the compensation phase

FIGURE 5.51
Frequency-response
example of lag-
compensation design.

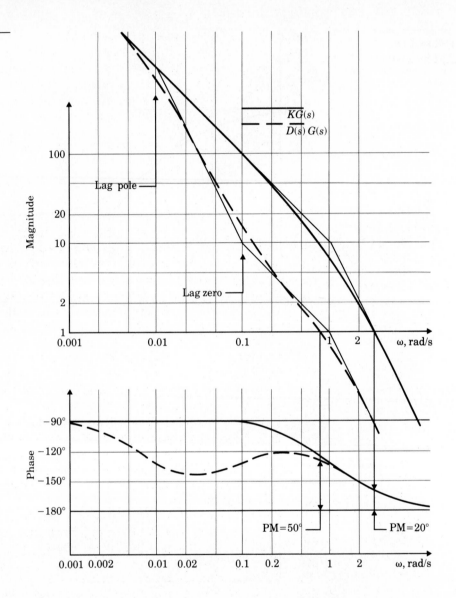

lag, the pole and zero of the compensation need to be substantially lower than the new crossover frequency. One possible choice is shown in Fig. 5.51: The lag zero is at 0.1 rad/s and the lag pole is at 0.01 rad/s. This selection of parameters produces a phase margin of 50°, thus satisfying the specifications. It can be seen from this example that the stabilization is achieved by lowering the crossover frequency to a region where $G(s)$ had more favorable phase characteristics. Furthermore, the reduction in the crossover frequency is dependent on the ratio (α) of the compensation zero breakpoint to the pole breakpoint and not on the absolute location of the pole and zero. The criterion for the pole and zero location selection is to make it low enough to minimize the effects of the phase lag from the compensation at the crossover frequency. Generally, however, the pole and zero are located no lower than necessary because the additional system root (see the root locus of a similar system design in Fig. 4.17(b) introduced by the lag will be in the same frequency range as the compensation zero and will have some effect on the output. ∎

One can interpret the beneficial effects of lag compensation in two ways: It reduces high-frequency gain for better phase margin, or it increases low-frequency gain for better error characteristics. The example design shown in Fig. 5.51 was illustrative of the first interpretation. However, given the same system with proportional feedback $K = 1$ so that the uncompensated frequency-response magnitude would fall on the dashed line in Fig. 5.51 for $\omega \geq 0.5$ rad/s, and told to meet the error specifications while preserving stability and bandwidth, a designer could accomplish the task by introducing the lag compensation shown in Fig. 5.51. This would simply increase the low-frequency gain by a factor of 10 with essentially no change in gain or phase at crossover. Thus we have an illustration of the second interpretation.

The last two design examples (shown in Figs. 5.48 and 5.51) meet an identical set of specifications for the same plant in very different ways. In the first case the specifications are met through the use of a lead network; a crossover frequency of 5 rad/s ($\omega_{BW} \cong 6$ rad/s) results. In the second case the same specifications are met through the use of a lag network, and a crossover frequency of approximately 0.8 rad/s ($\omega_{BW} \cong 1$ rad/s) results. If there was a rise-time or bandwidth specification, the choice of compensation might have been influenced by that specification; however, in this particularly simple example, any bandwidth or steady-state error specification could be met by simply raising the gain sufficiently high using either type of compensation.

In more realistic systems, typically there are dynamic elements representing the actuator and sensor as well as the system itself, and it is typically impossible to raise the crossover frequency much beyond the value that represents the speed of response of the components being used. Although linear analysis seems to suggest that almost any system can be "compensated," in fact, if one attempts to drive a set of components much faster than their natural frequency, there will be saturations in the system, the linearity assumptions are no longer valid, and the linear design represents little more than wishful thinking. With this background, we see that to simply increase the gain of a system with a lead compensator may not always be feasible. In this case, use of the lag compensator to increase the low-frequency gain while holding the crossover frequency constant may be the only viable option.

FIGURE 5.52
Frequency response of PID with $T_I/T_D = 20$.

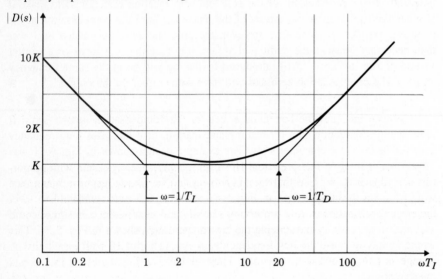

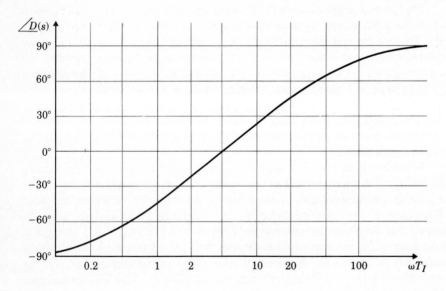

5.9.5 PID Compensation

By combining the derivative and integral feedback, Eqs. (5.28) and (5.31), we obtain PID control. Its transfer function is

$$D(s) = \frac{K}{s}\left[(T_D s + 1)\left(s + \frac{1}{T_I}\right)\right] \qquad (5.33)$$

and its frequency-response characteristics are shown in Fig. 5.52. This form is slightly different than that given by Eq. (3.16); however, the effect of the difference is inconsequential. This compensation is roughly equivalent to using both a lead and a lag compensator, sometimes referred to as a *lead-lag* compensator.

EXAMPLE 5.12   As an example of PID compensation design using frequency-response methods, consider the spacecraft attitude-control problem. A simplified design was presented in Section 5.7; however, here we have a more realistic situation that includes a sensor lag and a disturbing torque. Figure 5.53 defines the system. The design specifications are to have zero steady-state error, a phase margin of 65°, and as high a bandwidth as is reasonably possible, all using a PID controller.

First, let's take care of the steady-state error. For the spacecraft to be at a steady final value, the total input torque, $T_d + T_c$, must equal zero. Therefore if T_d is nonzero, T_c must equal $-T_d$. The only way this can be true with no error ($e = 0$) is for $D(s)$ to contain an integral term, hence the necessity for the "I" in PID. This could also be verified mathematically by use of the final-value theorem (see Problem 5.35).

FIGURE 5.53
PID design example.

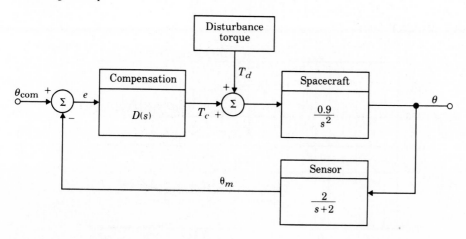

The frequency response of the spacecraft and sensor,

$$G(s) = \frac{0.9}{s^2} \frac{2}{s+2},$$ (5.34)

is shown in Fig. 5.54. The -2 (that is, -40 dB per decade) and -3 (-60 dB per decade) slopes show that the system would be unstable for any K if no derivative feedback were used. Therefore the "D" in PID is required to bring the slope to -1 (-20 dB per decade) at the crossover frequency, and the problem now is to pick

FIGURE 5.54
Compensation design for PID example.

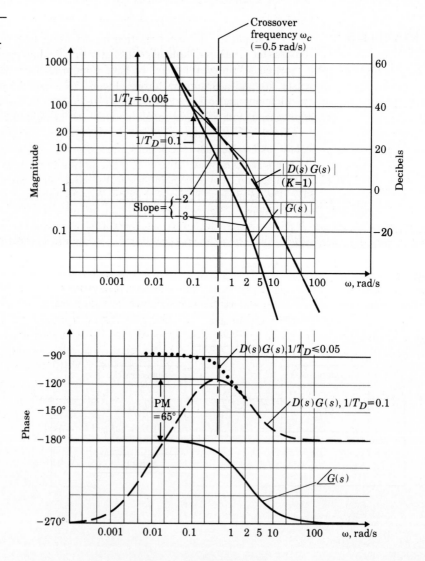

the three parameters in Eq. (5.33)—K, T_D, and T_I—so that the specifications are satisfied.

The easiest approach is to work first on the phase so that a 65° PM is achieved at a reasonably high frequency. This can be accomplished primarily by adjusting T_D, although T_I has a minor effect if sufficiently larger than T_D. Once the phase is adjusted, the crossover frequency is established and the gain K is easily determined.

By examining the phase of the PID in Fig. 5.52, we can consider what would happen to the compensated spacecraft system, $D(s)G(s)$, as T_D is varied. If $1/T_D \geq$ 2 rad/s, the phase lead from the PID would simply cancel the sensor phase lag and the composite phase would never exceed $-180°$, an unacceptable situation. If $1/T_D \leq 0.01$, the composite phase would approach $-90°$ for a range of frequencies and would exceed $-115°$ (potentially providing a PM of 65°) for an even wider range of frequencies. The compensated phase curve shown in Fig. 5.54 has $1/T_D = 0.1$, which is the largest value of $1/T_D$ that could provide the required PM of 65°. The phase would never cross the $-115°$ (65° PM) line for any $1/T_D > 0.1$. For $1/T_D = 0.1$, the crossover frequency ω_c that produces the 65° PM is 0.5 rad/s. For a value of $1/T_D \ll 0.05$, the phase essentially follows the dotted curve in Fig. 5.54, which indicates that the maximum possible ω_c is approximately 1 rad/s and is provided by $1/T_D = 0.05$. Therefore $0.05 < 1/T_D < 0.1$ is the only sensible range for $1/T_D$; anything less than 0.05 provides no significant increase in bandwidth, anything more than 0.1 could not meet the PM specification. Although the final choice is somewhat arbitrary, we have chosen $1/T_D = 0.1$ for our final design.

Our choice for $1/T_I$ is a factor of 20 lower than $1/T_D$; that is, $1/T_I = 0.005$. The factor of 20 was selected because anything less than 20 would have negatively impacted the phase at crossover, thus lowering the PM. Furthermore, it is generally desirable to maintain the compensated magnitude as large as possible at frequencies below ω_c because that provides a faster transient response and smaller errors. In order to bring this about, it is desirable to maintain $1/T_D$ and $1/T_I$ at the highest possible frequencies, hence the rationale for selecting the values of $1/T_D$ and $1/T_I$.

Determination of K is the only remaining task. Unlike the previous example where K was selected in order to meet a steady-state error specification, here we select K in order to yield a crossover frequency at the point where we have carefully tailored the phase to yield the required 65° PM. The basic idea for finding K was discussed in Section 5.6 and consists of plotting the compensated system amplitude with $K = 1$, finding its value at crossover, then setting $1/K$ equal to that value. This is done in Fig. 5.54 and shows that when $K = 1$, $|D(s)G(s)| = 20$ at the desired crossover frequency of 0.5 rad/s. Therefore

$$\frac{1}{K} = 20,$$

and

$$K = \frac{1}{20} = 0.05.$$

The compensation that satisfies all of the specifications is now complete:

$$D(s) = \frac{0.05}{s}[(10s + 1)(s + 0.005)].$$

It is interesting to note that this system would become unstable if the gain were lowered so that the crossover frequency became less than $\omega = 0.02$ rad/s, the region in Fig. 5.54 where the phase of the compensated system is less than $-180°$. This situation is referred to as a *conditionally stable* system. A root locus versus K for this and any conditionally stable system would show the portion of the locus corresponding to very low gains to be in the RHP.   ■

5.9.6 Summary of Compensation Characteristics

PD control adds phase lead at all frequencies above the breakpoint. If no change in gain of low-frequency asymptote, PD compensation will increase crossover frequency and speed of response. Increasing frequency-response magnitude at the higher frequencies will increase sensitivity to noise.

Lead compensation adds phase lead at a frequency band between the two breakpoints, which are usually selected to bracket the crossover frequency. If no change in gain of low-frequency asymptote, lead compensation will increase the crossover frequency and speed of response over the uncompensated system. If gain of low-frequency asymptote is reduced in order not to increase crossover frequency, the steady-state errors of the system will increase.

PI control increases frequency-response magnitude at frequencies below the breakpoint thereby decreasing steady-state errors. It also contributes phase lag below the breakpoint, which must be kept at a low enough frequency so that it does not degrade stability excessively.

Lag compensation increases frequency-response magnitude at frequencies below the two breakpoints thereby decreasing steady-state errors. With suitable adjustments in loop gain, it can alternatively be used to decrease the frequency-response magnitude at frequencies above the two breakpoints so that the magnitude 1 crossover occurs at a frequency that yields an acceptable phase margin. It also contributes phase lag between the two breakpoints, which must be kept at low enough frequencies so that the phase decrease does not degrade stability excessively.

5.10 Alternative Presentation of Data

Other presentations of frequency-response data have been developed over the years to aid both in understanding design and in easing the designer's work load.

Nichols Chart

A plot of $|G(j\omega)|$ versus $\angle G(j\omega)$ can be drawn by simply transferring the information directly from the separate magnitude and phase Bode plots: One point on the new curve thus results from a given value of frequency ω. The magnitude information is plotted on a log scale while the phase information is plotted with a linear scale, just as was the case with the Bode plots. This template was suggested by N. Nichols* and is usually referred to as a Nichols chart. The idea of plotting the magnitude versus phase of $G(j\omega)$ is similar to the plotting of the real and imaginary parts of $G(j\omega)$ that was the basis for the Nyquist plots shown in Sections 5.5 and 5.6. However, it is difficult to capture all the pertinent characteristics of $G(j\omega)$ on the linear scale of the Nyquist plot. The log scale for magnitude in the Nichols chart alleviates this difficulty and allows this kind of presentation to be used for design.

For any value of the complex transfer function $G(j\omega)$, Section 5.8 shows that there is a unique mapping to the unity feedback, closed-loop transfer function

$$F(j\omega) = \frac{G(j\omega)}{1 + G(j\omega)}.$$

The Nichols chart also contains contours of constant *closed-loop* magnitude and phase based on this relationship, as shown in Fig. 5.55. A designer can, therefore, graphically determine the bandwidth of a closed-loop unity-feedback system from the plot of the open-loop data on a Nichols chart by noting where the curve crosses the 0.7 (–3dB) contour of the closed-loop magnitude and determining the frequency of the corresponding data point. Likewise, a designer can determine the resonant-peak amplitude M_r by noting the value of the highest closed-loop magnitude contour that is tangent to the curve. The frequency associated with the magnitude and phase at the point of tangency is sometimes referred to as the resonant frequency ω_r. Similarly, a designer can determine the GM by observing the value of gain where the Nichols plot crosses the $-180°$ line and can determine the PM by observing the phase where the plot crosses the amplitude 1 line.

EXAMPLE 5.13   To illustrate, the compensated design example seen in Fig. 5.54 is shown in a Nichols chart in Fig. 5.56. In comparing the two figures, it is important to divide the magnitudes in Fig. 5.54 by a factor of 20 so that $|D(s)G(s)|$ is obtained, rather than the normalized values used in Fig. 5.54. Since the curve crosses the closed-loop magnitude contour of 0.7 at $\omega = 0.7$ rad/s, we see that the bandwidth of this system is 0.7 rad/s. Since the largest magnitude contour touched by the curve is 1.2, we also see that the resonant peak M_r is 1.2.   ∎

*James, Nichols, & Philips (1947).

FIGURE 5.55
Nichols chart.

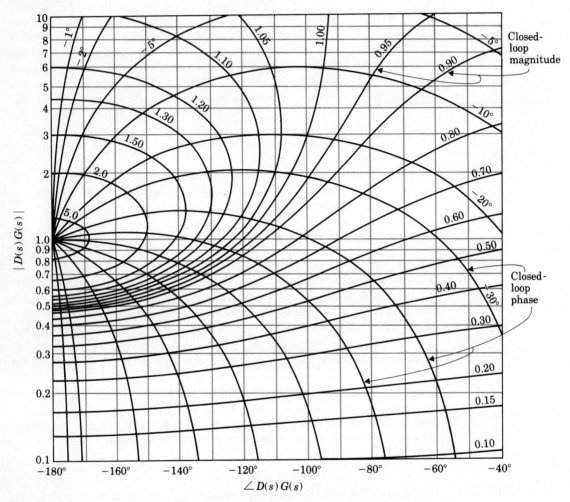

It can be proven that the contours of constant closed-loop magnitude and phase are circles when $G(j\omega)$ is presented in the Nyquist plot. These circles are referred to as the M and N circles, respectively.

This presentation of data was particularly valuable when a designer generated plots and performed calculations by hand. A change in gain, for example, could be evaluated by sliding the curve vertically on transparent paper over a standard

FIGURE 5.56
Example plot on Nichols chart for bandwidth and M_r determination.

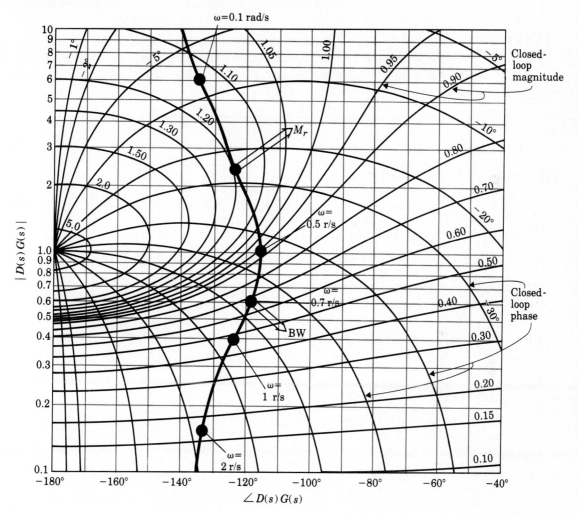

Nichols chart. The GM, PM, and bandwidth were then seen easily from the chart, thus allowing evaluations of several values of gain with a minimal amount of effort. For designers with access to computer-aids, however, calculations of bandwidth and many repetitive evaluations of gain or any other parameter are quickly accomplished with a few keystrokes. The primary value of the Nichols chart today is as an alternate presentation of the information in a Nyquist plot. For complex systems where the −1 encirclements need to be evaluated, the magnitude log scale of the Nichols presentation enables examination of a wider range of frequencies than a Nyquist plot and enables the designer to read the gain and phase margins directly.

EXAMPLE 5.14   Figure 5.57 shows a Nichols chart with the data from the same case whose Nyquist plot is shown in Fig. 5.35. Note that the phase margin for both crossover frequencies is seen to be 36° and the gain margin is seen to be 1.25 (= 1/0.8). It is clear

FIGURE 5.57
Nichols chart of the complex system whose Nyquist plot is in Fig. 5.35.

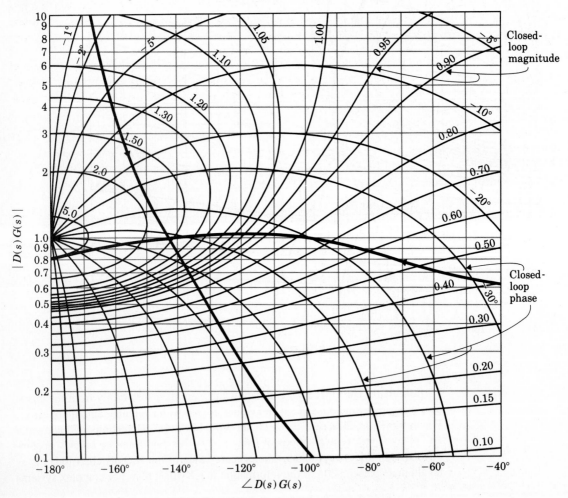

from this presentation of the data that the most critical portion of the curve is seen as it crosses the 180° line; hence the GM is the most relevant stability margin in this example.

■

Inverse Nyquist

This plot is simply the reciprocal of the complex quantity in the Nyquist plot described in Section 5.5 and used in Section 5.6 for the definition and discussion of stability margins. It is obtained most easily by computing the reciprocal of the magnitude from the Bode plot and plotting that quantity at an angle in the complex plane, which is the negative of the phase from the Bode plot. It can be used to find the PM and GM in the same way that the Nyquist plot was used. When $|G(j\omega)| = 1$, $|G^{-1}(j\omega)| = 1$ also, so the PM is measured from the negative real axis, but positive upward due to the phase sign reversal. However, when the phase $= -180°$ or $+180°$, the value of $|G^{-1}(j\omega)|$ is the GM directly; no calculation of an inverse is required as was the case for the Nyquist plot.

The inverse Nyquist plot of Example 5.5 from Section 5.5 defined in Fig. 5.20 is shown in Fig. 5.58 for the case where the gain is low and the system is stable. Note that the GM = 2 and the PM ≅ 20°. As an example of a more complex case,

FIGURE 5.58
Inverse Nyquist plot for Example 5.5.

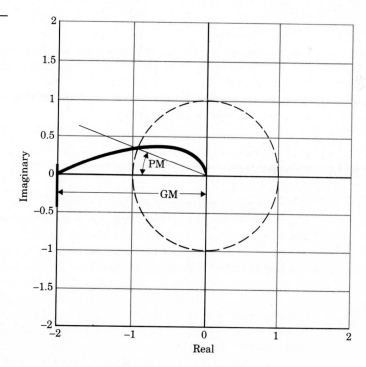

FIGURE 5.59
Inverse Nyquist plot
of the system whose
Nyquist plot is in
Fig. 5.35.

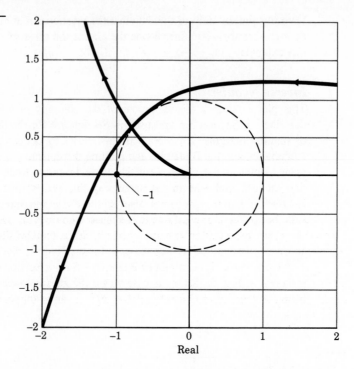

Fig. 5.59 shows an inverse Nyquist plot for the sixth-order case whose Nyquist plot was shown in Fig. 5.35 and whose Nichols chart is shown in Fig 5.57. Note here that the GM = 1.2 and the PM $\cong$ 35°. Had the two crossings of the unit circle not occurred at the same point, the crossing with the smallest PM would have been the appropriate one to use.

5.11   Sensitivity

In this section we will develop conditions on the Bode plot of the open-loop transfer function *DG* to ensure good performance with respect to sensitivity, steady-state errors, and sensor noise. One of the justifications for feedback, developed in Chapter 3, is to reduce the effect of disturbances and parameter changes on the performance of a control system. One aspect of the sensitivity issue has been the consideration of steady-state errors due to command inputs and disturbances. This has been an important design component in the different design methods. Design for acceptable steady-state errors can be thought of as the placement of a lower bound on the very low-frequency gain of the system. Another aspect of the sensitivity issue concerns the high-frequency portion of the system. So far, Chapter 3 and Sections 4.5 and 5.9 have briefly discussed the idea that alleviation of the effects of sensor noise dictate that the gain of the system at high frequencies be kept low. In fact, in the development of lead compensation, we placed an extra pole in the system specifically to reduce effects of sensor noise at the higher frequencies. It is not unusual for designers to place extra poles in the compensation, that is, to use

$$D(s) = K \frac{Ts + 1}{(\alpha Ts + 1)^2}$$

so that even more attenuation is introduced for noise reduction.

Yet another consideration affecting high-frequency gains is that many systems have high-frequency dynamic phenomena such as mechanical resonances that can sometimes impact the stability of a system. In very high performance designs, these high-frequency dynamics are included in the plant model, and a compensator is designed with a specific knowledge of those dynamics. A standard approach to designing for unknown high-frequency dynamics is to keep the high-frequency gain low, just as we did for sensor-noise reduction. The reason for this can be seen from the gain versus frequency of a typical system, shown in Fig. 5.60. The only way instability can result from high-frequency dynamics is if an unknown high-frequency resonance causes the magnitude to rise above 1. Conversely, if all

FIGURE 5.60
Effect of high-frequency plant uncertainty.

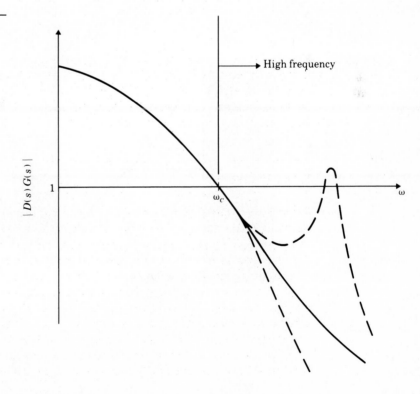

unknown high-frequency phenomena are guaranteed to remain below a magnitude of 1, stability can be guaranteed. The likelihood of an unknown resonance in the plant, G, rising above 1 can be reduced if the nominal high-frequency gain of DG is lowered by the addition of extra poles in D. When the stability of a system with resonances is ensured by tailoring the high-frequency magnitude so that it never exceeds 1, we refer to this process as *amplitude stabilization*. Of course, if the resonance characteristics are known exactly, a specially tailored compensation such as one with a notch at the resonant frequency can be used to reduce the gain at a specific frequency. This method of stabilization is referred to as *phase stabilization*. A drawback to phase stabilization is that the resonance information is often not available with adequate precision and therefore the method is more susceptible to errors in the plant model used in the design. Thus we see that sensitivity to plant uncertainty and sensor noise are both reduced by sufficiently low gain at high frequencies.

FIGURE 5.61
Design criteria for low sensitivity.

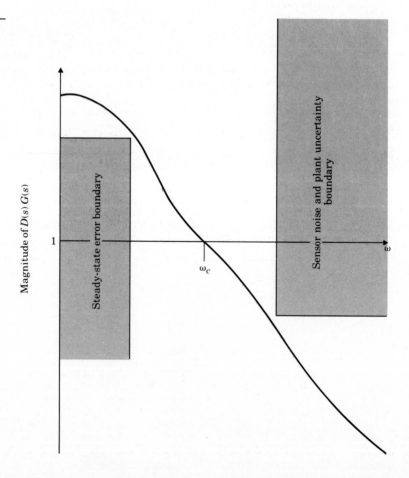

These two aspects of sensitivity—high- and low-frequency behavior—can be depicted graphically, as shown in Fig. 5.61. There is a minimum low-frequency gain allowable for acceptable steady-state error performance and a maximum high-frequency gain allowable for acceptable noise performance and for low probability of instabilities caused by plant-modeling errors. Between these two bounds the control engineer must achieve a gain crossover near the required bandwidth, which as we have seen, must be done at essentially a slope of -1.

□ 5.11.1 Sensitivity Function

In order to aid the designer in making a control system as insensitive as possible, special functions have been defined that exhibit the sensitivity of a closed-loop control system. We will initially define the functions for the SISO systems that are the subject of this text and then discuss briefly how they are extended to MIMO cases where they are used extensively in the design of insensitive or *robust* compensation.

Consider the unity feedback system in Fig. 5.62 with the plant G and the compensator D. One of the main objectives of the control system is to keep the tracking error ($e = r - y$) *small* for a reference input excitation and to keep the output (y) *small* for a disturbance input (d). The measurement noise is represented by n. From the block diagram we may write,

$$y = d + GD(r - y - n) \qquad (5.35)$$

$$(1 + GD)y = d + GD(r - n) \qquad (5.36)$$

or

$$y = (1 + GD)^{-1}d + (1 + GD)^{-1}GD(r - n) \qquad (5.37)$$

also for the tracking error we have,

$$e = r - (d + GDe) + GDn$$

$$(1 + GD)e = r - d + GDn \qquad (5.38)$$

or

$$e = (1 + GD)^{-1}(r - d) + (1 + GD)^{-1}GDn \qquad (5.39)$$

FIGURE 5.62
Feedback control system.

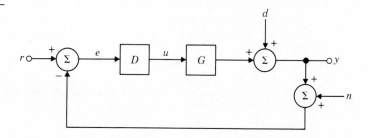

The *sensitivity function* is defined as,

$$S(s) = (1 + GD)^{-1} \tag{5.40}$$

which Eq. (5.37) shows is the transfer function from d to y. We also see from Eq. (5.39) that S is the transfer function from d to $-e$.

The *complementary sensitivity function* is defined as,*

$$T(s) = (1 + GD)^{-1}GD \tag{5.41}$$

and is the transfer function between the reference input r and the output y, that is, it is the *closed-loop* system transfer function. With some manipulation, the complementary sensitivity function can also be expressed as

$$T(s) = [1 + (GD)^{-1}]^{-1} \tag{5.42}$$

and the sum of the two sensitivity functions shown to obey

$$S(s) + T(s) = 1. \tag{5.43}$$

Therefore, with $n = 0$, we may then write Eq. (5.37) as

$$y = Sd + Tr \tag{5.44}$$

and (5.39) as

$$e = S(r - d). \tag{5.45}$$

Eq. (5.43) states that regardless of the compensation used, the sum of the sensitivity and the complementary sensitivity will always be unity.

The sensitivity function S is then the primary measure of performance so far as it relates to tracking performance (making e small for a given r) and also to disturbance rejection (making y small for a given d). Thus it is important to make $S(s)$ *small* in some sense. For physically realizable systems, the *loop gain*

$$L = |GD| \tag{5.46}$$

must become small for high frequencies. This means that

$$\lim_{s \to \infty} S(s) = 1. \tag{5.47}$$

Therefore it is only possible to make the sensitivity function small over low and midrange frequencies but not at very high frequencies. Figure 5.63(a) shows a typical plot of the sensitivity function S. The function is small at low frequencies, which is as desired for good tracking and disturbance rejection. At the midrange frequency, it approaches unity and crosses it before approaching unity again for high frequencies.

---

*The symbol T for the complementary sensitivity function is traditional, as is its use for the breakpoint in the lead and lag compensators discussed in Section 5.9. These two items are unrelated and the meaning of the symbol should be clear from the context.

FIGURE 5.63
Typical plots of (a) sensitivity function and (b) complementary sensitivity function.

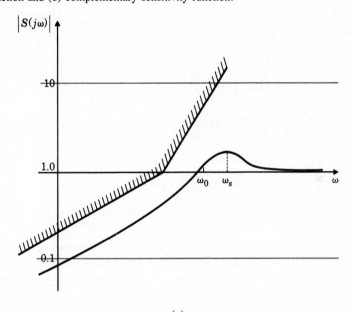

(a)

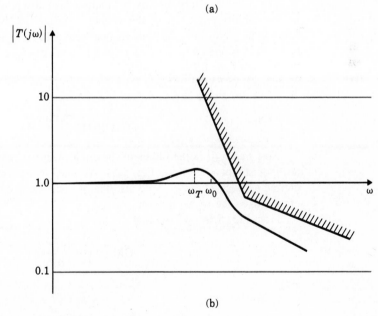

(b)

At the same time, we ideally wish to make the complementary sensitivity function one at all frequencies. However, this is not physically possible as

$$\lim_{s \to \infty} T(s) = 0. \tag{5.48}$$

Therefore it is only possible to make $T(s)$ close to unity at low and midrange frequencies, but the complementary sensitivity function must roll off at high frequencies as suggested by Eq. (5.43). Figure 5.63(b) shows a typical plot of the complementary sensitivity function. It is unity at low and midrange frequencies but rolls off where $S(s)$ approaches one.

The transfer function from the measurement noise n to the output y is, from Eq. (5.37) with $d = r = 0$,

$$y = -(1 + GD)^{-1}GDn. \tag{5.49}$$

Therefore, we also see that the complementary sensitivity function

$$T(s) = \frac{y}{-n} = (1 + GD)^{-1}GD. \tag{5.50}$$

We would like to make this transfer function as *small* as possible. This brings out the classical tradeoff in feedback control: Good tracking and disturbance rejection (S small and T large) must be balanced against minimization of the effect of sensor noise (S large and T small).

The complementary sensitivity function $T(s)$ also plays a key role in determining the stability properties of the system. Consider the Nyquist plot in Fig. 5.64 showing the phase and gain margins that were defined in Section 5.6. The sensitivity function is related to these margins because Eq. (5.40) shows that it represents the inverse of the distance from $G(j\omega)D(j\omega)$ to the -1 point. For convenience, we define α to be the distance from $G(j\omega)D(j\omega)$ to the -1 point. Therefore,

$$\alpha_{\min} = \frac{1}{\max_{\omega} |S(j\omega)|} = \min_{\omega} \frac{1}{|S(j\omega)|}. \tag{5.51}$$

Note that $\alpha_{\min}$ is the radius of the circle with the center at -1 and tangent to the Nyquist plot as shown in Fig. 5.64. As noted in Section 5.10, this is sometimes referred to as an M circle. Note also that it is the same quantity that Smith (1958) referred to as the *vector* margin.

From Fig. 5.64, we see that

$$\text{GM} = \frac{1}{(1 - \alpha_1)} \tag{5.52}$$

and

$$\text{PM} = 2\arcsin(\frac{\alpha_2}{2}) \tag{5.53}$$

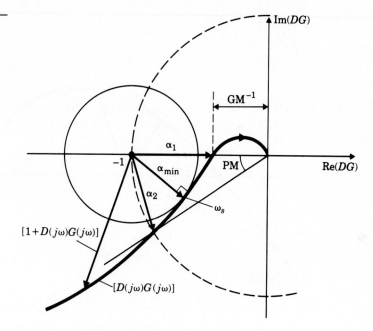

and, because α_1 and α_2 must both be greater than α_{min}, we see that the maximum value of the sensitivity function yields information about the phase and gain margins. One could use α_{min} in Eqs. (5.52) and (5.53) to yield conservative estimates of PM and GM.

In the MIMO case, the sensitivity function is defined as in Eq. (5.40), except GD becomes the matrix of transfer functions between the inputs and the outputs, and the value 1 is replaced with the identity matrix. Furthermore, rather than maximizing the sensitivity function as in Eq. (5.51), we judge stability by defining

$$\alpha_{min} = \min_{\omega} \frac{1}{\overline{\sigma}(S(j\omega))} = \frac{1}{\|S(j\omega)\|_{\infty}} \tag{5.54}$$

where $\overline{\sigma}(\,.\,)$ and $\|\,.\,\|_{\infty}$ denote the maximum singular value and the infinity norm, respectively (see Appendix C). For the SISO case, these quantities reduce to the definitions in Eq. (5.51). The definition of the sensitivity functions and the extensions of stability ideas based on these functions to the MIMO case may be done in terms of the recent H^{∞} *control theory.*

The complementary sensitivity function also may be related to the phase and gain margins. Consider the inverse Nyquist plot shown in Fig. 5.65. From Eq. (5.42) it is apparent that $T(j\omega)$ is the reciprocal of the distance between the -1

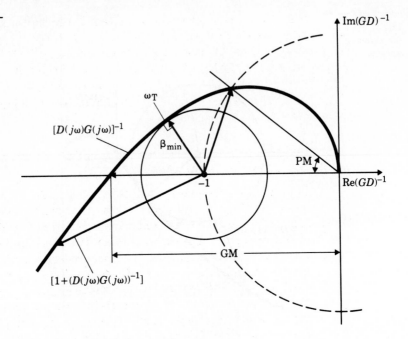

point and $(GD)^{-1}$. We can therefore define

$$\beta_{\min} = \frac{1}{\max_{\omega} |T(j\omega)|} = \min_{\omega} \frac{1}{|T(j\omega)|} = \min_{\omega} \frac{1}{\bar{\sigma}(T(j\omega))} = \frac{1}{\|T(j\omega)\|_{\infty}}. \quad (5.55)$$

The gain margin is

$$\text{GM} \geq 1 + \beta_{\min} \quad (5.56)$$

and the phase margin is

$$\text{PM} \geq 2 \arcsin(\frac{\beta_{\min}}{2}). \quad (5.57)$$

Note that $\beta_{\min}$ is the radius of the circle with center at -1 and tangent to the inverse Nyquist plot, that is, it defines the minimum distance from the -1 point. The quantity $\beta_{\min}$ is sometimes referred to as the *complex margin*.

Use of $\beta_{\min}$ to compute the GM and PM will obviously yield conservative estimates. The GM estimate will be exact only if the vector defining the minimum distance lies along the negative real axis. Likewise, the PM estimate will be exact only if the vector defining the minimum distance coincides with the point where GD crosses the unit circle.

The phase and gain margin estimates discussed in this section are attractive in the study of MIMO systems because they provide familiar measures (PM and GM) of the system's stability. The phase- and gain-margin estimates can be computed based on both the sensitivity and complementary sensitivity functions and the least conservative answer selected.

□ **5.11.2 Stability Robustness**

Most control system designs are based on a model of the plant; however, it is inevitable that the model we use is only an approximation of the true system dynamics. The difference between the model on which the design is based and the true system used in the actual control is referred to generally as *model uncertainty*. If the design performs well for substantial variations in the dynamics of the plant from the design values, we say the design is *robust*. We turn now to consider the effect of model uncertainty on the stability of control systems. Typical sources of uncertainty include unmodeled (high-frequency) dynamics, neglected nonlinearities, effects of deliberate reduced-order modeling, and plant-parameter changes due to environmental factors such as temperature, air speed, and age. Uncertainty can be represented in the form of either *additive* or *multiplicative perturbation*, as shown in Fig. 5.66. For the multiplicative case of Fig. 5.66(b), the true transfer function is

$$G(s) = G_0(s, \theta)[1 + l(s)],$$

where $G_0(s, \theta)$ is a parameterized model of the plant with *structured uncertainty* θ, which represents the plant-parameter variations. The expression $G_0(s, \theta)$ is a known function (known structure), but the values of the parameters θ are uncertain. The function $l(s)$ is an *unstructured uncertainty* and is entirely unknown, except that it is limited in magnitude by some function of frequency as

$$|l(j\omega)| \leq l_0(\omega), \tag{5.58}$$

where $l_0(\omega)$ is a known, real scalar function. The bound can be viewed as a frequency-dependent "radius" of uncertainty of the true plant $G(s)$ about some model $G_0(s, \theta)$ for a given θ. Figure 5.67 shows how such a bound can be calculated experimentally by simply comparing the actual plant with the model. In general, a good model will be well known at low frequencies, resulting in small $l(\omega)$ for $\omega \ll \omega_1$, and less well known at high frequencies where we have large $l(\omega)$. The curve of Fig. 5.67 is characteristic of unstructured uncertainty.

FIGURE 5.66
Representation of
uncertainty: (a) additive
perturbation and
(b) multiplicative
perturbation.

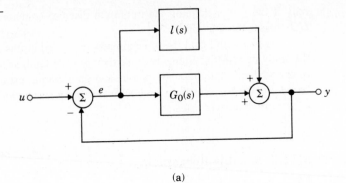

(a)

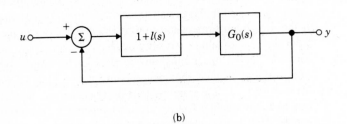

(b)

FIGURE 5.67
Frequency-dependent
uncertainty bound.

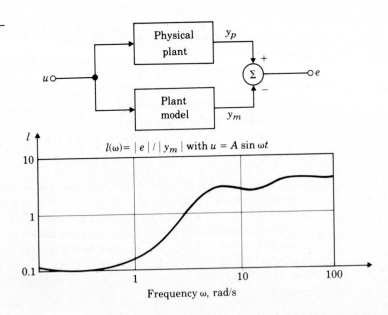

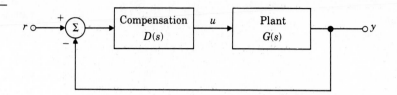

FIGURE 5.68
Compensated feedback
system.

Let us now assume that all structured and unstructured uncertainties can be lumped into a *stable* multiplicative perturbation, that is,

$$G(s) = G_0(s)[1 + l(s)],\qquad(5.59)$$

where $G_0(s)$ represents the nominal plant transfer function and $l(s)$ is the model error. The system with compensation $D(s)$ is as shown in Fig. 5.68, and the nominal closed-loop system $[l(s) \equiv 0]$ is designed to be stable.* A typical Nyquist plot of the compensated system is shown in Fig. 5.69. As $G_0(s)$ is perturbed toward $G(s)$, the Nyquist plot moves around in an envelope, as shown in the figure. The system remains stable so long as the number of encirclements of -1 remains unchanged, which will be true if, for all possible values of l,

$$|1 + D(j\omega)G(j\omega)| \neq 0,\qquad(5.60)$$

that is, so long as the perturbed Nyquist plot does not pass through the -1 point. The above stability condition can be rewritten as

$$|1 + D(j\omega)G_0(j\omega)(1 + \epsilon l)| \neq 0 \qquad 0 \leq \epsilon \leq 1$$
$$0 \leq \omega \leq \infty,\qquad(5.61)$$

which is true if and only if

$$|1 + D(j\omega)G_0(j\omega)(1 + \epsilon l)| > 0.\qquad(5.62)$$

If we take out a factor of DG_0, the open-loop gain, this can be expressed as

$$|DG_0[(DG_0)^{-1} + 1 + \epsilon l]| > 0,\qquad(5.63)$$

which is true if $|DG_0| \neq 0$ and

$$|(DG_0)^{-1} + 1 + \epsilon l| > 0.\qquad(5.64)$$

We would like to be able to express the requirement in terms of the open-loop

*$[D(s)G(s)]/[1 + D(s)G(s)]$ is stable.

FIGURE 5.69
Modification of Nyquist
plot by model
uncertainty.

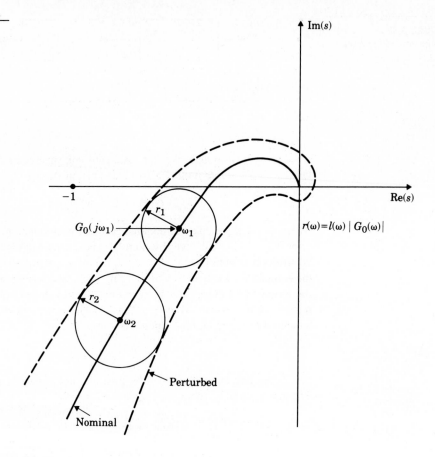

gain of the nominal system DG_0. We can do this if we notice that, for ϵ between 0 and 1,

$$\left|(DG_0)^{-1} + 1 + \epsilon l\right| > \left|1 + (DG_0)^{-1}\right| - |l| > 0, \qquad (5.65)$$

or

$$\left|1 + (DG_0)^{-1}\right| > |l|. \qquad (5.66)$$

The system is then guaranteed to be stable so long as Eq. (5.66) is satisfied. We define the *stability robustness measure* as

$$\delta_{\text{SR}}(\omega) = \left|1 + (DG_0)^{-1}\right|. \qquad (5.67)$$

Then as long as the model uncertainty remains below δ_{SR} for all frequencies, the system is guaranteed to remain stable in spite of the perturbations $l(s)$.* An alternative derivation of this result is possible using the "small-gain theorem" (see Problem 5.51). Note that δ_{SR} is simply the inverse of the closed-loop magnitude frequency response, that is, the inverse of the complementary sensitivity function $T(s)$. In terms of the complementary sensitivity function, we require that

$$|T(s)^{-1}| > |l| \qquad (5.68)$$

or

$$|T(s)| < \frac{1}{|l|} \qquad (5.69)$$

for stability. This means that model uncertainty dictates an upper bound on the magnitude of $T(s)$. The quantity l defines the minimum distance from the -1 point to the inverse Nyquist plot and, for the case where Eq. (5.69) is an equality, l would be the quantity $\beta_{\min}$ as in Fig. 5.65. For high frequencies, Fig. 5.67 shows that $|l(j\omega)| \gg 1$, therefore Eq. (5.68) suggests

$$|GD| < \frac{1}{|l|} \qquad (5.70)$$

that is, the (multiplicative) modeling error defines an upper bound on the loop gain at high frequencies.

The above discussion suggests the introduction of the following bounds on the sensitivity and complementary sensitivity functions depending on the tracking accuracy, bandwidth, and roll-off required:[†]

$$|S| < |W_1^{-1}| \qquad (5.71)$$

$$|T| < |W_2^{-1}|. \qquad (5.72)$$

Figure 5.70 shows these bounds in terms of the open-loop gain and is essentially the same idea as that shown in Fig. 5.61 and used in the discussion at the beginning of this section to relate the ideas from compensation design to a magnitude Bode plot.

---

*For the extension of this result, the MIMO case, see Doyle and Stein (1981).

[†] These bounds are used in H^{∞} control theory as weightings on the sensitivity and complementary sensitivity functions to design compensation.

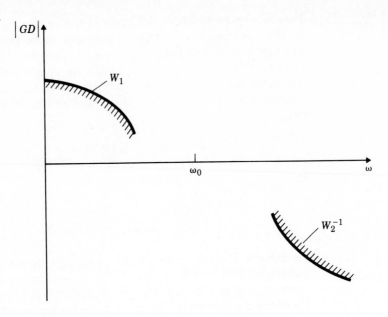

□ 5.12   Time Delay

The Laplace transform of pure time delay is $G_D(s) = e^{-sT}$ and was approximated by a rational function (Padé approximant) for root locus analysis. Although this same approximation could be used with frequency-response methods, an exact analysis of the delay is quite straightforward.

The frequency response of the delay is given by the magnitude and phase of $e^{-sT}|_{s=j\omega}$. The magnitude is

$$|G_D(j\omega)| = |e^{-j\omega T}| = |\cos \omega T - j \sin \omega T| = 1 \qquad \text{for all } \omega.$$

This result is expected, since a time delay merely shifts the signal in time and has no effect on its magnitude. The phase is

$$\angle G_D(j\omega) = -\omega T \qquad \text{radians},$$

and it grows increasingly negative in proportion to the frequency. This, too, is expected since a fixed time delay T becomes a larger fraction or multiple of a sine-wave period as the period drops because of the increasing frequency. A plot of $\angle G_D(j\omega)$ is drawn in Fig. 5.71 and shows how the phase lag is greater than $270°$ for values of ωT greater than about 5 rad. This implies that it would be virtually

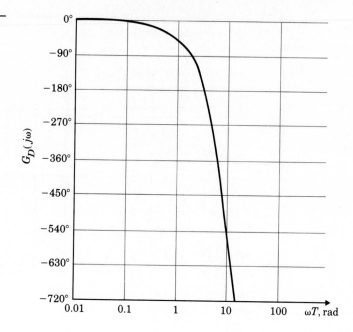

FIGURE 5.71
Phase lag due to pure
time delay.

impossible to stabilize a system (or to achieve a positive PM) with a crossover
frequency greater than $\omega = 5/T$, and it is typically difficult to achieve bandwidths
greater than $\omega \cong 3/T$. These characteristics essentially place a constraint on the
achievable bandwidth of any system with a time delay. See Problem 5.52 for an
illustration of the constraint.

□ 5.13 Obtaining a Pole-Zero Model from Frequency-Response Data

As has been pointed out earlier, it is relatively easy to obtain the frequency response
of a system experimentally. Sometimes it is desirable to obtain an approximate
model, in terms of a transfer function, directly from the frequency response. The
derivation of such a model can be done to various degrees of accuracy. The method
described below is usually adequate and is widely used in practice.

   To obtain a model from frequency-response data there are two choices. In
the first case, one can introduce a sinusoidal input, measure the gain and the
phase difference between output and input, and accept the curves plotted from this
data as the model. Using the methods given in previous sections, the design can
be done directly from this information. If we wish to use the frequency data to

FIGURE 5.72
An experimental frequency response.

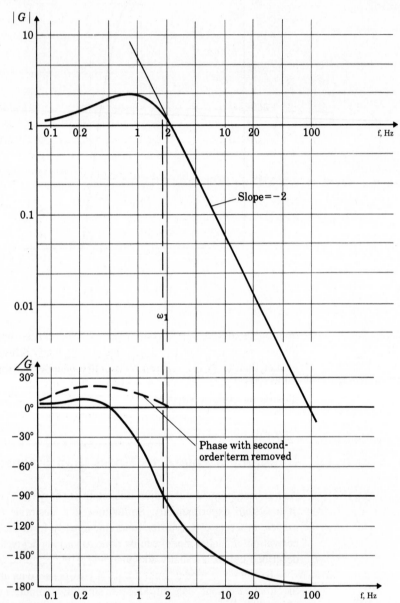

verify a mathematical model obtained by other means, then we need to extract an approximate transfer function from the plots. The method suggested is to fit straight lines to the data, estimate breakpoints (find the poles and zeros), and project the damping ratios of complex factors by the frequency overshoot or the phase, as found from Fig. 5.5. One systematic procedure is to begin at the highest frequency of interest and fit a first or second order term, subtract this from the data, move to a lower frequency on the residual, and repeat.

EXAMPLE 5.15   For example, consider the frequency response plotted in Fig. 5.72. Drawing an asymptote to the final slope of -2 or -40 dB per decade, we assume a break at the frequency where the phase is $-90°$, which occurs at $f_1 \cong 1.66$, or $\omega_1 = 2\pi f_1 = 10.4$ rad/s. We need to know the damping ratio to be able to subtract out this second-order pole. Actually the phase curve may be more help. Since the phase around the break frequency is symmetric, we draw a line at the slope of the phase curve at f_1 to find that the phase asymptote intersects at $f_0 \cong 0.71$ Hz corresponding to $f_1/f_0 \cong 2.34$, which in time corresponds to $\zeta \cong 0.5$, as seen on the normalized response curves in Fig. 5.5(b). The rest of the curve shows an asymptotic amplitude gain of about 6.0 dB, or a factor of $10^{6.0/20} = 2.0$. Since this is a gain rise, it occurs because of a lead compensation of the form

$$\frac{(s/z) + 1}{(s/p) + 1},$$

where $p/z = 2.0$. If we remove the phase of the second-order terms, we obtain a phase curve with a maximum phase of about $20°$, which also corresponds to a frequency separation of about 2. To locate the center of the lead network we must estimate the point of maximum phase on the lead term alone, which occurs at the geometric mean of the two break frequencies. This seems to occur at $f_2 \cong 0.3$ Hz or $\omega_2 = 1.88$ rad/s. Thus we have the relations

$$zp = (1.88)^2 = 3.55,$$

$$\frac{p}{z} = 2,$$

from which we can solve

$$2z^2 = 3.55,$$

$$z = 1.33,$$

$$p = 2.66.$$

Our final model is given by

$$\hat{G}(s) = \frac{(s/1.33) + 1}{[(s/2.66) + 1][(s/10.4)^2 + (s/10.4) + 1]}. \tag{5.73}$$

The actual data was plotted from

$$G(s) = \frac{(s/2) + 1}{[(s/4) + 1][(s/10)^2 + (s/10) + 1]}.$$

As can be seen, we found the second-order term quite easily, but the location of the lead network is off in center frequency by a factor of $4/2.66 \cong 1.5$. However, the subtraction of the second-order term from the composite curve was not done with great accuracy, just by reading the curves. Again, as with the transient response, the conclusion is that by a bit of approximate plotting one can obtain a crude model (usually within ± 3 dB in amplitude and $\pm 10°$ in phase) that can be used for control design. ∎

Refinements on these techniques with computer aids are rather obvious, and an interactive program for removing standard first- and second-order terms and accurately plotting the residual would greatly improve the speed and accuracy of the process. It is also common to have computer tools that can find the parameters of an assumed model structure by minimizing the sum of squares of the difference between the model frequency response with the experimental frequency response.

Summary

A method of determining the stability of a closed-loop system based on the frequency response of the system's open-loop transfer function has been described. The degree of stability has been quantified in terms of a gain margin and phase margin, which can be determined directly by inspection from the magnitude and phase of the open-loop transfer function. The frequency-response characteristics of several types of compensation have been described, and examples of design using these characteristics have been discussed. They show the ease of selecting specific values of design variables, a result of using frequency-response methods.

The advantages of design using frequency response have been shown to be the following:

Numerical values of the design parameters that meet specifications are more readily determined than with root locus design.

Experimental data of frequency response can be used directly.

Tracking-error reduction and disturbance rejection can be specified in terms of low-frequency gain.

Sensor noise rejection can be specified in terms of high-frequency attenuation.

It develops an understanding that will be useful for design of multi-input multi-output control systems.

Problems   Section 5.2

5.1 a) Show that α_o in Eq. (5.1) is given by

$$\alpha_o = G(s)\frac{U_o\omega}{s - j\omega}\Bigg|_{s=-j\omega} = -U_oG(-j\omega)\frac{1}{2j}$$

and

$$\alpha_o^* = G(s)\frac{U_o\omega}{s + j\omega}\Bigg|_{s=+j\omega} = U_oG(j\omega)\frac{1}{2j}$$

b) By assuming the output can be written as

$$y(t) = \alpha_o e^{-j\omega t} + \alpha_o^* e^{j\omega t}$$

derive the relations in Eqs. (5.3) and (5.4).

5.2 a) Calculate the magnitude and phase of

$$G(s) = \frac{1}{s + 1}$$

for $\omega = 0.1, 0.2, 0.5, 1, 2, 5, 10$ rad/sec.

b) Sketch the asymptotes for the $G(s)$ in **(a)** and compare with your computed results from **(a)**.

5.3 Draw the Bode plot for the following systems.

a) $G(s) = \dfrac{1}{(s + 1)^2(s^2 + s + 4)}.$

b) $G(s) = \dfrac{s}{(s + 1)(s + 10)(s^2 + 5s + 2500)}.$

c) $G(s) = \dfrac{4s(s + 10)}{(s + 50)(4s^2 + 5s + 4)}.$

d) $G(s) = \dfrac{10(s + 4)}{s(s + 1)(s^2 + 2s + 5)}.$

e) $G(s) = \dfrac{1000(s + 1)}{s(s + 2)(s^2 + 8s + 64)}.$

f) $G(s) = \dfrac{(s + 5)}{s(s + 1)(s^2 + s + 4)}.$

g) $G(s) = \dfrac{4000}{s(s + 40)}.$

h) $G(s) = \dfrac{100}{s(1 + 0.1s)(1 + 0.5s)}.$

i) $G(s) = \dfrac{1}{s(1 + s)(1 + 0.02s)}.$

5.4 A certain system is represented by the asymptotic Bode plot shown in Fig. 5.73. Find and sketch the response of this system to a unit step input (zero initial conditions).

5.5 Prove that a magnitude slope of -1 corresponds to -20 dB per decade.

FIGURE 5.73

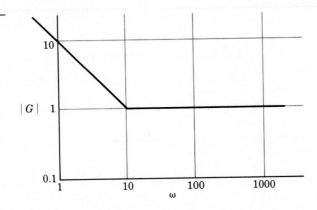

Section 5.3

5.6 Draw the Bode plots for a normalized second-order system with $\zeta = 0.5$ and an added zero. Do the plots lead one to expect extra peak overshoot in the transient response? How?

$$G(s) = \frac{(s/z) + 1}{s^2 + s + 1}.$$

Let $z = \alpha/2$ and plot for $\alpha = 0.1, 1,$ and 10.

5.7 Draw the Bode plot for a normalized second-order system with $\zeta = 0.5$ and an added pole. Do the plots lead one to expect additional rise time? How?

$$G(s) = \frac{1}{[(s/p) + 1](s^2 + s + 1)}.$$

Let $p = \alpha/2$ and plot for $\alpha = 0.1, 1,$ and 10.

5.8 For the closed-loop transfer function, $F(s) = \omega_n^2/(s^2 + 2\zeta\omega_n s + \omega_n^2)$, determine the ratio of the bandwidth ω_{BW} of $F(s)$ to ω_n for $\zeta = 0.2, 0.5,$ and 0.8.

5.9 A DC voltmeter schematic is shown in Fig. 5.74. The pointer is damped so that its maximum peak overshoot to a step input is 10%.

a) What is the undamped natural frequency?

b) What is the damped natural frequency?

c) What input frequency will produce the largest magnitude output?

d) This meter is now used to measure a 1-V AC input with a frequency of 2 rad/s. What is the amplitude the meter will indicate after initial transients have died out? What is the phase lag of the output with respect to the input? Use a Bode magnitude and phase plot analysis.

FIGURE 5.74

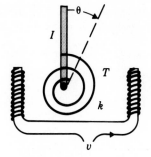

$I = 40 \times 10^{-6}\, kg \cdot m^2$

$k = 4 \times 10^{-6}\, kg \cdot m^2/s^2$

$T = \text{input torque}$

$\quad = K_m v$

$v = \text{input voltage}$

$K_m = 4 \times 10^{-6}\, N \cdot m/V$

Section 5.4

5.10 Determine the range of K for which each system is stable by making a Bode plot with $K = 1$ and imagining the magnitude plot sliding up or down until instability results. Qualitatively verify the stability rules by a very rough root-locus sketch.

a) $KG(s) = \dfrac{K(s + 2)}{s + 10}$.

b) $KG(s) = \dfrac{K}{(s + 10)(s + 2)^2}$.

c) $KG(s) = \dfrac{K(s + 10)(s + 1)}{(s + 100)(s + 2)^3}$.

5.11 Determine the range of K for which each system is stable by making a Bode plot with $K = 1$ and imagining the magnitude plot sliding up or down until instability results. Qualitatively verify the stability rules by a very rough root-locus sketch.

a) $KG(s) = \dfrac{K(s + 1)}{s^2(s + 10)}$.

b) $KG(s) = \dfrac{K(s + 1)}{s(s + 2)}$.

c) $KG(s) = \dfrac{K}{(s + 2)(s^2 + 9)}$.

d) $KG(s) = \dfrac{K(s + 1)^2}{s^3(s + 10)}$.

Section 5.5

5.12 a) Sketch the Nyquist plot for a $1/s^2$ open-loop system; that is, sketch.

$$\left. \frac{1}{s^2} \right|_{s=C_1}$$

where C_1 is a contour enclosing the entire RHP, as shown in Fig. 5.14.
Hint: Assume C_1 takes a small detour around the poles at $s = 0$, as shown in Fig. 5.23.

b) Repeat the problem for a $1/(s^2 + \omega_0^2)$ open-loop system.

5.13 Draw the Nyquist diagrams of the following systems.

a) $KG(s) = \dfrac{K(s + 2)}{s + 10}.$

b) $KG(s) = \dfrac{K}{(s + 10)(s + 2)^2}.$

c) $KG(s) = \dfrac{K(s + 10)(s + 1)}{(s + 100)(s + 2)^3}.$

Estimate the range of K for which each system is stable, and qualitatively verify your result by a rough root-locus sketch, if not already done for Problem 5.10.

5.14 Draw the Nyquist diagrams for the following systems.

a) $KG(s) = \dfrac{K(s + 1)}{s^2(s + 10)}.$

b) $KG(s) = \dfrac{K(s + 1)}{s(s + 2)}.$

c) $KG(s) = \dfrac{K}{(s + 2)(s^2 + 9)}.$

d) For **(b)** and **(c)**, choose the Nyquist contour to be to the left of the singularities on the imaginary axis and interpret the Nyquist plot in each case.

e) Estimate the range of K for which each system is stable and verify with a root locus, if not already done for Problem 5.11.

5.15 Draw the Nyquist plot for the system in Fig. 5.75. Determine the range of K for which the system is stable.

FIGURE 5.75

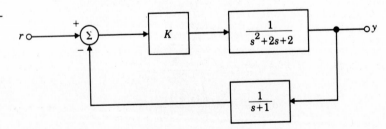

5.16 a) Sketch the phase for $\omega = 0.1$ to 100 rad/s for the minimum phase system

$$G(j\omega) = \left. \frac{s+1}{s+10} \right|_{s=j\omega}$$

and the nonminimum-phase system

$$G(j\omega) = -\left. \frac{s-1}{s+10} \right|_{s=j\omega} ,$$

noting that $\angle(j\omega - 1)$ decreases with ω rather than increasing. Does an RHP zero affect the relationship between the -1 encirclements on a Nyquist plot and the number of unstable closed-loop roots in Eq. (5.22)?

b) Sketch the phase for $\omega = 0.1$ to 100 rad/s for the unstable system

$$G(j\omega) = \left. \frac{s+1}{s-10} \right|_{s=j\omega} .$$

c) Check stability of the systems in (a) and (b) using the Nyquist criterion, and check your results qualitatively using a root-locus sketch.

5.17 Let

$$G(s) = \frac{100[(s/10)+1]}{s[(s/1)-1][(s/100)+1]}.$$

The Bode plot of $G(s)$ is provided in Fig. 5.76.

a) Sketch the Nyquist plot for $G(s)$.

b) Is the closed-loop system shown in Fig. 5.77 stable?

5.18 In Fig. 5.78

$$G(s) = \frac{25(s+1)}{s(s+2)(s^2+2s+16)}.$$

Carefully draw a Bode plot of the frequency response of $G(j\omega)$ and use this data to plot a Nyquist plot of $G(j\omega)$. Conclude whether or not the loop is stable. Sketch a root locus and verify your conclusion.

5.19 Consider the system given in Fig. 5.79.

a) Using the Routh criterion, determine closed-loop stability for all values of K.

b) Sketch the locus of the closed-loop poles as K varies from zero to infinity. (Include angles of departure or arrival and K and s values at breakpoints and crossings of the imaginary axis.)

c) Sketch the Nyquist plot of the closed-loop system and use it to verify the results of (a) and (b).

FIGURE 5.76
Bode plot for Problem 5.17.

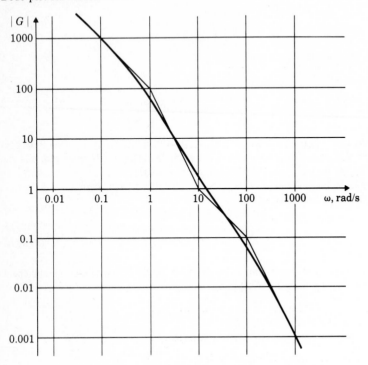

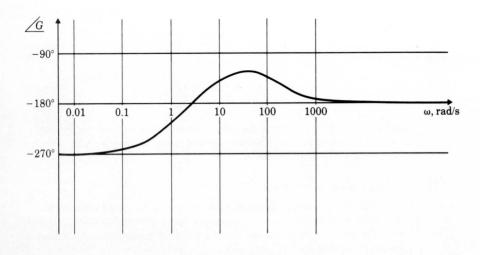

FIGURE 5.77

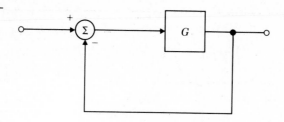

FIGURE 5.78

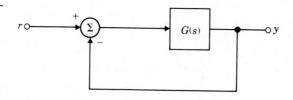

FIGURE 5.79

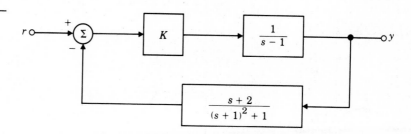

Section 5.6

5.20 Correlate the Nyquist plots, frequency responses, and step responses shown in Fig. 5.80.

5.21 a) Assuming an open loop

$$G(s) = \frac{\omega_n^2}{s(s + 2\zeta\omega_n)}$$

with unity feedback, the closed-loop system is

$$F(s) = \frac{\omega_n^2}{s^2 + 2\zeta\omega_n s + \omega_n^2}.$$

Set $\omega_n = 1$ and draw Bode plots for $\zeta = 0.1$, 0.4, and 0.7 to verify the values of PM in Fig. 5.32. By plotting G versus ω/ω_n, the value of ω_n does not enter

FIGURE 5.80

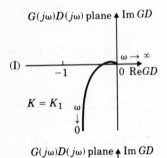

(I)

$K = K_1$

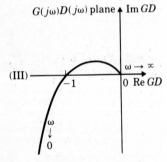

(II)

$K = K_2$

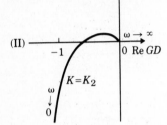

(III)

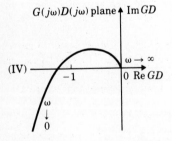

(IV)

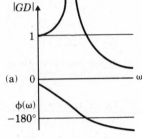

(a)

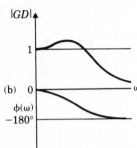

(b)

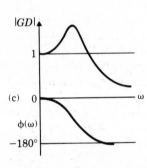

(c)

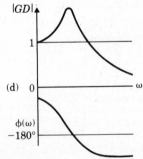

(d)

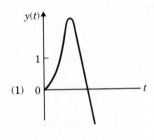

(1)

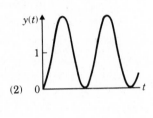

(2)

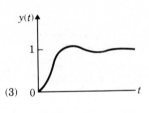

(3)

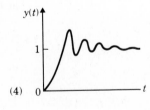

(4)

the problem. If the loop gain is some value other than 1, does the relationship between PM and the closed-loop system damping change?

b) Show that PM $= \tan^{-1}\left(2\zeta \left/ \sqrt{\sqrt{1 + 4\zeta^4} - 2\zeta^2}\right.\right)$.

5.22 For the unity feedback system with

$$G(s) = \frac{1}{s(s + 1)[(s^2/25) + 0.4(s/5) + 1]}$$

a) Draw the Bode plots (gain and phase) for $G(j\omega)$.
b) Indicate the gain margin when the gain is set for a phase margin of $45°$.
c) What is K_v when the gain is set for a $45°$ phase margin?
d) Sketch the root locus versus K and indicate the roots for a phase margin of $45°$.

5.23 Let

$$G(s) = \frac{1}{(s + 2)^2(s + 4)} \qquad \text{and} \qquad H(s) = \frac{1}{s + 1}.$$

FIGURE 5.81

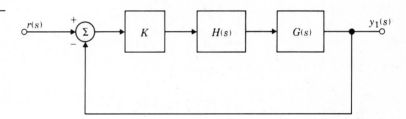

The root locus for the system depicted by the block diagram in Fig. 5.81 is shown in Fig. 5.82.

a) Using root-locus methods, determine the value of K at the stability boundary.
b) Using root-locus methods, determine the value of K that will produce roots with 0.707 damping.
c) What is the gain margin of the system if the gain is set to the value determined in **(b)**? Do *not* draw Bode or Nyquist plots.
d) Draw the Bode plots for the system and determine the gain margin that results for a $65°$ PM. Using the rule-of-thumb in Eq. (5.25), what damping ratio would you expect for this PM?
e) How does the root locus of the system shown in Fig. 5.83 differ from the one in Fig. 5.81?
f) How does the transfer function $Y_2(s)/R(s)$ differ from $Y_1(s)/R(s)$?

5.24 For the system shown in Fig. 5.84, use a Bode plot and root locus to determine the gain and frequency at which instability occurs. What two values of gain give a PM of $20°$? What is the vector margin at these two values of gain yielding identical values for the PM?

FIGURE 5.82
Root locus for Problem
5.23.

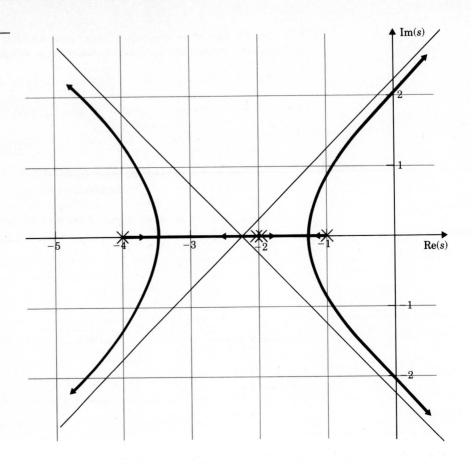

FIGURE 5.83

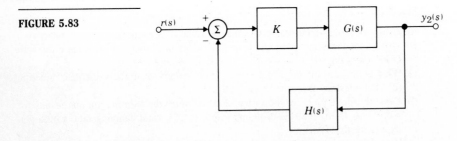

FIGURE 5.84

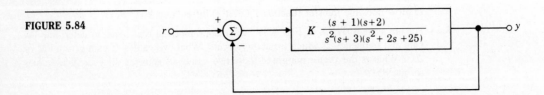

5.25 A magnetic-tape drive speed-control system is shown in Fig. 5.85. The speed sensor is slow enough that its dynamics must be included. The speed-measurement time constant is $\tau_m = 0.5$ s, the reel time constant $\tau_r = J/b = 4$ s, the output shaft-damping constant is $b = 1$ N $\cdot$ m $\cdot$ s, and the motor time constant is $\tau_1 = 1$ s.

a) Determine the necessary gain K if the steady-state speed error is required to be less than 7% of the reference-speed setting.

b) With the gain determined from **(a)**, use the Nyquist criterion to investigate the stability of the system.

c) Estimate the GM and PM of the system. Is this a good system design? Why?

FIGURE 5.85

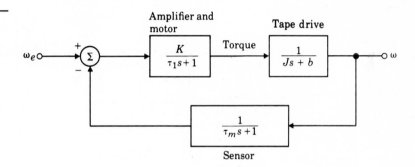

5.26 Apply the formal Nyquist criterion to determine the range of values of K (positive and negative) for which the system of Fig. 5.86 will be stable and to determine the number of roots in the RHP for those values of K for which it is unstable. Correlate with a root-locus sketch.

FIGURE 5.86

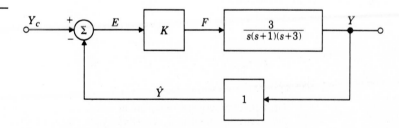

5.27 Repeat Problem 5.26 for the feedback system in Fig. 5.87.

FIGURE 5.87

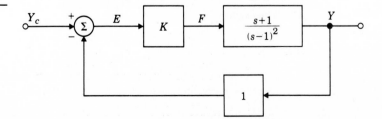

5.28 Repeat Problem 5.26 for the system shown in Fig. 5.88.

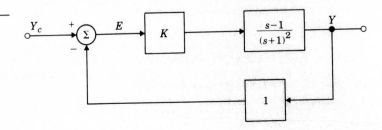

5.29 Two Nyquist plots for unity feedback systems that are open-loop stable are sketched in Figs. 5.89 and 5.90. The proposed operating gain is indicated as K_0, and arrows indicate increasing frequency. In each case, give your best estimates for:

  a) Phase margin.
  b) Damping ratio to step response.
  c) Range of gain for stability (if any).
  d) System type as 0, I, or II.

5.30 A ship's steering has the transfer function

$$G(s) = \frac{K_v[-(s/0.142) + 1]}{s[(s/0.325) + 1][(s/0.0362) + 1]}.$$

  a) Plot the log magnitude of $G(j\omega)$ versus log ω on log-log paper, with $K_v = 2$.
  b) Plot the phase of $G(j\omega)$ versus log ω using the same frequency axis as used for the magnitude.
  c) 1) Indicate the crossover frequency on the plot of **(a)**.
     2) Indicate the phase margin at crossover on the plot of **(b)**.
     3) Indicate the phase-crossover frequency where $\phi = -180°$ on the plot of **(b)**.

Section 5.7

5.31 The frequency response of a plant in a unity feedback configuration is sketched in Fig. 5.91. Assume the plant is open-loop stable and minimum phase.

  a) What is the velocity constant K_v for the system as drawn?
  b) What is the damping ratio of the complex poles at $\omega = 100$?
  c) What is the phase margin of the system as drawn? (Estimate to within $\pm 10°$.)

5.32 For the system

$$G(s) = \frac{100(s/a + 1)}{s(s + 1)(s/b + 1)},$$

where $b = 10a$, find a value of a that will yield good stability by plotting the frequency-response magnitude only.

FIGURE 5.89

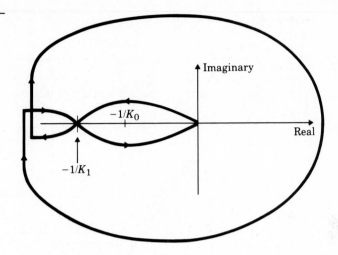

FIGURE 5.90

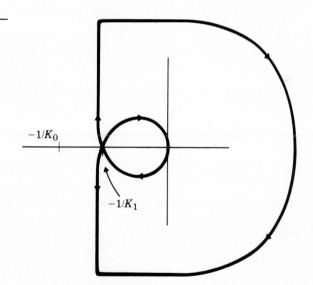

FIGURE 5.91

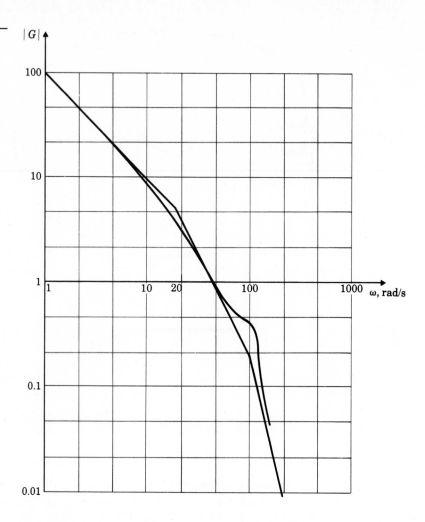

Section 5.8

5.33 a) Show that the magnitude

$$\left| \frac{1}{(j\omega/\omega_n)^2 + 2\zeta(j\omega/\omega_n) + 1} \right| = \frac{1}{2\zeta} \quad \text{at} \quad \omega = \omega_n.$$

b) Show that

$$|F(j\omega_c)| = \frac{1}{2\sin(\text{PM}/2)},$$

where F is the closed-loop transfer function and ω_c is the crossover frequency.

c) By using results from (a) and (b), show that

$$\zeta \approx \frac{\pi \mathrm{PM}^\circ}{360} \approx \frac{\mathrm{PM}^\circ}{100}$$

for small ζ and PM.

Section 5.9

5.34 For the system in Fig. 5.92

$$G(s) = \frac{5}{s[(s/1.4) + 1][(s/3) + 1]},$$

design a lead compensation $D(s)$ so that the phase margin is $\sim 40^\circ$. What is the approximate bandwidth of the system? (Maintain parameters of $D(s)$ so that its DC gain $= 1$.)

FIGURE 5.92

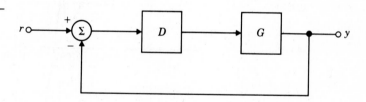

5.35 Referring to Fig. 5.53, derive the transfer function from T_d to θ, then apply the final-value theorem (assuming $T_d =$ constant) to determine whether $\theta(\infty)$ is nonzero:

a) When $D(s)$ has no integral term, that is, $\lim_{s \to 0} D(s) =$ constant;

b) When $D(s)$ has an integral term, that is,

$$D(s) = \frac{D'(s)}{s},$$

where $\lim_{s \to 0} D'(s) =$ constant.

5.36 Consider the type I unity feedback system with

$$G(s) = \frac{K}{s(s + 1)}.$$

Design a lead compensator such that

$$K_v = 12 \sec^{-1} \quad \text{and} \quad \mathrm{PM} \geq 40^\circ.$$

FIGURE 5.93

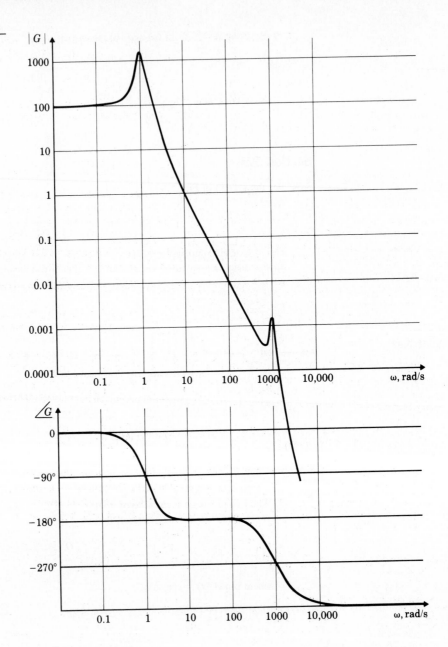

5.37 You are given the experimentally determined Bode plot shown in Fig. 5.93. Design a compensation that will yield a crossover frequency of 10 rad/s with a phase margin greater than 75°.

5.38 In one mode of operation, the autopilot of a jet transport is used to control altitude. For the purpose of designing the autopilot loop, only the long-period airplane dynamics are important. The linearized relationship of altitude *rate* to elevator angle is therefore second-order. Specifically,

$$\frac{\dot{h}(s)}{\delta(s)} = \frac{20(s + 0.01)}{s^2 + 0.01s + 0.0025} \frac{\text{ft/s}}{\text{deg}}.$$

The autopilot receives an electrical signal proportional to altitude from the altimeter. This signal is compared with a command signal (proportional to the altitude selected by the pilot) and the difference provides an error signal. The error signal is processed through a compensation network, and the result is used to command the elevator actuators. A block diagram of this system is shown in Fig. 5.94. You have been given the task of designing the compensation. Consider first $D(s) = K$.

a) Sketch the magnitude and phase versus frequency (Bode plots) of the open-loop system for $D(s) = K = 1$.
b) What value of K would provide a crossover frequency of 0.16 rad/s?
c) For the value of K in part (**b**), would this system be stable if the loop were closed?
d) What is the PM?
e) Sketch the Nyquist plot for K from part (**b**). Locate carefully the points where the phase angle is $-180°$ and where the magnitude is unity.
f) Sketch the root locus versus K. Locate the roots for your value of K from (**b**).
g) What steady-state error would result with K from part (**b**) if the command were a ramp of 1 ft/s? Next consider

$$D(s) = K\frac{Ts + 1}{\alpha Ts + 1}$$

h) Choose the parameters K, T, and α so the crossover frequency is 0.16 and the phase margin is greater than 50°. Indicate your work on the Bode plot of (**a**).
i) Sketch the root locus versus K. Locate the roots for your value of K from (**h**).
j) What steady-state error would result with the K from part (**h**) if the command were a ramp of 1 ft/s?
k) If the error in (**j**) is too large, what type of compensation would you add? Approximately, at what frequencies would you put its breakpoints?

FIGURE 5.94

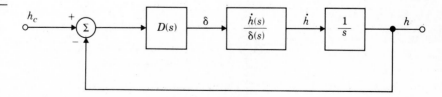

5.39 For the system with

$$G(s) = \frac{10}{s[(s/1.4) + 1][(s/3) + 1]},$$

design lag compensation so that PM ~40° while maintaining a DC gain of $D(s) = 1$. What is the approximate bandwidth of this system?

5.40 a) Design a compensation for the ship-steering control of Problem 5.30 so as to maintain a velocity constant K_v of 2, a phase margin $\geq 50°$, and unconditional stability (phase margin > 0 for all $\omega \leq \omega_c$).

b) Draw a root locus of your final design versus K_v and indicate the location of the closed-loop poles.

5.41 Consider the system frequency-response characteristics shown for Problem 5.31.

a) A lead compensation with $\alpha = 1/5$ is introduced with its zero at $1/T = 20$. How must the gain be changed to obtain crossover at $\omega = 31.6$ rad/s and what is the resulting value of K_v?

b) With the lead compensation in place, what is the required value of K for a lag network that will readjust the gain to the original K_v value of 100?

c) Place the pole of the lag at 3.16 rad/s and determine the zero location that will maintain the crossover frequency at $\omega = 31.6$ rad/s. Sketch the compensated frequency response on the same graph.

d) Estimate the PM of the compensated design.

Section 5.10

5.42 A feedback control system is shown in Fig. 5.95. The specification for the closed-loop system requires that the overshoot to a step input be less than 30%.

a) Determine the corresponding specification in the frequency domain M_r (resonant peak value) for the closed-loop transfer function.

b) Determine the resonant frequency ω_r for the maximum K that satisfies the specifications.

c) Determine the bandwidth of the closed-loop system resulting from **(b)**.
   Hint: Plot the open-loop magnitude versus phase (as shown in Fig. 5.56) on transparent paper, then slide it vertically on the Nichols chart (Fig. 5.55) to obtain the answers.

FIGURE 5.95

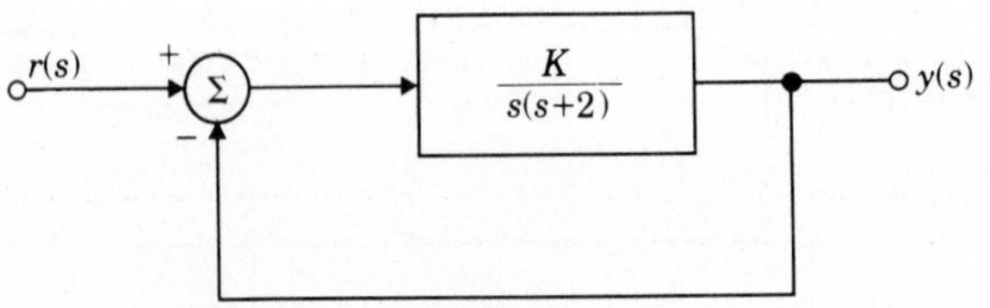

5.43 The Nichols plots of an uncompensated and compensated system are shown in Fig. 5.96.

a) What are the resonant peaks of the uncompensated and compensated systems?

b) What are the phase and gain margins of the uncompensated and compensated systems?

c) What are the bandwidths of the uncompensated and compensated systems?

d) What type of compensation is used?

FIGURE 5.96
Nichols plot for Problem 5.43.

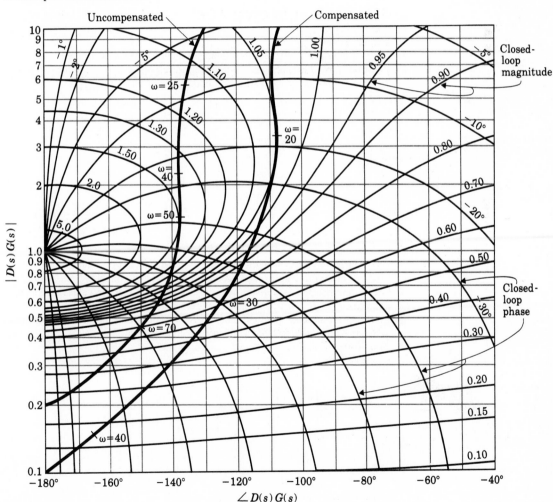

5.44 Consider the system in Fig. 5.86 on page 347.

a) Construct an inverse Nyquist plot $[Y(j\omega)/E(j\omega)]^{-1}$.

b) Show how the value of K for neutral stability (defined to be K_u) can be read directly from the inverse Nyquist plot.

c) For $K = 4$, 2, and 1, determine the gain margin and phase margin.

d) Construct a root-locus plot for the system and identify corresponding points in the two plots. To what ζ do the GM and PM of **(c)** correspond?

5.45 An unstable plant has a transfer function

$$\frac{Y}{F} = \frac{s+1}{(s-1)^2}.$$

A simple control loop is to be closed around it, as shown in Fig. 5.87.

a) Construct an inverse Nyquist plot of Y/F.

b) Choose a value of K to provide a phase margin of $45°$. What is the corresponding gain margin?

c) Can you infer from your plot the stability situation when K is negative?

d) Construct a root-locus plot and describe the correspondence between the two plots. To what ζ does the phase margin $45°$ correspond in this case?

FIGURE 5.97

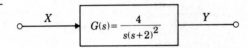

$$X \quad\quad G(s) = \frac{4}{s(s+2)^2} \quad\quad Y$$

5.46 Consider the system shown in Fig. 5.97.

a) Construct a Bode plot for the system.

b) Use your Bode plot to sketch an inverse Nyquist plot.

c) Consider closing a control loop around $G(s)$, as shown in Fig. 5.98. Using the inverse Nyquist plot as a guide, read from your Bode plot the values of GM and PM when $K = 0.7$, 1.0, 1.4, and 2. What value of K yields $30°$ PM?

d) Construct a root-locus plot and label the same values of K on the loci. For what value of ζ does each PM/GM correspond?

FIGURE 5.98

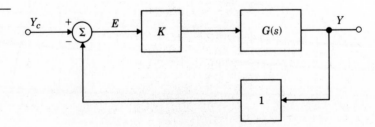

$$Y_c \quad + \quad \Sigma \quad \xrightarrow{E} \quad K \quad\quad G(s) \quad\quad Y$$
$$1$$

5.47 Repeat Problem 5.46 for the system shown in Fig. 5.99. (Use values of $K = 0.5$, 1, 2, and 4.)

FIGURE 5.99

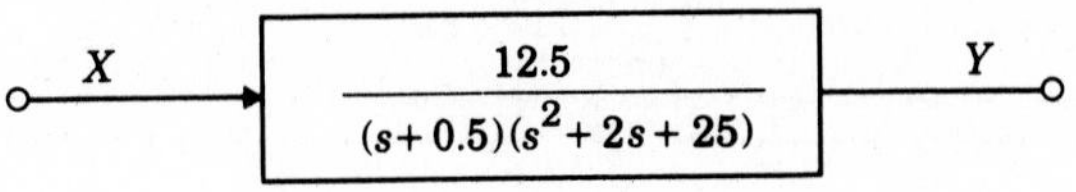

$$X \longrightarrow \boxed{\dfrac{12.5}{(s+0.5)(s^2+2s+25)}} \longrightarrow Y$$

5.48 For the open-loop system,

$$KG(s) = \frac{K(s + 1)}{s^2(s + 10)^2},$$

create the inverse Nyquist plot for $G(s)$ and determine the values of K at the stability boundary and that yield a PM of 30°.

5.49 Augment the system of Problem 5.46 to include a lead compensator as shown in Fig. 5.100.

a) Superimpose the new root locus, Bode, and inverse Nyquist plots on your old ones.

b) What value of K now produces PM = 30°? For this K what are the values of ζ and GM?

c) Compare the closed-loop system's lowest natural frequency, with and without the network, when K is set to produce a PM of 30° in each case.

FIGURE 5.100

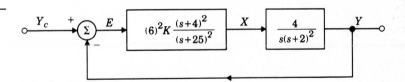

Section 5.11

5.50 Consider the system with the open-loop gain

$$G(s) = \frac{1}{s(s + 1)}.$$

a) Draw the Bode and Nyquist plots and find the GM, PM and vector margin (α_{min}) for the system.

b) Compute the sensitivity and complementary sensitivity functions and plot their magnitude frequency responses.

c) Find estimates of the GM and PM from **(b)** by determining α_{min} and using Eqs. (5.52) and (5.53). Compare the results to those of **(a)** and comment on their conservatism.

d) Draw the inverse Nyquist plot for the system. Identify the complex (vector) margin. Compute the approximate values of the PM and GM based on the complex margin and compare with (**a**).

FIGURE 5.101
Nonlinear feedback with inputs r_1 and r_2 and outputs u_1, y_1, u_2, and y_2.

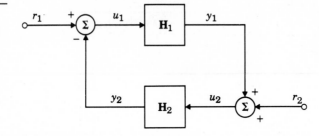

5.51 *Small-Gain Theorem.* An alternative derivation of the stability robustness results is possible via the small-gain theorem (Zames, 1966; Mees, 1981). Consider the system shown in Fig. 5.101, where $\mathbf{H}_1$ and $\mathbf{H}_2$ are not necessarily linear. The small-gain theorem states that if the inputs (r_1 and r_2) are bounded and if, in addition, the open-loop gain is less than 1, that is,

$$\|\mathbf{H}_2\|\|\mathbf{H}_1\| < 1,$$

then the feedback system is closed-loop stable, that is, u_1, u_2, y_1, and y_2 are all bounded. Now consider the typical feedback system shown in Fig. 5.102, where it is known that

$$\left|\frac{G - G_0}{G_0}\right| < l(\omega).$$

FIGURE 5.102
Typical feedback system with plant dynamics G and feedback dynamics D. The reference input is r and the disturbance input is d.

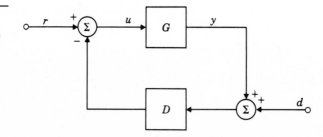

Prove that the system shown in Fig. 5.103 is equivalent to that of Fig. 5.102. Apply the small-gain theorem to the system shown in Fig. 5.103 to obtain the stability robustness result (Eq. 5.66) derived in this chapter. Why is it not useful to apply the theorem to the system in Fig. 5.102?

FIGURE 5.103
Equivalent feedback representation of Fig. 5.102.

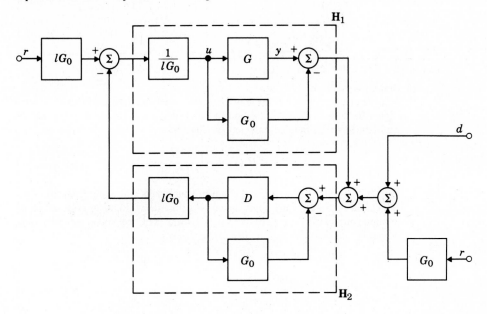

Section 5.12

5.52 Find the maximum possible bandwidth for the system below with a 0.2-s time delay
($T = 0.2$s), keeping PM $\geq 40°$.

$$G(s) = \frac{e^{-0.2s}}{s(s + 10)}.$$

a) Use one section of lead

$$D(s) = K\frac{TS + 1}{\alpha TS + 1},$$

where $\alpha = \frac{1}{100}$.

b) Use two sections of lead

$$D(s) = K\left(\frac{TS + 1}{\alpha TS + 1}\right)^2,$$

where $\alpha = \frac{1}{10}$.

5.53 Determine the range of K for which the following systems are stable:

a) $G(s) = K \dfrac{e^{-s}}{s}$.

b) $G(s) = K \dfrac{e^{-s}}{s(s + 2)}$.

5.54 A pure time delay, that is $x_{out}(t) = x_{in}(t - \tau)$, is represented by $H(s) = e^{-s\tau}$.

a) Graph $|H(j\omega)|$ and $\angle H(j\omega)$ versus ω.

b) $H(s)$ cannot be represented exactly as a ratio of polynomials in s. Therefore $H(s)$ cannot be described by a collection of poles and zeros. However, $H(s)$ may be approximated by a rational polynomial. Consider

$$H_1(s) = \frac{1 - (\tau s/2)}{1 + (\tau s/2)}.$$

Plot $|H_1(j\omega)|$ and $\angle H_1(j\omega)$, and compare your results with those from **(a)**. *Note:* $H_1(s)$ is a first-order Padé approximant to a time delay. Rational polynomial approximations to $H(s)$ may be constructed using RLC networks, whereas $H(s)$ cannot be realized with any lumped element network.

6

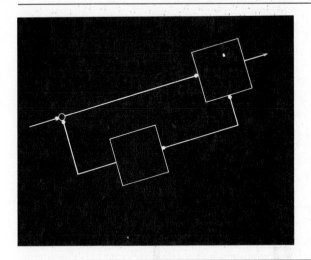

State-Space

Design

6.1 Introduction

The idea of *state space* comes from the state-variable method of describing differential equations. In this method, dynamic systems are described by a set of first-order differential equations in variables called the *state,* and the solution may be visualized as a trajectory in space. The method is particularly well suited to performing calculations by computer. State-space control design is the technique whereby the control engineer designs compensation by working directly with the state-variable description of the system.

Use of the state-space approach has often been referred to as *modern control design,* whereas use of transfer-function-based methods such as root locus and frequency response have been referred to as *classical control design.* However, since the state-space method of description for differential equations is over 100 years old and was introduced to control design in the late 1950s, it seems somewhat misleading to refer to it as "modern." We prefer to refer to the two approaches to design as state-space methods and transform methods.

Both the root-locus and frequency-response design techniques discussed in

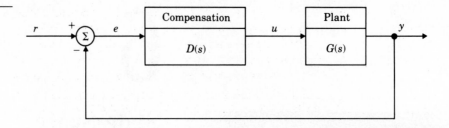

Chapters 4 and 5 are aimed at finding a compensation $D(s)$, shown in Fig. 6.1, that best satisfies the design specifications. The goal of the state-space design technique is precisely the same; however, because the method of describing the plant and the compensation is so different from the other techniques, it may seem at first that an entirely different problem is being solved. The examples given toward the end of this chapter should help convince the reader that, indeed, state-space design does result in the same sort of $D(s)$ that is obtained with the other methods.

Advantages of state-space design are especially apparent when engineers design controllers for systems with more than one control input or more than one sensed output; however, in this chapter, we will illustrate the ideas of state-space design using single-input, single-output systems. A further advantage of state-space design is that the system representation provides a complete (internal) description of the system, including possible internal oscillations or instabilities that might be hidden by inappropriate cancellations in the transfer-function (input/output) description.

6.2   System Description

The motion of any finite dynamic system can be expressed as a set of first-order ordinary differential equations. This is often referred to as the *state-variable representation*. For example, Newton's law for a single mass M moving in one dimension x under force F is

$$M\ddot{x} = F. \tag{6.1}$$

If we define one state variable as the position $x_1 = x$ and the other state variable as the velocity $x_2 = \dot{x}$, this equation can be written as

$$\dot{x}_1 = x_2,$$

$$\dot{x}_2 = \frac{F}{M}. \tag{6.2}$$

Furthermore, first-order linear differential equations can be concisely expressed using matrix notation. While linear algebra is a vast topic, and one with which control engineers need to be very familiar, we will use only the most basic relations and results here. A few of these results from the point of view of matrix algebra are collected in Appendix C for reference, study, and review.

Consider again the state equations of Eq. (6.2). If we collect the state into a column vector $\mathbf{x}$, and the coefficients of the state equations into a square matrix $\mathbf{F}$, and the coefficients of the input into the column matrix $\mathbf{G}$, these equations can be written in matrix form as

$$\begin{bmatrix} \dot{x}_1 \\ \dot{x}_2 \end{bmatrix} = \begin{bmatrix} 0 & 1 \\ 0 & 0 \end{bmatrix} \begin{bmatrix} x_1 \\ x_2 \end{bmatrix} + \begin{bmatrix} 0 \\ 1 \end{bmatrix} \frac{F}{M} \tag{6.3}$$

or

$$\dot{\mathbf{x}} = \mathbf{F}\mathbf{x} + \mathbf{G}u,$$

where $\mathbf{F}$ is the *system* matrix and $\mathbf{G}$ is the *input* matrix. If we take the output to be $y = x_1$, this too can be expressed in matrix form as

$$y = \begin{bmatrix} 1 & 0 \end{bmatrix} \begin{bmatrix} x_1 \\ x_2 \end{bmatrix} \tag{6.4}$$

or

$$y = \mathbf{H}\mathbf{x},$$

where $\mathbf{H}$ is a row vector, referred to as the *output* matrix. Collecting these matrix equations, we have the extremely compact notation,

$$\dot{\mathbf{x}} = \mathbf{F}\mathbf{x} + \mathbf{G}u + \mathbf{G}_1 w, \tag{6.5}$$

$$y = \mathbf{H}\mathbf{x} + Ju, \tag{6.6}$$

where the system input has been divided into control input u and the disturbance input w. Furthermore, we have added the term Ju to include the possibility that there is a direct path from the input u to the output y. J is called the *direct transmission* or the *feedthrough* term. Therefore, the transfer function of the control input u to system output y—which has been used heretofore as the system description and called $G(s)$—is now replaced by the state-variable description given by Eqs. (6.5) and (6.6). Figure 6.2 shows the two different methods of system representation for the case where there is no disturbance input ($\mathbf{G}_1 = 0$) and no direct transmission term ($J = 0$).

Similarly, the differential-equation models of more complex systems, such as those developed in earlier sections on mechanical, electrical, and electromechan-

364   Chapter 6 / State-Space Design

FIGURE 6.2
System representation:
(a) transform method and
(b) state-variable
method.

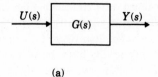

(a)

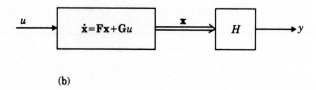

(b)

ical systems, can be described by state variables through selection of positions, velocities, capacitor voltages, and inductor currents as state variables. In Section 2.3, Eqs. (2.16) and (2.17) have already been manipulated into state-variable form through selection of suitable variables.

EXAMPLE 6.1   As an example of the formation of equations in state-variable form, consider the two-mass system shown in Fig. 6.3.

The equations of motion are (see Section 2.2)

$$M\ddot{y} + b(\dot{y} - \dot{d}) + k(y - d) = u,$$
$$m\ddot{d} + b(\dot{d} - \dot{y}) + k(d - y) = 0. \tag{6.7}$$

If we take the state as the position and velocity of each mass as follows:

$$x_1 = y$$
$$x_2 = \dot{y}$$
$$x_3 = d$$
$$x_4 = \dot{d},$$

FIGURE 6.3
A two-mass system.

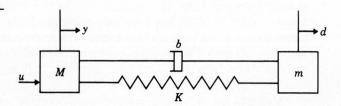

the equations in state-variable form are

$$\dot{x}_1 = x_2,$$

$$\dot{x}_2 = -\frac{k}{M}x_1 - \frac{b}{M}x_2 + \frac{k}{M}x_3 + \frac{b}{M}x_4 + \frac{u}{M},$$

$$\dot{x}_3 = x_4,$$

$$\dot{x}_4 = \frac{k}{m}x_1 + \frac{b}{m}x_2 - \frac{k}{m}x_3 - \frac{b}{m}x_4, \tag{6.8}$$

which, again, is the same form as Eq. (6.5) and, including the output y in (6.6), the state-space matrices are

$$\mathbf{F} = \begin{bmatrix} 0 & 1 & 0 & 0 \\ -\dfrac{k}{M} & -\dfrac{b}{M} & \dfrac{k}{M} & \dfrac{b}{M} \\ 0 & 0 & 0 & 1 \\ \dfrac{k}{m} & \dfrac{b}{m} & -\dfrac{k}{m} & -\dfrac{b}{m} \end{bmatrix}; \quad \mathbf{G} = \begin{bmatrix} 0 \\ \dfrac{1}{M} \\ 0 \\ 0 \end{bmatrix}; \quad \mathbf{G}_1 = \mathbf{0};$$

$$\mathbf{H} = [1 \quad 0 \quad 0 \quad 0]; \qquad\qquad J = 0. \qquad \blacksquare$$

EXAMPLE 6.2 Consider the equation of motion for the simple pendulum shown in Fig. 6.4, which is

$$\ddot{\theta} + \omega_0^2 \sin\theta = \frac{T_c}{ml^2}, \tag{6.9}$$

FIGURE 6.4
A simple pendulum.

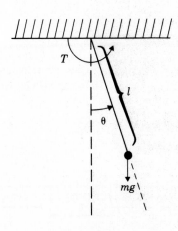

where $\omega_0^2 = g/l$. If we choose as the state variables $x_1 = \omega_0\theta$ and $x_2 = \dot{\theta}$, then the equations are

$$\dot{x}_1 = \omega_0 x_2,$$

$$\dot{x}_2 = -\omega_0^2 \sin\left(\frac{x_1}{\omega_0}\right) + \frac{T_c}{ml^2}. \tag{6.10}$$

These equations are nonlinear but are in state-variable form. Possible equilibrium state variables for the pendulum (obtained by setting $\dot{x}_1 = 0$, $\dot{x}_2 = 0$) are down ($x_1 = 0$) or up ($x_1 = \pi\omega_0$); that is,

$$x_1^* = 0 \quad \text{or} \quad \pi\omega_0,$$

$$x_2^* = 0.$$

If we assume the motion is close to the down position, we can linearize the equations about the solution $x_1^* = x_2^* = 0$, and hence, since $\sin(\theta) \approx \theta$ for small x,

$$\dot{x}_1 = \omega_0 x_2$$

$$\dot{x}_2 = -\omega_0 x_1 + \frac{T_c}{ml^2}, \tag{6.11}$$

where $|x_1|/\omega_0 \ll 1$. Once again, the result fits Eqs. (6.5) and (6.6), and the state-space matrices with measurement of θ are

$$\mathbf{F} = \begin{bmatrix} 0 & \omega_0 \\ -\omega_0 & 0 \end{bmatrix}; \quad \mathbf{G} = \begin{bmatrix} 0 \\ \dfrac{1}{ml^2} \end{bmatrix};$$

$$\mathbf{H} = \begin{bmatrix} \dfrac{1}{\omega_0} & 0 \end{bmatrix}; \quad J = 0. \qquad \blacksquare$$

EXAMPLE 6.3   Consider the spacecraft system shown in Fig. 2.6, with the dynamics of a sensor lag added to it. Its system description, using the transfer-function block diagram, is contained in Fig. 6.5. This same system can also be specified (see Problem 6.1)

FIGURE 6.5
Example system
description.

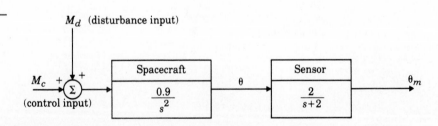

by equations in the form of Eqs. (6.5) and (6.6):

$$\begin{bmatrix} \dot{q} \\ \dot{\theta} \\ \dot{\theta}_m \end{bmatrix} = \begin{bmatrix} 0 & 0 & 0 \\ 1 & 0 & 0 \\ 0 & 2 & -2 \end{bmatrix} \begin{bmatrix} q \\ \theta \\ \theta_m \end{bmatrix} + \begin{bmatrix} 0.9 \\ 0 \\ 0 \end{bmatrix} M_c + \begin{bmatrix} 0.9 \\ 0 \\ 0 \end{bmatrix} M_d, \tag{6.12}$$

$$\dot{\mathbf{x}} = \qquad \mathbf{F} \qquad \mathbf{x} \quad + \quad \mathbf{G} \ u \ + \ \mathbf{G}_1 \ w,$$

$$y = \begin{bmatrix} 0 & 0 & 1 \end{bmatrix} \begin{bmatrix} q \\ \theta \\ \theta_m \end{bmatrix}, y = \mathbf{Hx}. \tag{6.13}$$

∎

EXAMPLE 6.4   Consider the longitudinal dynamics of an aircraft (Fig. 6.6). It obeys the equations of motion for any mechanical system given by Newton's laws [Eqs. (2.1) and (2.5)].

FIGURE 6.6
(a) Boeing 747 aircraft (*courtesy Boeing Commercial Airplane Co.*); (b) definition of aircraft coordinates.

(a)

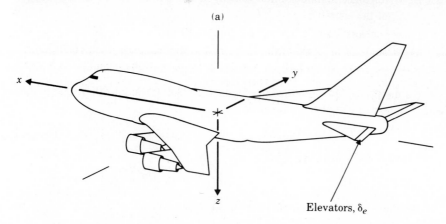

Elevators, δ_e

(b)

The forces are from aerodynamic lift and drag, the propulsion system, and gravity. Except for gravity, the forces are most easily expressed in a coordinate frame that is fixed to and rotating with the aircraft. Newton's laws need special expression in such a coordinate frame because in general it may be rotating with respect to the Earth's surface. It is therefore necessary to modify the equations to account for the rotation of the coordinate frame. For a detailed derivation of the equations, see Ashley (1974).

For linearized longitudinal motion, the results are

$$
\begin{bmatrix} \dot{u} \\ \dot{w} \\ \dot{q} \\ \dot{\theta} \end{bmatrix} = \begin{bmatrix} X_u & X_w & -W_0 & -g\cos\theta_0 \\ Z_u & Z_w & U_0 & -g\sin\theta_0 \\ M_u & M_w & M_q & 0 \\ 0 & 0 & 1 & 0 \end{bmatrix} \begin{bmatrix} u \\ w \\ q \\ \theta \end{bmatrix} + \begin{bmatrix} X_{\delta e} \\ Z_{\delta e} \\ M_{\delta e} \\ 0 \end{bmatrix} \delta e,
$$

where

u = forward velocity perturbation in the aircraft in x direction [see Fig. 6.6(b)],

w = velocity perturbation in the aircraft in z direction (also proportional to the angle of attack, $\alpha = w/U_0$),

q = angular rate about positive y axis, or pitch rate,

θ = pitch-angle perturbation with respect to horizontal,

X = normalized aerodynamic force derivatives in x direction,

Z = normalized aerodynamic force derivatives in z direction,

M = normalized aerodynamic moment derivatives,

W_0 = reference flight-condition velocity in z axis (usually $\cong 0$ for an aircraft flying horizontally),*

U_0 = reference flight-condition velocity in x axis (that is, the forward velocity of the aircraft),*

g = acceleration of gravity (32.2 ft/s^2 or 10 m/s^2),

θ_0 = angle of the x axis from horizontal in the reference flight condition,

δe = movable tail-section, or "elevator," angle for pitch control.

For the Boeing 747 in horizontal flight, $U = 830$ ft/s at $20,000$ ft (Mach 0.8) with a weight of $637,000$ lb, we have

$$
\begin{bmatrix} \dot{u} \\ \dot{w} \\ \dot{q} \\ \dot{\theta} \end{bmatrix} = \begin{bmatrix} -0.00643 & 0.0263 & 0 & -32.2 \\ -0.0941 & -0.624 & 820 & 0 \\ -.000222 & -0.00153 & -0.668 & 0 \\ 0 & 0 & 1 & 0 \end{bmatrix} \begin{bmatrix} u \\ w \\ q \\ \theta \end{bmatrix} + \begin{bmatrix} 0 \\ -32.7 \\ -2.08 \\ 0 \end{bmatrix} \delta e,
$$

We shall study this example in detail in Chapter 7.   ∎

*The $W_0 q$ and $U_0 q$ terms in the equations account for the acceleration of the coordinate frame due to aircraft rotational rates.

Equations (6.5) and (6.6) express the algebraic connection between input u and output y and state $\mathbf{x}$. The state for the specific cases above is of dimension 2, 3, and 4, but Eqs. (6.5) and (6.6) can be used to describe a system of any dimension—10, 200, or n for any integer n. Thus, we can express algebraic relations between input and state between state and output, and between one state representation and another by matrix algebra on the $\mathbf{F}$, $\mathbf{G}$, and $\mathbf{H}$; then these general expressions can be converted to specific cases by substituting the appropriate values of parameters.

Thus far, we have seen that the ordinary differential equations of physical dynamic systems can be manipulated into state-variable form. In fact, in the field of mathematics, where ordinary differential equations are studied, the state-variable form is called the *normal form* for the equations. There are several good reasons for the study of equations in this form, and some of them are listed here:

To study more general models. The differential equations do not have to be linear or stationary. Thus, by the study of the equations themselves, we develop methods that are very general. Having them in state-variable form gives us a compact, standard form for the study. Of course, in this book we study mainly linear, constant models (for the reasons given earlier).

To introduce the ideas of geometry into differential equations. In physics the plane of position versus velocity of a particle or rigid body is called the *phase plane*. The state is a generalization of that idea to more than two dimensions. While we cannot plot more than three dimensions, the concepts of distance as well as of "orthogonal" and "parallel" and other concepts from geometry can be useful in visualizing the solution of a differential equation as a path in state-space.

To connect internal and external descriptions. The state of a dynamic system often directly describes the flow of internal energy, as when we select position (potential energy), velocity (kinetic energy), capacitor voltage (electric energy), and inductor current (magnetic energy). In every case, the state is equivalent to this direct accounting and the internal energy can always be computed from the state. By a system of analysis that we will shortly describe, we can relate the state to the system inputs and outputs and thus connect the internal variables to the external inputs and to the sensed outputs. In contrast, the transfer function just relates the input to the output and does not show the internal behavior. The state form keeps this sometimes important connection.

6.2.1 Block Diagrams and State-Space

Perhaps the most effective way of understanding the state-space equations is via an analog computer, block-diagram representation. The structure of the representation uses integrators as the central element, which are quite suitable for first-order, state-space representation of equations of motion for a system. Even though the

FIGURE 6.7
An integrator.

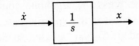

use of analog computers has declined, it is still a useful concept for state-space design and in the circuit design of analog compensation.

The analog computer is a device composed of electric components designed to simulate ordinary differential equations. The basic dynamic component of the analog computer is the *integrator*, constructed from an operational amplifier with a capacitor feedback and a resistor feed-forward, as shown in Fig. 2.16(b). Since an integrator is a device whose input is the derivative of its output, as shown in Fig. 6.7, if, in an analog-computer simulation, we identify the outputs of the integrators as the state, we will then automatically have the equations in state-variable form. Conversely, if a system is described by state variables, we can construct an analog-computer simulation of that system by taking one integrator for each state and connecting its input according to the given equation for that state as expressed in the state equations. The analog-computer diagram is a picture of the state equations.

The components of a typical analog computer used to accomplish these functions are shown in Fig. 6.8. Notice that the operational amplifier has a sign change that gives it a negative gain. Implementation of a simulation of a differential equation with an analog computer is best illustrated by an example, and we will start with a third-order differential equation. Consider the system represented by the equation

$$\dddot{y} + 6\ddot{y} + 11\dot{y} + 6y = 6u. \tag{6.14}$$

If we assume that $\dddot{y}$ is "available," one integration will give $\ddot{y}$, another will give $\dot{y}$, and another will give y itself. With these variables at hand, the equation for y

FIGURE 6.8
Elements of an analog computer.

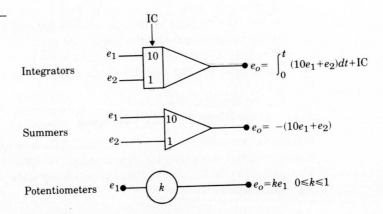

FIGURE 6.9
Block diagram of a
system to solve $\ddot{y} +
6\ddot{y} + 11\dot{y} + 6y = 6u$
using only integrators
as dynamic elements.

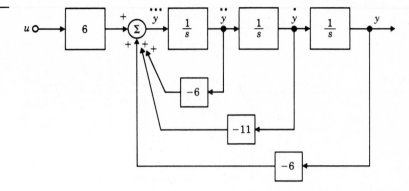

given by Eq. (6.14) can be implemented and the simulation is complete. The block
diagram of the resulting realization is shown in Fig. 6.9. The state variables, the
outputs of the integrators, are $x_1 = y$, $x_2 = \dot{y}$, and $x_3 = \ddot{y}$. In these terms, the
equations of motion are

$$\dot{x}_1 = x_2,$$
$$\dot{x}_2 = x_3,$$
$$\dot{x}_3 = -6x_1 - 11x_2 - 6x_3 + 6u. \tag{6.15}$$

As we noted above, the analog computer based on the operational amplifier
uses integrators that have a negative gain, so a modification of the block diagram
is necessary to take that into account. The realization of this system using analog-
computer components is shown in Fig. 6.10.

Notice here that the outputs of the integrators are $-\ddot{y}$, $\dot{y}$, and $-y$. Also note
that an amplifier (sign changer) is needed to get the correct sign of $\dot{y}$. We frequently
use a block diagram having only integrators as dynamic elements to help visualize
the meaning of certain state-space selections and to display the relations between
input, state, output, and transfer function. The translation of such a block diagram
into specific analog-computer components, including correct signs and gains, is
done only if an actual simulation is to be run.

□ **6.2.2 State Transformations**

A particular definition of the state vector $\mathbf{x}$ can be redefined by any nonsingular
linear transformation of that state, $\mathbf{p} = \mathbf{Tx}$. The new state vector $\mathbf{p}$ will describe
the identical system as $\mathbf{x}$ with suitable changes to $\mathbf{F}$, $\mathbf{G}$, $\mathbf{G}_1$, $\mathbf{H}$ and J.

For example, if, in the spacecraft example of Fig. 6.5, we define a new state
to be

$$\mathbf{p} = \begin{bmatrix} \theta_m \\ \theta \\ q \end{bmatrix}; \quad \text{that is} \quad \mathbf{T} = \begin{bmatrix} 0 & 0 & 1 \\ 0 & 1 & 0 \\ 1 & 0 & 0 \end{bmatrix}$$

FIGURE 6.10
Example of an
analog-computer
simulation of $\ddot{y} + 6\ddot{y} + 11\dot{y} + 6y = 6u$.

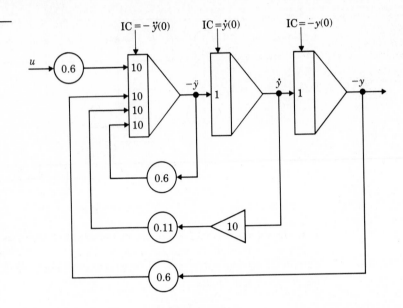

instead of the state definition given in Eq. (6.12), we see that the following revisions of the state-space matrices,

$$\mathbf{F}' = \begin{bmatrix} -2 & 2 & 0 \\ 0 & 0 & 1 \\ 0 & 0 & 0 \end{bmatrix}, \qquad \mathbf{G}' = \mathbf{G}'_1 = \begin{bmatrix} 0 \\ 0 \\ 0.9 \end{bmatrix}, \qquad \mathbf{H}' = [1 \quad 0 \quad 0],$$

will result in identical system dynamics.

For more complicated systems where the revised matrices cannot be determined by inspection, we can derive formulas for the revised matrices (see Problem 6.2):

$$\mathbf{F}' = \mathbf{TFT}^{-1}; \qquad \mathbf{G}' = \mathbf{TG}; \qquad \mathbf{G}'_1 = \mathbf{TG}_1; \qquad \mathbf{H}' = \mathbf{HT}^{-1}. \qquad (6.16)$$

This kind of transformation is sometimes used to arrive at a "canonical" form of the equations, which is well suited to a particular analysis. One commonly used transformation converts $\mathbf{F}$ to a diagonal matrix (if possible); therefore each diagonal element represents a pole of the system transfer function, and the state vector describes the normal modes. Other canonical forms will be described in subsequent sections.

6.2.3 System Transfer Functions

In order to relate the state-variable equations to our earlier consideration of frequency response and poles and zeros, let's now go back to the nonhomogeneous formula, Eq. (6.5), and consider the problem in the frequency domain. Taking the Laplace transform of

$$\dot{\mathbf{x}} = \mathbf{F}\mathbf{x} + \mathbf{G}u, \tag{6.17}$$

we obtain

$$s\mathbf{X}(s) - \mathbf{x}(0) = \mathbf{F}\mathbf{X}(s) + \mathbf{G}U(s), \tag{6.18}$$

which is now an algebraic equation. If we collect the terms involving $\mathbf{X}(s)$ on the left-hand side, we find (remember that in matrix multiplication it is very important to keep the order)*

$$(s\mathbf{I} - \mathbf{F})\mathbf{X}(s) = \mathbf{G}U(s) + \mathbf{x}(0).$$

If we premultiply both sides by the inverse of $(s\mathbf{I} - \mathbf{F})$, then

$$\mathbf{X}(s) = (s\mathbf{I} - \mathbf{F})^{-1}\mathbf{G}U(s) + (s\mathbf{I} - \mathbf{F})^{-1}\mathbf{x}(0). \tag{6.19}$$

The output of the system is

$$Y(s) = \mathbf{H}X(s) + JU(s),$$

which means

$$Y(s) = \mathbf{H}(s\mathbf{I} - \mathbf{F})^{-1}\mathbf{G}U(s) + \mathbf{H}(s\mathbf{I} - \mathbf{F})^{-1}\mathbf{x}(0) + JU(s). \tag{6.20}$$

If we assume that the initial conditions are zero, the input-output relationship is

$$G(s) = \frac{Y(s)}{U(s)} = \mathbf{H}(s\mathbf{I} - \mathbf{F})^{-1}\mathbf{G} + J. \tag{6.21}$$

This G is called the *transfer function* of the system.[†]

EXAMPLE 6.5   Consider the system with state-variable description

$$\mathbf{F} = \begin{bmatrix} -5 & -6 \\ 1 & 0 \end{bmatrix}, \qquad \mathbf{G} = \begin{bmatrix} 1 \\ 0 \end{bmatrix},$$

$$\mathbf{H} = [0 \quad 1], \qquad J = 0.$$

The block diagram corresponding to these matrices is given in Fig. 6.11. To com-

---

*The matrix $\mathbf{I}$ is the identity, a matrix of 1s on the main diagonal and zeros everywhere else and therefore $\mathbf{I}\mathbf{x} = \mathbf{x}$.

[†] Note that the initial-condition assumption is important and is always a factor when dealing with transfer functions.

FIGURE 6.11
Block diagram of a
second-order system.

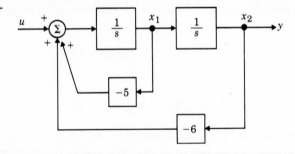

pute the transfer function, form

$$sI - F = \begin{bmatrix} s + 5 & 6 \\ -1 & s \end{bmatrix}$$

and compute

$$(sI - F)^{-1} = \frac{\begin{bmatrix} s & -6 \\ 1 & s + 5 \end{bmatrix}}{s(s + 5) + 6}.$$

Therefore

$$G(s) = \frac{[0 \quad 1]\begin{bmatrix} s & -6 \\ 1 & s + 5 \end{bmatrix}\begin{bmatrix} 1 \\ 0 \end{bmatrix}}{s(s + 5) + 6}$$

$$= \frac{[1 \quad s + 5]\begin{bmatrix} 1 \\ 0 \end{bmatrix}}{s(s + 5) + 6}$$

$$= \frac{1}{(s + 2)(s + 3)}. \qquad (6.22)$$

∎

Since Eq. (6.21) expresses the transfer function in terms of the state-space matrices **F**, **G**, **H**, and *J*, we should be able to express poles and zeros in terms of these matrices too, and we can.

6.2.4 Poles, Zeros, and Eigenvalues from the State-Space Description

A pole of the transfer function $G(s)$ is a value of frequency s such that, if $s = \lambda_i$, then the system responds as $K_i e^{\lambda_i t}$, *with no forcing function.* λ_i is called a *natural frequency* of the system. In state-space we have, from Eq. (6.5) with no forcing function,

$$\dot{\mathbf{x}} = \mathbf{Fx}, \tag{6.23}$$

and we assume

$$\mathbf{x}(0) = \mathbf{x}_0, \tag{6.24}$$

where $\mathbf{x}_0$ is a constant initial-condition vector. If we also assume that $\mathbf{x}(t) = e^{\lambda_i t}\mathbf{x}_0$, it follows that

$$\dot{\mathbf{x}}(t) = \lambda_i e^{\lambda_i t}\mathbf{x}_0 = \mathbf{Fx} = \mathbf{F}e^{\lambda_i t}\mathbf{x}_0 \tag{6.25}$$

or

$$\mathbf{Fx}_0 = \lambda_i \mathbf{x}_0. \tag{6.26}$$

We can rewrite Eq. (6.26) as

$$(\lambda_i \mathbf{I} - \mathbf{F})\mathbf{x}_0 = 0. \tag{6.27}$$

Equations (6.26) and (6.27) are well known in matrix algebra as equations of eigenvalues λ_i and eigenvectors $\mathbf{x}_0$ of the matrix $\mathbf{F}$. For a nonzero $\mathbf{x}_0$, the above equation is satisfied if and only if

$$\det[\lambda_i \mathbf{I} - \mathbf{F}] = 0. \tag{6.28}$$

These equations show that the *poles* of the transfer function are the eigenvalues of system matrix $\mathbf{F}$.

EXAMPLE 6.6   Consider the system with poles at -2 and -3 and a transfer function of

$$G(s) = \frac{1}{s^2 + 5s + 6} = \frac{1}{(s + 2)(s + 3)}.$$

We saw earlier that a state-variable description of this transfer function is

$$\mathbf{F} = \begin{bmatrix} -5 & -6 \\ 1 & 0 \end{bmatrix}, \qquad \mathbf{G} = \begin{bmatrix} 1 \\ 0 \end{bmatrix},$$
$$\mathbf{H} = [0 \quad 1], \qquad J = 0.$$

The poles of the system are found from Eq. (6.28) by solving the *characteristic equation*

$$\det[s\mathbf{I} - \mathbf{F}] = 0,$$
$$\det\begin{bmatrix} s + 5 & 6 \\ -1 & s \end{bmatrix} = 0, \tag{6.29}$$

or

$$s(s + 5) + 6 = (s + 2)(s + 3) = 0,$$

and we confirm that the poles of the system are the eigenvalues of $\mathbf{F}$.   ∎

The eigenvalues of quite large systems can be computed using the *QR* algorithm* (Francis, 1961).

The zeros of the system can also be determined from the state-variable description matrices (**F**, **G**, **H**, and *J*). A *zero* is a value of frequency *s* such that if the input is exponential at the zero frequency, or

$$u(t) = u_0 e^{st}, \tag{6.30}$$

then the output is identically zero; that is,

$$y(t) \equiv 0, \tag{6.31}$$

which means that the system has a nonzero input signal but nothing comes out! In state-space description, we would have

$$u = u_0 e^{st}; \qquad \mathbf{x}(t) = \mathbf{x}_0 e^{st}; \qquad y(t) \equiv 0. \tag{6.32}$$

Thus, we have that

$$\dot{\mathbf{x}} = s e^{st} \mathbf{x}_0 = \mathbf{F} e^{st} \mathbf{x}_0 + \mathbf{G} u_0 e^{st} \tag{6.33}$$

or

$$[s\mathbf{I} - \mathbf{F}, -\mathbf{G}] \begin{bmatrix} \mathbf{x}_0 \\ u_0 \end{bmatrix} = \mathbf{0} \tag{6.34}$$

and

$$y = \mathbf{H}x + Ju = \mathbf{H} e^{st} \mathbf{x}_0 + J u_0 e^{st} \equiv 0. \tag{6.35}$$

Combining Eqs. (6.34) and (6.35), we have that

$$\begin{bmatrix} s\mathbf{I} - \mathbf{F} & -\mathbf{G} \\ \mathbf{H} & J \end{bmatrix} \begin{bmatrix} \mathbf{x}_0 \\ u_0 \end{bmatrix} = \begin{bmatrix} \mathbf{0} \\ 0 \end{bmatrix}. \tag{6.36}$$

A *zero* of the state-space system is a value of *s* where the above equation has a nontrivial solution. With one input and one output, the matrix is square and a solution to Eq. (6.36) is equivalent to a solution to

$$\det \begin{bmatrix} s\mathbf{I} - \mathbf{F} & -\mathbf{G} \\ \mathbf{H} & J \end{bmatrix} = 0. \tag{6.37}$$

EXAMPLE 6.7   Consider a transfer function with a zero at -2

$$G(s) = \frac{s + 2}{s(s + 3)},$$

---

*This algorithm is available at most computer centers and is carefully documented in the software package EISPAK (Garbow et al., 1977).

which corresponds to the state-variable description

$$\mathbf{F} = \begin{bmatrix} -3 & 1 \\ 0 & 0 \end{bmatrix}; \qquad \mathbf{G} = \begin{bmatrix} 1 \\ 2 \end{bmatrix};$$

$$\mathbf{H} = [1 \quad 0]; \qquad J = 0.$$

The zeros are such that [from Eq. (6.37)]

$$\det \begin{bmatrix} s+3 & -1 & -1 \\ 0 & s & -2 \\ 1 & 0 & 0 \end{bmatrix} = 0,$$

or

$$2 + s = 0,$$

$$s = -2. \qquad \blacksquare$$

These results can be combined to express the whole transfer function in a compact form (see Appendix C):

$$G(s) = \frac{\det \begin{bmatrix} s\mathbf{I} - \mathbf{F} & -\mathbf{G} \\ \mathbf{H} & J \end{bmatrix}}{\det[s\mathbf{I} - \mathbf{F}]}, \qquad (6.38)$$

since both numerator and denominator polynomials can be expressed in terms of determinants. For a numerically stable algorithm to compute transfer functions, see Emami-Naeini and Van Dooren (1982). Given the transfer function, we can compute the frequency response as $G(j\omega)$, and, as we have seen, Eqs. (6.28) and (6.37) can be used to find the poles and zeros from which the transient response follows, as we saw in Section 2.7.

6.3   Control Law Design for Pole Placement

One of the attractive features of the state-space design method is that the procedure consists of two independent steps. One step assumes that we have all the elements of the state vector at our disposal for feedback purposes. In general, of course, this would be an unreasonable assumption, since a well-trained control designer would know that, using other design methods, it would not be necessary to provide that many sensors. The assumption that all state variables are available merely allows us to proceed with the first design step, namely, the control law. The next step is to design an *estimator* (sometimes called an *observer*), which estimates the entire state vector when provided with the measurements of the system indicated by the output [Eq. (6.6)]. The final step will consist of combining the control law and the estimator, where the control law calculations are based on the estimated state

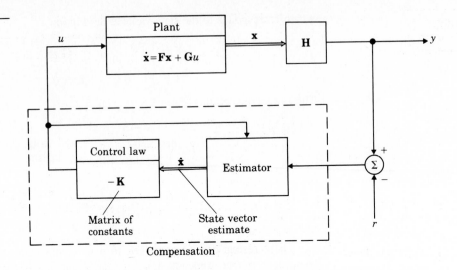

FIGURE 6.12
Schematic diagram of state-space design elements.

variables rather than the actual state. In Section 6.6, we show that this substitution is reasonable and that the combined control law and estimator gives closed-loop dynamics (as determined by the closed-loop pole locations) that are unchanged from those determined when designing the control law and estimator separately. Figure 6.12 shows how the control law and the estimator fit together and how the combination takes the place of what we have been previously referring to as dynamic *compensation*. The figure shows the command input r introduced in the same relative position as previously; however, in a later section we will show how a different location will usually provide superior control.

The first step, then, is to find the control law. Assuming temporarily that there is no command input $(r = 0)$,* the control law is the simple feedback of a linear combination of all the state variables—that is,

$$u = -\mathbf{K}\mathbf{x} = -[K_1 K_2 \cdots K_n] \begin{bmatrix} x_1 \\ x_2 \\ \vdots \\ x_n \end{bmatrix}. \qquad (6.39)$$

Therefore the system has a constant matrix in the state-vector feedback path, as shown in Fig. 6.13. For an nth-order system, there will be n feedback gains $K_1, \ldots, K_n$, and since there are n roots of the system, there are enough degrees of freedom to select arbitrarily any desired set of root locations by proper

---

*This assumption will be removed in Section 6.7.

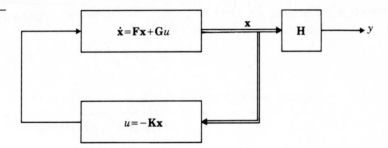

FIGURE 6.13
Assumed system for the control-law design.

choice of the K's. This is not always so in transform design methods where the number of design parameters in the compensation is usually less than the order of the system being controlled.

Substituting the feedback law, Eq. (6.39), into the system, Eq. (6.5) yields

$$\dot{\mathbf{x}} = \mathbf{Fx} - \mathbf{GKx} + \mathbf{G}_1 w. \tag{6.40}$$

Therefore, the homogeneous portion, which determines the system's natural behavior, is

$$\dot{\mathbf{x}} = (\mathbf{F} - \mathbf{GK})\mathbf{x}, \tag{6.41}$$

and the characteristic equation of this closed-loop system is

$$\det[s\mathbf{I} - (\mathbf{F} - \mathbf{GK})] = 0. \tag{6.42}$$

When evaluated, this yields an nth-order polynomial in s containing the n gains $K_1, \ldots, K_n$. The control-law design then consists of picking the gains $\mathbf{K}$ so that the roots of Eq. (6.42) are in desirable locations.

Selection of desirable locations for the roots usually requires some iteration by the designer. Some of the issues in their selection will be discussed in the following examples and will be further discussed in Section 6.4. For now, we will assume that the desired locations are known, say,

$$s = s_1, s_2, \ldots, s_n.$$

Then the desired (control) characteristic equation is

$$\alpha_c(s) = (s - s_1)(s - s_2)\cdots(s - s_n) = 0. \tag{6.43}$$

Hence, the required elements of $\mathbf{K}$ are obtained by matching coefficients in Eqs. (6.42) and (6.43), thus forcing the system-characteristic equation to be identical with the desired formula, Eq. (6.43). An example should help clarify this pole placement idea.

EXAMPLE 6.8   Suppose we have an undamped oscillator with frequency ω_0 and a state-space description given by

$$\begin{bmatrix} \dot{x}_1 \\ \dot{x}_2 \end{bmatrix} = \begin{bmatrix} 0 & 1 \\ -\omega_0^2 & 0 \end{bmatrix} \begin{bmatrix} x_1 \\ x_2 \end{bmatrix} + \begin{bmatrix} 0 \\ 1 \end{bmatrix} u, \qquad (6.44)$$

and we wish to relocate the poles of the system so that they are both at $-2\omega_0$. In other words, we wish to double the natural frequency and increase the damping from $\zeta = 0$ to $\zeta = 1$. From Eq. (6.43), we find that

$$\alpha_c(s) = (s + 2\omega_0)^2 = s^2 + 4\omega_0 s + 4\omega_0^2, \qquad (6.45)$$

and from Eq. (6.42),

$$\det[s\mathbf{I} - (\mathbf{F} - \mathbf{GK})] = \det\left(\begin{bmatrix} s & 0 \\ 0 & s \end{bmatrix} - \left\{ \begin{bmatrix} 0 & 1 \\ -\omega_0^2 & 0 \end{bmatrix} - \begin{bmatrix} 0 \\ 1 \end{bmatrix} [K_1 \quad K_2] \right\} \right)$$

or

$$s^2 + K_2 s + \omega_0^2 + K_1 = 0. \qquad (6.46)$$

Equating the coefficients with like powers of s in Eqs. (6.45) and (6.46) yields

$$K_2 = 4\omega_0,$$
$$\omega_0^2 + K_1 = 4\omega_0^2, \qquad (6.47)$$

and therefore

$$K_1 = 3\omega_0^2,$$
$$K_2 = 4\omega_0.$$

Or, more concisely, the control law is

$$\mathbf{K} = [K_1 \quad K_2] = [3\omega_0^2 \quad 4\omega_0]. \qquad (6.48)$$

Figure 6.14 shows the response of the closed-loop system to the initial condition $x_1 = 0.3$, $x_2 = -0.5$, and $\omega_0 = 1$. It shows a very well-damped response, as would be expected from two roots at $s = -2$.   ∎

The calculation of the gains using the technique illustrated in this example becomes rather tedious when the order of the system is larger than 3. There are, however, special "canonical" forms of the state-variable equations where the algebra for finding the gains is especially simple. One such canonical form useful in control-law design is the *control canonical form*. Consider the third-order system*

$$\dddot{y} + a_1\ddot{y} + a_2\dot{y} + a_3 y = b_1\ddot{u} + b_2\dot{u} + b_3 u, \qquad (6.49)$$

---

*This development is exactly the same for higher-order systems.

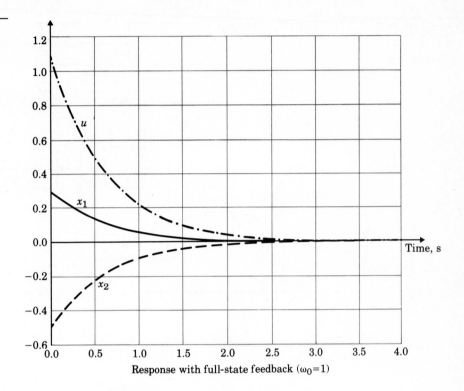

FIGURE 6.14
Undamped oscillator
response with full-state
feedback ($\omega_0 = 1$).

Response with full-state feedback ($\omega_0=1$)

which corresponds to the transfer function

$$H(s) = \frac{Y(s)}{U(s)} = \frac{b_1 s^2 + b_2 s + b_3}{s^3 + a_1 s^2 + a_2 s + a_3} = \frac{b(s)}{a(s)}. \qquad (6.50)$$

Suppose we introduce an auxiliary variable (referred to as the *partial state*) ξ, which relates to $a(s)$ and $b(s)$ as shown in Fig. 6.15(a). The transfer function of u to ξ is

$$\frac{\xi(s)}{U(s)} = \frac{1}{a(s)} \qquad (6.51)$$

or

$$\dddot{\xi} + a_1 \ddot{\xi} + a_2 \dot{\xi} + a_3 \xi = u. \qquad (6.52)$$

It is easy to draw a block diagram corresponding to Eq. (6.52) if we rearrange the equation as follows:

$$\dddot{\xi} = -a_1 \ddot{\xi} - a_2 \dot{\xi} - a_3 \xi + u. \qquad (6.53)$$

FIGURE 6.15
Derivation of control
canonical form.

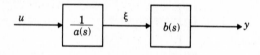

(a)

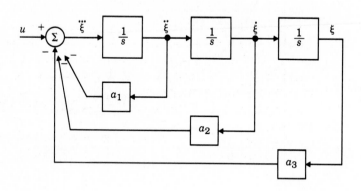

(b)

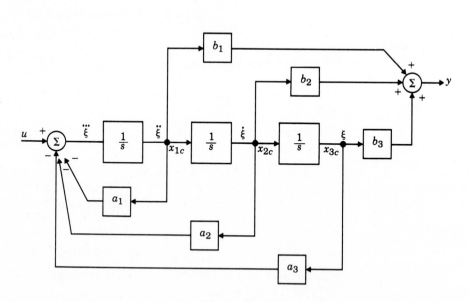

(c)

The summation indicated is shown in Fig. 6.15(b), where each ξ on the right-hand side is obtained by sequential integrations of $\dddot{\xi}$. To form the output, we go back to Fig. 6.15(a) and note that

$$Y(s) = b(s)\xi(s), \tag{6.54}$$

which means

$$y = b_1\ddot{\xi} + b_2\dot{\xi} + b_3\xi. \tag{6.55}$$

We again pick off the outputs of the integrators, multiply them by $\{b_i\}$'s, and form the right-hand side of Eq. (6.49) by using a summer to yield the output, as shown in Fig. 6.15(c). In this case, all the feedback loops return to the point of application of the input, or "control," variable, and hence the form is referred to as the *control canonical form*. Reduction of the structure by Mason's rule or by elementary operations verifies that this structure has the transfer function given by $H(s)$.

Taking the state as the outputs of the three integrators numbered, by convention, from the left,

$$x_1 = \ddot{\xi}, \qquad x_2 = \dot{\xi}, \qquad x_3 = \xi,$$

we obtain

$$\dot{x}_1 = \dddot{\xi} = -a_1x_1 - a_2x_2 - a_3x_3 + u,$$
$$\dot{x}_2 = x_1,$$
$$\dot{x}_3 = x_2, \tag{6.56}$$

or, in our matrix notation,

$$\mathbf{F}_c = \begin{bmatrix} -a_1 & -a_2 & -a_3 \\ 1 & 0 & 0 \\ 0 & 1 & 0 \end{bmatrix}; \qquad \mathbf{G}_c = \begin{bmatrix} 1 \\ 0 \\ 0 \end{bmatrix};$$
$$\mathbf{H}_c = [\, b_1 \quad b_2 \quad b_3 \,]; \qquad J_c = 0. \tag{6.57}$$

The special structure of the system matrix is referred to as the *(upper) companion form*. The characteristic equation of such a system is $a(s) = s^3 + a_1s^2 + a_2s + a_3$ and has the property that its coefficients are the elements in the first row of $\mathbf{F}_c$. If we now form the closed-loop system matrix, $\mathbf{F}_c - \mathbf{G}_c\mathbf{K}$, we find that

$$\mathbf{F}_c - \mathbf{G}_c\mathbf{K} = \begin{bmatrix} -a_1 - K_1 & -a_2 - K_2 & -a_3 - K_3 \\ 1 & 0 & 0 \\ 0 & 1 & 0 \end{bmatrix}. \tag{6.58}$$

By inspection [comparing Eq. (6.58) with Eq. (6.57)], the characteristic equation [Eq. (6.58)] is

$$s^3 + (a_1 + K_1)s^2 + (a_2 + K_2)s + (a_3 + K_3) = 0. \tag{6.59}$$

Therefore, if the desired pole locations result in the characteristic equation given by Eq. (6.43),

$$\alpha_c(s) = s^3 + \alpha_1 s^2 + \alpha_2 s + \alpha_3 = 0, \qquad (6.60)$$

the necessary feedback gains can be found by equating the coefficients in Eqs. (6.59) and (6.60). This result is

$$K_1 = -a_1 + \alpha_1, \qquad K_2 = -a_2 + \alpha_2, \qquad K_3 = -a_3 + \alpha_3. \qquad (6.61)$$

We now have the basis for a design procedure: Given an arbitrary $(\mathbf{F}, \mathbf{G})$ and a desired characteristic polynomial $\alpha_c(s)$, we transform $(\mathbf{F}, \mathbf{G})$ to control canonical form $(\mathbf{F}_c, \mathbf{G}_c)$ and solve for the control gains by inspection via Eq. (6.61). Since this gain is for the state in the control form, we must transform the gain back to the original state. (See Appendix 6C for a development of this method.)

The fact that one can shift the poles of a system by state feedback to any desired locations is a rather remarkable result. However, we must pause and ask ourselves whether this can always be done. Equivalently, we ask ourselves if one can always transform $(\mathbf{F}, \mathbf{G})$ to the control form $(\mathbf{F}_c, \mathbf{G}_c)$. The answer is always "yes" if the system has a property called *controllability*. Roughly speaking, a system is "controllable" if the state can be moved in any desired direction by suitable choice of control signals. Equivalently, if the system is controllable, the roots of the system can be moved to any desired locations by state feedback. In rare instances, the system may be "uncontrollable," in which case no possible control will yield arbitrary root locations. These systems have certain modes, or subsystems, that are unaffected by the control. This usually means that parts of the system are physically disconnected from the input. For example, the steering wheel of a car cannot be used to control the engine speed. Engine speed is not "controllable" by the steering wheel but it is controllable by the throttle.

If there are no repeated roots of the system (that is, if the roots are "distinct"), it is always possible to choose a *realization* (selection of state) so that each state variable represents a natural mode of the system. This means that the system can be represented as

$$\dot{\mathbf{x}} = \begin{bmatrix} \lambda_1 & & & \bigcirc \\ & \lambda_2 & & \\ & & \ddots & \\ \bigcirc & & & \lambda_n \end{bmatrix} \mathbf{x} + \begin{bmatrix} g_1 \\ g_2 \\ \vdots \\ g_n \end{bmatrix} u, \qquad (6.62)$$

which we refer to as the *diagonal form*. This form explicitly exhibits the criterion for controllability. All the elements in the input matrix $\mathbf{G}$ must be nonzero if the system is controllable. Otherwise, the modes corresponding to zero entries in $\mathbf{G}$ are not influenced by control, and the associated state variable will conform to their natural behavior. This would, of course, be a sad state of affairs if the uncontrollable modes happen to be unstable. A good physical understanding of the system being controlled would prevent any attempt to design a controller for an uncontrollable system. There are mathematical tests for controllability in Appendix 6B, which are applicable to any state-space description that essentially evaluates **G** in Eq. (6.62), tests whether all elements are nonzero, and provides additional help in discovering this property. However, no mathematical test can replace the control engineer's physical understanding of the system. This is especially true since often the physical situation is such that every mode is controllable to some degree, and, while the mathematical tests indicate the system is controllable, certain modes are so weakly controllable that control design is ineffective for all practical purposes.

Airplane control is a good example of weak controllability of certain modes. In steady cruise, pitch plane motion $\mathbf{x}_p$ is primarily affected by the elevator δe and weakly affected by rolling motion $\mathbf{x}_r$. Rolling motion is mostly affected by the ailerons δa. The state-space description of these facts is

$$\begin{bmatrix} \dot{\mathbf{x}}_p \\ \dot{\mathbf{x}}_r \end{bmatrix} = \begin{bmatrix} \mathbf{F}_p & \boldsymbol{\epsilon} \\ 0 & \mathbf{F}_r \end{bmatrix} \begin{bmatrix} \mathbf{x}_p \\ \mathbf{x}_r \end{bmatrix} + \begin{bmatrix} \mathbf{G}_p & 0 \\ 0 & \mathbf{G}_r \end{bmatrix} \begin{bmatrix} \delta e \\ \delta a \end{bmatrix},$$

where ϵ represents the weak coupling from rolling motion to pitching motion and may result from trimming* during steady turning flight.

A mathematical test of controllability for this system would conclude that pitch plane motion (and therefore altitude) is controllable by the ailerons as well as by the elevator! But physically, it is unreasonable to attempt to control an airplane's altitude by rolling the aircraft with the ailerons.

In addition to the transformation method described, we have an alternative method for solving for the feedback gains of a controllable system: Ackermann's formula. This formula organizes the process of converting to $(\mathbf{F}_c, \mathbf{G}_c)$ and back again in a very convenient fashion. This rather elegant formula was derived by Ackermann (1972), and a proof can be found in Appendix 6C. The relation is

$$\mathbf{K} = [0 \quad \cdots \quad 0 \quad 1]\mathscr{C}^{-1}\alpha_c(\mathbf{F}), \tag{6.63}$$

where

$$\mathscr{C} = [\mathbf{G} \quad \mathbf{FG} \quad \mathbf{F}^2\mathbf{G} \quad \cdots \quad \mathbf{F}^{n-1}\mathbf{G}]$$

is the *controllability matrix*,† n is the order of the system or the number of state variables, and $\alpha_c(\mathbf{F})$ is defined as

$$\alpha_c(\mathbf{F}) = \mathbf{F}^n + \alpha_1\mathbf{F}^{n-1} + \alpha_2\mathbf{F}^{n-2} + \cdots + \alpha_n\mathbf{I}, \tag{6.64}$$

---

*A reference condition where the body axis accelerations are zero.

† We should mention that the matrix should not be inverted explicitly, as it may be ill-conditioned. Rather Eq. (6.63) should be converted to solving linear equations by a numerically stable method such as gaussian elimination with pivoting. For large problems, other methods should be used (see Problem 6.32).

where the α_i's are the coefficients of the desired characteristic polynomial [Eq. (6.43)]:

$$\alpha_c(s) = s^n + \alpha_1 s^{n-1} + \cdots + \alpha_n. \tag{6.65}$$

Note that Eq. (6.64) is a matrix equation.

EXAMPLE 6.9   Applying this formula to the undamped oscillator example, we find that

$$\alpha_1 = 4\omega_0, \qquad \alpha_2 = 4\omega_0^2$$

and therefore

$$\alpha_c(\mathbf{F}) = \begin{bmatrix} -\omega_0^2 & 0 \\ 0 & -\omega_0^2 \end{bmatrix} + 4\omega_0 \begin{bmatrix} 0 & 1 \\ -\omega_0^2 & 0 \end{bmatrix} + 4\omega_0^2 \begin{bmatrix} 1 & 0 \\ 0 & 1 \end{bmatrix} = \begin{bmatrix} 3\omega_0^2 & 4\omega_0 \\ -4\omega_0^3 & 3\omega_0^2 \end{bmatrix}.$$

Furthermore,

$$\mathscr{C} = [\mathbf{G} \quad \mathbf{FG}] = \begin{bmatrix} 0 & 1 \\ 1 & 0 \end{bmatrix},$$

$$\mathscr{C}^{-1} = \begin{bmatrix} 0 & 1 \\ 1 & 0 \end{bmatrix},$$

and finally

$$\mathbf{K} = [K_1 \quad K_2] = [0 \quad 1] \begin{bmatrix} 0 & 1 \\ 1 & 0 \end{bmatrix} \begin{bmatrix} 3\omega_0^2 & 4\omega_0 \\ -4\omega_0^3 & 3\omega_0^2 \end{bmatrix}.$$

Therefore

$$\mathbf{K} = [3\omega_0^2 \quad 4\omega_0],$$

which is the same result as was obtained previously.   ∎

Implementation of Ackermann's formula for the design of single-input, single-output systems and more complicated systems is usually accomplished with a computer aid, such as MATLAB, CTRL-C, or MATRIX$_x$.

Generally speaking, Ackermann's formula starts to break down for systems of higher order than 6 or 7. The reason is that forming $\mathscr{C}$ is not a good idea from a numerical point of view. Even if the system is completely controllable, the columns of $\mathscr{C}$ tend to become nearly linearly dependent with finite precision arithmetic. Careful selection of units and scaling on the state variables and input should help the computation. For results of experiments with various feedback gain calculation techniques the reader is referred to Laub et al. (1985). Robust algorithms for pole placement are reported in Kautsky, Nichols and Van Dooren (1985), Misra and Patel (1989), and Minimis and Paige (1982).

6.4 Selection of Pole Locations for Good Design

The first step in the pole-placement design approach described in the last section is to decide on the closed-loop pole locations. When selecting pole locations, it is always useful to keep in mind that the control effort required is related to how far the open-loop poles are moved by the feedback. Furthermore, open-loop zeros attract poles, so considerable control effort is required to move a pole away from a nearby zero. Therefore a pole-placement philosophy that aims to fix only the undesirable aspects of the open-loop response will typically allow smaller control actuators than one that arbitrarily picks all the poles in some location without regard to the original open-loop poles. Sometimes, closed-loop pole locations are specified, as in aircraft flying qualities specifications. In this section, we discuss three techniques to aid in the pole-selection process. The first two approaches deal with pole selection without explicit regard for their effect on control effort. The third method (called optimal control, or symmetric root locus) was developed to specifically address the issue of achieving a balance between good system response and control effort.

Dominant Second-Order

The second-order transient with complex poles at radius ω_n and damping ζ was discussed in Chapter 2. The rise time, overshoot, and settling time can be deduced directly from the pole locations. The closed-loop poles for a system can be chosen as a desired pair of dominant second-order poles, with the rest of the poles selected to have real parts corresponding to sufficiently damped modes so the system will mimic a second-order response with reasonable control effort. It is also necessary to be sure that the zeros are far enough in the LHP so as not to have any appreciable effect on the second-order behavior. A system with several lightly damped high-frequency vibration modes plus two "rigid-body" low-frequency modes lends itself to this philosophy. Here one can pick the low-frequency modes to achieve a desired ω_n and ζ and increase the damping of the high-frequency modes, holding their frequency constant in order to minimize control effort.

Prototype Design

For higher-order systems, which cannot be usefully approximated as a perturbation on a second-order system, the situation is more complex. Furthermore, the variety of responses is too great for a catalog to be very useful.* What can be helpful is a selection of higher-order prototype responses with very particular characteristics against which systems can usefully be compared. There are several sets of these, of which the ITAE and the Bessel responses seem most relevant to control.

*For one such catalog, see Elgerd and Stephens (1959).

One prototype set of transient responses was worked out by Graham and Lathrop (1953) to minimize the integral of the time multiplied by the absolute values of the error (ITAE)—that is, to minimize

$$\mathcal{I} = \int_0^\infty t|e|\, dt.$$

The evaluation was done numerically, and the resulting pole locations and step responses are shown in Table 6.1(a) and Fig. 6.16, respectively, for a nominal cutoff frequency $\omega_0 = 1$ rad/s.

For those applications where overshoot must be avoided altogether, the Bessel filter prototype can be considered. These transfer functions are given by $1/B_n(s)$, where $B_n(s)$ is the nth-degree Bessel polynomial. The pole locations are given in Table 6.1(b), and the step responses are shown in Fig. 6.17.

Comparing the transient responses shown in Fig. 6.16 with those shown in Fig. 6.17, it is seen that the ITAE has some overshoot, whereas the Bessel prototype has almost no overshoot. However, the ITAE responses have a faster rise time compared with Bessel prototypes. Further insight into the two prototype designs can be gained from studying their frequency responses shown in Fig. 6.18. It shows that, given the same ω_0, the ITAE has a higher bandwidth for the same attenuation

TABLE 6.1
Prototype response poles

(a) ITAE transfer functions	k	Pole locations for $\omega_0 = 1$ rad/s*
	1	$s + 1$
	2	$s + 0.7071 \pm j0.7071^\dagger$
	3	$(s + 0.7081)(s + 0.5210 \pm j1.068)$
	4	$(s + 0.4240 \pm j1.2630)(s + 0.6260 \pm j0.4141)$
	5	$(s + 0.8955)(s + 0.3764 \pm j1.2920)$ $\cdot(s + 0.5758 \pm j0.5339)$
	6	$(s + 0.3099 \pm j1.2634)(s + 0.5805 \pm j0.7828)$ $\cdot(s + 0.7346 \pm j0.2873)$

(b) Bessel transfer functions	k	Pole locations for $\omega_0 = 1$ rad/s*
	1	$s + 1$
	2	$s + 0.8660 \pm j0.5000^\dagger$
	3	$(s + 0.9420)(s + 0.7455 \pm j0.7112)$
	4	$(s + 0.6573 \pm j0.8302)(s + 0.9047 \pm j0.2711)$
	5	$(s + 0.9264)(s + 0.5906 \pm j0.9072)$ $\cdot(s + 0.8516 \pm j0.4427)$
	6	$(s + 0.5385 \pm j0.9617)(s + 0.7998 \pm j0.5622)$ $\cdot(s + 0.9093 \pm j0.1856)$

* Pole locations for other values of ω_0 can be obtained by substituting s/ω_0 for s everywhere.

$\dagger$ The factors $(s + a + jb)(s + a - jb)$ are written as $(s + a \pm jb)$ to conserve space.

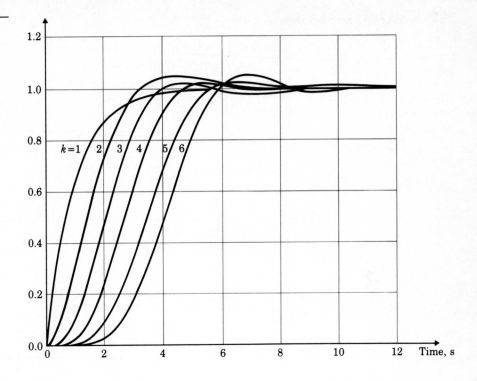

FIGURE 6.16
Step response of ITAE prototypes for $\omega_0 = 1$ rad/s.

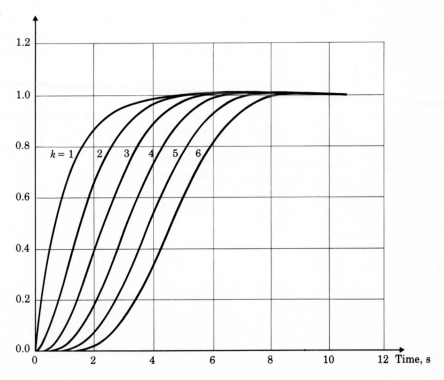

FIGURE 6.17
Step responses of the Bessel prototype systems for $\omega_0 = 1$ rad/s.

389

FIGURE 6.18
Frequency response
magnitude of ITAE and
Bessel prototype systems.

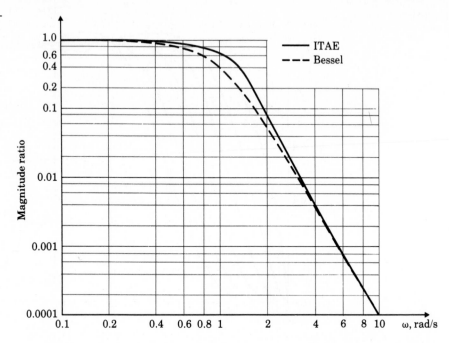

at higher frequencies. Therefore, by selecting ω_0 to give the same bandwidths, one can achieve reduced sensitivity to sensor noise at high frequencies with the ITAE.

Note that all the poles in a particular prototype response have similar natural frequencies. Therefore, direct application of a prototype design to a system with a large spread in the frequencies of its open-loop modes will likely result in excessive control usage. An alternative approach would be to use a prototype pole constellation for the lower-frequency portions of the system and simply add some damping, if required, to the higher-frequency portions. Examples of these approaches are found in Section 6.6.

The Symmetric Root Locus

A most effective and widely used design technique of linear control system design is the optimal linear quadratic regulator (LQR) design. The simplified version of the LQR problem is to find the control such that the performance index

$$\mathcal{J} = \int_0^\infty [\rho z^2(t) + u^2(t)]\, dt \tag{6.66}$$

is minimized for the system

$$\dot{\mathbf{x}} = \mathbf{F}\mathbf{x} + \mathbf{G}u,$$
$$z = \mathbf{H}_1\mathbf{x}, \tag{6.67}$$

where ρ in Eq. (6.66) is a weighting factor of the designer's choice. It can be shown (see Chapter 5, Bryson and Ho, 1969) that the control that minimizes the $\mathscr{J}$ in Eq. (6.66) is given by linear-state feedback

$$u = -\mathbf{K}\mathbf{x}, \tag{6.68}$$

where the *optimal* value of $\mathbf{K}$ is obtained as follows. It is shown in Kailath (1980) that the optimal closed-loop poles are given by the *stable* roots (those in the LHP) of the equation

$$1 + \rho G_0(-s)G_0(s) = 0, \tag{6.69}$$

where G_0 is the open-loop transfer function,

$$G_0(s) = \frac{z(s)}{u(s)} = \mathbf{H}_1(s\mathbf{I} - \mathbf{F})^{-1}\mathbf{G} = \frac{N(s)}{D(s)}. \tag{6.70}$$

Note that this is a root-locus problem, as discussed in Chapter 4, with respect to the parameter ρ, which weights the "tracking error" z^2 versus the control effort u^2 in the performance index [Eq. (6.66)]. Note also that s and $-s$ affect the equation in an identical manner; therefore, for any root s_0 of the equation, there will also be a root at $-s_0$. We thus call the resulting root locus a symmetric root locus (SRL), and the locus in the LHP will have a mirror image in the RHP; that is, there is symmetry with respect to the imaginary axis. Selection of the optimal roots may thus be accomplished by first selecting the matrix $\mathbf{H}_1$, which defines the "tracking error" and then choosing ρ, which balances the importance of tracking error against control effort. Notice that the output selected for this purpose does *not* need to be the output measured and used for feedback.

Solving Eq. (6.69) results in the desired closed-loop roots, which are then used in a pole-placement calculation such as Ackermann's formula [Eq. (6.63)] to obtain $\mathbf{K}$.

As with all root loci for real transfer functions G_0, the locus is also symmetric with respect to the real axis, and thus there is the special *double symmetry* with respect to both the real and imaginary axes. The SRL equation can be written in the standard form,

$$1 + \rho\frac{N(-s)N(s)}{D(-s)D(s)} = 0, \tag{6.71}$$

and it can be obtained by reflecting the open-loop poles and zeros of the plant across the imaginary axis (which doubles the number of poles and zeros), and then sketching the locus. Note that the locus could be either a $0°$ or $180°$ locus, depending on the sign of $G_0(-s)G_0(s)$. A quick way to determine which type of locus ($0°$ or $180°$) is to pick the one that has *no* part on the imaginary axis. This is because the optimal closed-loop system is guaranteed to be stable, and

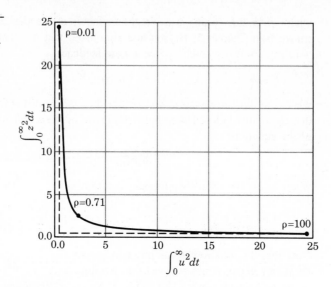

thus no segment of the locus can lie on the imaginary axis. Examination of the SRL can provide the designer with pole locations determined so as to achieve varying balances (different ρ's) between fast response and low control. Figure 6.19 shows the design-tradeoff curve for the double integrator plant [Eqs. (6.3) and (6.4)] for various values of ρ ranging from .01 to 100. The tradeoff is between control energy ($\int u^2 dt$) and the rate of decay of the output ($\int z^2 dt$). The curve has two asymptotes (dashed lines) corresponding to low (large ρ) and high (small ρ) penalty on the control usage. In practice, usually a value of ρ corresponding to a point close to the knee of the tradeoff curve is chosen. This is because it provides a reasonable compromise between the use of control and the speed of response. For the double-integrator plant, the value of $\rho = 0.71$ corresponds to the knee of the curve. In this case the closed-loop poles have a damping ratio of $\zeta = 0.707$! Figure 6.20 shows the associated Nyquist plot, which has a phase margin PM $= 65°$ and infinite gain margin.* It is interesting to consider the limiting behavior of the optimal closed-loop poles as a function of the root-locus parameter (i.e., ρ) although, in practice, neither case would be used.

* It has been proved (Anderson and Moore, 1971) that the Nyquist plot for an LQR design avoids a circle of radius one centered at the -1 point. This leads to extraordinary phase and gain margin properties.

FIGURE 6.20
Nyquist plot for LQR
design.

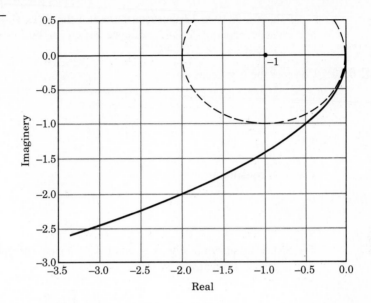

"*Expensive control" case ($\rho \to 0$)* Equation (6.66) penalizes output deviations as well as the use of control energy. If the control is expensive, the optimal control does not move any of the open-loop poles except for those that are in the right-half plane (RHP). The poles in the RHP are simply moved to their mirror images in the LHP. The optimal control does this to stabilize the system using minimum control effort and makes no attempt to move any of the poles of the system in the LHP. The closed-loop pole locations are simply the starting points on the SRL in the LHP. The optimal control does not speed up the response of the system in this case. For the double-integrator plant, the vertical dashed line in Fig. 6.19 corresponds to the "expensive control" case.

"*Cheap control" case ($\rho \to \infty$)* In this case, control energy is no object and may be used by the optimal control law. The control law then moves some of the closed-loop pole locations right on top of the zeros in the LHP. The rest are moved to infinity along the SRL asymptotes. If the system is nonminimum phase, some of the closed-loop poles are moved to mirror images of these zeros in the LHP. The rest of the poles go to infinity and do so along a Butterworth filter pole pattern. The optimal control law provides the fastest possible response time. The feedback gain matrix K becomes unbounded in this case. For the double-integrator plant, the horizontal dashed line in Fig. 6.19 corresponds to the "cheap control" case.

Examples 6.10–6.12 illustrate the use of the SRL while examples of the use of dominant second-order and prototype design can be found in Section 6.6.

EXAMPLE 6.10   Consider the first-order system

$$\dot{y} = -ay + u,$$

$$G(s) = \frac{1}{s + a}.$$

The SRL equation is

$$1 + \rho \frac{1}{(-s + a)(s + a)} = 0.$$

The SRL is as shown in Fig. 6.21 and is a 0° locus. The optimal (stable) pole can be determined explicitly in this case as

$$s = -\sqrt{a^2 + \rho}.$$

This result states that the closed-loop root location that minimizes the cost function of Eq. (6.66) lies on the real axis at the distance given above, which is always to the left of the open-loop root.   ∎

EXAMPLE 6.11   Consider the unstable second-order system that corresponds to the inverted pendulum equation of motion:

$$\dot{\mathbf{x}} = \begin{bmatrix} 0 & 1 \\ a^2 & 0 \end{bmatrix} \mathbf{x} + \begin{bmatrix} 0 \\ -1 \end{bmatrix} u.$$

FIGURE 6.21
Symmetric root locus for Example 6.10.

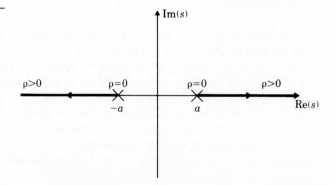

Select an output equally weighted on position and velocity; that is,

$$y = [\,1 \quad 1\,]\mathbf{x}.$$

We calculate from Eq. (6.70):

$$G_0(s) = \frac{s + 1}{s^2 - a^2},$$

and the symmetric loci are as shown as in Fig. 6.22. Two stable roots would be chosen for a given value of ρ and used for pole-placement and control-law design.   ∎

FIGURE 6.22
Symmetric root locus for
Example 6.11A.

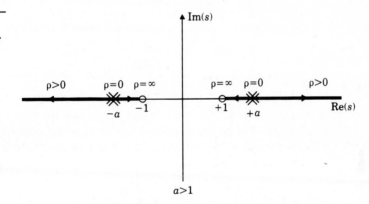

$a > 1$

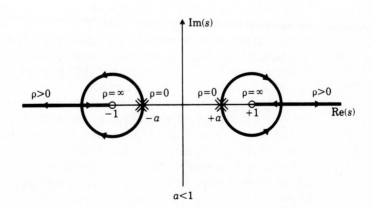

$a < 1$

FIGURE 6.23
Symmetric root locus for
Example 6.12.

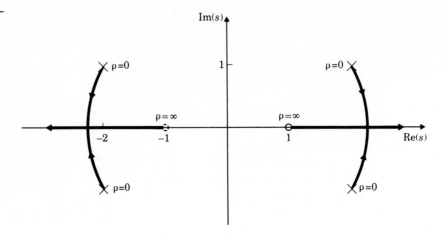

EXAMPLE 6.12   Consider the second-order nonminimum phase system with

$$G_0(s) = \frac{(s - 1)}{(s + 2 \pm j1)}.$$

The symmetric root locus is as shown in Fig. 6.23 (it is a $0°$ locus). For the expensive control case, the optimal closed-loop roots are at $-2 \pm j1$, that is, do not move. For the cheap control case, one of the poles is moved to -1, which is the location of the reflection of the zero in the RHP. The other pole moves to minus infinity along the real axis.   ∎

6.4.1 Comments on the Methods

These three methods of pole selection are alternatives that the designer can use for an initial design by pole placement. Note that the first two methods suggest closed-loop pole selection without regard to the effect on the control effort required to achieve that response. In some cases, therefore, the resulting control effort may be unreasonably high. The third method (SRL), on the other hand, selects poles that result in some balance between system errors and control effort. The designer can easily examine the relationship between shifts in that balance (by changing ρ) and system root locations, time response, and feedback gains. However, the SRL method does not provide direct specification of the transient shape in the way the other methods do. Whatever the initial pole-selection method, some modification is then made to achieve the desired balance of bandwidth, overshoot, sensitivity, control effort, or other design requirements. Further insight into pole selection will be gained from Example 6.16, given in Section 6.6.

6.5 Estimator Design

The control law designed in the last section assumed that all the state variables are available for feedback. In most cases, all the state variables are not measured. The cost of the required sensors may be prohibitive, or it may be physically impossible to measure all the state variables—for example, in a nuclear power plant. The purpose of this section is to demonstrate how to reconstruct all the state variables of a system from measurements of some of them. If the estimate of the state is denoted by $\hat{\mathbf{x}}$, it would be nice if the true state in the control law given by Eq. (6.39) could be replaced by its estimate, $u = -\mathbf{K}\hat{\mathbf{x}}$. This is indeed possible, as we shall see in Section 6.6.

6.5.1 Full-Order Estimators

One method of estimating all the state variables that one may consider is to construct a model of the plant dynamics:

$$\dot{\hat{\mathbf{x}}} = \mathbf{F}\hat{\mathbf{x}} + \mathbf{G}u, \qquad (6.72)$$

where $\hat{\mathbf{x}}$ is the estimate of the actual state $\mathbf{x}$. We know $\mathbf{F}$, $\mathbf{G}$, and $u(t)$, and hence this estimator is satisfactory if we can obtain the correct initial condition $\mathbf{x}(0)$ and set $\hat{\mathbf{x}}(0)$ equal to it. Figure 6.24 depicts this "open-loop" estimator. However, it is precisely the lack of information on $\mathbf{x}(0)$ that requires the construction of an estimator. Otherwise, the estimated state could track the true state exactly. Thus, if the initial condition was off, the estimated state would have a continually growing error or an error that goes to zero too slowly to be of any use. Furthermore, small errors in our knowledge of the system $(\mathbf{F}, \mathbf{G})$ would also cause the estimate to slowly diverge from the true state.

To study the dynamics of this estimator, we define the error in the estimate as

$$\tilde{\mathbf{x}} \triangleq \mathbf{x} - \hat{\mathbf{x}}. \qquad (6.73)$$

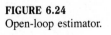

FIGURE 6.24
Open-loop estimator.

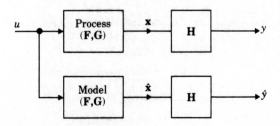

The dynamics of this error system are given by

$$\dot{\tilde{\mathbf{x}}} = \mathbf{F}\tilde{\mathbf{x}}, \qquad \tilde{\mathbf{x}}(0) = \mathbf{x}(0) - \hat{\mathbf{x}}(0). \tag{6.74}$$

The error converges to zero for a stable system (**F** with all poles in LHP) because **x** and $\hat{\mathbf{x}}$ are converging to zero, but the state estimates are not converging to the true state in a meaningful way.

In order to speed up the process and provide a useful state estimate, we feed back the difference between the measured and estimated outputs and correct the model continuously with this error signal. This scheme is shown in Fig. 6.25, and the equation for it is

$$\dot{\hat{\mathbf{x}}} = \mathbf{F}\hat{\mathbf{x}} + \mathbf{G}u + \mathbf{L}(y - \mathbf{H}\hat{\mathbf{x}}), \tag{6.75}$$

where **L** is a proportional gain,

$$\mathbf{L} = [l_1, \, l_2, \, \ldots, \, l_n]^{\mathrm{T}}, \tag{6.76}$$

to be chosen to achieve satisfactory error characteristics. The dynamics of the error can be obtained by subtracting the $\hat{\mathbf{x}}$ equation (Eq. 6.75) from the state equation (6.5), which results in

$$\dot{\tilde{\mathbf{x}}} = (\mathbf{F} - \mathbf{LH})\tilde{\mathbf{x}} + \mathbf{G}_1 w. \tag{6.77}$$

The characteristic equation of the error is now given by

$$\det[s\mathbf{I} - (\mathbf{F} - \mathbf{LH})] = 0. \tag{6.78}$$

If **L** can (we hope) be chosen so that $\mathbf{F} - \mathbf{LH}$ has stable and reasonably fast roots, $\tilde{\mathbf{x}}$ will decay to zero if there is no disturbing input w, irrespective of $\tilde{\mathbf{x}}(0)$. This means that $\hat{\mathbf{x}}(t)$ will converge to $\mathbf{x}(t)$, regardless of the value of $\hat{\mathbf{x}}(0)$, and, furthermore, the dynamics of the error can be chosen to be faster than the open-loop dynamics **F**. Note that Eq. (6.77) is only forced by one term, w. This is a consequence of assuming **F**, **G**, and **H** to be identical in the plant and estimator. Therefore, for

FIGURE 6.25
Closed-loop estimator.

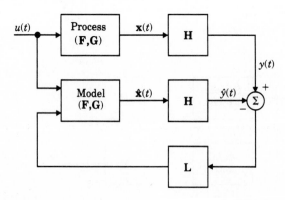

$w = 0$, the estimation error $\tilde{\mathbf{x}}$ converges to zero and remains there, independent of any known forcing function $u(t)$ on the plant and its effect on the state $\mathbf{x}(t)$. If we do not have a very accurate model of the plant $(\mathbf{F}, \mathbf{G}, \mathbf{H})$, the dynamics of the error are no longer governed by Eq. (6.77). However, $\mathbf{L}$ can typically be chosen so that the error system is stable and the error is acceptably small, even with small modeling errors and disturbing inputs. It is important to realize that the nature of the plant and estimator are quite different. The plant could be a chemical process, whereas the estimator is an electronic unit or computer implementing the estimated state differential equation (Eq. 6.75).

The selection of $\mathbf{L}$ can be approached in exactly the same fashion as the selection of $\mathbf{K}$ is in the control-law design. If we specify the desired location of the estimator error roots as

$$s_i = \beta_1, \beta_2, \ldots, \beta_n,$$

the desired estimator characteristic equation is

$$\alpha_e(s) \triangleq (s - \beta_1)(s - \beta_2) \cdots (s - \beta_n) = 0, \tag{6.79}$$

and one can solve for $\mathbf{L}$ by comparing coefficients in Eqs. (6.78) and (6.79). However, this can only be done if the system is "observable." Roughly speaking, *observability* refers to the ability to deduce information about all the modes of the system by monitoring the sensed outputs. Unobservability results from some mode or subsystem being disconnected physically from the output and therefore not appearing in the measurements. For example, if only derivatives of certain state variables are measured, and these state variables do not affect the dynamics, a constant of integration is obscured. This situation occurs with a $1/s^2$ plant if only velocity is measured, for then it is impossible to deduce the initial value of the position. On the other hand, for an oscillator a velocity measurement is sufficient to estimate position because the acceleration, and consequently the velocity observed, are affected by position. The mathematical test for determining observability is developed in Appendix 6A. The result is that the *observability matrix*,

$$\mathbb{O} = \begin{bmatrix} \mathbf{H} \\ \mathbf{HF} \\ \vdots \\ \mathbf{HF}^{n-1} \end{bmatrix},$$

must be full-rank.

EXAMPLE 6.13   Again, consider the undamped oscillator with frequency ω_0:

$$\dot{\mathbf{x}} = \begin{bmatrix} 0 & 1 \\ -\omega_0^2 & 0 \end{bmatrix} \mathbf{x} + \begin{bmatrix} 0 \\ 1 \end{bmatrix} u,$$

$$y = [\,1 \quad 0\,]\mathbf{x}.$$

Suppose we wish to place the estimator error equation at $-10\omega_0$ (five times as fast as the state feedback controller poles); then

$$\alpha_e(s) = (s + 10\omega_0)^2 = s^2 + 20\omega_0 s + 100\omega_0^2.$$

From Eq. (6.78),

$$\det[s\mathbf{I} - (\mathbf{F} - \mathbf{LH})] = s^2 + l_1 s + l_2 + \omega_0^2.$$

Comparing coefficients in the above two equations, we find that

$$\mathbf{L} = \begin{bmatrix} l_1 \\ l_2 \end{bmatrix} = \begin{bmatrix} 20\omega_0 \\ 99\omega_0^2 \end{bmatrix}. \qquad (6.80)$$

FIGURE 6.26
Undamped-oscillator initial-condition response of y and $\hat{y}$.

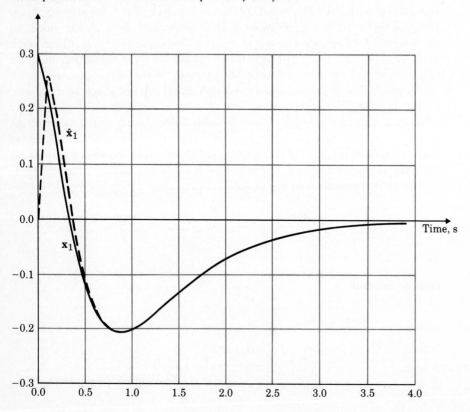

This completes the design of the estimator (Eq. 6.75) for this example. The overall system equations are obtained by substituting $u = -K\hat{x}$ and $y = Hx$ in Eqs. (6.5) and (6.75). Combining these two equations in matrix notation results is

$$\begin{bmatrix} \dot{x} \\ \dot{\hat{x}} \end{bmatrix} = \begin{bmatrix} F & -GK \\ LH & F - LH - GK \end{bmatrix} \begin{bmatrix} x \\ \hat{x} \end{bmatrix},$$

$$y = [H \quad 0] \begin{bmatrix} x \\ \hat{x} \end{bmatrix}. \tag{6.81}$$

The response of the above closed-loop system with $\omega_0 = 1$ to an initial condition $x_0 = [0.3, -0.5]^T$ is shown in Figs. 6.26 and 6.27, where K is obtained from Eq. (6.48) and L from Eq. (6.80). Note that the state variable estimates track the actual ones after an initial transient, and the error between them decays approximately

FIGURE 6.27
Undamped-oscillator initial-condition response of $\dot{y}$ and $\hat{\dot{y}}$.

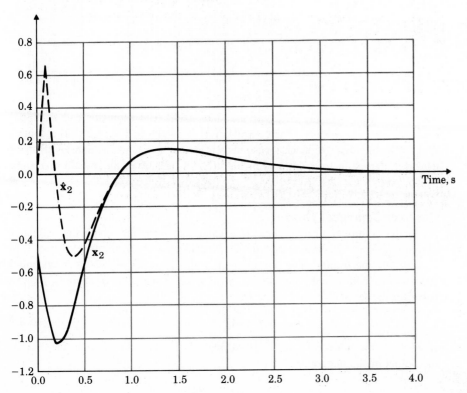

FIGURE 6.28
Comparison of control
efforts for state and
estimated state feedback.

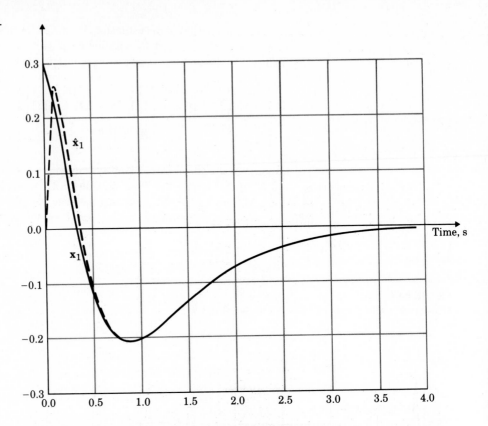

five times faster than the decay of the state itself. The estimator had an initial
condition of $\hat{\mathbf{x}}_0 = [0, 0]^T$. The control effort is shown in Fig. 6.28. The control
effort with full-state feedback is also shown for comparison. The control effort of
the controller-estimator is initially much different because of the initial uncertainty
in the state.   ∎

Observer Canonical Form

As in the controller design, there is a canonical form for which the estimator
design equations are particularly simple. Consider again the third-order system in
Eqs. (6.49) and (6.50). (The development is the same for higher-order systems.)
Assume that both u and y signals are "available." Then from Eq. (6.49), we can
solve for the terms that have no differentiation as follows:

$$b_3 u - a_3 y = \dddot{y} + a_1 \ddot{y} + a_2 \dot{y} - b_2 \dot{u} - b_1 \ddot{u}. \qquad (6.82)$$

If we form the left-hand side as shown in Fig. 6.29(a) and integrate the results,
the output of the integrator is

$$\ddot{y} + a_1 \dot{y} - b_1 \dot{u} + a_2 y - b_2 u.$$

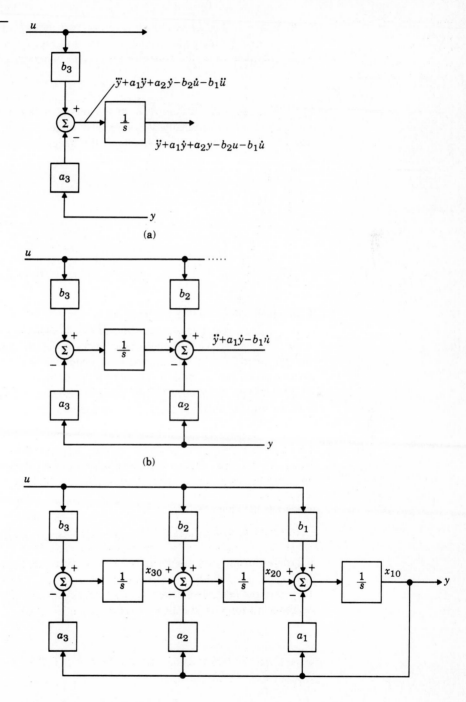

FIGURE 6.29
Derivation of observer canonical form.

Since both u and y are available, the last two terms can be nulled by adding $-a_2 y + b_2$ to the integrator output, as shown in Fig. 6.29(b). If we integrate the results one more time, the output of the second integrator is

$$\dot{y} + a_1 y - b_1 u.$$

Again, we can add $-a_1 y + b_1 u$ to the output of the integrator, as shown in Fig. 6.29(c), to get $\dot{y}$. Finally, $\dot{y}$ is integrated to yield y, and the realization is complete. If we take the state as outputs of the integration going from right to left, then

$$\dot{\mathbf{x}}_{\circ} = \mathbf{F}_{\circ}\mathbf{x}_{\circ} + \mathbf{G}_{\circ}u,$$
$$y = \mathbf{H}_{\circ}\mathbf{x}_{\circ}, \tag{6.83}$$

where

$$\mathbf{F}_{\circ} = \begin{bmatrix} -a_1 & 1 & 0 \\ -a_2 & 0 & 1 \\ -a_3 & 0 & 0 \end{bmatrix}, \qquad \mathbf{G}_{\circ} = \begin{bmatrix} b_1 \\ b_2 \\ b_3 \end{bmatrix}, \qquad \mathbf{H}_{\circ} = [\, 1 \quad 0 \quad 0\,]. \tag{6.84}$$

In this realization, all the feedback loops come from the output or "observed" signal, and we call this structure the *observer canonical form*. It is a "direct" canonical form because the gains of the blocks in the structure are obtained directly from the coefficients of the numerator and denominator polynomials in the transfer function $H(s)$, just as with the control canonical form.

One of the advantages of the observer canonical form is that the estimator gains can be obtained from it by inspection. The estimator error closed-loop matrix is

$$\mathbf{F}_{\circ} - \mathbf{L}\mathbf{H}_{\circ} = \begin{bmatrix} -a_1 - l_1 & 1 & 0 \\ -a_2 - l_2 & 0 & 1 \\ -a_3 - l_3 & 0 & 0 \end{bmatrix}, \tag{6.85}$$

which has the characteristic equation

$$s^3 + (a_1 + l_1)s^2 + (a_2 + l_2)s + (a_3 + l_3) = 0, \tag{6.86}$$

and the estimator gain can be found by comparing the coefficients of Eq. (6.86) with $\alpha_e(s)$.

As in the controller design, an alternative method of computing $\mathbf{L}$ is to use Ackermann's formula in estimator form, which is

$$\mathbf{L} = \alpha_e(\mathbf{F})\mathbb{O}^{-1} \begin{bmatrix} 0 \\ 0 \\ \vdots \\ 1 \end{bmatrix}, \tag{6.87}$$

TABLE 6.2
Duality between Control and Estimation

Control	Estimation
F	F^T
G	H^T
H	G^T

where

$$\mathbb{O} = \begin{bmatrix} H \\ HF \\ \vdots \\ HF^{n-1} \end{bmatrix}$$

is the *observability matrix*.

Duality

The reader may already have seen some resemblance between the estimation and control problems. In fact, the two problems are mathematically equivalent. This property is called *duality,* and Table 6.2 shows the duality relations between the estimation and control problems. For example, Ackermann's control formula (Eq. 6.63) becomes the estimator formula (Eq. 6.87) if the substitutions of Table 6.2 are made. Furthermore, duality allows one to use the same design tools for estimator problems as for control problems, with proper substitutions. The two canonical forms, that is, the control and observer forms are dual, as seen from comparison of the triples (F_c, G_c, H_c) and (F_o, G_o, H_o). Basically, the role of input and output are interchanged. If in the observer form, starting from the output, all signal flows are reversed—the summers are changed to nodes and nodes are changed to summers—we obtain the control canonical form. Using the same rules, one can obtain the observer form from the control form.

6.5.2 Reduced-Order Estimators

The estimation design in the previous section reconstructs the entire state vector using measurements of some of the state variables. If the sensors have no noise, there seems to be a certain redundancy, and one may become curious as to the necessity of estimating state variables that are measured directly. Can we reduce the complexity of the estimator using the state variables that are measured directly and exactly? Yes. However, it is better to implement a full-order estimator if there is significant noise on the measurements because the estimator filters the measurements in addition to estimating unmeasured state variables.

The reduced-order estimator reduces the number of required integrators by the number of sensed outputs. To derive this estimator (Gopinath, 1971), we partition the state vector into two parts: one part x_a, which is directly measured ($x_a = y$), and another part $\mathbf{x}_b$, representing the remaining state variables that need to be estimated. If we partition the system matrices accordingly, the complete description of the system is given by

$$\begin{bmatrix} \dot{x}_a \\ \dot{\mathbf{x}}_b \end{bmatrix} = \begin{bmatrix} \mathbf{F}_{aa} & \mathbf{F}_{ab} \\ \mathbf{F}_{ba} & \mathbf{F}_{bb} \end{bmatrix} \begin{bmatrix} x_a \\ \mathbf{x}_b \end{bmatrix} + \begin{bmatrix} \mathbf{G}_a \\ \mathbf{G}_b \end{bmatrix} u,$$

$$y = \begin{bmatrix} \mathbf{I} & 0 \end{bmatrix} \begin{bmatrix} x_a \\ \mathbf{x}_b \end{bmatrix}. \tag{6.88}$$

The dynamics of the unmeasured state variables are given by

$$\dot{\mathbf{x}}_b = \mathbf{F}_{bb}\mathbf{x}_b + \underbrace{\mathbf{F}_{ba}x_a + \mathbf{G}_b u}_{\text{known input}}, \tag{6.89}$$

where the rightmost two terms are known and can be considered as an input into the $\mathbf{x}_b$ dynamics. Since $x_a = y$, the measured dynamics are given by the scalar equation

$$\dot{x}_a = \dot{y} = \mathbf{F}_{aa}y + \mathbf{F}_{ab}\mathbf{x}_b + \mathbf{G}_a u. \tag{6.90}$$

If we collect the known terms of Eq. (6.90) on one side,

$$\underbrace{\dot{y} - \mathbf{F}_{aa}y - \mathbf{G}_a u}_{\text{known measurement}} = \mathbf{F}_{ab}\mathbf{x}_b, \tag{6.91}$$

we obtain a relationship between known quantities on the left-hand side, considered as measurements, and unknown state variables on the right. Therefore, Eqs. (6.90) and (6.91) have the same relationship to the state $\mathbf{x}_b$ that the original equation (Eq. 6.88) had to the entire state $\mathbf{x}$. Following this line of reasoning, we can establish the following substitutions in the original estimator equations to obtain a (reduced-order) estimator of $\mathbf{x}_b$:

$$\mathbf{x} \leftarrow \mathbf{x}_b,$$
$$\mathbf{F} \leftarrow \mathbf{F}_{bb},$$
$$\mathbf{G}u \leftarrow \mathbf{F}_{ba}y + \mathbf{G}_b u,$$
$$y \leftarrow \dot{y} - \mathbf{F}_{aa}y - \mathbf{G}_a u,$$
$$\mathbf{H} \leftarrow \mathbf{F}_{ab}.$$

Therefore, the reduced-order estimator equations are obtained by making the substitutions above in the full-order estimator (Eq. 6.75):

$$\dot{\hat{\mathbf{x}}}_b = \mathbf{F}_{bb}\hat{\mathbf{x}}_b + \underbrace{\mathbf{F}_{ba}y + \mathbf{G}_b u}_{\text{input}} + \mathbf{L}(\underbrace{\dot{y} - \mathbf{F}_{aa}y - \mathbf{G}_a u}_{\text{measurement}} - \mathbf{F}_{ab}\hat{\mathbf{x}}_b). \tag{6.92}$$

If we define the estimator error as

$$\tilde{\mathbf{x}}_b \triangleq \mathbf{x}_b - \hat{\mathbf{x}}_b, \tag{6.93}$$

the dynamics of the error are given by subtracting Eq. (6.92) from Eq. (6.89):

$$\dot{\tilde{\mathbf{x}}}_b = (\mathbf{F}_{bb} - \mathbf{L}\mathbf{F}_{ab})\tilde{\mathbf{x}}_b. \tag{6.94}$$

Its characteristic equation is given by

$$\det[s\mathbf{I} - (\mathbf{F}_{bb} - \mathbf{L}\mathbf{F}_{ab})] = 0. \tag{6.95}$$

We design the dynamics of this estimator by selecting $\mathbf{L}$ so that Eq. (6.95) matches a reduced order $\alpha_e(s)$. Equation (6.92) can be rewritten as

$$\dot{\hat{\mathbf{x}}}_b = (\mathbf{F}_{bb} - \mathbf{L}\mathbf{F}_{ab})]\hat{\mathbf{x}}_b + (\mathbf{F}_{ba} - \mathbf{L}\mathbf{F}_{aa})y + (\mathbf{G}_b - \mathbf{L}\mathbf{G}_a)u + \mathbf{L}\dot{y}. \tag{6.96}$$

The fact that we must form the derivative of the measurements in Eq. (6.96) appears to present a practical difficulty. It is known that differentiation amplifies noise, so if y is noisy, the use of $\dot{y}$ is unacceptable. To get around this difficulty, we define the new controller state as

$$\mathbf{x}_c \triangleq \hat{\mathbf{x}}_b - \mathbf{L}y. \tag{6.97}$$

Then, in terms of this new state, the implementation of the reduced-order estimator is given by

$$\dot{\mathbf{x}}_c = (\mathbf{F}_{bb} - \mathbf{L}\mathbf{F}_{ab})\hat{\mathbf{x}}_b + (\mathbf{F}_{ba} - \mathbf{L}\mathbf{F}_{aa})y + (\mathbf{G}_b - \mathbf{L}\mathbf{G}_a)u, \tag{6.98}$$

and $\dot{y}$ no longer appears directly. A block-diagram representation of the reduced-order estimator is shown in Fig. 6.30.

FIGURE 6.30
Reduced-order estimator structure.

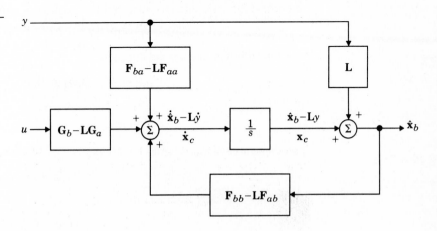

The reduced-order estimator gain can also be found from Ackermann's formula (Eq. 6.87), with appropriate substitutions from Eq. (6.95):

$$\mathbf{L} = \alpha_e(\mathbf{F}_{bb})\mathbb{O}_b^{-1} \begin{bmatrix} 0 \\ 0 \\ \vdots \\ 1 \end{bmatrix}, \qquad (6.99)$$

where

$$\mathbb{O}_b = \begin{bmatrix} \mathbf{F}_{ab} \\ \mathbf{F}_{ab}\mathbf{F}_{bb} \\ \mathbf{F}_{ab}\mathbf{F}_{bb}^2 \\ \vdots \\ \mathbf{F}_{ab}\mathbf{F}_{bb}^{n-2} \end{bmatrix}.$$

We should point out that the conditions for the existence of the reduced-order estimator are the same as for the full-order estimator—namely, observability of $(\mathbf{F}, \mathbf{H})$.

EXAMPLE 6.14   Again, consider the undamped oscillator

$$\begin{bmatrix} \dot{x}_1 \\ \dot{x}_2 \end{bmatrix} = \begin{bmatrix} 0 & 1 \\ -\omega_0^2 & 0 \end{bmatrix} \begin{bmatrix} x_1 \\ x_2 \end{bmatrix} + \begin{bmatrix} 0 \\ 1 \end{bmatrix} u, \qquad y = [1 \quad 0] \begin{bmatrix} x_1 \\ x_2 \end{bmatrix}.$$

Suppose we would like to place the roots of the reduced-order estimator-error characteristic equations at $-10\omega_0$. The partitioned matrices are

$$\begin{bmatrix} \mathbf{F}_{aa} & \mathbf{F}_{ab} \\ \mathbf{F}_{ba} & \mathbf{F}_{bb} \end{bmatrix} = \begin{bmatrix} 0 & 1 \\ -\omega_0^2 & 0 \end{bmatrix}; \qquad \begin{bmatrix} \mathbf{G}_a \\ \mathbf{G}_b \end{bmatrix} = \begin{bmatrix} 0 \\ 1 \end{bmatrix}.$$

From Eq. (6.95), we find the characteristic equation in terms of L:

$$s - (0 - L) = 0.$$

We compare it with the desired equation,

$$\alpha_e(s) = s + 10\omega_0 = 0,$$

which yields

$$L = 10\omega_0.$$

The estimator equation is, from Eq. (6.98),

$$\dot{x}_c = -10\omega_0\hat{x}_2 - \omega_0^2 y + u,$$

and the state estimate, from Eq. (6.97), is

$$\hat{x}_2 = x_c + 10\omega_0 y.$$

The control law is from Eqs. (6.39) and (6.48) where x_1 is measured directly and x_2 is used for x_2. The response of the closed-loop system to an initial condition $\mathbf{x}_0 = [0.3, -0.5]^T$ is shown in Fig. 6.31, and the control effort is shown in Fig. 6.32. The estimator had an initial condition $x_{c0} = 0$ and $\omega_0 = 1$. Note the similarity of the initial-condition response and control effort to that of the full-order estimator in Figs. 6.26, 6.27, and 6.28. ∎

6.5.3 Estimator-Pole Selection

The selection of estimator pole locations can be based on the techniques discussed in Section 6.4. As a rule of thumb, the estimator poles can be chosen by a factor of 2 to 6 times faster than the controller poles. This is to ensure a faster decay of the estimator errors compared with the desired dynamics, thus causing the controller poles to dominate the total response. If sensor noise is large enough to be a major concern, one may decide to choose the estimator poles to be slower than 2 times the controller poles, which would yield a system with lower bandwidth and more noise-smoothing. However, the total system response in this case could be strongly influenced by the location of the estimator poles. If the estimator poles were slower

FIGURE 6.31
Initial-condition response of reduced-order controller.

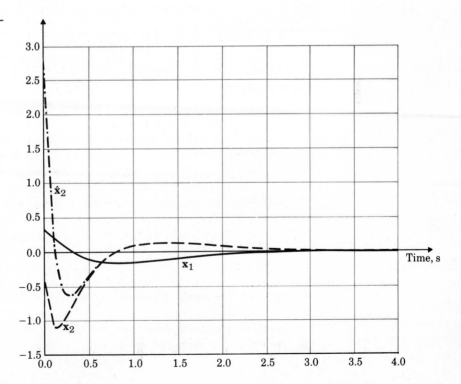

FIGURE 6.32
Control effort for
reduced-order controller.

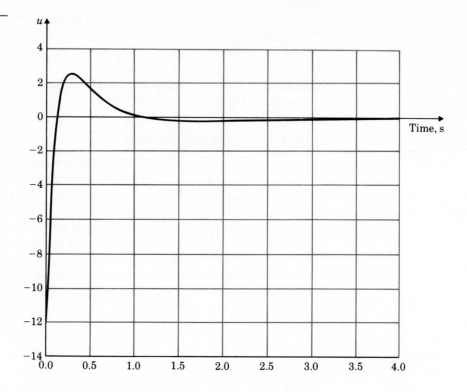

than the controller poles, the system response to disturbances would be dominated by the estimator dynamic characteristics.

In comparison with the selection of controller poles, there is a much different relationship to *control effort* to be concerned with. As in the controller, there is a feedback term in the estimator that grows in magnitude as the speed of response increases. This feedback is in the form of an electronic signal or a digital word in a computer, but its growth causes no difficulty. In the controller, increasing the speed of response, and thus the control effort, implies the use of a larger actuator, which increases size, weight, and cost. The important consequence of increasing the speed of response of an estimator is that the bandwidth of the estimator becomes higher, thus causing more sensor noise to pass on to the control actuator. Thus, similar to controller design, the best estimator design is a balance between good transient response with low enough bandwidth so that sensor noise does not significantly affect actuator activity.

In optimal estimation theory (Bryson and Ho, 1969), it has been shown that the best balance is dependent on the ratio of sensor noise to disturbance input noise (w in Eq. 6.5). This is best understood by reexamining the estimator formula (Eq. 6.75),

$$\dot{\hat{\mathbf{x}}} = \mathbf{F}\hat{\mathbf{x}} + \mathbf{G}u + \mathbf{L}(y - \mathbf{H}\hat{\mathbf{x}}), \tag{6.100}$$

and how it interacts with the system equations (Eq. 6.5),

$$\dot{\mathbf{x}} = \mathbf{F}\mathbf{x} + \mathbf{G}u + \mathbf{G}_1 w, \tag{6.101}$$

and the measurement equation,

$$y = \mathbf{H}\mathbf{x} + v, \tag{6.102}$$

where v is the measurement noise. The question to be addressed is how much the state estimate revision ($\dot{\hat{\mathbf{x}}}$ in Eq. 6.100) should be based on the system model ($\mathbf{F}\hat{\mathbf{x}} + \mathbf{G}u$) versus the measurement error [$\mathbf{L}(y - \mathbf{H}\hat{\mathbf{x}})$]. If the sensor noise v is relatively large, it is better to base the estimate more heavily on the model, and thus $\mathbf{L}$ should be small, which produces a slow estimator. If the input disturbance noise is relatively large, the system model in the estimator is less useful, and it is more important to believe the measurements. Thus $\mathbf{L}$ should be large and the estimator fast. The *optimal* solution to this balance is achieved by solving an estimator symmetric root-locus equation that is almost identical to the SRL equation established for the optimal control formulation, Eq. (6.69) (Kailath, 1980):

$$1 + qG_e(-s)G_e(s) = 0, \tag{6.103}$$

where q is the ratio of input disturbance noise to sensor noise and

$$G_e(s) = \mathbf{H}(s\mathbf{I} - \mathbf{F})^{-1}\mathbf{G}_1.$$

Note that $G_e(s)$ is similar to the $G_0(s)$ in Eq. (6.69). It differs in that $G_e(s)$ is the transfer function from the input disturbance ($\mathbf{G}_1$ instead of $\mathbf{G}$ in Eq. 6.101) to the measurement, whereas G_0 is the transfer function from the control input to output for minimization in Eq. (6.66), that is, $\mathbf{H}_1$ instead of $\mathbf{H}$.

The use of the estimator SRL, Eq. (6.103), is identical to the use of the controller SRL. A root locus versus q is generated, thus yielding sets of optimal estimator poles. The designer then picks the set of (stable) poles that seems best considering all aspects of the problem. An important advantage of using the SRL technique is that the arbitrariness is reduced to one degree of freedom (q) instead of the many degrees of freedom required to select the poles directly in a higher-order system.

A final comment concerns the reduced-order estimator. Because of the presence of a direct transmission term (see Fig. 6.30), the reduced-order estimator has a much higher bandwidth from sensor to control compared with the full-order estimator. Therefore, if sensor noise is a significant factor, the reduced-order estimator is less attractive, since the potential savings in complexity is more than offset by the increased sensitivity to noise.

6.6   Compensator Design: Combined Control Law and Estimator

If we take the control (Section 6.3) and implement it using the estimated state variables, the design for a regulator (no reference input to be tracked) is complete. However, since we designed the control law assuming that all the state variables were available, we may wonder (and need to compute) what effect using $\hat{\mathbf{x}}$ in place of $\mathbf{x}$ has on the system dynamics. The plant equation with feedback is now

$$\dot{\mathbf{x}} = \mathbf{F}\mathbf{x} - \mathbf{G}\mathbf{K}\hat{\mathbf{x}}, \tag{6.104}$$

which can be rewritten in terms of the state error $\tilde{\mathbf{x}}$ as

$$\dot{\mathbf{x}} = \mathbf{F}\mathbf{x} - \mathbf{G}\mathbf{K}(\mathbf{x} - \tilde{\mathbf{x}}). \tag{6.105}$$

The overall system dynamics are obtained by combining Eq. (6.105) with the error equation (Eq. 6.77), which, in overall state form, is

$$\begin{bmatrix} \dot{\mathbf{x}} \\ \dot{\tilde{\mathbf{x}}} \end{bmatrix} = \begin{bmatrix} \mathbf{F} - \mathbf{G}\mathbf{K} & \mathbf{G}\mathbf{K} \\ 0 & \mathbf{F} - \mathbf{L}\mathbf{H} \end{bmatrix} \begin{bmatrix} \mathbf{x} \\ \tilde{\mathbf{x}} \end{bmatrix}. \tag{6.106}$$

The characteristic equation of the closed-loop system is

$$\det \begin{bmatrix} s\mathbf{I} - \mathbf{F} + \mathbf{G}\mathbf{K} & -\mathbf{G}\mathbf{K} \\ 0 & s\mathbf{I} - \mathbf{F} + \mathbf{L}\mathbf{H} \end{bmatrix} = 0. \tag{6.107}$$

which, because it is block triangular (see Appendix C), can be written as

$$\det[s\mathbf{I} - \mathbf{F} + \mathbf{G}\mathbf{K}]\det[s\mathbf{I} - \mathbf{F} + \mathbf{L}\mathbf{H}] = \alpha_c(s)\alpha_e(s) = 0. \tag{6.108}$$

In other words, the poles of the combined system consist of the union of the control roots and the estimator roots. This means that the design of the control law and the estimator can be carried out independently; yet, when they are used together the roots are unchanged. This is a special case of the *separation principle* (see Gunkel and Franklin, 1963), which holds in much more general contexts and allows for separate design of control law and estimator in certain stochastic cases.

   To compare the state-variable method of design with the transform methods discussed in Chapters 4 and 5, we note from Fig. 6.33 that the dashed portion corresponds to a compensator. The state equation for this compensator is obtained by including the feedback law $u = -\mathbf{K}\hat{\mathbf{x}}$ (since it is part of the compensator) in the estimator equation (Eq. 6.75):

$$\dot{\hat{\mathbf{x}}} = (\mathbf{F} - \mathbf{G}\mathbf{K} - \mathbf{L}\mathbf{H})\hat{\mathbf{x}} + \mathbf{L}y, \qquad u = -\mathbf{K}\hat{\mathbf{x}}, \tag{6.109}$$

which is the same form as Eq. (6.17):

$$\dot{\mathbf{x}} = \mathbf{F}\mathbf{x} + \mathbf{G}u, \tag{6.110}$$

which was shown in Eq. (6.29) to have a characteristic equation

$$\det[s\mathbf{I} - \mathbf{F}] = 0. \tag{6.111}$$

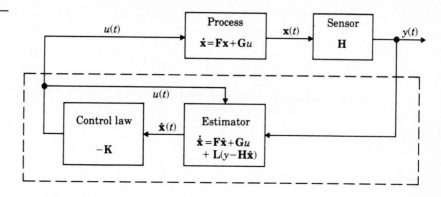

Therefore, the characteristic equation of the compensator is found by comparing Eqs. (6.109) and (6.110) and substituting the equivalent matrices in Eq. (6.111):

$$\det[s\mathbf{I} - \mathbf{F} + \mathbf{GK} + \mathbf{LH}] = 0. \tag{6.112}$$

Note that the roots of Eq. (6.112) were never specified nor used during the state-space design technique. (Note also that the compensator is not guaranteed to be *stable*.) The transfer function representing the dynamic compensator is obtained by inspecting Eq. (6.21) and substituting in the similar matrices from Eq. (6.108). This yields

$$D(s) = \frac{U(s)}{Y(s)} = -\mathbf{K}[s\mathbf{I} - \mathbf{F} + \mathbf{GK} + \mathbf{LH}]^{-1}\mathbf{L}. \tag{6.113}$$

The same development can be carried out for the reduced-order estimator. In that case, the control law is

$$u = -[\,K_a \quad \mathbf{K}_b\,]\begin{bmatrix} x_a \\ \hat{\mathbf{x}}_b \end{bmatrix} = -K_a y - \mathbf{K}_b \hat{\mathbf{x}}_b. \tag{6.114}$$

Substituting Eq. (6.114) into Eq. (6.98) and using Eq. (6.97), we obtain

$$\dot{\mathbf{x}}_c = \mathbf{A}_r \mathbf{x}_c + \mathbf{B}_r y,$$
$$u = \mathbf{C}_r \mathbf{x}_c + D_r y, \tag{6.115}$$

where

$$\mathbf{A}_r = \mathbf{F}_{bb} - \mathbf{LF}_{ab} - (\mathbf{G}_b - \mathbf{L}G_a)\mathbf{K}_b,$$
$$\mathbf{B}_r = \mathbf{A}_r \mathbf{L} + \mathbf{F}_{ba} - \mathbf{LF}_{aa} - (\mathbf{G}_b - \mathbf{L}G_a)K_a,$$
$$\mathbf{C}_r = -\mathbf{K}_b,$$
$$D_r = -K_a - \mathbf{K}_b \mathbf{L}. \tag{6.116}$$

The dynamic compensator now has the transfer function

$$D(s) = \frac{U(s)}{Y(s)} = C_r(s\mathbf{I} - \mathbf{A}_r)^{-1}\mathbf{B}_r + D_r. \tag{6.117}$$

When we compute $D(s)$ for a specific case, we will find that they are very similar to the classical compensators of Chapters 4 and 5, in spite of the fact that they are arrived at by entirely different means. Examples 6.15 and 6.16 illustrate state-space design.

EXAMPLE 6.15   Consider the double integrator plant

$$G(s) = \frac{1}{s^2}$$

with state-variable Eqs. (6.3) and (6.4):

$$\dot{\mathbf{x}} = \begin{bmatrix} 0 & 1 \\ 0 & 0 \end{bmatrix} \mathbf{x} + \begin{bmatrix} 0 \\ 1 \end{bmatrix} u,$$

$$y = [\,1 \quad 0\,]\mathbf{x}.$$

If we place the controller roots at $s = -(\sqrt{2}/2) \pm j(\sqrt{2}/2)(\omega_n = a, \zeta = 0.7)$, then

$$\alpha_c(s) = s^2 + \sqrt{2}s + 1. \tag{6.118}$$

The state feedback gain is then found to be

$$\mathbf{K} = [\,1 \quad \sqrt{2}\,].$$

If the estimator-error roots are at $\omega_n = 5$, $\zeta = 0.5$, the desired estimator characteristic polynomial is

$$\alpha_e(s) = s^2 + 5s + 25 = s + 2.5 \pm j4.3, \tag{6.119}$$

and the estimator feedback gain matrix is

$$\mathbf{L} = \begin{bmatrix} 5 \\ 25 \end{bmatrix}.$$

The compensator transfer function given by Eq. (6.113) is

$$D(s) = \frac{-40.4(s + 0.619)}{s + 3.21 \pm j4.77}, \tag{6.120}$$

which looks very much like a *lead compensator* in that it has a zero on the real axis to the right of its poles; but, rather than one real pole, it has two complex poles. The zero provides the derivative feedback and the double pole provides some smoothing of sensor noise. The effect of the compensation on the system closed-loop roots can be evaluated in exactly the same way that compensation was evaluated in Chapters 4 and 5 using root-locus or frequency-response tools. The gain of 40.4 in Eq. (6.120) is a result of the root selection inherent in Eqs. (6.118) and (6.119). If we replace this specific value of gain with a variable value K, the characteristic equation of the closed-loop system becomes

$$1 + \frac{K(s + 0.619)}{(s + 3.21 \pm j4.77)s^2} = 0. \tag{6.121}$$

The root-locus technique allows us to evaluate the roots of this equation versus K, and this is done in Fig. 6.34. Note that the loci go through the roots selected in Eqs. (6.118) and (6.119), and, when $K = 40.4$, the four roots of the closed-loop system are equal to those specified.

The frequency-response plots seen in Fig. 6.35 show that the compensation designed using state-space accomplishes the same results that one would strive for using frequency-response design. Specifically, the uncompensated phase margin of $0°$ is increased to $53°$ in the compensated case, and the gain of $K = 40.4$ produces a crossover frequency of 1.35 rad/s. Both these values are roughly consistent with the controller closed-loop roots, with $\omega_n = 1$ rad/s and $\zeta = 0.7$, as one should expect since these slow roots are more dominant in the system response than the estimator.

Consider designing a reduced-order estimator for the system. If we choose to place the single estimator pole at -5, then, from Eq. (6.95),

$$L = 5.$$

The scalar compensator equations are

$$\dot{x}_c = -6.41x_c - 33.1y,$$
$$u = -1.41x_c - 8.07y,$$

where

$$x_c = \hat{x}_2 - 5y.$$

The compensator is shown in Fig. 6.36 and has the transfer function given by Eq. (6.117):

$$D(s) = \frac{-8.07(s + 0.62)}{s + 6.4}.$$

FIGURE 6.34
Root locus for the
combined control
estimator, with process
gain as the parameter.

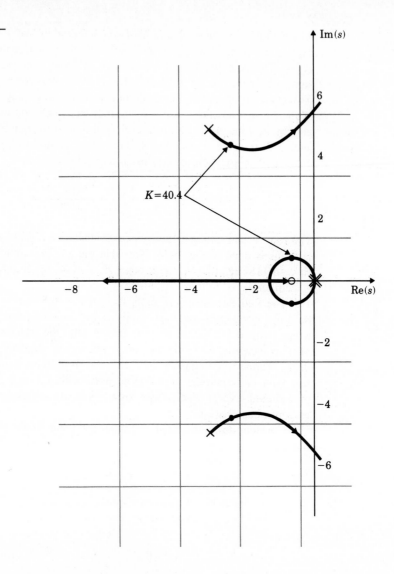

The reduced-order compensator is precisely the *lead network* that was described in Chapters 4 and 5. This is a pleasant discovery, as it shows that classical and state-variable techniques can result in exactly the same type of compensation. The root locus of Fig. 6.37 shows that the closed-loop poles are at the assigned locations, and the frequency response of the compensated system seen in Fig. 6.38 shows a phase margin of about 55°. As with the full-order estimator, analysis by the other methods confirms the selected root locations. ∎

FIGURE 6.35
Frequency response for $1/s^2$ example.

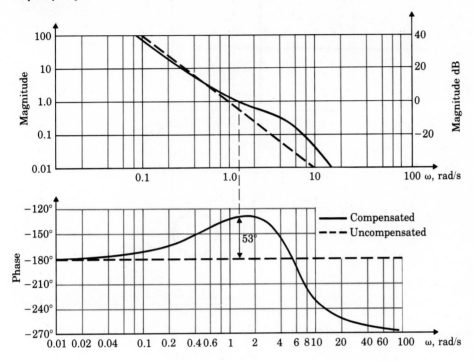

FIGURE 6.36
Simplified block diagram
of a reduced-order
controller, a lead
network.

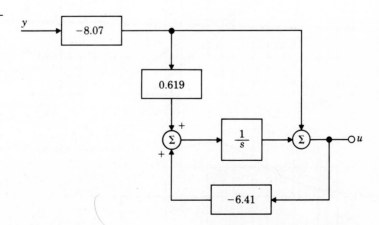

FIGURE 6.37
Root locus of a reduced-order controller and $1/s^2$ process. Root locations at $K = 8.07$ are shown by the dots.

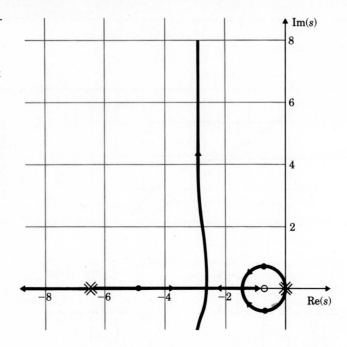

FIGURE 6.38
Frequency response of the $1/s^2$ example with a reduced-order estimator

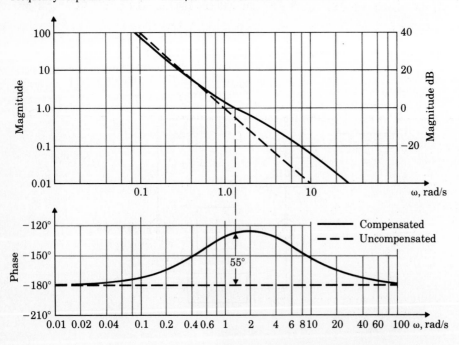

EXAMPLE 6.16   Consider the third-order system

$$G(s) = \frac{10}{s(s + 2)(s + 8)}.$$

A block diagram representation of this system in observer canonical form is shown in Fig. 6.39. The corresponding state-variable matrices are

$$\mathbf{F} = \begin{bmatrix} -10 & 1 & 0 \\ -16 & 0 & 1 \\ 0 & 0 & 0 \end{bmatrix}, \qquad \mathbf{G} = \begin{bmatrix} 0 \\ 0 \\ 10 \end{bmatrix},$$

$$\mathbf{H} = [1 \quad 0 \quad 0], \qquad J = 0.$$

Suppose we want the state feedback poles to be at the third-order ITAE locations, with $\omega_0 = 2$. From Table 6.1(a), we let s be replaced by $s/2$, clear fractions, and compute

$$\alpha_c(s) = (s + 1.42)(s + 1.04 \pm j2.14)$$
$$= s^3 + 3.5s^2 + 8.6s + 8.$$

By comparing coefficients of α_c with the control characteristic equation

$$\det[s\mathbf{I} - \mathbf{F} + \mathbf{GK}] = \alpha_c(s),$$

the state feedback gain is found to be

$$\mathbf{K} = [-46.4 \quad 5.76 \quad -0.65].$$

If we wish to place the full-order estimator poles at the third-order ITAE locations, with $\omega_0 = 6$, we compute

$$\alpha_e(s) = (s + 4.25)(s + 3.13 \pm j6.41)$$
$$= s^3 + 10.5s^2 + 77.4s + 216.$$

FIGURE 6.39
Third-order example in observer canonical form.

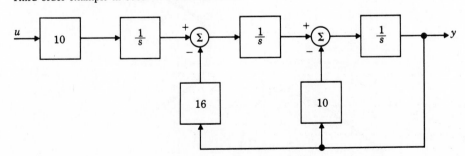

By comparing coefficients in the estimator characteristic equation

$$\det[s\mathbf{I} - \mathbf{F} + \mathbf{L}\mathbf{H}] = \alpha_e(s),$$

the estimator gain is found to be

$$\mathbf{L} = \begin{bmatrix} 0.5 \\ 61.4 \\ 216 \end{bmatrix}.$$

The compensator transfer function as given by Eq. (6.113) is

$$D(s) = -190 \frac{(s + 0.432)(s + 2.10)}{(s - 1.88)(s + 2.94 \pm j8.32)}.$$

Figure 6.40 shows the root locus of the system. It verifies that the roots are in the ITAE locations specified, in spite of the peculiar compensation that results.

Note that the compensator has an unstable root at $s = +1.88$ but that all system closed-loop roots (controller and estimator) are stable. Compensators are not

FIGURE 6.40
Root locus for ITAE pole assignment.

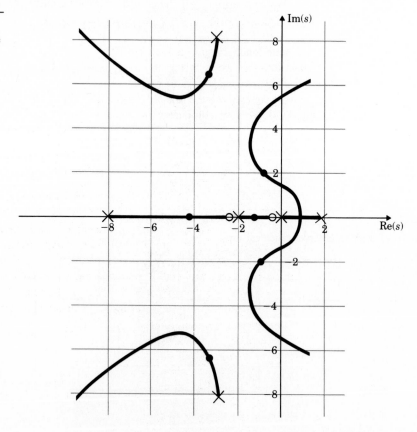

typically designed this way because of the difficulty in testing the compensator by itself in a bench checkout. In some cases, however, better control can be achieved with an unstable compensator, and its inconvenience in checkout may be worthwhile.

Figure 6.40 also shows that the system is *conditionally stable* in that the system is unstable for small values of gain. Therefore, if the gain of any part of the system was to take on a smaller than nominal value during startup, the system would be unstable. This type of situation is considered less desirable than one where the system is stable for all values of gain lower than the nominal value.

If, instead of the full-order estimator, we design a reduced-order estimator, with estimator poles at the second-order ITAE locations and with $\omega_0 = 6$, that is, at $-4.2 \pm j4.28$, the desired estimator characteristic polynomial is

$$\alpha_e(s) = s^2 + 8.4s + 36.$$

Following the partitions,

$$\left[\begin{array}{c|c} F_{aa} & F_{ab} \\ \hline \mathbf{F}_{ba} & \mathbf{F}_{bb} \end{array} \right] = \left[\begin{array}{c|cc} -10 & 1 & 0 \\ \hline -16 & 0 & 1 \\ 0 & 0 & 0 \end{array} \right], \qquad \left[\begin{array}{c} G_a \\ \hline \mathbf{G}_b \end{array} \right] = \left[\begin{array}{c} 0 \\ \hline 0 \\ 10 \end{array} \right],$$

and solving for the error characteristic polynomial,

$$\det[s\mathbf{I} - \mathbf{F}_{bb} + \mathbf{L}\mathbf{F}_{ab}] = \alpha_e(s),$$

we find that

$$\mathbf{L} = \left[\begin{array}{c} 8.4 \\ 36 \end{array} \right].$$

The compensator transfer function is given by Eq. (6.117) and is computed to be

$$D(s) = 21.416 \frac{(s - 0.718)(s + 1.87)}{s + 0.95 \pm j6.17}.$$

The associated root locus for this system is shown in Fig. 6.41. Note that this is a non-minimum phase compensator and a zero-degree root locus.

As an alternative to placing poles using the ITAE criterion, consider using the symmetric root locus for this problem. The symmetric root locus for the system is as shown in Fig. 6.42. If we assume that $\mathbf{G}_1 = G$ and $\mathbf{H}_1 = H$, the SRL is the same for the controller and estimator, and one only needs to be generated. Suppose we select $-2 \pm j1.56$ and -8.04 as the desired state feedback poles and $-4 \pm j4.9$ and -9.169 as the estimator poles. The feedback gain is

$$\mathbf{K} = [\,-0.275 \quad 0.218 \quad 0.204\,]$$

and the estimator gain is

$$\mathbf{L} = \left[\begin{array}{c} 7.17 \\ 97.5 \\ 368 \end{array} \right]$$

FIGURE 6.41
Root locus for an ITAE
reduced-order controller.

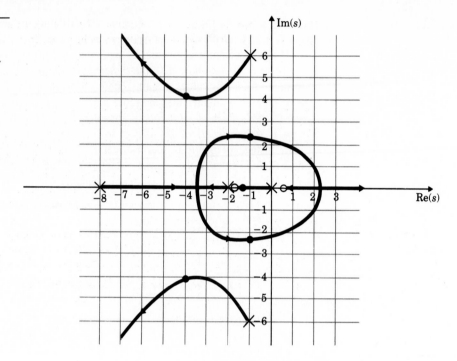

FIGURE 6.42
Symmetric root locus.

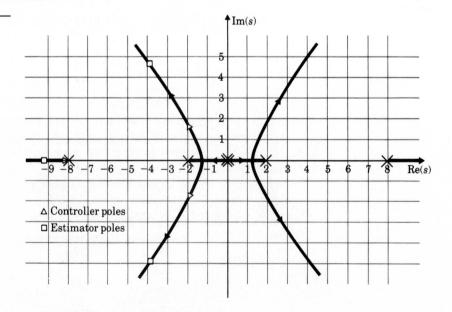

Notice that the feedback gains are much smaller than those required for ITAE placement. The resulting compensator transfer function is given by Eq. (6.113),

$$D(s) = \frac{-94.5(s + 7.98)(s + 2.52)}{(s + 4.28 \pm j6.42)(s + 10.6)},$$

and the root locus of the closed-loop system is shown in Fig. 6.43. It shows the same closed-loop locations as those selected from the SRL, as it should. Note that the compensator is now minimum phase. This improved design comes about in large part because the root at $s = -8$ is virtually unchanged. It doesn't need to be changed for good performance; in fact, the only feature in need of repair in the original $G(s)$ is the root at $s = 0$. The SRL technique essentially discovered that the best use of control effort is to shift the two low-frequency roots at $s = 0, -2$ and to leave the $s = -8$ root alone.

Armed with this knowledge, let's go back, with a better selection of poles, and investigate the use of pole placement for this example. Initially we used the third-order ITAE locations, which produced three poles with a natural frequency around 2 rad/s. This violated the principle that open-loop poles should not be moved unless they're a problem. (The pole at $s = -8$ was moved to $s = -1.4$.) Let's try it

FIGURE 6.43
Root locus for optimum pole assignment.

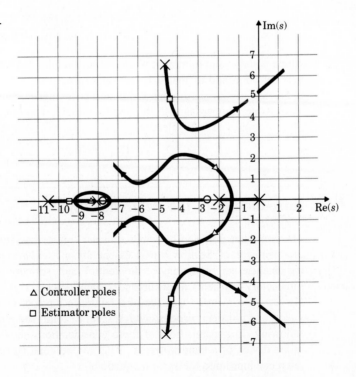

again, but we'll only use ITAE (second-order) to shift the slow poles, and we'll leave the fast pole at $s = -8$; that is,

$$\alpha_c(s) = (s + 1.41 \pm j1.41)(s + 8).$$

We find that the required feedback gain is

$$\mathbf{K} = [\,-0.469 \quad 0.234 \quad 0.0828\,],$$

which has a smaller magnitude than the case where the pole at -8 was moved. It is also interesting to note that the SRL yields a feedback gain that is much smaller in magnitude than either pole-placement alternatives. If we select

$$\alpha_e(s) = (s + 4.24 \pm j4.24)(s + 8),$$

the estimator gain is found to be

$$\mathbf{L} = \begin{bmatrix} 6.48 \\ 87.9 \\ 288 \end{bmatrix},$$

and the compensator transfer function is

$$D(s) = \frac{-41.4(s + 2.78)(s + 8)}{(s + 4.13 \pm j5.29)(s + 9.05)},$$

which is stable and minimum phase. This example illustrates the value of judicious pole selection and of the SRL technique. The poor pole selection inherent to the initial use of the third-order ITAE results in higher control effort and produces an unstable compensator. Both these undesirable features are eliminated by use of SRL or improved pole selection. ∎

6.7   Introduction of the Reference Input

The controller obtained by combining the control law studied in Section 6.3 and the estimator of Section 6.6 is essentially a *regulator design*. This means that the characteristic equations of the control and estimator are chosen so as to give satisfactory transients to disturbances such as $w(t)$, that is, disturbance rejection. However, no mention is made of a reference input or of design considerations to yield good transient response with respect to command changes, that is, command following. In general, both of these considerations should be taken into account in the design of a control system. This can be done by proper introduction of the reference input into the system equations.

Let's repeat the plant and controller equations for the full-order estimator (the reduced-order case is the same in concept, differing only in detail):

$$\left.\begin{aligned} \dot{\mathbf{x}} &= \mathbf{Fx} + \mathbf{G}u \\ y &= \mathbf{Hx} \end{aligned}\right\} \text{ plant;} \tag{6.122}$$

$$\left.\begin{aligned} \dot{\hat{\mathbf{x}}} &= (\mathbf{F} - \mathbf{GK} - \mathbf{LH})\hat{\mathbf{x}} + \mathbf{L}y \\ u &= -\mathbf{K}\hat{\mathbf{x}} \end{aligned}\right\} \text{ controller.} \tag{6.123}$$

Figure 6.44(a) and (b) shows the system with two possibilities for introducing the command input r. In general terms, they are similar to the issue of whether the compensation should be put in the feedback or feed-forward paths. The response of the system to command inputs is different, depending on the configuration, because the zeros of the transfer function are different. The closed-loop roots are identical, however, as can be easily verified by letting $r = 0$ and noting that the systems are then identical.

The difference in the response of the two configurations can be seen quite easily. Consider the effect of a step input in r. For the configuration in Fig. 6.44(a), the step will excite the estimator in precisely the same way that it excites the plant; thus the estimator error will remain zero during and after the step. This means that the estimator dynamics are not excited in the configuration of Fig. 6.44(a), and therefore the transfer function from r to y must have zeros at the estimator

FIGURE 6.44
Possible locations for the
introduction of command
input: (a) compensation
in the feedback path;
(b) compensation in
the feed-forward path.

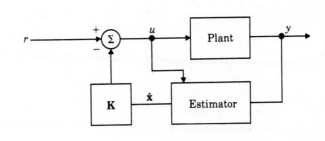

(a)

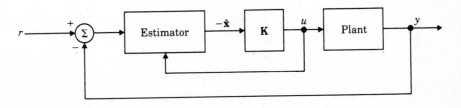

(b)

roots in order to cancel the estimator roots. Therefore, a step command will excite system behavior consistent with the control roots above, that is, with roots of $\det[s\mathbf{I} - \mathbf{F} + \mathbf{GK}] = 0$.

In the configuration seen in Fig. 6.44(b), a step command in r only enters directly into the estimator, thus causing an estimation error that settles out with the estimator dynamic characteristics. Thus a step command will excite system behavior consistent with both control plus estimator roots, that is, the roots of $\det[s\mathbf{I} - \mathbf{F} + \mathbf{GK}]\det[s\mathbf{I} - \mathbf{F} + \mathbf{LH}] = 0$. It is therefore typically superior to command the system by the configuration shown in Fig. 6.44(a).

A more detailed treatment of this issue is covered in the next section. The idea is developed that the command input can be introduced in both locations, and therefore the zeros, as well as the poles, of the transfer function can be arbitrarily selected.

□ 6.7.1 Servodesign

Given a reference input $r(t)$, the most general linear manner to introduce r to the system equations is to add terms proportional to it in the controller equations. We can do this by adding a term Nr to the control equation for u and a term $\mathbf{M}r$ to the estimator equation. Note that N is a scalar and $\mathbf{M}$ is an $n \times 1$ vector in this case. With these additions, the controller equations become

$$\dot{\hat{\mathbf{x}}} = (\mathbf{F} - \mathbf{GK} - \mathbf{LH})\hat{\mathbf{x}} + \mathbf{L}y + \mathbf{M}r,$$
$$u = -\mathbf{K}\hat{\mathbf{x}} + Nr. \tag{6.124}$$

Since $r(t)$ is an external signal, it is clear that neither $\mathbf{M}$ nor N affects the characteristic equation of the combined controller-estimator system. In transfer-function terms, the selection of $\mathbf{M}$ and N will only affect the zeros of transmission from r to y and as a consequence can have a significant effect on the transient response but no effect on stability. The question then is: How can we choose $\mathbf{M}$ and N to obtain satisfactory transient response? We should point out that we assigned the poles of the system by feedback gains $\mathbf{K}$ and $\mathbf{L}$ and are now adding zeros by feed-forward control gains $\mathbf{M}$ and N.

The selection of $\mathbf{M}$ and N can be based on the following four choices:

1. Select $\mathbf{M}$ and N so that the state estimator-error equation is autonomous, that is, independent of r.

2. Select $\mathbf{M}$ and N so that only the tracking error, $e = (r - y)$, is used in the control.

3. Select $\mathbf{M}$ and N to give the designer maximum freedom in setting the dynamic and static response of the system.

4. Select $\mathbf{M}$ and N so as to minimize

$$\min \mathscr{J} = \int_0^\infty e^T e \, dt,$$

the integral-squared tracking error. The algebra for the selection of **M** and N via the fourth method is rather involved and is omitted (see Emami-Naeini and Franklin, 1982). This method is a member of a class of computer-selected controller parameters. In the sequel, we shall discuss Cases 1–3.

Case 1 From the viewpoint of estimator performance, the first of these methods is quite attractive. If $\hat{\mathbf{x}}$ is to generate a good estimate of $\mathbf{x}$, then surely $\tilde{\mathbf{x}}$ should be free of external excitation (that is, it should be "uncontrollable" from r). The computation of **M** and N is quite easy in this case. The estimator-error equation is found by subtracting Eq. (6.75) from Eq. (6.40), with the plant output substituted into the estimator equation and the control law substituted into the plant equation. The result is

$$\dot{\mathbf{x}} - \dot{\hat{\mathbf{x}}} = \mathbf{Fx} + \mathbf{G}(-\mathbf{K}\hat{\mathbf{x}} + Nr) - (\mathbf{F} - \mathbf{GK} - \mathbf{LH})\hat{\mathbf{x}} + \mathbf{L}y - \mathbf{M}r,$$
$$\dot{\tilde{\mathbf{x}}} = (\mathbf{F} - \mathbf{LH})\tilde{\mathbf{x}} + \mathbf{G}Nr - \mathbf{M}r. \tag{6.125}$$

If r is not to appear in Eq. (6.125), we should choose

$$\mathbf{M} = \mathbf{G}N. \tag{6.126}$$

Since N is a scalar, **M** is fixed to within a constant factor. Note that with this choice of **M** we can write the controller equation as

$$u = -\mathbf{K}\hat{\mathbf{x}} + Nr,$$
$$\dot{\hat{\mathbf{x}}} = (\mathbf{F} - \mathbf{LH})\hat{\mathbf{x}} + \mathbf{G}u + \mathbf{L}y, \tag{6.127}$$

which is the configuration of Fig. 6.44(a). In this form, if the true control is subject to saturation, the same control limits can be applied in Eq. (6.127) to the control entering the equation in $\hat{\mathbf{x}}$, and the nonlinearity does not influence the $\tilde{\mathbf{x}}$ equation. This is essential for proper estimator performance and can be used as an anti-integrator-windup technique important in practical problems. We will return to the selection of N, the gain factor of the reference input, after discussing the alternative methods of selecting **M**.

Case 2 The second approach suggested above is that the tracking error be used. This solution is sometimes forced on the control designer because the sensor only measures the output error. As an example, many thermostats have output that is the difference between the temperature to be controlled and the setpoint temperature; that is, there is no absolute indication of the reference temperature available to the controller. Also, some radar tracking systems have a reading that is proportional to the pointing error, and this error signal alone must be used for feedback control. In these situations, we must select **M** and N so that Eq. (6.125) is driven by the error only. This requirement is satisfied if we select

$$N = 0, \qquad \mathbf{M} = -\mathbf{L}. \tag{6.128}$$

Then the estimator equation is

$$\dot{\hat{\mathbf{x}}} = (\mathbf{F} - \mathbf{G}\mathbf{K} - \mathbf{L}\mathbf{H})\hat{\mathbf{x}} + \mathbf{L}(y - r). \qquad (6.129)$$

The compensator for this case, for low-order cases, is a *lead* network in the forward path. The design can have a considerable amount of overshoot because of the zero of the network. However, this goes along with what we found in Chapter 2: The addition of a zero tends to increase the overshoot. This design corresponds exactly to the compensators designed by transform methods.

Case 3 The third method of selecting **M** and N provides the designer with maximum flexibility in satisfying transient-response constraints. This will be accomplished if **M** and N can be selected to add arbitrary zeros of transmission from r to y. First we will discuss briefly the relation between the state-variable system matrices and the zeros of the corresponding closed-loop transfer function. Consider the case where the system uses state feedback and no estimator is involved. Suppose we also wish to include a reference input in the control law as $u = -\mathbf{K}\mathbf{x} + Nr$. The equations for the closed-loop system are

$$\dot{\mathbf{x}} = \mathbf{F}\mathbf{x} + \mathbf{G}(-\mathbf{K}\mathbf{x} + Nr),$$
$$y = \mathbf{H}\mathbf{x}, \qquad (6.130)$$

which may be written as

$$\dot{\mathbf{x}} = (\mathbf{F} - \mathbf{G}\mathbf{K})\mathbf{x} + \mathbf{G}Nr,$$
$$y = \mathbf{H}\mathbf{x}. \qquad (6.131)$$

The system has a zero at s if the equation

$$\begin{bmatrix} s\mathbf{I} - \mathbf{F} + \mathbf{G}\mathbf{K} & -\mathbf{G} \\ \mathbf{H} & 0 \end{bmatrix} \begin{bmatrix} \mathbf{X}(s) \\ R(s) \end{bmatrix} = 0 \qquad (6.132)$$

has a nontrivial solution. This is true if the square matrix on the left loses rank at s, that is, if

$$\det \begin{bmatrix} s\mathbf{I} - \mathbf{F} + \mathbf{G}\mathbf{K} & -\mathbf{G} \\ \mathbf{H} & 0 \end{bmatrix} = 0. \qquad (6.133)$$

The value of this determinant is unchanged if we multiply the last column by **K** and add it to the rest:

$$\det \begin{bmatrix} s\mathbf{I} - \mathbf{F} & -\mathbf{G} \\ \mathbf{H} & 0 \end{bmatrix} = 0. \qquad (6.134)$$

But now the matrix is independent of the feedback gains **K**. We have proved the important result that the zeros of the system are *not* changed by linear state-variable feedback. Equation (6.134) yields the (open-loop) zeros of the plant, which remain invariant under state feedback.

Consider the controller of Eq. (6.123). If there is a zero of transmission from r to u, there is necessarily a zero of transmission from r to y unless there is a coinciding point of infinite transmission—which is to say there is a pole at the location of the zero. It is therefore sufficient to treat the controller alone. The equations for a zero from r to u (we let $y = 0$ since we only care about the effect of r) from Eq. (6.124) is given by

$$\det \begin{bmatrix} s\mathbf{I} - \mathbf{F} + \mathbf{GK} + \mathbf{LH} & -\mathbf{M} \\ -\mathbf{K} & N \end{bmatrix} = 0. \tag{6.135}$$

If we divide the last column by the (nonzero) scalar N and add $\mathbf{K}$ times the last column to the rest, we find the feed-forward zeros are at the values of s such that

$$\det \begin{bmatrix} s\mathbf{I} - \mathbf{F} + \mathbf{GK} + \mathbf{LH} - \frac{\mathbf{M}}{N}\mathbf{K} & \frac{\mathbf{M}}{N} \\ 0 & 1 \end{bmatrix} = 0,$$

$$\det [\, s\mathbf{I} - \mathbf{F} + \mathbf{GK} + \mathbf{LH} - \tfrac{\mathbf{M}}{N}\mathbf{K} \,] = \gamma(s) = 0. \tag{6.136}$$

Equation (6.136) is now in exactly the form of the equation we found in the estimator design for the selection of $\mathbf{L}$ to yield desired locations for the estimator poles. Here we have to select $\mathbf{M}/N$ for a desired zero polynomial, $\gamma(s)$, in the transfer function from the reference input to the control. Thus, there is a substantial amount of freedom to influence the transient response by selection of $\mathbf{M}$. We can add an arbitrary nth-order polynomial to the transfer function from r to u and hence from r to y; that is, we can assign n zeros in addition to all the poles that we assigned previously. If the roots of $\gamma(s)$ are not canceled by the poles of the system, they will be included in zeros of transmission from r to y.

There are two considerations that can guide us in the choice of $\mathbf{M}/N$, that is, in the location of the zeros. The first is dynamic response. We have seen in Chapter 2 that the zeros influence the step response significantly. The second consideration is steady-state error control. The relationship between steady-state accuracy and closed-loop poles and zeros was derived in Chapter 3. If the system is type I, the steady-state error to a step input is zero and the error to a unit ramp input is

$$e_\infty = \frac{1}{K_v}, \tag{6.137}$$

where K_v is the velocity constant coefficient. Furthermore, it was shown in Chapter 3 that if the *closed-loop* poles are at $\{-p_i\}$ and the *closed-loop* zeros are at $\{-z_i\}$, then (for a type I system)

$$\frac{1}{K_v} = \sum \frac{1}{p_i} - \sum \frac{1}{z_i}. \tag{6.138}$$

Equation (6.138) forms the basis for a partial selection of $\gamma(s)$, and hence $\mathbf{M}$ and N. The choice is based on two observations:

1. If $p_i - z_i \ll 1$, then the effect of this pole-zero pair on the dynamic response will be small, since the pole is almost canceled by the zero, and, in any transient, the residue of the pole at p_i will be very small.

2. Even though $p_i - z_i$ is small, it is possible for $(1/p_i) - (1/z_i)$ to be substantial and thus have a significant influence on K_v according to Eq. (6.138).

Application of these two guidelines to the selection of $\gamma(s)$, and hence of $\mathbf{M}$ and N, results in a *lag* network design. We will illustrate this by an example at the end of this section.

It is now interesting to go back and examine the implication of the first two rules in terms of zeros. Under the first rule, we let $\mathbf{M} = \mathbf{G}N$. Substituting this in Eq. (6.136) yields for the controller feed-forward zeros

$$\det[s\mathbf{I} - \mathbf{F} + \mathbf{LH}] = 0. \tag{6.139}$$

But this is exactly the equation from which $\mathbf{L}$ was selected to make the characteristic polynomial of the estimator equation equal to $\alpha_e(s)$. This means that we have created zeros in exactly the same locations as the poles of the estimator. Therefore we have placed n zeros right on top of the n poles of the estimator (this pole-zero cancellation causes "uncontrollability" of the estimator modes), and the overall transfer function poles consist of the state feedback controller poles above. The second rule selects $\mathbf{M} = -\mathbf{L}$ and $N = 0$. If these are substituted in Eq. (6.135), we have that the feed-forward zeros are given by

$$\det\begin{bmatrix} s\mathbf{I} - \mathbf{F} + \mathbf{GK} + \mathbf{LH} & \mathbf{L} \\ -\mathbf{K} & 0 \end{bmatrix} = 0. \tag{6.140}$$

If we postmultiply the last column by $\mathbf{H}$ and subtract it from the rest, and premultiply the last row by $\mathbf{G}$ and add it to the rest, we find that

$$\det\begin{bmatrix} s\mathbf{I} - \mathbf{F} & \mathbf{L} \\ -\mathbf{K} & 0 \end{bmatrix} = 0. \tag{6.141}$$

If we postmultiply the first set of columns by $(s\mathbf{I} - \mathbf{F})^{-1}\mathbf{L}$ and subtract it from the last column, we obtain

$$\det\begin{bmatrix} s\mathbf{I} - \mathbf{F} & 0 \\ -\mathbf{K} & -\mathbf{K}(s\mathbf{I} - \mathbf{F})^{-1}\mathbf{L} \end{bmatrix} = 0, \tag{6.142}$$

which is

$$\det(s\mathbf{I} - \mathbf{F}) \det\left[\mathbf{K}(s\mathbf{I} - \mathbf{F})^{-1}\mathbf{L}\right] = 0,$$

$$\mathbf{K} \operatorname{adj}(s\mathbf{I} - \mathbf{F})\mathbf{L} = 0, \tag{6.143}$$

where adj denotes the matrix adjugate. Thus the added zeros are those obtained by replacing the input matrix by **L** and the output by **K**. This means that if we wish to use error control, we have to accept the presence of these additional zeros. For low-order cases this results, as we said before, in a lead network compensator in a unity feedback topology.

Let us now summarize our findings with respect to the introduction of the reference input. When the reference input signal is included in the controller, the overall transfer function of the closed-loop system is

$$H(s) = \frac{Y(s)}{R(s)} = \frac{K^* \gamma(s) b(s)}{\alpha_e(s) \alpha_c(s)}, \tag{6.144}$$

where K^* is the total system gain and $\gamma(s)$ and $b(s)$ are monic polynomials. The polynomial $\alpha_c(s)$ is selected by the designer and results in a control gain **K** such that $\det[s\mathbf{I} - \mathbf{F} + \mathbf{GK}] = \alpha_c(s)$. The polynomial $\alpha_e(s)$ is selected by the designer and results in estimator gains **L** such that $\det[s\mathbf{I} - \mathbf{F} + \mathbf{LH}] = \alpha_e(s)$. Therefore, the designer has complete freedom in assigning the poles of the closed-loop system. The polynomial $\gamma(s)$ can be selected by the designer so that $\gamma(s) = \alpha_e(s)$, from which **M**/N is given by Eq. (6.126); or else $\gamma(s)$ may be accepted as given by Eq. (6.141), so that error control is used; or finally $\gamma(s)$ may be given arbitrary coefficients by selection of **M**/N from Eq. (6.136). It is important to point out that the plant zeros represented by $b(s)$ are *not* moved by this technique and remain as part of the closed-loop transfer function unless α_c is selected to cancel some of these zeros.

We now turn to the selection of the gain N in the case where we select method 1 or 3 for **M** (recall $N = 0$ for case 2, error control). For these cases, the zeros of the polynomial $\gamma(s)$ are a function of the ratio **M**/$N = \bar{\mathbf{M}}$. How should we select N? One selection that seems reasonable is to pick N such that the overall closed-loop DC gain is unity.* The overall system equations are

$$\begin{bmatrix} \dot{\mathbf{x}} \\ \dot{\tilde{\mathbf{x}}} \end{bmatrix} = \begin{bmatrix} \mathbf{F} - \mathbf{GK} & \mathbf{GK} \\ 0 & \mathbf{F} - \mathbf{LH} \end{bmatrix} \begin{bmatrix} \mathbf{x} \\ \tilde{\mathbf{x}} \end{bmatrix} + \begin{bmatrix} \mathbf{G} \\ \mathbf{G} - \bar{\mathbf{M}} \end{bmatrix} N r,$$

$$y = \begin{bmatrix} \mathbf{H} & 0 \end{bmatrix} \begin{bmatrix} \mathbf{x} \\ \tilde{\mathbf{x}} \end{bmatrix}. \tag{6.145}$$

The closed-loop system has unity DC gain if

$$\begin{bmatrix} \mathbf{H} & 0 \end{bmatrix} \begin{bmatrix} \mathbf{F} - \mathbf{GK} & \mathbf{GK} \\ 0 & \mathbf{F} - \mathbf{LH} \end{bmatrix}^{-1} \begin{bmatrix} \mathbf{G} \\ \mathbf{G} - \bar{\mathbf{M}} \end{bmatrix} N = 1. \tag{6.146}$$

---

*A reasonable alternative is to select N such that, when r and y are both unchanging, the DC gain from r to u is the *negative* of the DC gain from y to u. The consequences of this choice are that our controller can be structured as error control plus generalized derivative control, and if the system is capable of type I behavior, that capability will be realized and the two choices are the same (see Problem 6.40).

If we solve for N,*

$$N = \frac{1}{H(F - GK)^{-1}G[1 - K(F - LH)^{-1}(G - \bar{M})]}, \qquad (6.147)$$

where $\bar{M}$ is the outcome of selection of zero locations via either Eq. (6.136) or (6.126). The techniques in this section can be readily extended to reduced-order estimators.

EXAMPLE 6.17 The generic servomechanism modeled for small deviations from an operating point and scaled for time and amplitude is described by the transfer function

$$G(s) = \frac{1}{s(s + 1)}$$

or state equations

$$\dot{x}_1 = x_2,$$
$$\dot{x}_2 = -x_2 + u.$$

Suppose that dynamic response requirements are met by a closed-loop characteristic equation

$$\alpha_c(s) = (s + 2)^2 + 4 = s^2 = 4s + 8,$$

and steady-state error requirements are met by $K_v = 10$. The state feedback gain

$$\mathbf{K} = [8 \quad 3]$$

results in the desired poles. However, as the reader may verify, this design has $K_v = 2$, and this value is not changed by estimator design if the reference input is applied according to choice 1 (the autonomous estimator). If the lead network option (that is, error control) is used, a zero is introduced at a location unknown beforehand, and the effect on K_v is not under direct design control. However, if we use the third option, we can satisfy both dynamic response and steady-state requirements. First we must select the estimator pole $-p_3$ and the zero z_3 so that Eq. (6.138) is satisfied. To keep $p_3 - z_3$ small and yet have $1/p_3 - 1/z_3$ be large, we arbitrarily set p_3 small compared with the control dynamics. For example, we let

$$-p_3 = -0.1.$$

*We have used the fact that

$$\begin{bmatrix} A & C \\ 0 & B \end{bmatrix}^{-1} = \begin{bmatrix} A^{-1} & -A^{-1}CB^{-1} \\ 0 & B^{-1} \end{bmatrix}.$$

Notice that this is the opposite of the usual philosophy on estimation design, where fast response is the requirement. Using Eq. (6.138),

$$\frac{1}{K_v} = \frac{1}{p_1} + \frac{1}{p_2} + \frac{1}{p_3} - \frac{1}{z_3},$$

where $-p_1 = -2 + j2$, $-p_2 = -2 - j2$, and $-p_3 = -0.1$. We want to solve for z_3 so that $K_v = 10$:

$$\frac{1}{K_v} = \frac{4}{8} + \frac{1}{0.1} - \frac{1}{z_3} = \frac{1}{10}$$

or

$$-z_3 = -\frac{1}{10.4} = -0.096.$$

We thus design a reduced-order estimator to have a pole at -0.1 and choose $\mathbf{M}/N$ such that $\gamma(s)$ has a zero at -0.09615. A block diagram of the resulting system is shown in Fig. 6.45. The reader can readily verify that this system has the overall transfer function

$$H(s) = \frac{Y(s)}{R(s)} = \frac{8.3(s + 0.09)}{(s^2 + 4s + 8)(s + 0.1)}, \tag{6.148}$$

which has $K_v = 10$, as prescribed. The compensation shown in Fig. 6.45(a) is nonclassical in the sense that it has two inputs (e and y) and one output. If we resolve the equations to provide pure error compensation by finding the transfer function from e and u, which would give the overall system Eq. (6.148), we obtain the system shown in Fig. 6.45(b). This compensation is a classical *lag-lead network*. The root locus of the system of Fig. 6.45(b) is shown in Fig. 6.46, which shows the pole-zero pattern near the origin characteristic of the lag network. The Bode plot in Fig. 6.47 shows the phase lag at low frequencies and phase lead at high frequencies. The step response of the system is shown in Fig. 6.48. ∎

□ 6.8 Polynomial Solution

An alternative to the state-space methods discussed so far is to postulate a general-structure dynamic controller with two inputs (r and y) and one output (u) and to solve for the transfer function of the controller to give a specified overall r-to-y transfer function. The situation is as shown in Fig. 6.49. We model the plant as a transfer function

$$\frac{Y(s)}{U(s)} = \frac{b(s)}{a(s)}, \tag{6.149}$$

FIGURE 6.45
Servomechanism with
assigned zeros:
a lag network.

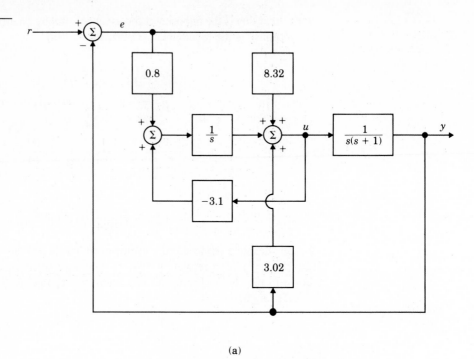

(a)

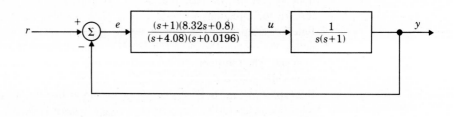

(b)

rather than by state equations. The controller is modeled similarly as

$$U(s) = -\frac{c_y(s)}{d(s)}Y(s) + \frac{c_r(s)}{d(s)}R(s), \qquad (6.150)$$

where $d(s)$, $c_y(s)$, and $c_r(s)$ are polynomials. To complete the design, we require
that the closed-loop transfer function be given by

$$H(s) = \frac{Y(s)}{R(s)} = \frac{c_r(s)b(s)}{\alpha_c(s)\alpha_e(s)}. \qquad (6.151)$$

FIGURE 6.46
Root locus of lag-lead
compensation.

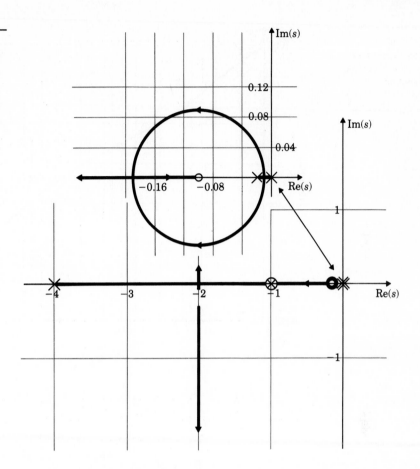

From Eqs. (6.149) and (6.150), we have that

$$a(s)Y(s) = b(s)\left[-\frac{c_y(s)}{d(s)}Y(s) + \frac{c_r(s)}{d(s)}R(s)\right], \qquad (6.152)$$

which can be rewritten as

$$[a(s)d(s) + b(s)c_y(s)]Y(s) = b(s)c_r(s)R(s). \qquad (6.153)$$

Comparing Eq. (6.151) with Eq. (6.153), we immediately see that the design can be accomplished if we can solve the equation (commonly called the Diophantine equation)

$$a(s)d(s) + b(s)c_y(s) = \alpha_c(s)\alpha_e(s) \qquad (6.154)$$

for given arbitrary a, b, α_c, and α_e. If $a(s)$ is of degree n and $b(s)$ is of degree n

FIGURE 6.47
Frequency response of
lag-lead compensation:
(a) magnitude and
(b) phase.

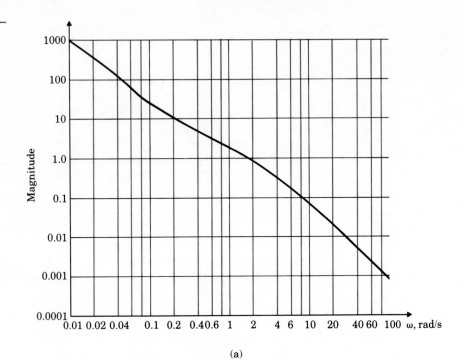

(a)

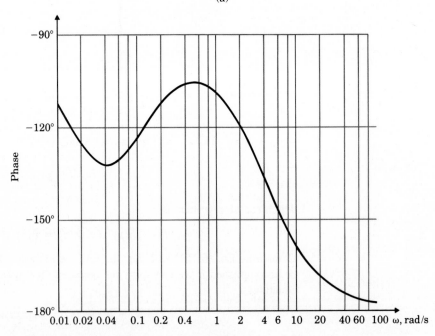

(b)

FIGURE 6.48
Step response.

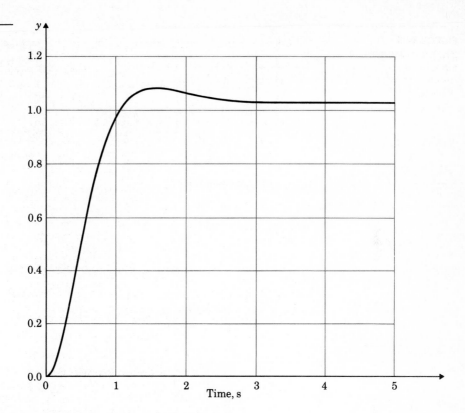

or less, then a solution for $d(s)$ of degree n and $c_y(s)$ of degree $n - 1$ exists for arbitrary $\alpha_c \alpha_e$ of degree $2n$, if and only if $a(s)$ and $b(s)$ have no common factors.

Notice that $c_r(s)$ does not enter in this analysis. However, one can select $c_r(s)$ to assign zeros in the transfer function from $R(s)$ to $Y(s)$. One choice is to select $c_r(s)$ to cancel $\alpha_e(s)$ so that the overall transfer function is

$$H(s) = \frac{Y(s)}{R(s)} = \frac{K^* b(s)}{\alpha_c(s)}.$$

This topic was treated in the state-space setting in the previous section.

EXAMPLE 6.18   Again, consider the third-order example, Example 6.16. For the case of a full-order estimator, we asked for

$$\alpha_c(s)\alpha_e(s) = s^6 + 14s^5 + 122.75s^4 + 585.2s^3 + 1505.64s^2$$
$$+ 2476.8s + 1728.$$

FIGURE 6.49
Direct transfer function
formulation.

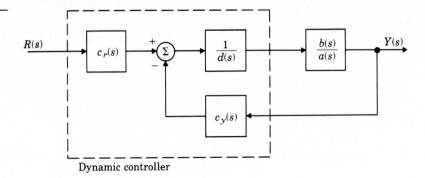

Dynamic controller

Using Eq. (6.154) with $b(s) = 1$ then,

$$(d_0 s^3 + d_1 s^2 + d_2 s + d_3)(s^3 + 10s^2 + 16s)$$
$$+ (c_0 s^2 + c_1 s + c_2) \equiv \alpha_c(s)\alpha_e(s).$$

Equating coefficients of powers of s in the above equation, we immediately see that $d_0 = 1$ and the coefficients satisfy*

$$\begin{bmatrix} 1 & 0 & 0 & 0 & 0 & 0 & 0 \\ 10 & 1 & 0 & 0 & 0 & 0 & 0 \\ 16 & 10 & 1 & 0 & 0 & 0 & 0 \\ 0 & 16 & 10 & 1 & 0 & 0 & 0 \\ 0 & 0 & 16 & 10 & 1 & 0 & 0 \\ 0 & 0 & 0 & 16 & 0 & 1 & 0 \\ 0 & 0 & 0 & 0 & 0 & 0 & 1 \end{bmatrix} \begin{bmatrix} d_0 \\ d_1 \\ d_2 \\ d_3 \\ c_0 \\ c_1 \\ c_2 \end{bmatrix} = \begin{bmatrix} 1 \\ 14 \\ 122.75 \\ 585.2 \\ 1505.64 \\ 2476.8 \\ 1728 \end{bmatrix}.$$

The solution is

$$d_0 = 1; \qquad d_1 = 4; \qquad d_2 = 66.75; \qquad d_3 = -146.3;$$
$$c_0 = 1901; \qquad c_1 = 4818; \qquad c_2 = 1728.$$

Hence, the controller transfer function is

$$D(s) = \frac{c_y(s)}{d(s)} = \frac{190.1s^2 + 481.8s + 172.8}{s^3 + 4s^2 + 66.75s - 146.3},$$

and these are exactly the coefficients of the controller that was obtained using the state-variable techniques. ∎

The reduced-order compensator can also be derived using a polynomial solution. The equations are the same as above, except that both $d(s)$ and $c_y(s)$ are of

*The matrix on the left-hand side is called a *Sylvester matrix* and is nonsingular if and only if $a(s)$ and $b(s)$ have no common factor; see Kailath (1980).

degree $n - 1$. Again, working Example 6.16,

$$\alpha_c(s)\alpha_e(s) = s^5 + 11.9s^4 + 74s^3 + 206s^2 + 377s + 288,$$

and with $b(s) = 1$ we should solve

$$(d_0s^2 + d_1s + d)(s^3 + 10s + 16s) + (c_0s^2 + c_1s + c_2) \equiv \alpha_c(s)\alpha_e(s).$$

Equating coefficients of like powers of s in the above equation, we obtain

$$\begin{bmatrix} 1 & 0 & 0 & 0 & 0 & 0 \\ 10 & 1 & 0 & 0 & 0 & 0 \\ 16 & 10 & 1 & 0 & 0 & 0 \\ 0 & 16 & 10 & 1 & 0 & 0 \\ 0 & 0 & 16 & 0 & 1 & 0 \\ 0 & 0 & 0 & 0 & 0 & 1 \end{bmatrix} \begin{bmatrix} d_0 \\ d_1 \\ d_2 \\ c_0 \\ c_1 \\ c_2 \end{bmatrix} = \begin{bmatrix} 1 \\ 11.9 \\ 74 \\ 206 \\ 377 \\ 288 \end{bmatrix},$$

which has the solution

$$d_0 = 1; \qquad d_1 = 1.9; \qquad d_2 = 39;$$
$$c_0 = -214; \qquad c_1 = -247.2; \qquad c_2 = 288.$$

The resulting controller is

$$D(s) = \frac{c_y(s)}{d(s)} = \frac{-21.4s^2 - 24.72s + 28.8}{s^2 + 1.9s + 39},$$

which again is exactly the same as that derived using the state-variable techniques.

6.9   Integral Control

In the state-space design method discussed up to now, no mention has been made of integral control, and no design examples have produced a compensation containing an integral term. In fact, state-space design naturally produces a generalization of proportional and derivative feedback, as has been pointed out by the design examples, but it does not produce integral control unless special steps are taken in the design process.

One way to introduce an integral term is to augment the state vector with the desired integral. More specifically, for the system

$$\dot{\mathbf{x}} = \mathbf{F}\mathbf{x} + \mathbf{G}u,$$
$$y = \mathbf{H}\mathbf{x}.$$

We can feed back the integral of y as well as the state $\mathbf{x}$ by augmenting the plant state with the extra state x_I, which obeys the differential equation

$$\dot{x}_I = \mathbf{H}\mathbf{x}(= y).$$

FIGURE 6.50
Integral control structure.

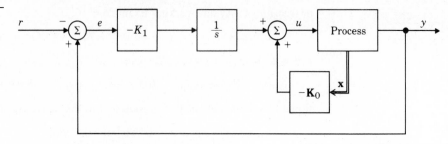

Thus

$$x_I = \int y \, dt.$$

The augmented state equations become

$$\begin{bmatrix} \dot{x}_I \\ \dot{\mathbf{x}} \end{bmatrix} = \begin{bmatrix} 0 & \mathbf{H} \\ 0 & \mathbf{F} \end{bmatrix} \begin{bmatrix} x_I \\ \mathbf{x} \end{bmatrix} + \begin{bmatrix} 0 \\ \mathbf{G} \end{bmatrix} u,$$

and the feedback law is

$$u = -[K_1 \quad \mathbf{K}_0] \begin{bmatrix} x_I \\ \mathbf{x} \end{bmatrix},$$

or simply

$$u = -\mathbf{K} \begin{bmatrix} x_I \\ \mathbf{x} \end{bmatrix}.$$

With this revised definition of the system, the design techniques from Section 6.3 can be applied in a similar fashion; they will result in the control structure shown in Fig. 6.50.

EXAMPLE 6.19   Suppose the system is given by

$$\frac{Y(s)}{U(s)} = \frac{1}{s + 3}, \qquad \text{that is,} \qquad F = -3, \, G = 1, \, H = 1,$$

and we wish to have integral control with closed-loop roots at $\omega_n = 5$ and $\zeta = 0.5$, which is equivalent to asking for a desired characteristic equation:

$$\alpha_c(s) = s^2 + 5s + 25 = 0.$$

FIGURE 6.51
Integral control example.

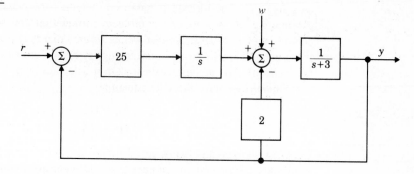

The augmented system description is

$$\begin{bmatrix} \dot{x}_I \\ \dot{x} \end{bmatrix} = \begin{bmatrix} 0 & 1 \\ 0 & -3 \end{bmatrix} \begin{bmatrix} x_I \\ x \end{bmatrix} + \begin{bmatrix} 0 \\ 1 \end{bmatrix} u.$$

Therefore, we can find $\mathbf{K}$ from

$$\det \left[s\mathbf{I} - \begin{bmatrix} 0 & 1 \\ 0 & -3 \end{bmatrix} + \begin{bmatrix} 0 \\ 1 \end{bmatrix} \mathbf{K} \right] = s^2 + 5s + 25,$$

or

$$s^2 + (3 + K_0)s + K_1 = s^2 + 5s + 25.$$

Therefore

$$\mathbf{K} = [K_1 \quad K_0] = [25 \quad 2].$$

The system is shown with feedbacks in Fig. 6.51 along with a disturbing input w. Notice that if a reference command signal $r(t)$ is introduced, it is the system error, $e = r - y$, that must be integrated and not the output alone. This system will behave according to the desired closed-loop roots ($\omega_n = 5$, $\zeta = 0.5$) and will exhibit the characteristics of integral control: Zero steady-state error to a step r and zero steady-state error to a constant disturbance w. ∎

☐ 6.9.1 Internal Model Control

In the previous section, integral control was introduced in a direct way and the structure of the implementation was selected to achieve integral action with respect to reference and disturbance inputs. We now present a more analytical approach to giving a control system the ability to track (with zero error) a nondecaying input and to reject (with zero error) a nondecaying disturbance such as a step, a ramp, or a sinusoid. The method is based on including the equations satisfied by these external signals as part of the problem formulation and solving the problem of control in an *error space* so that we are assured that the error approaches zero even if the output is following a nondecaying, or even a growing, command (such as a ramp signal). The method is illustrated in detail for signals of order 2.

Suppose we have the state equations

$$\dot{\mathbf{x}} = \mathbf{Fx} + \mathbf{G}u + \mathbf{G}_1 w,$$
$$y = \mathbf{Hx} + Ju, \qquad (6.155)$$

and we wish to do a pole-placement design and at the same time design a controller so the closed-loop system can track a second-order input with zero steady-state error and reject second-order disturbances without error. We can define the reference input as satisfying the differential equation

$$\ddot{r} = \alpha_1 \dot{r} + \alpha_2 r. \qquad (6.156)$$

In a similar fashion, we take the disturbance to satisfy

$$\ddot{w} = \alpha_1 \dot{w} + \alpha_2 w. \qquad (6.157)$$

The (tracking) error is defined as

$$e = y - r. \qquad (6.158)$$

The problem of tracking r and rejecting w can be seen as an exercise in designing a control law to provide *regulation of the error,* which is to say such that the error e tends to zero as time gets large. The control must also be *structurally stable* or *robust,* in the sense that regulation of e to zero in the steady-state occurs in the presence of perturbation of system parameters. Note that in practice one never has a perfect model and that the values of parameters are virtually always subject to change.

One way to formulate the tracking problem is to differentiate the error equation and introduce the error as a state. Thus, we have from Eq. (6.158) the formula

$$\ddot{e} = \ddot{y} - \ddot{r}$$
$$= \mathbf{H}\ddot{\mathbf{x}} + J\ddot{u} - \alpha_1 \dot{r} - \alpha_2 r. \qquad (6.159)$$

To obtain the overall state vector, the plant-state vector is replaced by the state in error space defined by

$$\xi \triangleq \ddot{\mathbf{x}} - \alpha_1 \dot{\mathbf{x}} - \alpha_2 \mathbf{x}, \qquad (6.160)$$

and the control vector is replaced by the control vector in error space

$$\mu \triangleq \ddot{u} - \alpha_1 \dot{u} - \alpha_2 u. \qquad (6.161)$$

The error-state formula, from Eq. (6.159), is

$$\ddot{e} - \alpha_1 \dot{e} - \alpha_2 e = H\xi + J\mu. \qquad (6.162)$$

The state equation for ξ is given by*

$$\dot{\xi} = \dddot{x} - \alpha_1 \ddot{x} - \alpha_2 \dot{x}$$
$$= F\xi + G\mu. \tag{6.163}$$

Equations (6.160) and (6.161) now describe the overall system state; in state-variable form, these are

$$\dot{z} = Az + B\mu, \tag{6.164}$$

where $\mathbf{z} = [e \quad \dot{e} \quad \xi]^T$ and

$$A = \begin{bmatrix} 0 & 1 & 0 \\ \alpha_2 & \alpha_1 & H \\ 0 & 0 & F \end{bmatrix} \qquad B = \begin{bmatrix} 0 \\ J \\ G \end{bmatrix}. \tag{6.165}$$

The error system (A, B) can be given arbitrary dynamics by state feedback if it is controllable. If the plant (F, G) is controllable and does not have a zero at the roots of

$$\alpha_r(s) = s^2 - \alpha_1 s - \alpha_2,$$

the error system (A, B) is controllable. Therefore there exists a control law of the form

$$\mu = -[K_2 \quad K_1 \quad K_0] \begin{bmatrix} e \\ \dot{e} \\ \xi \end{bmatrix} = -Kz, \tag{6.166}$$

so that the error system has arbitrary dynamics by pole placement. We now need to express this control law in terms of the actual process state $\mathbf{x}$ and the actual control. In terms of u and $\mathbf{x}$ we have the control law,

$$(u + K_0 x)^{(2)} = \sum_{i=1}^{2} \alpha_i (u + K_0 x)^{(2-i)} - \sum_{i=1}^{2} K_i e^{(2-i)}.$$

EXAMPLE 6.20 Consider a servomechanism to follow the data track on a computer-disk memory system. Because the data track is not exactly a centered circle, the radial servo must follow a ("runout") sinusoid input of radian frequency ω_0. The (normalized) parameters are

$$F = \begin{bmatrix} 0 & 1 \\ 0 & -1 \end{bmatrix}; \qquad G = \begin{bmatrix} 0 \\ 1 \end{bmatrix}; \qquad H = [1 \quad 0];$$

$$J = 0; \qquad G_1 = \begin{bmatrix} 0 \\ 0 \end{bmatrix};$$

---

*Notice that this concept can be extended to more complex equations in r and to multivariable systems.

FIGURE 6.52
Structure of the compensator for the servomechanism to track exactly the sinusoid of frequency ω_0.

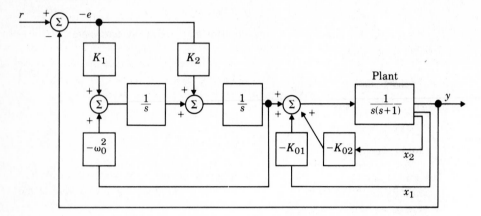

and the reference input satisfies $\ddot{r} = -\omega_0^2 r$. The error state matrices are

$$
\mathbf{A} = \begin{bmatrix} 0 & 1 & 0 & 0 \\ -\omega_0^2 & 0 & 1 & 0 \\ 0 & 0 & 0 & 1 \\ 0 & 0 & 0 & -1 \end{bmatrix} ; \qquad \mathbf{B} = \begin{bmatrix} 0 \\ 0 \\ 0 \\ 1 \end{bmatrix}.
$$

The characteristic equation of $\mathbf{A} - \mathbf{BK}$ is

$$
s^4 + (1 + K_{02})s^3 + (\omega_0^2 + K_{01})s^2 + [K_1 + \omega_0^2(1 + K_{02})]s + K_2 + K_{01}\omega_0^2 = 0,
$$

from which the gain may be selected if pole assignment is satisfactory for the design (see Problem 6.44). The compensator implementation has the structure shown in Fig. 6.52, which clearly shows the presence of the oscillator with frequency ω_0 (known as the *internal model* of the input generator) in the controller.*

Let's now consider the case of tracking constant inputs, that is, $\dot{r} = 0$. In terms of u and $\mathbf{x}$, we have the control law

$$
\dot{u} = -K_1 e - \mathbf{K}_0 \dot{\mathbf{x}}. \tag{6.167}
$$

Here we only need to integrate to reveal the control law and the action of integral control,

$$
u = -K_1 \int^t e\, d\tau - \mathbf{K}_0 \mathbf{x}. \tag{6.168}
$$

---

*This is a special case of the internal-model principle, which requires a model of the exogenous signal in the controller for robust tracking and disturbance rejection. See Francis and Wonham (1976).

A block diagram of the system is seen in Fig. 6.50; it clearly shows the presence of a pure integrator in the controller. ∎

From the nature of the pole-placement problem, the state z in Eq. (6.164) will tend to zero for all perturbations in the system parameters so long as $A - BK$ remains stable. The nature of the design is such that it creates structurally robust "blocking" zeros (that is, those which do *not* change as the parameters change) from r and w to e, thereby completely blocking transmission of signals at roots of $\alpha_r(s)$ between the external signals and the system output error e.

EXAMPLE 6.21   Let's repeat the example of the previous section:

$$H(s) = \frac{1}{s + 3},$$

with state-variable description

$$F = -3, \qquad G = 1, \qquad H = 1.$$

The error system is

$$\begin{bmatrix} \dot{e} \\ \dot{z} \end{bmatrix} = \begin{bmatrix} 0 & 1 \\ 0 & -3 \end{bmatrix} \begin{bmatrix} e \\ z \end{bmatrix} + \begin{bmatrix} 0 \\ 1 \end{bmatrix} u.$$

If the desired characteristic equation is taken to be

$$\alpha_c(s) = s^2 + 5s + 25,$$

the pole-placement equation for K is

$$\det[sI - A + BK] = \alpha_c(s).$$

In detail, it is

$$s^2 + (3 + K_1)s + K_0 = s^2 + 5s + 25,$$

which gives

$$K = [25 \qquad 2] = [K_0 : K_1],$$

and the system is implemented as shown in Fig. 6.51. If we compute the transfer function from r to e,

$$\frac{E(s)}{R(s)} = \frac{-s(s + 5)}{s^2 + 5s + 25},$$

which clearly shows the "blocking" zero at $s = 0$. The overall transfer function is

$$H(s) = \frac{Y(s)}{R(s)} = \frac{25}{s^2 + 5s + 25}.$$

FIGURE 6.53
Robust servomechanism with zero assignment.

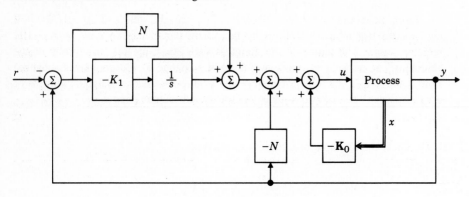

The structure of Fig. 6.50 permits the engineer to add the feed-forward of the reference input, which provides additional freedom in zero assignment. If we add a term proportional to r in Eq. (6.168), then

$$u = -K_1 \int^t e(\tau)\,d\tau - \mathbf{K}_0 \mathbf{x} + Nr,$$

and this has the effect of creating a zero at $-K_1/N$. The location of this zero can be chosen to improve the transient response of the system. For actual implementation, the above equation can be rewritten in terms of e:

$$u = -K_1 \int^t e(\tau)\,d\tau - \mathbf{K}_0 \mathbf{x} + N(y - e). \tag{6.169}$$

The system is as shown in Fig 6.53. For our example, the overall transfer function

FIGURE 6.54
Robust servomechanism.

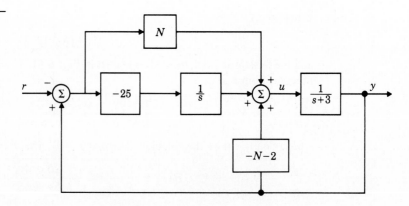

now becomes

$$H(s) = \frac{Y(s)}{R(s)} = \frac{Ns + 25}{s^2 + 5s + 25},$$

and its implementation is shown in Fig. 6.54. Notice that the DC gain is unity for any value of N and that by choice of N we can place the zero anywhere we like to improve the dynamic response. The step response of the system is as shown in Fig. 6.55 for two values of N. In computer-aided design Emami-Naeini and Franklin (1982) show how to optimize the choice of N with respect to a performance criterion.   ■

FIGURE 6.55
Step responses.

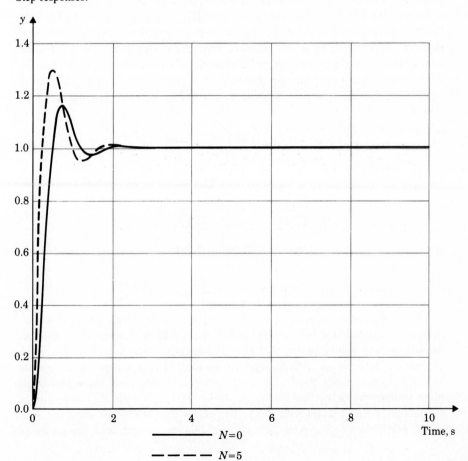

□ 6.10 Design for Systems with Pure Time Delay

In any linear system consisting of lumped elements, the response of the system appears immediately after an excitation of the system. In some instances in feedback systems—for example, in process control, whether in systems controlled by a human operator in the loop or in computer-controlled systems—there is a pure time delay or transportation lag in the system [see Fig. 6.56(a)]. As a result of the distributed nature of these systems, the response remains identically zero until after some time period λ. The transfer function of a pure transportation lag is $e^{-\lambda s}$. We can represent an overall transfer function of a (SISO) system with time delay as

$$G_I(s) = G(s)e^{-\lambda s}, \qquad (6.170)$$

where $G(s)$ has no pure time delay. Since $G_I(s)$ does not have a finite state description, standard use of state-variable methods is impossible. However, O. J. M. Smith (1958) showed how to construct a feedback structure that effectively takes the delay outside the loop and allows a feedback design based on $G(s)$ alone, which does have a state description. The result is a closed-loop transfer function with delay λ and otherwise the response of a closed-loop design based on no delay. Consider the feedback structure shown in Fig. 6.56(b). The transfer function is

$$\frac{Y(s)}{R(s)} = \frac{D(s)'e^{-\lambda s}G(s)}{1 + D(s)'e^{-\lambda s}G(s)}. \qquad (6.171)$$

Let us postulate a modified compensator, $D(s)$, such that the transfer function is

$$\frac{Y(s)}{R(s)} = \frac{D(s)G(s)}{1 + G(s)D(s)}e^{-\lambda s}. \qquad (6.172)$$

If we equate Eqs. (6.171) and (6.172), we will find

$$D(s) = \frac{D(s)'}{1 + D(s)'[-G(s) + G(s)e^{-\lambda s}]} \qquad (6.173)$$

and the compensator is as shown in Fig. 6.56(c). So with this structure we can design the compensator $D(s)$ in the usual way, as if there were no delay, and then implement it as shown in Fig. 6.56(c). The resulting closed-loop system would exhibit the behavior of a finite state closed-loop system except for a time delay of λ. Conceptually, the Smith compensator is feeding back the output of the plant without the delay, that is, the output of $G(s)$, and with $e^{-\lambda s}$ appearing as the second block in series with $G(s)$. The reader may verify this using block diagram manipulation. It can be demonstrated that the compensator, shown by the

FIGURE 6.56
A Smith regulator for systems with time delay.

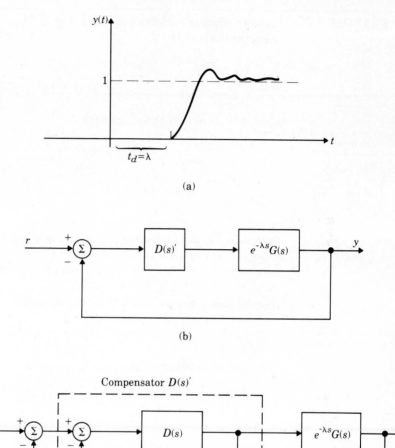

(a)

(b)

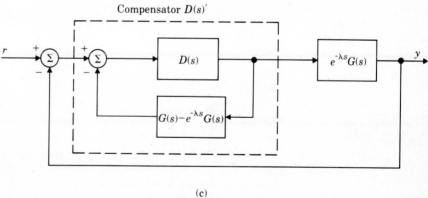

(c)

dashed lines in Fig. 6.56(c), is equivalent to an ordinary regulator in cascade with a compensator that provides significant *phase lead* (see Åström, 1977). It would be difficult to implement such compensators in analog systems; however, they can be implemented satisfactorily using digital compensators (see Chapter 8) and are currently being used by process-control industries.

EXAMPLE 6.22   Consider the heat exchanges shown in Fig. 6.57. A linear model for the system can be represented by

$$G(s) = \frac{e^{-5s}}{(10s + 1)(60s + 1)}.$$

A suitable set of state-space equations is

$$\dot{\mathbf{x}}(t) = \begin{bmatrix} -0.017 & 0.017 \\ 0 & -0.1 \end{bmatrix} \mathbf{x}(t) + \begin{bmatrix} 0 \\ 0.1 \end{bmatrix} u(t - 5),$$

$$y = [\,1 \quad 0\,]\mathbf{x},$$

$$\lambda = 5.$$

Suppose we wish to place the closed-loop poles at

$$\alpha_c(s) = s + 0.05 \pm j0.087;$$

then the state feedback gain is (ignoring the delay)

$$\mathbf{K} = [\,5.2 \quad -0.17\,].$$

If we pick the estimator poles three times faster,

$$\alpha_e(s) = s + 0.15 \pm j0.26,$$

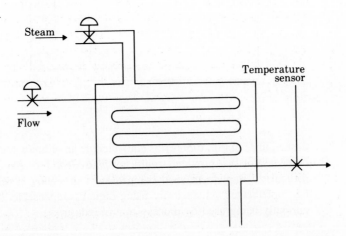

FIGURE 6.57
A heat exchanger.

then the estimator gain matrix for a full-order estimator is

$$\mathbf{L} = \begin{bmatrix} 0.18 \\ 4.2 \end{bmatrix}.$$

If we choose to cancel the estimator poles with the feed-forward zeros and want unity DC gain, then

$$N = 6.0 \quad \text{and} \quad \mathbf{M} = \mathbf{G}N.$$

The controller transfer function resulting from all this is

$$D(s) = \frac{U(s)}{Y(s)} = \frac{-0.25(s + 1.8)}{s + 0.14 \pm j0.27}.$$

The open-loop and closed-loop step responses of the system and the control effort are shown in Figs. 6.58 and 6.59, and the root locus of the system (without the delay) is shown in Fig. 6.60. Note that the time delay in Figs. 6.58 and 6.60 is quite small compared with the response of the system, and is barely noticeable in this case. ∎

□ 6.11 Lyapunov Stability

It should be obvious by now that stability plays a major role in control-system design. We have already considered bounded-input–bounded-output stability based on impulse response, Routh's criterion based on the coefficients of the characteristic equation, and Nyquist's criterion based on the frequency response. In this section, we consider a very important technique for the analysis (and design) of nonlinear dynamic systems based on the state-variable form of ordinary differential equations.

A. M. Lyapunov* considered stability of general systems described by ordinary differential equations in state-variable form. To illustrate a few points of his very useful theory, we consider first the general set of state equations,

$$\dot{\mathbf{x}} = \mathbf{f}(\mathbf{x}). \tag{6.174}$$

If the function $\mathbf{f}(\mathbf{x}) = \mathbf{F}\mathbf{x}$, then we have the linear case considered throughout this chapter. We assume that the equations have been written so that $\mathbf{x} = 0$ is an

---

*A. M. Lyapunov studied dynamic systems in Russia, and his fundamental work, *On the General Problem of Stability of Motion,* was published in 1892 by the Kharkov Mathematical Society.

FIGURE 6.58
Step responses for a heat exchanger.

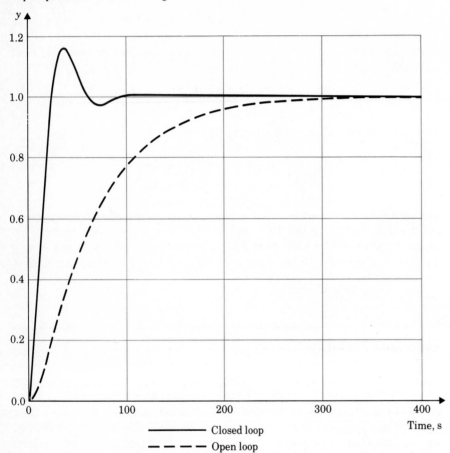

Closed loop

Open loop

equilibrium point, which is to say that $\mathbf{f}(0) = 0$. This equilibrium point is said to be stable in the sense of Lyapunov if we are able to select a nonzero initial condition that will result in a trajectory that remains small. The equilibrium is said to be asymptotically stable if it is Lyapunov-stable, and furthermore the state goes to zero as time goes to infinity. More formally, the system described by Eq. (6.174) has a stable equilibrium at $\mathbf{x} = 0$ if, for every ε there is a δ such that if $\|\mathbf{x}(0)\| < \delta$ then $\|\mathbf{x}(t)\| < \epsilon$ for all t.* Pictures of stable and unstable responses are sketched in Fig. 6.61.

*$\|\mathbf{x}\|^2 = \mathbf{x}^T\mathbf{x} = \sum_{i=1}^{n} x_i^2$ is a measure of the length of the vector $\mathbf{x}$.

FIGURE 6.59
Control effort for a heat exchanger.

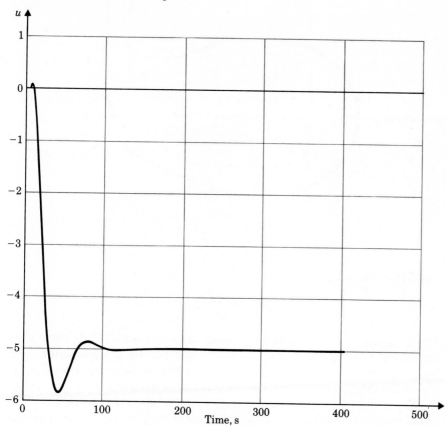

For constant linear systems, we see at once that $\dot{\mathbf{x}} = \mathbf{Fx}$ is stable if none of the characteristic roots of $\mathbf{F}$ are in the right-half plane and if any roots on the imaginary axis are simple. (A multiple root on the imaginary axis would have a response that grows in time and could not be stable.) Furthermore, the response of a linear constant system is asymptotically stable if all the characteristic values of $\mathbf{F}$ are inside the left-half plane. Comparing with the results on BIBO stability (see Section 3.5.1), we see that linear constant systems that are BIBO-stable are asymptotically stable in the sense of Lyapunov, and, furthermore, we can have stable systems in the sense of Lyapunov that are not BIBO-stable. (The capacitor is a case of this.)

Lyapunov showed that if one has a nonlinear system, such as in Eq. (6.174), and it is expanded into the form

$$\dot{\mathbf{x}} = \mathbf{Fx} + \mathbf{g(x)}, \tag{6.175}$$

FIGURE 6.60
Root locus for a heat ex-
changer.

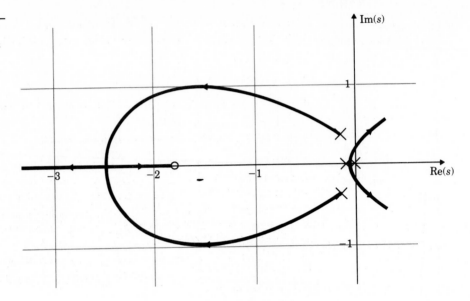

where $\mathbf{g(x)}$ contains all the higher powers of $\mathbf{x}$ to the extent that

$$\lim_{\|x\| \to 0} \frac{\|\mathbf{g(x)}\|}{\|\mathbf{x}\|} = 0$$

(that is, $\mathbf{g(x)}$ goes to zero faster than $\mathbf{x}$ does), then the system will be stable if all
the roots of $\mathbf{F}$ are strictly inside the left-half plane and will be unstable if at least
one root is in the right-half plane. For systems with roots in the left-half plane
and on the imaginary axis, the stability depends on the terms in the function $\mathbf{g}$.
In order to improve these results, Lyapunov introduced a function that has many
of the properties of energy. He then proved that if the "energy" is always getting

FIGURE 6.61
Definition of stability in
the sense of Lyapunov.

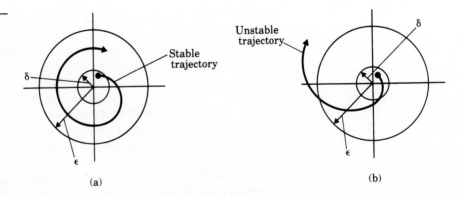

smaller, it must finally run out and the system come to rest. The argument goes like this: For a state $\mathbf{x}$, consider a scalar function $V(\mathbf{x})$ having the property that

1. $V(0) = 0$;
2. $V(\mathbf{x}) > 0$, $\|\mathbf{x}\| \neq 0$;
3. V is continuous and has continuous derivatives with respect to all components of $\mathbf{x}$;
4. $\dot{V}(\mathbf{x}) \leq 0$ along trajectories of the equation.

Properties 1 and 2 mean that V, like energy, is positive if any state is different from zero but equals zero when the state is zero. Property 3 ensures that V is a smooth function and generally has the shape of a bowl near the equilibrium. Property 4 guarantees that the trajectory moves so as to never climb higher on the bowl than it starts out. If property 4 is made stronger so that $\dot{V} < 0$ for $\|\mathbf{x}\| \neq 0$, then the trajectory must be drawn to the origin. The stability theorem is: Given the system of equations $\dot{\mathbf{x}} = \mathbf{f}(\mathbf{x})$, if there exists a Lyapunov function for this equation, the origin is a stable equilibrium point. If, in addition, $\dot{V} < 0$, the stability is asymptotic.

For the linear constant systems of main concern here, the quadratic function is adequate. Consider the function $V = \mathbf{x}^T\mathbf{P}\mathbf{x}$, where $\mathbf{P}$ is a symmetric positive matrix. For example, if $\mathbf{P} = \mathbf{I}$, then V is the sum of squares of the components of $\mathbf{x}$ as $V = \sum_{i=1}^{n} x_i^2$. In general, if $\mathbf{P}$ is positive, it means that we can find a matrix $\mathbf{T}$ such that $\mathbf{P} = \mathbf{T}^T\mathbf{T}$ and $V = \sum_{i=1}^{n} z_i^2$, where $\mathbf{z} = \mathbf{T}\mathbf{x}$. For such a $\mathbf{P}$ the quadratic function satisfies conditions 1, 2, and 3 of a Lyapunov function. A well-known test for a symmetric matrix to be positive is that the determinants of the n principal minors of the matrix are positive. We need to consider condition 4, the derivative condition. To compute the derivative of V, we use the chain rule as follows:

$$\dot{V} = \frac{d}{dt}\mathbf{x}^T\mathbf{P}\mathbf{x}$$
$$= \dot{\mathbf{x}}^T\mathbf{P}\mathbf{x} + \mathbf{x}^T\mathbf{P}\dot{\mathbf{x}}$$
$$= \mathbf{x}^T(\mathbf{F}^T\mathbf{P} + \mathbf{P}\mathbf{F})\mathbf{x}$$
$$= -\mathbf{x}^T\mathbf{Q}\mathbf{x},$$

where

$$\mathbf{F}^T\mathbf{P} + \mathbf{P}\mathbf{F} = -\mathbf{Q}. \tag{6.176}$$

Lyapunov showed that, for any positive $\mathbf{Q}$, the solution $\mathbf{P}$ of the Lyapunov equation (Eq. 6.176) is positive if and only if all the characteristic roots of $\mathbf{F}$ have negative real parts. In other words, if we are given the system matrix $\mathbf{F}$, we can select a positive $\mathbf{Q}$ such as $\mathbf{I}$, solve the Lyapunov equation (which is just a system of linear equations in $n(n-1)/2$ unknowns) and test to see if $\mathbf{P}$ is positive by looking at the determinants of the n principal minors. From this we can determine the stability of the equations without either solving them or finding the characteristic roots, a much harder problem. An example may help illustrate this theory.

EXAMPLE 6.23   Consider

$$\mathbf{F} = \begin{bmatrix} -\alpha & \beta \\ -\beta & -\alpha \end{bmatrix}; \qquad \alpha > 0.$$

Let $\mathbf{Q} = \mathbf{I}$. The Lyapunov equation is

$$\begin{bmatrix} -\alpha & -\beta \\ \beta & -\alpha \end{bmatrix}\begin{bmatrix} p & q \\ q & r \end{bmatrix} + \begin{bmatrix} p & q \\ q & r \end{bmatrix}\begin{bmatrix} -\alpha & \beta \\ -\beta & -\alpha \end{bmatrix} = \begin{bmatrix} -1 & 0 \\ 0 & -1 \end{bmatrix}.$$

Multiplying out and equating coefficients,

$$-\alpha p - \beta q - \alpha p - \beta q = -1,$$
$$-\alpha q - \beta r + \beta p - \alpha q = 0,$$
$$\beta q - \alpha r + \beta q - \alpha r = -1.$$

These equations are readily solved by $p = r = 1/2\alpha$, $q = 0$ so that

$$\mathbf{P} = \begin{bmatrix} 1/2\alpha & 0 \\ 0 & 1/2\alpha \end{bmatrix},$$

and the determinants are $1/2\alpha > 0$ and $1/4\alpha^2 > 0$.

Thus $\mathbf{P}$ is positive, and we conclude that the system is stable if $\alpha > 0$. ∎

For systems with more state variables, solution of the Lyapunov equation (Eq. 6.176) can be burdensome, but the result is an alternative to computing the characteristic values.

Testing for stability by considering the linear part is referred to as *Lyapunov's first*, or *indirect, method;* using the idea of the Lyapunov function directly on the nonlinear equations themselves is called the *second*, or *direct method*. As an example of the direct method, consider the position feedback system modeled in Fig. 6.62.

We assume that the actuator, which is perhaps only an amplifier in this case, has a significant nonlinearity sketched as a saturation but possibly more complex.

FIGURE 6.62
A model of an elementary position feedback system with a nonlinear actuator.

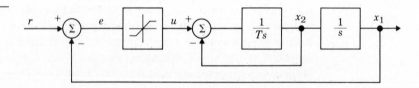

Here we will assume only that $u = f(e)$, where the function lies in the first and third quadrants so that $\int_0^e f(e)\,de > 0$. We also assume that $T > 0$, so that the system is open-loop stable. The equations of motion are

$$\dot{e} = -x_2,$$

$$\dot{x}_2 = \frac{-1}{T}x_2 + \frac{f(e)}{T}. \tag{6.177}$$

For a Lyapunov function, consider something like kinetic plus potential energy:

$$V = \frac{T}{2}x_2^2 + \int_0^e f(\sigma)\,d\sigma.$$

Clearly, $V = 0$ if $x_2 = e = 0$ and, because of the assumptions on f, $V > 0$ if $x_2^2 + e^2 \neq 0$. To check to see if this is a Lyapunov function, we compute $\dot{V}$ as follows:

$$\dot{V} = Tx_2\dot{x}_2 + f(e)\dot{e}$$
$$= Tx_2\left[-\frac{1}{T}x_2 + \frac{f(e)}{T}\right] + f(e)(-x_2)$$
$$= -x_2^2.$$

Hence, $\dot{V} \leq 0$ and the origin is stable. In fact, $\dot{V}$ is always decreasing if $x_2 \neq 0$, and, from Eq. (6.177), the system has no trajectory with $x_2 = 0$, except $x_2 = e = 0$; so in fact we can conclude that the system is asymptotically stable for every f that satisfies $\int f\,d\sigma > 0$.

As we mentioned, the study of the stability of nonlinear systems is vast, and we have only touched on some important points and methods. Further material for study can be found in LaSalle and Lefschetz (1961), Kalman and Bertram (1960), and Vidyasagar (1978).

Summary

In this chapter, we have considered systems described by state variables. Analysis techniques and design methods to obtain compensation for stabilization and improved dynamic behavior have been presented. We have seen that, if a system is controllable, a unique control gain $\mathbf{K}$ can be found such that the closed-loop characteristic polynomial $\alpha_c(s)$ of $\mathbf{F} - \mathbf{GK}$ can be given arbitrary poles. (Complex poles must appear in conjugate pairs.) The selection of poles via ITAE and Bessel responses and by the symmetric root-locus method have been discussed.

We have seen how an estimator can be designed such that the estimated-error characteristic polynomial $\alpha_e(s)$ of $\mathbf{F} - \mathbf{LH}$ can be given arbitrary poles. We have seen how the combination of the controller and estimator can be interpreted to yield a classical compensator similar to those designed in Chapters 4 and 5. The separation principle allows for independent design of the controller and estimator. Duality between the control and estimation problems has been pointed out.

The idea of tracking a reference input has also been discussed. Introduction of the reference signal allows for assigning new zeros, placing zeros on top of estimator poles, or using error control. The general idea is that state feedback allows for assigning poles, and feed-forward control allows for assigning new zeros.

We have seen that the same type of compensator can be obtained using polynomials as using state variables. However, this technique is guided by insight into the problem gained using state-variable techniques. The use of integral control in robust servomechanisms has been discussed and the Smith regulator for systems with pure time delay was presented briefly. Finally, we have touched on the notions of stability for nonlinear systems as set forth by Lyapunov.

Appendix 6A:
Solution of State Equations

In this appendix, we consider the solution of state-variable equations. This material is not necessary to understand the design method of pole placement but will give a deeper insight into the method of state variables. It is instructive to consider first the unforced, or *homogeneous*, system, which has the form

$$\dot{\mathbf{x}} = \mathbf{F}(t)\mathbf{x}, \qquad \mathbf{x}(0) = \mathbf{x}_0. \tag{6A.1}$$

If the elements of $\mathbf{F}(t)$ are continuous functions of time, the above equation has a *unique* solution for any initial state vector $\mathbf{x}_0$. There is a useful representation for the solution of this equation in terms of a matrix, called the *transition matrix*. Let $\phi_i(t, t_0)$ be the solution to the special initial condition

$$\mathbf{x}(0) = \mathbf{e}_i = \begin{bmatrix} 0 \\ 0 \\ 1 \\ 0 \\ \vdots \\ 0 \end{bmatrix} \leftarrow i\text{th row}. \tag{6A.2}$$

If $\mathbf{x}(t_0)$ is the actual initial condition at t_0, we can express it in the decomposed form

$$\mathbf{x}(t_0) = \mathbf{x}_{01}\mathbf{e}_1 + \mathbf{x}_{02}\mathbf{e}_2 + \cdots + \mathbf{x}_{0n}\mathbf{e}_n. \tag{6A.3}$$

Because Eq. (6A.1) is linear, the state $\mathbf{x}(t)$ can also be expressed as a sum of the solutions to the special initial conditions ϕ_i, as follows:

$$\mathbf{x}(t) = \mathbf{x}_{01}\phi_1(t, t_0) + \mathbf{x}_{02}\phi_2(t, t_0) + \cdots + \mathbf{x}_{0n}\phi_n(t, t_0), \tag{6A.4}$$

or, in matrix notation,

$$\mathbf{x}(t) = [\boldsymbol{\phi}_1(t, t_0), \boldsymbol{\phi}_2(t, t_0), \ldots, \boldsymbol{\phi}_n(t, t_0)] \begin{bmatrix} \mathbf{x}_{01} \\ \mathbf{x}_{02} \\ \vdots \\ \mathbf{x}_{0n} \end{bmatrix}. \quad (6A.5)$$

So we define the *transition matrix** to be

$$\boldsymbol{\Phi}(t, t_0) = [\boldsymbol{\phi}_1(t, t_0), \boldsymbol{\phi}_2(t, t_0), \ldots, \boldsymbol{\phi}_n(t, t_0)] \quad (6A.6)$$

and write the solution as

$$\mathbf{x}(t) = \boldsymbol{\Phi}(t, t_0)\mathbf{x}(t_0), \quad (6A.7)$$

where, as the name implies, the transition matrix provides the transition between the state at time t_0 to the state time t. Furthermore, from Eq. (6A.7) we have that

$$\frac{d}{dt}\mathbf{x}(t) = \frac{d}{dt}[\boldsymbol{\Phi}(t, t_0)]\mathbf{x}(t_0), \quad (6A.8)$$

and from Eq. (6A.1) and (6A.8) we have

$$\frac{d}{dt}[\mathbf{x}(t)] = \mathbf{F}\mathbf{x}(t) = \mathbf{F}\boldsymbol{\Phi}(t, t_0)\mathbf{x}(t_0). \quad (6A.9)$$

Therefore

$$\frac{d}{dt}[\boldsymbol{\Phi}(t, t_0)] = \mathbf{F}\boldsymbol{\Phi}(t, t_0) \quad (6A.10)$$

and also

$$\boldsymbol{\Phi}(t_0, t_0) = \mathbf{I}. \quad (6A.11)$$

The transition matrix can be shown to have many interesting properties. Among them are the following:

1. $\boldsymbol{\Phi}(t_2, t_0) = \boldsymbol{\Phi}(t_2, t_1)\boldsymbol{\Phi}(t_1, t_0);$
2. $\boldsymbol{\Phi}^{-1}(t, \tau) = \boldsymbol{\Phi}(\tau, t);$
3. $\dfrac{d}{d\tau}\boldsymbol{\Phi}(t, \tau) = -\boldsymbol{\Phi}(t, \tau)\mathbf{F}(\tau);$
4. $\det \boldsymbol{\Phi}(t, t_0) = e^{\int_{t_0}^{t} \text{trace}\mathbf{F}(\tau)d\tau}. \quad (6A.12)$

The second property implies that $\boldsymbol{\Phi}(t, \tau)$ is always invertible. What this means is that the solution is always unique so that, given a particular value of state at time

---

*This is also referred to as the *fundamental matrix* of the differential equation.

τ not only can we compute the future state from $\mathbf{\Phi}(t, \tau)$ but also past values from $\mathbf{\Phi}^{-1}(t, \tau)$.

For the homogeneous case, with a forcing function input $u(t)$, the equation is

$$\dot{\mathbf{x}}(t) = \mathbf{F}(t)\mathbf{x}(t) + \mathbf{G}(t)u(t), \tag{6A.13}$$

and the solution is*

$$\mathbf{x}(t) = \mathbf{\Phi}(t, t_0)\mathbf{x}_0 + \int_{t_0}^{t} \mathbf{\Phi}(t, \tau)\mathbf{G}(\tau)u(\tau)\, d\tau. \tag{6A.14}$$

We can verify this simply by substituting the supposed solution (Eq. 6A.14) into the differential equation (Eq. 6A.13):

$$\frac{d}{dt}\mathbf{x}(t) = \frac{d}{dt}\mathbf{\Phi}(t, t_0)\mathbf{x}_0 + \frac{d}{dt}\int_{t_0}^{t} [\mathbf{\Phi}(t, \tau)\mathbf{G}(\tau)u(\tau)]\, d\tau. \tag{6A.15}$$

The second term from calculus (using the Leibnitz formula) is

$$\frac{d}{dt}\int_{t_0}^{t} [\mathbf{\Phi}(t, \tau)\mathbf{G}(\tau)u(\tau)]\, d\tau$$

$$= \int_{t_0}^{t} \mathbf{F}(t)\mathbf{\Phi}(t, \tau)\mathbf{G}(\tau)u(\tau)\, d\tau + \mathbf{\Phi}(t, t)\mathbf{G}(t)u(t). \tag{6A.16}$$

Using the basic relations, we have that

$$\frac{d}{dt}\mathbf{x}(t) = \mathbf{F}(t)\mathbf{\Phi}(t, t_0)\mathbf{x}_0 + \mathbf{F}(t)\int_{t_0}^{t} \mathbf{\Phi}(t, \tau)\mathbf{G}(\tau)u(\tau)\, d\tau + \mathbf{G}(t)u(t)$$

$$= \mathbf{F}(t)\mathbf{x}(t) + \mathbf{G}(t)u(t), \tag{6A.17}$$

which shows that the proposed solution satisfies the system equation.

For the time-invariant case,

$$\mathbf{\Phi}(t, t_0) = e^{\mathbf{F}(t-t_0)} = \mathbf{\Phi}(t - t_0), \tag{6A.18}$$

where

$$e^{\mathbf{F}t} \triangleq \left(\mathbf{I} + \mathbf{F}t + \frac{\mathbf{F}^2 t^2}{2!} + \cdots \right) = \sum_{k=1}^{\infty} \frac{\mathbf{F}^k t^k}{k!} \tag{6A.19}$$

is an invertible $n \times n$ exponential matrix, and, by letting $t = 0$, we see that

$$e^0 = \mathbf{I}. \tag{6A.20}$$

The state solution is now

$$\mathbf{x}(t) = e^{\mathbf{F}(t-t_0)}\mathbf{x}_0 + \int_{t_0}^{t} e^{\mathbf{F}(t-\tau)}\mathbf{G}u(\tau)\, d\tau \tag{6A.21}$$

---

*The first term is the natural response due to the initial condition $\mathbf{x}_0$, and the second term is the superposition integral that applies because the system is linear.

and

$$y(t) = \mathbf{H}e^{\mathbf{F}(t-t_0)}\mathbf{x}_0 + \mathbf{H}\int_{t_0}^{t} e^{\mathbf{F}(t-\tau)}\mathbf{G}u(\tau)\,d\tau + Ju(t). \tag{6A.22}$$

Suppose $\mathbf{x}(t_0) = \mathbf{x}_0 \equiv 0$, then the output is given by the convolution integral

$$y(t) = \int_{t_0}^{t} h(t - \tau)u(\tau)\,d\tau, \tag{6A.23}$$

where $h(t)$ is the *impulse response*. In terms of the state-variable matrices,

$$h(t) = \mathbf{H}e^{\mathbf{F}t}\mathbf{G} + J\delta(t). \tag{6A.24}$$

While there is no uniformly best way to compute the transition matrix, there are several methods that can be used to compute approximations to it (see Moler and van Loan, 1978, and Franklin, Powell, and Workman, 1990). Three of these are matrix exponential series, inverse Laplace transform, and diagonalization of the system matrix. In the first technique, we use Eq. (6A.19):

$$e^{\mathbf{F}t} = \mathbf{I} + \mathbf{F}t + \mathbf{F}^2\frac{t^2}{2!} + \mathbf{F}^3\frac{t^3}{3!} + \cdots, \tag{6A.25}$$

and the series should be completed in a reliable fashion. For the second method, we notice that if we define

$$\boldsymbol{\Phi}(s) = (s\mathbf{I} - \mathbf{F})^{-1}, \tag{6A.26}$$

then we can compute $\boldsymbol{\Phi}(s)$ from the $\mathbf{F}$ matrix and matrix algebra. Given this matrix, we use the inverse Laplace transform to compute

$$\boldsymbol{\Phi}(t) = \mathcal{L}^{-1}[\boldsymbol{\Phi}(s)] = \mathcal{L}^{-1}[(s\mathbf{I} - \mathbf{F})^{-1}]. \tag{6A.27}$$

The last method we mentioned operates on the system matrix. If the system matrix can be diagonalized, that is, if we can find a transformation $\mathbf{T}$ so that

$$\boldsymbol{\Lambda} = \mathbf{T}^{-1}\mathbf{F}\mathbf{T}, \tag{6A.28}$$

where $\mathbf{F}$ is reduced to the similar but diagonal matrix,

$$\boldsymbol{\Lambda} = \text{diag}\{\lambda_1, \lambda_2, \ldots, \lambda_n\}, \tag{6A.29}$$

then from the series (Eq. 6A.19), we see that we need only compute scalar exponentials, since

$$e^{\mathbf{F}t} = \mathbf{T}^{-1}\text{diag}\{e^{\lambda_1 t}, \ldots, e^{\lambda_n t}\}\mathbf{T}. \tag{6A.30}$$

Appendix 6B:
Controllability and Observability

Controllability and observability are important structural properties of a dynamic system. These concepts were identified and studied first by Kalman (1960) and later by Kalman, Ho, and Narendra (1961). These properties have continued under much study during the last three decades. We will discuss only a few of the known results for linear constant systems with one input and one output. We have encountered these concepts already in connection with control law and estimator designs. We suggested in Section 6.3 that if the square matrix given by

$$\mathcal{C} = [\mathbf{G} : \mathbf{FG} : \mathbf{F}^2\mathbf{G} : \cdots : \mathbf{F}^{n-1}\mathbf{G}] \tag{6B.1}$$

is nonsingular, by transformation of the state we can convert the given description into the control canonical form. A control law can then be constructed such that the closed-loop system can be given an arbitrary characteristic equation. We begin our discussion of controllability by making the statement (the first of three):

Definition I: The system $(\mathbf{F}, \mathbf{G})$ is controllable if for any given nth-order polynomial $\alpha_c(s)$ there exists a (unique) control law $u = -\mathbf{Kx}$ such that the characteristic polynomial of $(\mathbf{F} - \mathbf{GK})$ is $\alpha_c(s)$.

From the results of Ackermann's formula (Appendix 6C), we have the mathematical test for controllability: $(\mathbf{F}, \mathbf{G})$ are a controllable pair if and only if the rank of $\mathcal{C}$ is n. The above definition based on pole placement is a frequency-domain concept. Controllability can be equivalently defined in the time domain.

Definition II: The system $(\mathbf{F}, \mathbf{G})$ is controllable if there exists a (piecewise continuous) control signal $u(t)$ such that the state of the system can be taken from any initial state $\mathbf{x}_0$ to any desired final state $\mathbf{x}_f$ in a finite time interval.

We will now show that the system is controllable by this definition if and only if $\mathscr{C}$ is full-rank. Assume that the system is controllable but

$$\text{rank}[\mathbf{G} : \mathbf{FG} : \mathbf{F}^2\mathbf{G} : \cdots : \mathbf{F}^{n-1}\mathbf{G}] < n. \tag{6B.2}$$

We can then find a vector $\mathbf{V}$ such that

$$\mathbf{V}[\mathbf{G} : \mathbf{FG} : \mathbf{F}^2\mathbf{G} : \cdots : \mathbf{F}^{n-1}\mathbf{G}] = 0 \tag{6B.3}$$

or

$$\mathbf{VG} = \mathbf{VFG} = \mathbf{VF}^2\mathbf{G} = \cdots = \mathbf{VF}^{n-1}\mathbf{G} = 0. \tag{6B.4}$$

The Cayley-Hamilton theorem states that $\mathbf{F}$ satisfies its own characteristic equations, namely,

$$-\mathbf{F}^n = a_1\mathbf{F}^{n-1} + a_2\mathbf{F}^{n-2} + \cdots + a_n\mathbf{I}. \tag{6B.5}$$

Therefore

$$-\mathbf{VF}^n\mathbf{G} = a_1\mathbf{VF}^{n-1}\mathbf{G} + a_2\mathbf{VF}^{n-2}\mathbf{G} + \cdots + a_n\mathbf{VG} = 0. \tag{6B.6}$$

By induction, $\mathbf{VF}^{n+k}\mathbf{G} = 0$ for $k = 0, 1, 2, \ldots$, or $\mathbf{VF}^m\mathbf{G} = 0$ for $m = 0, 1, 2, \ldots$, and thus

$$\mathbf{V}e^{\mathbf{F}t}\mathbf{G} = \mathbf{V}\left[\mathbf{I} + \mathbf{F}t + \frac{1}{2!}\mathbf{F}^2t^2 + \cdots\right]\mathbf{G} = 0 \tag{6B.7}$$

for all t. However, the zero initial-condition response ($\mathbf{x}_0 = 0$) is, from Appendix 6A,

$$\mathbf{x}(t) = \int_0^t e^{\mathbf{F}(t-\tau)}\mathbf{G}u(\tau)\,d\tau \tag{6B.8}$$

or

$$\mathbf{x}(t) = e^{\mathbf{F}t}\int_0^t e^{-\mathbf{F}\tau}\mathbf{G}u(\tau)\,d\tau. \tag{6B.9}$$

Using Eq. (6B.7) then,

$$\mathbf{Vx}(t) = \int_0^t \mathbf{V}e^{\mathbf{F}(t-\tau)}\mathbf{G}u(\tau)\,d\tau = 0 \tag{6B.10}$$

for all $u(t)$ and $t > 0$. This implies that all points reachable from the origin are orthogonal to $\mathbf{V}$. This restricts the reachable space and therefore contradicts the definition of controllability. Thus if $\mathscr{C}$ is singular, $(\mathbf{F}, \mathbf{G})$ are not controllable by definition II.

Next, assume that $\mathscr{C}$ is full-rank, but the system is uncontrollable by definition II. This means that there exists a nonzero $\mathbf{V}$ such that

$$\mathbf{V}\int_0^t e^{\mathbf{F}(t_f-\tau)}\mathbf{G}u(\tau)\,d\tau = 0, \tag{6B.11}$$

since the whole state space is not reachable. But Eq. (6B.11) implies that

$$\mathbf{V}e^{\mathbf{F}(t_f - \tau)}\mathbf{G} = 0, \qquad 0 \le \tau \le t_f. \tag{6B.12}$$

If we set $\tau = t_f$, we see that $\mathbf{VG} = 0$. Also, differentiating Eq. (6B.12) and letting $\tau = t_f$ gives $\mathbf{VFG} = 0$. Continuing this process, we find that

$$\mathbf{VG} = \mathbf{VFG} = \mathbf{VF}^2\mathbf{G} = \cdots = \mathbf{VF}^{n-1}\mathbf{G} = 0, \tag{6B.13}$$

which contradicts the assumption that $\mathscr{C}$ is full-rank.

We have shown that the system is controllable by definition II if and only if the rank of $\mathscr{C}$ is n, exactly the same condition we found for pole assignment.

Our final definition is closest to the structural character of controllability:

Definition III: The system $(\mathbf{F}, \mathbf{G})$ is controllable if every mode of $\mathbf{F}$ is connected to the control input.

Because of the generality of modal structure of systems, we will only treat the case of systems for which $\mathbf{F}$ can be transformed to diagonal form. (The double integration plant does *not* qualify.) Suppose we have a diagonal $\mathbf{F}_d$ matrix and the corresponding input matrix $\mathbf{G}_d$ with elements g_i. The structure is as shown in Fig. 6.63. By definition, for a controllable system the input must be connected to each mode so that the g_i's are all nonzero. However, this is not enough if the poles (λ_i's) are not distinct. Suppose, for instance, $\lambda_1 = \lambda_2$. The first two state equations are then

$$\dot{x}_{1d} = \lambda_1 x_{1d} + g_1 u,$$
$$\dot{x}_{2d} = \lambda_1 x_{2d} + g_2 u. \tag{6B.14}$$

If we define a new state, $\xi = g_2 x_{1d} - g_1 x_{2d}$, the equation for ξ is

$$\dot{\xi} = g_2 \dot{x}_{1d} - g_1 \dot{x}_{2d} = g_2 \lambda_1 x_{1d} + g_2 g_1 u - g_1 \lambda_1 x_{2d} - g_1 g_2 u = \lambda_1 \xi,$$

$$\tag{6B.15}$$

FIGURE 6.63
Block diagram of a
system with a diagonal
$\mathbf{F}$ matrix.

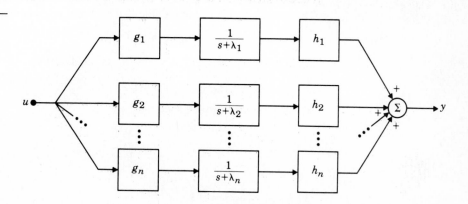

which does not include the control u; hence ξ is not controllable. The point is that if any two poles are equal *in a diagonal* $\mathbf{F}_d$ *system with only one input*, we effectively have a hidden mode that is not connected to the control, and the system is not controllable. This is because the two state variables move exactly together and we cannot control x_{1d} and x_{2d} *independently*. Therefore, even in such a simple case, we have two conditions for controllability:

1. All eigenvalues of $\mathbf{F}_d$ are distinct.

2. No element of $\mathbf{G}_d$ is zero.

Let us consider the controllability matrix of this diagonal system. By direct computation,

$$
\mathcal{C} = \begin{bmatrix} g_1 & g_1\lambda_1 & \cdots & g_1\lambda_1^{n-1} \\ g_2 & g_2\lambda_2 & \cdots & \vdots \\ \vdots & \vdots & \cdots & \vdots \\ g_n & g_n\lambda_n & \cdots & g_n\lambda_n^{n-1} \end{bmatrix}
$$

$$
= \begin{bmatrix} g_1 & & 0 \\ & g_2 & \\ & & \ddots \\ 0 & & g_n \end{bmatrix} \begin{bmatrix} 1 & \lambda_1 & \lambda_1^2 & \cdots & \lambda_1^{n-1} \\ 1 & \lambda_2 & \lambda_2^2 & \cdots & \vdots \\ \vdots & \vdots & \vdots & \cdots & \vdots \\ 1 & \lambda_n & \lambda_n^2 & \cdots & \lambda_n^{n-1} \end{bmatrix}. \tag{6B.16}
$$

The controllability matrix is the product of two matrices, and $\mathcal{C}$ is nonsingular if and only if both factors are invertible. The first term has a determinant that is the product of g_i's, and the second term (called the *Vandermonde matrix*) is nonsingular if and only if the λ_i's are distinct. Thus definition III is also equivalent to nonsingular $\mathcal{C}$.

As our final remark on the topic of controllability, we present a mathematical test that is an alternative to testing the rank (or determinant) of $\mathcal{C}$. This is the Popov-Hautus-Rosenbrock (PHR) test (see Rosenbrock, 1970, and Kailath, 1980). The system $(\mathbf{F}, \mathbf{G})$ is controllable if the system of equations

$$
\mathbf{V}^T[s\mathbf{I} - \mathbf{F} : \mathbf{G}] = \mathbf{0}^T \tag{6B.17}
$$

has only the trivial solution $\mathbf{V}^T = \mathbf{0}^T$, or, equivalently,

$$
\text{rank}[s\mathbf{I} - \mathbf{F} : \mathbf{G}] = n, \tag{6B.18}
$$

or if there is nonzero $\mathbf{V}^T$ such that*

$$
\mathbf{V}^T\mathbf{F} = s\mathbf{V}^T,
$$
$$
\mathbf{V}^T\mathbf{G} = 0. \tag{6B.19}
$$

---

*$\mathbf{V}^T$ is a left eigenvector of $\mathbf{F}$.

This test is equivalent to the rank-of-$\mathscr{C}$ test. It is easy to show that if such a $\mathbf{V}$ exists, $\mathscr{C}$ is singular, for if a nonzero $\mathbf{V}$ exists such that $\mathbf{V}^T\mathbf{G} = 0$ by Eq. (6B.19a), then

$$\mathbf{V}^T\mathbf{FG} = s\mathbf{V}^T\mathbf{G} = 0. \tag{6B.20}$$

Multiplying by $\mathbf{FG}$, we find

$$\mathbf{V}^T\mathbf{F}^2\mathbf{G} = s\mathbf{V}^T\mathbf{FG} = 0, \tag{6B.21}$$

and so on. Thus, we derive that $\mathbf{V}^T\mathscr{C} = 0^T$ has a nontrivial solution, $\mathscr{C}$ is singular, and the system is not controllable. To show that a nontrivial $\mathbf{V}^T$ exists if $\mathscr{C}$ is singular requires a bit more work.

We have given two pictures of uncontrollability. Either a mode is physically disconnected from the input, or else two parallel subsystems have identical characteristic roots. The control engineer should be aware of the existence of a third simple situation, illustrated in Fig. 6.64—namely, a *pole-zero cancellation*. Here the problem is that the mode at $s = 1$ appears to be connected to the input but is masked by the zero at $s = 1$ in the preceding member; the result is an uncontrollable system. This can be confirmed in several ways. First let us look at the controllability matrix. The system matrices are

$$\mathbf{F} = \begin{bmatrix} -1 & 0 \\ 1 & 1 \end{bmatrix} \qquad \mathbf{G} = \begin{bmatrix} -2 \\ 1 \end{bmatrix}$$

FIGURE 6.64
Examples of an uncontrollable system.

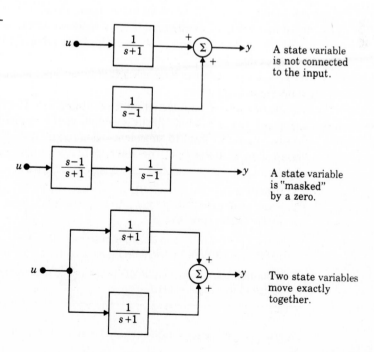

A state variable is not connected to the input.

A state variable is "masked" by a zero.

Two state variables move exactly together.

and

$$\mathscr{C} = [\,\mathbf{G} \quad \mathbf{FG}\,] = \begin{bmatrix} -2 & 2 \\ 1 & -1 \end{bmatrix},$$

which is clearly singular. If we compute the transfer function from u to x_1, we find

$$H(s) = \frac{s-1}{s+1}\frac{1}{s-1}$$

$$= \frac{1}{s+1}.$$

Since the natural mode at $s = 1$ disappears from the input-output description, it is not connected to the input. Finally, if we consider the PHR test,

$$[\,s\mathbf{I} - \mathbf{F} : \mathbf{G}\,] = \begin{bmatrix} s+1 & 0 & -2 \\ -1 & s-1 & 1 \end{bmatrix}$$

and let $s = 1$, we must test the rank of

$$\begin{bmatrix} 2 & 0 & -2 \\ -1 & 0 & 1 \end{bmatrix},$$

which is clearly less than 2, which means, again, the system is uncontrollable. In conclusion, we have three definitions for controllability: pole assignment, state reachability, and mode coupling to the input. These definitions are equivalent, and the tests for any of these properties are found in terms of the rank of the controllability matrix or the rank of the matrix pencil $[s\mathbf{I} - \mathbf{F} : \mathbf{G}]$. If $\mathscr{C}$ is nonsingular, we can assign the closed-loop poles arbitrarily by state feedback, we can move the state to any point in the state space in a finite time, and every mode is connected to the control input.*

So far, we have discussed only controllability. The concept of observability is parallel to that of controllability, and all of the results we have discussed thus far may be transformed to statements about observability by invoking duality, as discussed in Section 6.5.1. The observability definitions dual to controllability are

---

Definition I: The system $(\mathbf{F}, \mathbf{H})$ is observable if, for any nth-order polynomial $\alpha_e(s)$, there exists an estimator gain $\mathbf{L}$ such that the characteristic equation of the state estimator error is $\alpha_e(s)$.

Definition II: The system $(\mathbf{F}, \mathbf{H})$ is observable if, for any $\mathbf{x}(0)$, there is a finite time τ such that $\mathbf{x}(0)$ can be determined (uniquely) from $u(\tau)$ and $y(\tau)$.

Definition III: The system $(\mathbf{F}, \mathbf{H})$ is observable if every dynamic mode in $\mathbf{F}$ is connected to the output via $\mathbf{H}$.

---

*We have shown the latter for diagonal $\mathbf{F}$ only, but the result is true in general.

As seen in the discussion for controllability, a mathematical test can be developed for observability. The system is observable if the matrix

$$\mathbb{O} = \begin{bmatrix} \mathbf{H} \\ \mathbf{HF} \\ \vdots \\ \mathbf{HF}^{n-1} \end{bmatrix}$$

is nonsingular. If we take the transpose of $\mathbb{O}$ and let $\mathbf{H}^T = \mathbf{G}$ and $\mathbf{F}^T = \mathbf{F}$, we find the controllability matrix of $(\mathbf{F}, \mathbf{G})$, another manifestation of duality.

Appendix 6C:
Ackermann's Formula for Pole Placement

We have the plant and state-variable equation

$$\dot{\mathbf{x}} = \mathbf{F}\mathbf{x} + \mathbf{G}u \qquad (6C.1)$$

The objective is to find a state feedback control law

$$u = -\mathbf{K}\mathbf{x}, \qquad (6C.2)$$

so that the closed-loop characteristic polynomial is

$$\alpha_c(s) = \det[s\mathbf{I} - \mathbf{F} + \mathbf{G}\mathbf{K}]. \qquad (6C.3)$$

First, we have to select $\alpha_c(s)$, which determines where the poles are to be shifted, and then solve for $\mathbf{K}$ such that Eq. (6C.3) is satisfied. Our technique is based on the transformation of the plant equation to the control canonical form.

We consider the affect of an arbitrary nonsingular transformation of the state

$$\mathbf{x} = \mathbf{T}\bar{\mathbf{x}}, \qquad (6C.4)$$

where $\bar{\mathbf{x}}$ is the new transformed state. The equations of motion in the new coordinates are, from Eq. (6C.4),

$$\dot{\mathbf{x}} = \mathbf{T}\dot{\bar{\mathbf{x}}} = \mathbf{F}\mathbf{x} + \mathbf{G}u$$

$$= \mathbf{F}\mathbf{T}\bar{\mathbf{x}} + \mathbf{G}u; \qquad (6C.5)$$

$$\dot{\bar{\mathbf{x}}} = \mathbf{T}^{-1}\mathbf{F}\mathbf{T}\bar{\mathbf{x}} + \mathbf{T}^{-1}\mathbf{G}u = \bar{\mathbf{F}}\bar{\mathbf{x}} + \bar{\mathbf{G}}u. \qquad (6C.6)$$

The controllability matrix for the original state,

$$\mathscr{C}_x = [\mathbf{G} \quad \mathbf{F}\mathbf{G} \quad \mathbf{F}^2\mathbf{G} \quad \cdots \quad \mathbf{F}^{n-1}\mathbf{G}], \qquad (6C.7)$$

provides a useful transformation matrix. We can also define the controllability matrix for the transformed state:

$$\mathscr{C}_{\bar{x}} = [\bar{\mathbf{G}} \quad \bar{\mathbf{F}}\bar{\mathbf{G}} \quad \bar{\mathbf{F}}^2\bar{\mathbf{G}} \quad \cdots \quad \bar{\mathbf{F}}^{n-1}\bar{\mathbf{G}}]. \qquad (6C.8)$$

The two controllability matrices are related by

$$\mathcal{C}_{\bar{x}} = [\mathbf{T}^{-1}\mathbf{G} \ \mathbf{T}^{-1}\mathbf{F}\mathbf{T}\mathbf{T}^{-1}\mathbf{G} \cdots] = \mathbf{T}^{-1}\mathcal{C}_x \tag{6C.9}$$

and the transformation matrix

$$\mathbf{T} = \mathcal{C}_x \mathcal{C}_{\bar{x}}^{-1}. \tag{6C.10}$$

We can draw some important conclusions from Eqs. (6C.9) and (6C.10). From Eq. (6C.9), we see that if $\mathcal{C}_x$ is nonsingular, then for any nonsingular $\mathbf{T}$, $\mathcal{C}_{\bar{x}}$ is also nonsingular. This means that a similarity transformation on the state does not change the controllability properties of a system. From Eq. (6C.10), we can go the other way. Suppose we would like to find a transformation to take $(\mathbf{F}, \mathbf{G})$ into the control canonical form. As we shall see shortly, $\mathcal{C}_{\bar{x}}$ in that case is *always* nonsingular. From Eq. (6C.9), we see that a nonsingular $\mathbf{T}$ will always exist if and only if $\mathcal{C}_x$ is nonsingular. We conclude:

We can always transform $(\mathbf{F}, \mathbf{G})$ into the control canonical form if and only if the system is controllable.

Let's take a closer look at the control canonical form and treat the third-order case, although the results are true for any n:

$$\bar{\mathbf{F}} = \mathbf{F}_c = \begin{bmatrix} -a_1 & -a_2 & -a_3 \\ 1 & 0 & 0 \\ 0 & 1 & 0 \end{bmatrix}, \qquad \bar{\mathbf{G}} = \mathbf{G}_c = \begin{bmatrix} 1 \\ 0 \\ 0 \end{bmatrix}. \tag{6C.11}$$

The controllability matrix, by direct computation, is

$$\mathcal{C}_{\bar{x}} = \mathcal{C}_c = \begin{bmatrix} 1 & -a_1 & a_1^2 - a_2 \\ 0 & 1 & -a_1 \\ 0 & 0 & 1 \end{bmatrix}. \tag{6C.12}$$

Since this matrix is upper triangular with 1s along the diagonal, it is always invertible, as was mentioned earlier. Also note that the last row of $\mathcal{C}_{\bar{x}}$ is the unit vector with all 0s, except for the last element, which is unity. We shall use this fact in the sequel.

As we pointed out in Section 6.3, the design of a control law for the state $\bar{\mathbf{x}}$ is trivial if the equations of motion happen to be in control canonical form. The characteristic equation is

$$s^3 + a_1 s^2 + a_2 s + a_3 = 0, \tag{6C.13}$$

and the characteristic equation for the closed-loop system comes from

$$\mathbf{F}_{cL} = \mathbf{F}_c - \mathbf{G}_c \mathbf{K}_c \tag{6C.14}$$

and has the coefficients,

$$s^3 + (a_1 + K_{c1})s^2 + (a_2 + K_{c2})s + (a_3 + K_{c3}) = 0. \tag{6C.15}$$

To obtain the desired closed-loop pole location, the coefficients of this equation are to match

$$\alpha_c(s) = s^3 + \alpha_1 s^2 + \alpha_2 s + \alpha_3, \tag{6C.16}$$

and we see that

$$a_1 + K_{c1} = \alpha_1, \qquad a_2 + K_{c2} = \alpha_2, \qquad a_3 + K_{c3} = \alpha_3, \tag{6C.17}$$

or, in vector form,

$$\mathbf{a} + \mathbf{K}_c = \alpha, \tag{6C.18}$$

where $\mathbf{a}$ and α are row vectors containing the coefficients of the characteristic polynomials of the open-loop and closed-loop systems, respectively.

We need to find a relationship between these polynomial coefficients and the matrix $\mathbf{F}$. The requirement is achieved by the Cayley-Hamilton theorem, which states that a matrix satisfies its own characteristic polynomial. For $\mathbf{F}_c$, we have

$$\mathbf{F}_c^n + a_1 \mathbf{F}_c^{n-1} + a_2 \mathbf{F}_c^{n-2} + \cdots + a_n \mathbf{I} = 0. \tag{6C.19}$$

Suppose we form the polynomial $\alpha_c(\mathbf{F})$, which is the *closed-loop* characteristic polynomial with the matrix $\mathbf{F}$ substituted for the complex variable s:

$$\alpha_c(\mathbf{F}_c) = \mathbf{F}_c^n + \alpha_1 \mathbf{F}_c^{n-1} + \alpha_2 \mathbf{F}_c^{n-2} + \cdots + \alpha_n \mathbf{I}. \tag{6C.20}$$

If we solve Eq. (6C.19) for $\mathbf{F}_c^n$ and substitute in Eq. (6C.20), we find that

$$\alpha_c(\mathbf{F}_c) = (-a_1 + \alpha_1)\mathbf{F}_c^{n-1} + (-a_2 + \alpha_2)\mathbf{F}_c^{n-2} + \cdots + (-a_n + \alpha_n)\mathbf{I}. \tag{6C.21}$$

But, because $\mathbf{F}_c$ has such a special structure, we observe that if we multiply it by the transpose of the nth unit vector, $\mathbf{e}_n^\mathsf{T} = [0 \quad 0 \quad \cdots \quad 0 \quad 1]$,

$$\mathbf{e}_n^\mathsf{T} \mathbf{F}_c = [0 \quad \cdots \quad 1 \quad 0] = \mathbf{e}_{n-1}^\mathsf{T}, \tag{6C.22}$$

as we can see from Eq. (6C.11). If we multiply this vector again by $\mathbf{F}_c$,

$$(\mathbf{e}_n^\mathsf{T} \mathbf{F}_c)\mathbf{F}_c = [0 \quad \cdots \quad 1 \quad 0]\mathbf{F}_c = [0 \quad 0 \quad \cdots \quad 1 \quad 0 \quad 0] = \mathbf{e}_{n-2}^\mathsf{T}, \tag{6C.23}$$

and continue the process, the successive unit vectors are generated until

$$\mathbf{e}_n^\mathsf{T} \mathbf{F}_c^{n-1} = [1 \quad 0 \quad \cdots \quad 0] = \mathbf{e}_1^\mathsf{T}. \tag{6C.24}$$

Therefore, if we multiply Eq. (6C.21) by $\mathbf{e}_n^\mathsf{T}$, we find

$$\mathbf{e}_n^\mathsf{T} \alpha_c(\mathbf{F}_c) = (-a_1 + \alpha_1)\mathbf{e}_1^\mathsf{T} + (-a_2 + \alpha_2)\mathbf{e}_2^\mathsf{T} +$$
$$\cdots + (-a_n + \alpha_n)\mathbf{e}_n^\mathsf{T} = [K_{c1} K_{c2} \cdots K_{cn}] = \mathbf{K}_c, \tag{6C.25}$$

where we use Eq. (6C.18), which relates $\mathbf{K}_c$ to the $\mathbf{a}$'s and α's.

We now have a compact expression for the gains of the system in control canonical form as represented in Eq. (6C.25). But we really need the expression for **K**, that is, the gain on the original state. If $u = -\mathbf{K}_c\bar{\mathbf{x}}$, then $u = -\mathbf{K}_c\mathbf{T}^{-1}\mathbf{x}$, so that

$$\mathbf{K} = \mathbf{K}_c\mathbf{T}^{-1} = \mathbf{e}_n^T\alpha_c(\mathbf{F}_c)\mathbf{T}^{-1} = \mathbf{e}_n^T\alpha_c(\mathbf{T}^{-1}\mathbf{F}\mathbf{T})\mathbf{T}^{-1} = \mathbf{e}_n^T\mathbf{T}^{-1}\alpha_c(\mathbf{F}), \quad (6C.26)$$

where in the last step we use the fact that $(\mathbf{T}^{-1}\mathbf{F}\mathbf{T})^k = \mathbf{T}^{-1}\mathbf{F}^k\mathbf{T}$ and that α_c is a polynomial, that is, a sum of powers of $\mathbf{F}_c$. But from Eq. (6C.9), we see that

$$\mathbf{T}^{-1} = \mathscr{C}_c\mathscr{C}_x^{-1}, \quad (6C.27)$$

so that

$$\mathbf{K} = \mathbf{e}_n^T(\mathscr{C}_c\mathscr{C}_x^{-1})\alpha_c(\mathbf{F}). \quad (6C.28)$$

Now we use the observation made earlier that the last row of $\mathscr{C}_c$, which is $\mathbf{e}_n^T\mathscr{C}_c$, is again $\mathbf{e}_n^T$. We finally obtain Ackermann's formula:

$$\mathbf{K} = \mathbf{e}_n^T\mathscr{C}_x^{-1}\alpha_c(\mathbf{F}). \quad (6C.29)$$

We note again that forming the explicit inverse of $\mathscr{C}_x$ is not advisable for numerical accuracy. Thus, we need to solve $\mathbf{b}^T$ so that

$$\mathbf{e}_n^T\mathscr{C}_x^{-1} = \mathbf{b}^T. \quad (6C.30)$$

We solve the linear set of equations

$$\mathbf{b}^T\mathscr{C}_x = \mathbf{e}_n^T \quad (6C.31)$$

and then compute

$$\mathbf{K}_x = \mathbf{b}^T\alpha_c(\mathbf{F}). \quad (6C.32)$$

Problems Section 6.2

6.1 Verify Eqs. (6.12) and (6.13) by writing the differential equations representing the two blocks in Fig. 6.5 and combining the result into the proper form. Then use Eq. (6.21) to determine the transfer function from Eq. (6.12) (setting $M_d = 0$).

6.2 For the redefinition of state variables described in Section 6.2.2, derive Eq. (6.16). Note that, since $\mathbf{p} = \mathbf{Tx}$, we also have that $\dot{\mathbf{p}} = \mathbf{T\dot{x}}$ and $\mathbf{x} = \mathbf{T}^{-1}\mathbf{p}$.

6.3 Consider the system in Fig. 6.65.

FIGURE 6.65

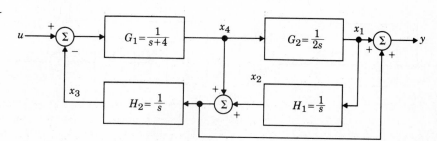

a) Find the transfer function from u and y.

b) Write the state-variable equations using the state indicated.

6.4 Using the indicated state variables, find the state-variable representation for each of the systems shown in Fig. 6.66. Find the transfer function for each system using block-diagram manipulation and matrix algebra.

6.5 An electric-field transducer that can be used as a microphone or a speaker can be described by the following equations:

$$\text{Mechanical: } M\frac{d^2x}{dt^2} + B\frac{dx}{dt} + K(x-l) + \frac{\varepsilon_0 A}{2}\left(\frac{v}{x}\right)^2 = f(t),$$

$$\text{Electrical: } \frac{d}{dt}\frac{\varepsilon_0 Av}{x} = \frac{V_s - v}{R},$$

where

M = mass of the moving plate,
B = damping coefficient,
K = spring constant,
x = separation of plates,
l = natural length of spring,
A = area of plate,
v = voltage across plates,
f = force of plate (input),
V_s = bias voltage,
R = resistance.

FIGURE 6.66

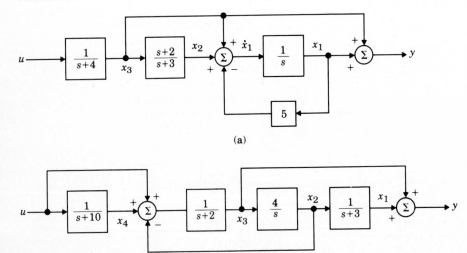

(a)

(b)

The output is

$$y = R\frac{d}{dt}\frac{\varepsilon_0 Av}{x}$$

a) Using state variables $x_1 = x$, $x_2 = \dot{x}$, and $x_3 = v/x$, write the equations in state-variable form.

b) Find the equation that describes equilibrium points for x and v if $f(t) = 0$.

c) Let the equilibrium points be X and V; find a set of linearized equations about X and V.

Section 6.3

6.6 Consider the system in Fig. 6.67.

  a) Write a set of equations that describes this system in the standard form $\dot{x} = \mathbf{Fx} + \mathbf{G}u$ and $y = \mathbf{Hx}$.

  b) Design a control law of the form

$$u = -[K_1 \quad K_2]\begin{bmatrix} x_1 \\ x_2 \end{bmatrix}$$

  that will place the closed-loop poles at $s = -2 \pm 2j$.

6.7 Consider the plant described by

$$\dot{\mathbf{x}} = \begin{bmatrix} 0 & 1 \\ 7 & -4 \end{bmatrix}\mathbf{x} + \begin{bmatrix} 1 \\ 2 \end{bmatrix}u,$$

$$y = [1 \quad 3]\mathbf{x}.$$

  a) Draw the block diagram for the plant at the integrator level.

  b) Find the transfer function using matrix algebra.

  c) Find the closed-loop characteristic equation if the feedback is

   1. $u = -[K_1 \quad K_2]\mathbf{x}$.

   2. $u = -Ky$.

6.8 Consider the transfer function

$$G(s) = \frac{Y(s)}{U(s)} = \frac{s + 1}{s^2 + 5s + 6}.$$

  a) By writing

$$G(s) = \left(\frac{1}{s + 3}\right)\left(\frac{s + 1}{s + 2}\right),$$

FIGURE 6.67

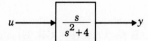

FIGURE 6.68

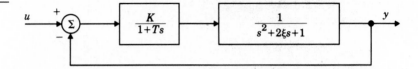

find a state-space realization of $G(s)$ as a cascade of first order systems [see Fig. 2.30(a)].

b) Using a partial fraction expansion of $G(s)$, find a realization of $G(s)$ as a parallel combination of two first order systems [see Fig. 2.30(b)].

c) Find a control canonical form realization.

6.9 The equations of motion for an inverted pendulum on a cart are

$$\ddot{\theta} = \theta + u; \qquad \ddot{x} = -\beta\theta - u.$$

Find the feedback gain K such that the poles of the closed-loop system are at $s = -1, -1, -1 \pm j1$. Let $x = [\theta, \dot{\theta}, x, \dot{x}]^T$ be the state vector.

6.10 Consider the feedback system in Fig. 6.68. Find the relationship between K, T, and ξ such that the closed-loop transfer function minimizes the ITAE criterion for a step input.

Section 6.5

6.11 For each of the transfer functions below, write the state equations in both controller and observer canonical form. In each case, draw a block diagram at the integrator level and give the appropriate **F**, **G**, and **H**.

a) $\dfrac{s^2 - 2}{s^2(s^2 - 1)}$ (control of an inverted pendulum by a force on the cart).

b) $\dfrac{3s + 4}{s^2 + 2s + 2}$.

6.12 A certain process has transfer function $G(s) = 4/(s^2 - 4)$.

a) Find **F**, **G**, and **H** for this system in *observer* canonical form.

b) If $u = -\mathbf{Kx}$, compute **K** so the closed-loop poles are located at $s = -2 \pm j2$.

c) Compute **L** so the estimator-error poles are located at $s = -4 \pm j4$.

d) Give the transfer function u/y of the resulting controller.

6.13 Consider the system

$$\mathbf{F} = \begin{bmatrix} -2 & 1 \\ 1 & 0 \end{bmatrix}, \qquad \mathbf{G} = \begin{bmatrix} 1 \\ 0 \end{bmatrix}, \qquad \mathbf{H} = [1 \quad 2],$$

and assume that we are using feedback of the form $u = -\mathbf{Kx} + r$, where r is the reference input signal.

a) Show that (**F**, **H**) is observable.

b) Show that there exists a **K** such that (**F** − **GK**, **H**) is unobservable.

c) Compute a **K** as in **(b)** such that $\mathbf{K} = [1, K_2]$—that is, find K_2.

d) Compute the open-loop and closed-loop transfer functions. What is the unobservability due to?

6.14 The normalized equations of motion for an inertial navigator are

$$\begin{bmatrix} \dot{x}_1 \\ \dot{x}_2 \\ \dot{x}_3 \end{bmatrix} = \begin{bmatrix} 0 & -1 & 0 \\ 1 & 0 & 1 \\ 0 & 0 & 0 \end{bmatrix} \begin{bmatrix} x_1 \\ x_2 \\ x_3 \end{bmatrix} + \begin{bmatrix} 0 \\ 0 \\ 1 \end{bmatrix} u,$$

where

$x_1 =$ east-velocity error,
$x_2 =$ platform tilt about north axis,
$x_3 =$ north-gyro drift,
$u =$ gyro drift rate of change.

Design a reduced-order estimator with $y = x_1$ as the measurement and the observer-error poles placed at $-0.1, -0.1$. Be sure to provide all the relevant estimator equations.

6.15 The equations of motion for a station-keeping satellite are

$$\ddot{x} - 2\omega\dot{y} - 9\omega^2 x = 0, \qquad \ddot{y} + 2\omega\dot{x} + 4\omega^2 y = u,$$

where

$x =$ radial position perturbation,
$y =$ azimuthal position perturbation,
$u =$ engine thrust in y direction.

a) Is the system observable from y?

b) Design a full-order observer with its poles placed at $s = -2\omega, -3\omega, -3\omega \pm j3\omega$. Choose $\mathbf{x} = [x, \dot{x}, y, \dot{y}]^T$ as the state vector and y as the measurement.

6.16 The equations of motion of the simple pendulum in Fig. 6.69 are

$$\ddot{\theta} + \omega^2 \theta = u.$$

FIGURE 6.69

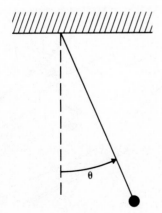

a) Write the equations of motion in state-space form.
b) Design an estimator (or observer) that reconstructs both state variables of the pendulum given measurements of $\dot{\theta}$. Assume $\omega = 5$ rad/s and pick the estimator roots to be at $s = -10 \pm 10j$.
c) Write the transfer function of the estimator between the measured value of $\dot{\theta}$ and the estimated value of θ (or $\hat{\theta}$).
d) Design a controller (that is, determine **K**), so that the roots are at $s = -4 \pm 4j$.
e) Find the transfer function for the compensator.

6.17 The longitudinal motions of a helicopter near hover (Fig. 6.70) can be modeled by the third-order system,

$$\begin{bmatrix} \dot{q} \\ \dot{\theta} \\ \dot{u} \end{bmatrix} = \begin{bmatrix} -0.4 & 0 & -0.01 \\ 1 & 0 & 0 \\ -1.4 & 9.8 & -0.02 \end{bmatrix} \begin{bmatrix} q \\ \theta \\ u \end{bmatrix} + \begin{bmatrix} 6.3 \\ 0 \\ 9.8 \end{bmatrix} \delta,$$

where

q = pitch rate,
θ = pitch angle of fuselage,
u = horizontal velocity,
δ = rotor tilt angle.

Suppose our sensor measures the velocity as $y = u$.

a) Find the open-loop poles.
b) Is the system controllable?
c) Assign the poles of the system at $s = -2, s = -1 \pm j1$ and determine the feedback gain K.

FIGURE 6.70

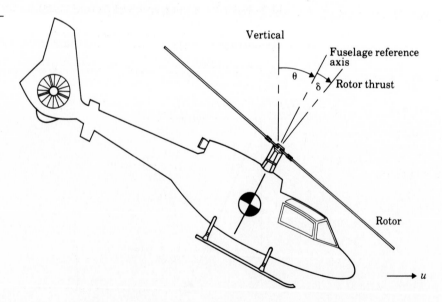

d) Design a full-order estimator to place the estimator poles at -1.5. and $-0.5 \pm j1$.

e) Design a reduced-order estimator to place the poles at -1. What are the advantages and disadvantages of this reduced-order estimator?

f) Compute the compensator transfer function using the controller in **(c)** combined with the estimator in **(d)**. Repeat for the estimator in **(e)**.

6.18 Consider the system with the transfer function

$$G(s) = \frac{9}{s^2 - 9}.$$

a) Find $(\mathbf{F}_0, \mathbf{G}_0, \mathbf{H}_0)$ for this system in observer canonical form.

b) Is $(\mathbf{F}_0, \mathbf{G}_0)$ controllable?

c) Compute $\mathbf{K}$ so that the closed-loop poles are assigned at $s = -3 \pm j3$.

d) Is the system observable?

e) Design a full-order estimator such that the estimator-error poles are located at $s = -6 \pm j6$.

f) Suppose the system has a zero such that

$$G_1(s) = \frac{9(s + 1)}{s^2 - 9}.$$

Prove that if $u = -\mathbf{K}\mathbf{x} + r$, there is a feedback gain $\mathbf{K}$ such that the system is unobservable.

6.19 Explain how the controllability, observability, and stability properties of a linear system are related.

6.20 Consider the electric network shown in Fig. 6.71.

a) Write the internal (state) equations for the network. The input $u(t)$ is a current and the output y is a voltage. Let $x_1 = i_L$ and $x_2 = v_c$.

b) Let $R^2 = L/C$. Is the system controllable?

c) Let $R^2 = L/C$. Is the system observable?

6.21 A block diagram of a feedback system is shown in Fig. 6.72. The state is

$$\mathbf{x} = \begin{bmatrix} \mathbf{x}_p \\ \mathbf{x}_f \end{bmatrix},$$

FIGURE 6.71

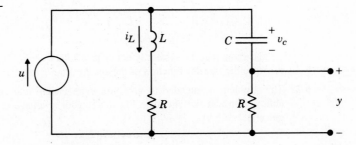

FIGURE 6.72

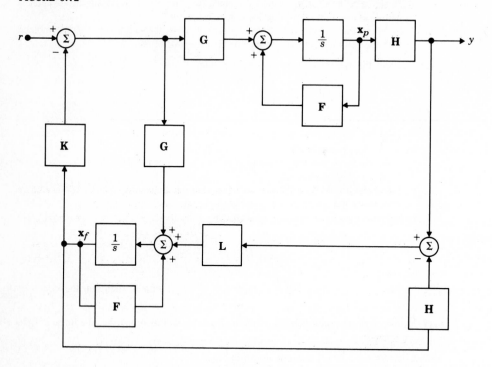

the input is r, and the output is y. The dimensions of the matrices are

$$\mathbf{F} = n \times n \qquad \mathbf{L} = n \times 1$$
$$\mathbf{G} = n \times 1 \qquad \mathbf{x} = 2n \times 1$$
$$\mathbf{H} = 1 \times n \qquad r = 1 \times 1$$
$$\mathbf{K} = 1 \times n \qquad y = 1 \times 1.$$

a) Write the equations in state-variable form.
b) Let $\mathbf{x} = \mathbf{TZ}$, where

$$\mathbf{T} = \begin{bmatrix} I & 0 \\ I & -I \end{bmatrix},$$

and show that the system is not controllable.
c) Write the transfer function of the system from r to y.

6.22 This problem is intended to give you more insight into controllability and observability. Consider the circuit in Fig. 6.73, with a voltage source of $u(t)$ and an output $y(t)$.

FIGURE 6.73

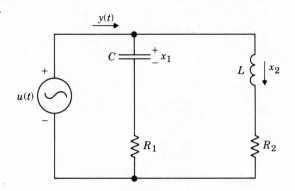

a) Using the capacitor voltage and inductor current as state variables, write the state and output equations of the system (single-input and single-output).

b) Find the conditions relating R_1, R_2, C, and L such that the system is uncontrollable. Find a similar set of conditions such that the system is unobservable.

c) Interpret these conditions physically in terms of the time constants of the system.

d) Find the transfer function of the system. Show that there is a pole-zero cancellation when the system is uncontrollable or unobservable.

6.23 Two pendulums, coupled by a spring, are to be controlled by two equal and opposite forces u, which are applied to the pendulum bobs shown in Fig. 6.74. The equations of motion are

$$ml^2\ddot{\theta}_1 = -ka^2(\theta_1 - \theta_2) - mgl\theta_1 - u,$$
$$ml^2\ddot{\theta}_2 = -ka^2(\theta_2 - \theta_1) - mgl\theta_2 + u.$$

a) Show that the system is uncontrollable. Can you associate a physical meaning with the controllable and uncontrollable modes?

b) Is there any way at all that the system can be made controllable?

FIGURE 6.74

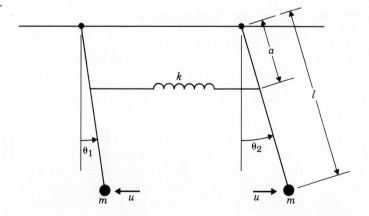

FIGURE 6.75

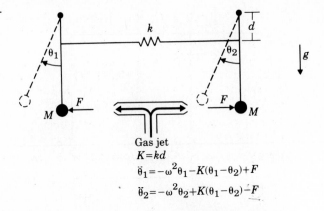

Gas jet
$K = kd$
$$\ddot{\theta}_1 = -\omega^2\theta_1 - K(\theta_1 - \theta_2) + F$$
$$\ddot{\theta}_2 = -\omega^2\theta_2 + K(\theta_1 - \theta_2) - F$$

6.24 Consider the system in Fig. 6.75.

a) Write the equations in state-variable form, using $[\theta_1, \theta_2, \dot{\theta}_1, \dot{\theta}_2]^T$ as the state vector and F as the single input.

b) Show that all the state variables are observable with $y = \theta_1$.

c) Show that the characteristic equation for the system is the product of two oscillators by first converting the system equations to the new state variables

$$\begin{bmatrix} y_1 \\ y_2 \\ \dot{y}_1 \\ \dot{y}_2 \end{bmatrix} = \begin{bmatrix} \theta_1 + \theta_2 \\ \theta_1 - \theta_2 \\ \dot{\theta}_1 + \dot{\theta}_2 \\ \dot{\theta}_1 - \dot{\theta}_2 \end{bmatrix}.$$

Hint: If **A** and **D** are invertible matrices, then

$$\begin{bmatrix} A & 0 \\ 0 & D \end{bmatrix}^{-1} = \begin{bmatrix} A^{-1} & 0 \\ 0 & D^{-1} \end{bmatrix}.$$

d) Deduce that the "spring mode" can be controlled with F but that the "pendulum mode" is uncontrollable with F.

6.25 The linearized equations of a satellite are

$$\dot{\mathbf{x}} = \mathbf{Fx} + \mathbf{Gu},$$
$$\mathbf{y} = \mathbf{Hx},$$

where

$$\mathbf{F} = \begin{bmatrix} 0 & 1 & 0 & 0 \\ 3\omega^2 & 0 & 0 & 2\omega \\ 0 & 0 & 0 & 1 \\ 0 & -2\omega & 0 & 0 \end{bmatrix}, \quad \mathbf{G} = \begin{bmatrix} 0 & 0 \\ 1 & 0 \\ 0 & 0 \\ 0 & 1 \end{bmatrix},$$

$$\mathbf{H} = \begin{bmatrix} 1 & 0 & 0 & 0 \\ 0 & 0 & 1 & 0 \end{bmatrix}, \quad \mathbf{u} = \begin{bmatrix} u_1 \\ u_2 \end{bmatrix}, \quad \mathbf{y} = \begin{bmatrix} y_1 \\ y_2 \end{bmatrix}.$$

u_1, u_2 are the radial and tangential thrusts, respectively. x_1 and x_3 are the radial and angular deviations from the reference (circular) orbit, respectively. y_1, y_2 are the radial and angular measurements.

a) Show that the system is controllable using both controls.
b) Show that the system is controllable with only one control. Which one is it?
c) Show that the system is observable using both measurements.
d) Show that the system is observable using only one measurement. Which one is it?

6.26 Consider the system shown in Fig. 6.76, employing series, parallel, and feedback configurations.

a) Suppose we have controllable-observable realizations for each subsystem:

$$\dot{\mathbf{x}}_i = \mathbf{F}_i \mathbf{x}_i + \mathbf{G}_i \mathbf{u}_i,$$

$$y_i = \mathbf{H}_i \mathbf{x}_i,$$

$$i = 1, 2.$$

FIGURE 6.76
(a) Series; (b) parallel;
and (c) feedback.

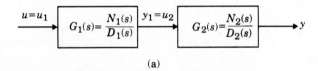

(a)

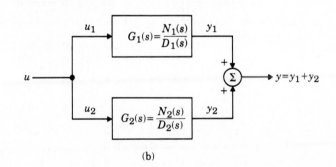

(b)

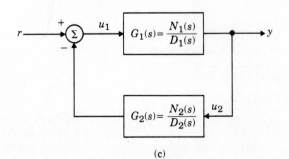

(c)

Write down a state-space realization for each system.

b) Determine under what conditions each system is controllable and/or observable.

6.27 *Output controllability.* In some practical control systems, the control engineer may be interested in controlling the *output y* rather than the *state* **x**. A system is said to be *output controllable* if at any time one is able to transfer the output from zero to any desired output y^* using an appropriate control signal u^* (in a finite time). Derive necessary and sufficient conditions for a continuous system $(\mathbf{F},\mathbf{G},\mathbf{H})$ to be *output controllable*. Are output and state controllability related?

6.28 Consider the system

$$\dot{\mathbf{x}} = \begin{bmatrix} 0 & 4 & 0 & 0 \\ -1 & -4 & 0 & 0 \\ 5 & 7 & 1 & 15 \\ 0 & 0 & 3 & -3 \end{bmatrix} \mathbf{x} + \begin{bmatrix} 0 \\ 0 \\ 1 \\ 0 \end{bmatrix} u.$$

a) Find the eigenvalues of this system. (Note the triangular structure.)

b) Find the controllable and uncontrollable modes of this system.

c) For each of the uncontrollable modes, give a **V** such that

$$\mathbf{V}^T\mathbf{G} = 0, \qquad \mathbf{V}^T\mathbf{F} = \lambda\mathbf{V}^T.$$

d) Show that there is an *infinite* number of feedback gains **K** that will relocate the modes of the system to $-5, -3, -2, -2$.

e) Find the *unique* **K** that achieves these locations *and* prevents initial conditions on the uncontrollable part from ever affecting the controllable part of the system.

6.29 A state-space model for a paper machine in a certain application had

$$\mathbf{F} = \begin{bmatrix} 0.174 & 0 & 0 & 0 & 0 \\ 0.157 & 0.645 & 0 & 0 & 0 \\ 0 & 1 & 0 & 0 & 0 \\ 0 & 0 & 1 & 0 & 0 \\ 0 & 0 & 0 & 1 & 0 \end{bmatrix},$$

$$\mathbf{G} = \begin{bmatrix} -0.207 \\ -0.005 \\ 0 \\ 0 \\ 0 \end{bmatrix}, \qquad \mathbf{H} = [1 \ 0 \ 0 \ 0 \ 0].$$

a) Draw a block diagram of the system at the integrator level.

b) A student has computed det $\mathscr{C} = 2.3 \times 10^{-7}$ and claims that the system is uncontrollable. Is she right or wrong? Why?

c) Is the realization observable?

6.30 Consider the system $\ddot{y} + 3\dot{y} + 2y = \dot{u} + u$.

a) Give $\mathbf{F}_c$, $\mathbf{G}_c$, and $\mathbf{H}_c$ in controller canonical form.

b) Sketch the eigenvectors of $\mathbf{F}_c$ in the (x_1, x_2) plane and the completely observable ($\mathbf{x}_0$) and completely unobservable ($\mathbf{x}_{\bar{0}}$) state variables in *orthogonal* form.

c) Express $\mathbf{x}_0$ and $\mathbf{x}_{\bar{0}}$ in terms of the observability matrix $\mathcal{O}$.

d) Repeat the problem for controllability instead of observability.

6.31 *Staircase algorithm* (Van Dooren et al., 1978). Any realization $(\mathbf{F},\mathbf{G},\mathbf{H})$ can be transformed by an *orthogonal similarity transformation* to $(\overline{\mathbf{F}}, \overline{\mathbf{G}}, \overline{\mathbf{H}})$, where $\overline{\mathbf{F}}$ is an upper-Hessenberg matrix (one nonzero diagonal above the main diagonal):

$$\overline{\mathbf{F}} = \mathbf{T}^{\mathsf{T}}\mathbf{F}\mathbf{T} = \begin{bmatrix} \ddots & \alpha_1 & & & \\ & \ddots & \ddots & & \mathbf{O} \\ & & \ddots & \ddots & \\ & * & & \ddots & \alpha_{n-1} \\ & & & & \ddots \end{bmatrix}; \qquad \overline{\mathbf{G}} = \mathbf{T}^{\mathsf{T}}\mathbf{G} = \begin{bmatrix} 0 \\ 0 \\ \vdots \\ 0 \\ g_1 \end{bmatrix};$$

$$\overline{\mathbf{H}} = \mathbf{H}\mathbf{T} = [h_1, \dots, h_n]; \qquad g_1 \neq 0; \qquad \mathbf{T}^{-1} = \mathbf{T}^{\mathsf{T}}.$$

a) Prove that if $\alpha_i = 0$ and $\alpha_{i+1}, \dots, \alpha_{n-1} \neq 0$, then the controllable and uncontrollable modes of the system can be identified.

b) How would you use this technique to identify the observable/unobservable modes of $(\mathbf{F}, \mathbf{G}, \mathbf{H})$?

c) What is the advantage of this approach over transforming the system to any other form for determining the controllable and uncontrollable forms?

d) How can we use this approach to determine a basis for the controllable/uncontrollable subspaces, as in Problem 6.30? This algorithm can be used to design a numerically stable algorithm for pole-placement (see Minimis and Paige, 1982). The name of the algorithm comes from the multi-input version in which the α_i's are the blocks that make $\overline{\mathbf{F}}$ resemble a staircase.

6.32 To control the inverted pendulum by a DC drive motor with current U we have the equations

$$\ddot{\theta} = \theta + V + U,$$
$$\dot{V} = \theta - V - U,$$

where θ is the angle of the pendulum and V is the velocity of the cart.

a) We wish to control θ by feedback to U of the form

$$U = -K_1\theta - K_2\dot{\theta} - K_3V.$$

Find the gains so that the resulting poles are located at $-1, -1 \pm j\sqrt{3}$.

b) Assume that θ and V are measured. Construct an estimator for θ and $\dot{\theta}$ in the form $\dot{\hat{\mathbf{x}}} = \mathbf{F}\hat{\mathbf{x}} + \mathbf{L}(y - \hat{y})$, where $\mathbf{x} = [\theta \quad \dot{\theta}]^{\mathsf{T}}$ and $y = \theta$. Treat V and U as both known. Select $\mathbf{L}$ so the error system has poles at $-2, -2$.

6.33 Consider the motor control of $G(s) = 10/[s(s + 1)]$.

a) Let $y = x_1$ and $\dot{x}_1 = x_2$. Complete the state equation by writing $\dot{\mathbf{x}}$.

b) Find K_1 and K_2 so $u = -K_1x_1 - K_2x_2$ gives closed-loop poles at $\omega_n = 3$, $\zeta = 0.5$.

c) Find L_1 and L_2 of a state estimator so that the state-error equations have characteristic equation with $\omega_1 = 15, \zeta_1 = 0.5$.

d) What is the transfer function of the controller obtained by combining **(b)** and **(c)**?

e) Sketch the root-locus of the resulting closed-loop system as plant gain (nominally 10) is varied.

6.34 In controlling an upside-down pendulum (as with a rocket), one aspect requires that we control the unstable equations (normalized)

$$\ddot{x} = x + u.$$

a) Let $u = -Kx$, and sketch the root locus with respect to K.

b) Let

$$U(s) = K \frac{s + a}{s + 10} X(s).$$

Select a and K so that the system will display a rise time of about 2 s and no more than 25% overshoot. Sketch the root locus versus K for your design.

c) Sketch the Bode plot of the uncompensated plant, both amplitude and phase.

d) Sketch the Bode plot of the closed-loop design and estimate the phase margin.

e) Design an estimator for x and $\dot{x}$ using measurement of $x = y$ and selecting the $\mathbf{L}$ matrix so that the $\tilde{\mathbf{x}}$ equation has characteristic roots at $\zeta = 0.5, \omega_n = 8$. (Note that there is no K in this part.)

f) Draw the block diagram of your estimator, and show where $\hat{x}$ and $\dot{x}$ appear.

6.35 Consider the pendulum problem with $g/l = 4$:

$$\ddot{\theta} + 4\theta = T'_c + T'_d.$$

Assume there is a potentiometer at 0 that measures the output angle θ, but that it includes a *constant*, but unknown, bias b—that is, $z = \theta + b$.

a) Taking the "augmented" state vector to be

$$\begin{bmatrix} \theta \\ \dot{\theta} \\ b \end{bmatrix},$$

write the system equations in state form. Indicate $\mathbf{F}$, $\mathbf{G}$, $\mathbf{H}$, and $\mathbf{G}_1$.

b) Show, using state-variable methods, that the characteristic equation is $s(s^2 + 4) = 0$.

c) Show b is observable with z and write the estimator equations for

$$\begin{bmatrix} \hat{\theta} \\ \dot{\hat{\theta}} \\ \hat{b} \end{bmatrix}.$$

Pick estimator gain L to place the roots of the estimator-error characteristic equation at $-2, -2, -2$.

d) Using full-state feedback of the estimated state, derive a control law to place the closed-loop poles at $-2 \pm 2j, -1$.

e) Draw a block diagram of the six-state system, that is, estimator, plant, and control law, using six integrator blocks.

6.36 A certain fifth-order system is found to have a characteristic equation with roots at $[0, -1, -2, -1 \pm j1]$. A decomposition into controllable and uncontrollable parts discloses that the controllable part has a characteristic equation with roots at $[0, -1 \pm j1]$. A decomposition into observable and nonobservable parts discloses that the observable modes are at $[0, -1, -2]$.

FIGURE 6.77

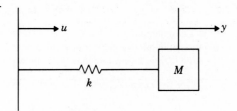

a) Where are the zeros of $b(s) = \mathbf{H} \operatorname{adj}(s\mathbf{I} - \mathbf{F})\mathbf{G}$ for this system?
b) Where are the poles of the reduced transfer function of this system?

Section 6.6

6.37 A simplified model for the control of a flexible robotic arm is shown in Fig. 6.77, where $k/M = 900$ rad/s.

a) Write the equations of motion in state-space form. (Output is y, the mass position; input is u, the position of the end of the spring.)
b) Design an estimator with roots at $s = -100 \pm 100j$.
c) Could both state variables of the system be estimated if only a measurement of $\dot{y}$ was available?
d) Design a controller with roots at $s = -20 \pm 20j$.
e) Would it be reasonable to design a controller for the system with roots at $s = -200 \pm 200j$? State your reasons.
f) Write down the equations representing the compensator. Include a command input for y.

6.38 The equations of motion of two cascaded tanks (see Fig. 6.78) linearized about nominal levels are

$$\delta \dot{h}_1 + \sigma \delta h_1 = \delta u,$$
$$\delta \dot{h}_2 + \sigma \delta h_2 = \sigma \delta h_1,$$

FIGURE 6.78

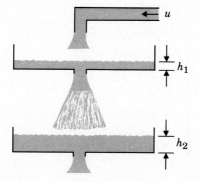

where

$$\delta h_1 = \text{deviation of depth in tank 1 from nominal level,}$$
$$\delta h_2 = \text{deviation of depth in tank 2 from nominal level,}$$
$$\delta u = \text{deviation in fluid inflow rate to tank 1 (control).}$$

a) *Level controller for two cascaded tanks.* Using state-feedback of form

$$\delta u = -K_1\delta h_1 - K_2\delta h_2,$$

choose K_1 and K_2 to place the closed-loop eigenvalues at

$$s = -2\sigma(1 \pm j).$$

b) *Level estimator for two cascaded tanks.* Only the deviation in level of tank 2 is measured (that is, $z = \delta h_2$). Using this measurement, design the logic for an estimator that will give continuous, smooth estimates of the deviation in levels of tank 1 and tank 2, such that the estimate errors have eigenvalues $-3\sigma(1 \pm j)$.

c) *Estimator/controller for two cascaded tanks.* Combining the estimator of **(b)** with the controller of **(a)** (that is, using estimated-state feedback), sketch a block diagram of the controlled system (at the detailed individual-integrator level).

6.39 The lateral motions of a ship that is 100 m long, moving at a constant velocity of 10 m/s, are described by

$$\begin{bmatrix} \dot{\beta} \\ \dot{r} \\ \dot{\psi} \end{bmatrix} = \begin{bmatrix} -0.0895 & -0.286 & 0 \\ -0.0439 & -0.272 & 0 \\ 0 & 1 & 0 \end{bmatrix} \begin{bmatrix} \beta \\ r \\ \psi \end{bmatrix} + \begin{bmatrix} 0.0145 \\ -0.0122 \\ 0 \end{bmatrix} \delta,$$

where β is the sideslip angle, r is the yaw rate, and ψ the heading angle.

a) Determine the transfer function from δ to ψ and the characteristic roots of the uncontrolled ship.

b) Using complete state feedback,

$$\delta = -K_1\beta - K_2 r - K_3(\psi - \psi_d),$$

where ψ_d is the desired heading, determine K_1, K_2, and K_3 to place the closed-loop roots at $s = -0.2, -0.2 \pm j0.2$.

c) Design an estimator using a measurement of ψ (obtained from a gyrocompass). Place the roots of the estimator-error equation at $s = -0.4, -0.4 \pm 0.4j$.

d) Write out the complete equations that correspond to the "compensation" of this system. In other words, specify the box labeled D_c in Fig. 6.79.

FIGURE 6.79

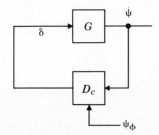

6.40 A reasonable alternative to the selection of feed-forward gain N in Eq. (6.147) is to select N such that when r and y are both unchanging, the DC gain from r to u is the *negative* of the DC gain from y to u. Derive a formula for N based on such a selection. Show that if the plant is type I, this choice is the same as that given by Eq. (6.147).

Section 6.9

6.41 Assume the linearized and time-scaled equation of the ball-bearing levitation device is $\ddot{x} - x = u + w$. Here w is a constant bias due to the power amplifier. Introduce integral error control and select three control gains $\mathbf{K}$ so that the closed-loop poles are at -1 and $-1 \pm j$ and the steady-state error to w and to a position command will be zero. Let $y = x$ and $r \triangleq y_{\text{ref}}$ be a constant. Draw a block diagram of your design showing locations of the K_i. Assume that both $\dot{x}$ and x can be measured.

6.42 A plant is described by the matrices

$$\mathbf{F} = \begin{bmatrix} -1 & 0 \\ 0 & -2 \end{bmatrix}; \qquad \mathbf{G} = \begin{bmatrix} 1 \\ 1 \end{bmatrix}; \qquad \mathbf{H} = [\, 1 \quad 3 \,].$$

a) Find $\mathbf{K}$ such that, if $u = -\mathbf{K}x + Nr$, the closed-loop system has poles at $-2 \pm j2$.
b) Find N so that, if $r = r_\infty = $ constant, $y = y_\infty = r_\infty$; that is, there is zero steady-state error. Show that this property (zero steady-state error) is not robust with respect to changes in $\mathbf{F}$.
c) Add to the plant equation in integrator $\dot{\eta} = e = y - r$ and select gains $\mathbf{K}, K_1$ so that, if $u = -\mathbf{K}x - K_1\eta$, the closed-loop poles are at $-2, -1 \pm j\sqrt{3}$. Show that this system has zero steady-state error and that this property is robust with respect to changes in $\mathbf{F}$ so long as the closed-loop system remains stable.

6.43 Consider a system with

$$\mathbf{F} = \begin{bmatrix} -2 & 1 \\ 0 & -3 \end{bmatrix}; \qquad \mathbf{G} = \begin{bmatrix} 1 \\ 1 \end{bmatrix}; \qquad \mathbf{H} = [\, 1 \quad 3 \,].$$

a) Use feedback of the form $u(t) = -\mathbf{K}x(t) + Nr(t)$, N being a nonzero scalar, to move poles to $-3 \pm j3$.
b) Choose N so that if r is a constant, the system has zero steady-state error, that is, $y(\infty) = r$.
c) Show that if $\mathbf{F}$ changes to $\mathbf{F} + \delta\mathbf{F}$, $\delta\mathbf{F}$ being an arbitrary 2×2 matrix, subject to $\mathbf{F} + \delta\mathbf{F} - \mathbf{GK}$ being stable, then the above choice of N will no longer make $y(\infty) = r$. Therefore the system is not robust under changes in the system parameters $\mathbf{F}$.
d) The system can be made robust by augmenting it with an integrator and using unity feedback, that is, $\dot{\xi} = r - y$, where ξ is the state of the integrator. To see this, first use state feedback of the form $u = -\mathbf{K}x - K_e\xi$ so that the poles of the augmented system are at $-3, -2 \pm j\sqrt{3}$.
e) Show that the resulting system will have $y(\infty) = r$ no matter how the matrices $[\mathbf{F}, \mathbf{G}, \mathbf{K}, K_e]$ are changed so long as the closed-loop system remains asymptotically stable.

6.44 Consider the servomechanism to follow the data track on a computer-disk memory system. The data track is not exactly a centered circle. Therefore, the radial servo must follow a sinusoid input of radian frequency ω_0. The matrices are

$$F = \begin{bmatrix} 0 & 1 \\ 0 & -1 \end{bmatrix}; \qquad G = \begin{bmatrix} 0 \\ 1 \end{bmatrix}; \qquad H = [1 \quad 3];$$

and $\ddot{r} = -\omega_0^2 r$. Place the poles of the error system at

$$\alpha_c(s) = (s + 1 \pm j\sqrt{3})(s + \sqrt{3} \pm j1)$$

and the pole of the reduced-order estimator at

$$\alpha_e(s) = s + 2.$$

Draw a block diagram of the system and clearly show the presence of the oscillator with frequency ω_0 (the internal model) in the controller. Verify the presence of the blocking zeros at $\pm j\omega_0$.

Section 6.10

6.45 Consider the system

$$G(s) = \frac{1}{s(s + 1)(s + 2)}.$$

The Smith compensator for $e^{-Ts}G(s)$ is given by

$$D_c(s) = \frac{1}{1 + (1 - e^{-sT})G(s)}.$$

Investigate the frequency-response properties of the compensator for $T = 5$. *(Problem given by Åström, 1977.)*

7

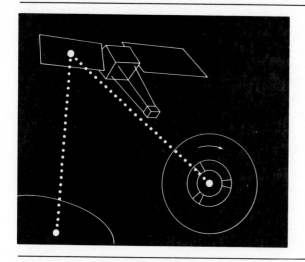

Control-System

Design:

Principles and

Case Studies

7.1 Introduction

In Chapters 4, 5, and 6 we have presented techniques for the analysis and design of feedback systems based on root locus, frequency response, and pole placement using state-variable models. In this chapter we return to the theme of Chapter 3 to reconsider the overall problem of control-system design, but now with these more sophisticated tools in hand. After a discussion of the general, problem setting, several characteristic problems are presented in something of a case-study format to illustrate the use of all the tools we have developed for control-system analysis and design.

7.1.1 An Outline of Control-System Design

Control engineering is an important part of the design process of many dynamic systems. As suggested in Chapter 3, the deliberate use of feedback can stabilize an otherwise unstable system, reduce the error due to disturbance inputs, and reduce the sensitivity of a closed-loop transfer function to small variations in internal system parameters. In those situations where feedback control is required, it is possible to outline an approach to control design that will often lead to a satisfactory solution. However, before giving our outline, it is important to emphasize that the purpose of control is to aid the product or process—the mechanism, the robot, the chemical plant, the aircraft, or whatever—in doing the job. It is interesting that others engaged in the design process are increasingly taking into account the contribution of control early on in their planning so that more and more systems simply will not work without feedback. This is especially significant in the design of high-performance aircraft, where control has taken its place along with structures and aerodynamics as an essential element for the craft even to fly at all. It is impossible to give a description of such overall design in this book, but recognition of its existence places in perspective not only the specific task of control design but also the central role this task can play in an enterprise.

Control design begins with a proposed product or process whose satisfactory dynamic performance depends upon feedback for stability, disturbance regulation, or reduction of the effects of parameter variations. We will give an outline of the design process that is general enough to be useful whether the product is an electronic amplifier or a large space structure. Obviously, to be so widely applicable, our outline will be vague with respect to physical details and specific only with respect to the feedback-control problem.

STEP 1: *Understand the process and translate dynamic performance requirements into time, frequency, and pole-zero specifications.* The importance of understanding the process, what it is intended to do, how much system error is permissible, how to describe the class of command and disturbance signals to be expected, and what the physical capabilities and limitations are: all of these can hardly be overemphasized. Regrettably, in instructional texts, one tends to view the "process" as a linear, time-invariant transfer function capable of responding to inputs of arbitrary size, and to overlook the fact that this is a gross caricature valid only for small signals, short times, and particular environmental conditions. The designer must not confuse the caricature with reality, but must be able to use the simplified model for its intended purpose and to return to an accurate physical model or the actual system for real verification of the design performance.

A typical result of this step is a specification that the system have: a step response inside some constraint boundaries such as shown in Fig. 7.1(a), an open-loop frequency response satisfying constraints such as those sketched in Fig. 7.1(b), or closed-loop poles to the left of some constraint boundary such as that sketched in Fig. 7.1(c).

STEP 2: *Select a sensor.* In sensor selection one must consider which variables are important to control and which variables can physically be measured. For example, in a jet engine there are critical internal temperatures that must be controlled but that cannot be measured directly in an operational engine. One must select sensors

FIGURE 7.1
(a) Time-response
specifications;
(b) frequency-response
specifications; and
(c) pole-zero
specifications.

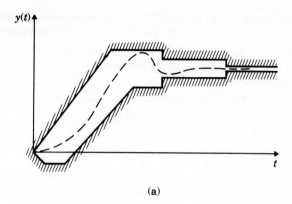

(a)

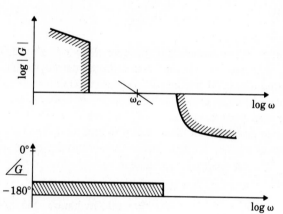

(b)

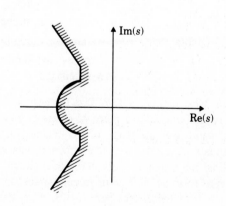

(c)

that indirectly allow a good estimate to be made of these critical variables. Also important is the consideration of sensors for the disturbance. Sometimes, especially in chemical processes, it is beneficial to directly sense a load disturbance, and improved performance can be obtained if this information is measured and fed forward to the controller.

Factors that influence sensor selection are

Technology	Electric or magnetic, mechanical, electromechanical, electrooptical, piezoelectric
Functional performance	Linearity, bias, accuracy, dynamic range, noise
Physical properties	Weight, size, strength
Quality factors	Reliability, durability, maintainability
Cost	Expense, availability, facilities for testing and maintenance

STEP 3: *Select an actuator*. In order to control a dynamic system, one must obviously be able to influence the response; the device that does this is the actuator. Before a specific actuator is considered, one must sometimes consider which variables can be influenced. For example, in a flight vehicle many configurations of movable surfaces are possible, and the influence these have on the performance and controlability of the craft can be profound. The locations of jets or other torque devices are also a major part of the control design of spacecraft.

Having selected a particular variable to control, other considerations enter, such as:

Technology	Electric, hydraulic, pneumatic, thermal, other
Functional performance	Maximum force possible, extent of the linear range, maximum speed possible, power, efficiency
Physical properties	Weight, size, strength
Quality factors	Reliability, durability, maintainability
Cost	Expense, availability, facilities for testing and maintenance

STEP 4: *Make a linear model*. Take the best choice for process, actuator, and sensor, identify the equilibrium point of interest, and construct a small-signal dynamic model valid over the range of frequencies included in the specifications of step 1. Validate the model with experimental data where possible. In order to be able to use all the available tools, the model should be expressed in state-variable and pole-zero form as well as in frequency-response form. Many modern computer-design software packages perform the transformation from one of these forms to the others.

STEP 5: *Try a simple PID or lead-lag design.* Sketch a frequency (Bode plot) response and a root locus versus plant gain to form an initial estimate of the complexity of the design problem. If the plant-actuator-sensor model is stable and minimum phase, the Bode plot is probably the most useful; otherwise the root locus shows very important information with respect to behavior in the right-half plane. In any case, try to meet the specifications with a simple controller of the lead-lag variety, including integral control if steady-state error response requires it.

Be sure not to overlook feed-forward of the disturbances if the information is available. Also consider the effects of sensor noise and consider the alternatives of a lead network or a speed (velocity) sensor to see which gives a better design.

Reevaluate the specifications, the physical configuration of the process, and the actuator and sensor selections in the light of the current design and return to step 1 if improvement seems feasible. For example, in many motion-control problems, after the first-pass design is tested there are vibrational modes that prevent the design from meeting the initial specifications of the problem. It may be possible to meet the specifications much more easily by altering the structure through the addition of stiffening members or by passive damping than to meet them by control strategies alone. An alternative solution may be to move a sensor so it is at a node of a vibration mode, thus providing no feedback of the motion. Also, some actuator technologies (hydraulic) have many more low-frequency vibrations than others (electric). The important point is that all parts of the design, not only the control logic, need to be considered to help meet the specifications in the most cost-effective way.

If the design now seems satisfactory, go to step 7; otherwise try step 6.

STEP 6: *Try an optimal design.* If the trial-and-error compensators do not give entirely satisfactory performance, a design based on optimal control should be considered. Draw the symmetric root locus; select locations for the control poles that meet the response specifications; select locations for the estimator poles that represent a compromise between sensor and process noise. Plot the corresponding open-loop frequency response and the root locus to evaluate the stability margins of this design and to estimate its robustness to parameter changes. Modify the pole locations until a "best" compromise is obtained.

Compare the optimal design that has satisfactory frequency response with the simple design of step 5. Select the better one and go to step 7.

STEP 7: *Build a computer model and compute (simulate) the performance of the design.** When the best compromise between process modification, actuator selection, and sensor choice is obtained, a computer model of the system should be run that

---

*Extensive computer software is available to assist in this difficult but critical step. Names of just a few of these packages are MODEL-C (Systems Control Technology), SIMNON (Lund Institute of Technology, Lund, Sweden), ACSL (Mitchell and Gauthier), DSL (IBM), EASY5 (Boeing Computer Services), MATRIX$_x$, (Integrated Systems, Inc.).

includes important nonlinearities such as actuator saturation, noise, and parameter variations that are expected during operation of the system. Such a simulation of the design will confirm stability and robustness and allow one to predict the true performance to be expected from the system. As part of this simulation one can often include parameter optimization whereby the computer can be used to "tune" the free parameters for best performance. In the early stages, a simple model is simulated; as design progresses, more complete and detailed models are studied. If the results of the simulation prove the design satisfactory, go to step 8; otherwise return to step 1.

STEP 8: *Build a prototype*. As the final test before production, a prototype can be built and tested. At this point the quality of the model will be verified, unsuspected vibration and other modes will be discovered, and ways to improve the design can be suggested. After these tests, one might want to reconsider the sensor, actuator, and process and conceivably could return to step 1—unless time, money, or ideas have run out.

This outline is an approximation of good practice; other engineers will have variations on these themes. In some cases, engineers may wish to carry out the steps in a different order, to omit a step, or to add one. The step of simulation and prototype construction varies widely, depending on the nature of the system. For systems where a prototype is difficult to test and rework (e.g., a satellite) or where a failure is dangerous (e.g., active stabilization of a high-speed centrifuge or landing a human on the moon), most of the design verification will be done by a simulation of some sort. It may be in the form of a digital numerical simulation, a laboratory-scale model, or a full-size laboratory model with a simulated environment. For systems that are easy to build and modify (e.g., an automotive feedback fuel system), the simulation step is often skipped entirely. Design verification and refinement are accomplished by working with prototypes.

Implicit in the process of design is the well-known fact that designs within a given class often draw upon experience gained from earlier models, and thus good designs evolve rather than appear in the best form the first time. To illustrate the method, a few cases will be discussed. For easy reference, a summary of the steps is repeated here.

1. Understand the process and its performance requirements.
2. Select the best sensor, considering location, technology, and noise.
3. Select the best actuator, considering location, technology, and power.
4. Make a linear model of the process, actuator, and sensor.
5. Make a simple trial design based on lead-lag or PID ideas. If satisfied, go to step 7.
6. Make a trial pole-placement design based on optimal control or other criteria.

7. Simulate the design, including effects of nonlinearities, noise, and parameter variations. If the performance is not satisfactory, go to step 1.

8. Build a prototype and test it. If not satisfied, go to step 1.

7.2 Design of a Satellite's Attitude Control

Our first example is taken from the space program and is suggested by the need to control the pointing direction, or attitude, of a satellite in orbit about the earth. We will go through each step in our design outline and touch on some of the factors that might be considered for such a system.

STEP 1: *Understand the process and its performance specifications.* A sketch of the satellite is shown in Fig. 7.2. We imagine that the vehicle has an astronomical survey mission that requires accurate pointing of a scientific sensor package that must be maintained in the quietest possible environment, isolated from the vibrations and electrical noise of the main service body, which contains power

FIGURE 7.2
A satellite problem.

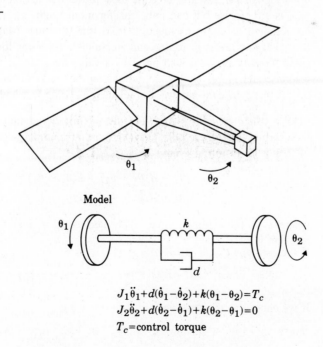

Model

$$J_1\ddot{\theta}_1 + d(\dot{\theta}_1 - \dot{\theta}_2) + k(\theta_1 - \theta_2) = T_c$$
$$J_2\ddot{\theta}_2 + d(\dot{\theta}_2 - \dot{\theta}_1) + k(\theta_2 - \theta_1) = 0$$
$$T_c = \text{control torque}$$

supplies, thrusters, and communication gear. The resulting structure is modeled as two masses connected by a flexible boom.

Disturbance torques due to solar pressure, micrometeorites, and orbit perturbations are computed to be negligible. The pointing requirement can be met by dynamics with a transient settling time of 14 s and an overshoot of 15%. The satellite will be overhead for about 10 min, and 14 s is considered adequate speed of response for the mission.

STEP 2: *Select a sensor.* In order to orient the scientific package, it is necessary to measure the attitude angles of the vehicle. For this purpose, a system based on gathering an image of a specific star and keeping it centered on the focal plane of a telescope is selected. This sensor gives a relatively noisy but very accurate (on the average) reading proportional to θ, the angle of deviation from the desired angle. To stabilize the control, a rate gyro is included to give a clean reading of $\dot{\theta}$, since a lead network on the star tracker would have too much noise. Furthermore, the rate gyro could stabilize any large motions without requiring the star tracker to acquire the target.

STEP 3: *Select an actuator.* Major considerations in the actuator selection are precision, reliability, weight, power requirements, and life-span. Alternatives for applying torque are cold-gas jets, reaction wheels or gyros, magnetic torquers, and gravity gradient. The jets have the most power and are the least accurate. Reaction wheels are precise but can only transfer momentum, so jets or magnetic torquers are required to dump momentum from time to time. Magnetic torquers provide relatively low levels of torque and are suitable for some low-altitude satellite missions. Gravity gradient also provides a very small torque that limits the speed of response and places severe restrictions on the shape of the satellite. For purposes of this mission, cold-gas jets are selected as being fast and adequately accurate.

STEP 4: *Make a linear model.* For the satellite we assume two masses connected by a spring with torque constant k and viscous damping constant d. The equations of motion from Fig. 7.2 are

$$J_1\ddot{\theta}_1 + d(\dot{\theta}_1 - \dot{\theta}_2) + k(\theta_1 - \theta_2) = T_c,$$
$$J_2\ddot{\theta}_2 + d(\dot{\theta}_2 - \dot{\theta}_1) + k(\theta_2 - \theta_1) = 0, \tag{7.1}$$

where T_c is the control torque. With inertias $J_1 = 1$ and $J_2 = 0.1$, the transfer function is

$$G(s) = \frac{10ds + 10k}{s^2(s^2 + 11ds + 11k)}. \tag{7.2}$$

If we choose as the state vector,

$$\mathbf{x}^T = [\theta_2 \quad \dot{\theta}_2 \quad \theta_1 \quad \dot{\theta}_1],$$

then, from Eq. (7.1), the equations are (with $T_c \equiv u$)

$$\dot{\mathbf{x}} = \begin{bmatrix} 0 & 1 & 0 & 0 \\ -\dfrac{k}{J_2} & -\dfrac{d}{J_2} & \dfrac{k}{J_2} & \dfrac{d}{J_2} \\ 0 & 0 & 0 & 1 \\ \dfrac{k}{J_1} & \dfrac{d}{J_1} & -\dfrac{k}{J_1} & -\dfrac{d}{J_1} \end{bmatrix} \mathbf{x} + \begin{bmatrix} 0 \\ 0 \\ 0 \\ \dfrac{1}{J_1} \end{bmatrix} u,$$

$$y = [\,1 \quad 0 \quad 0 \quad 0\,]\mathbf{x}. \tag{7.3}$$

Physical analysis of the boom leads us to assume that the k and d parameters vary as a result of temperature fluctuations but are bounded by

$$0.09 \le k \le 0.4,$$

$$0.04\sqrt{\frac{k}{10}} \le d \le 0.2\sqrt{\frac{k}{10}}. \tag{7.4}$$

As a result, the vehicle resonance ω_n can vary between 1 and 2 rad/s, and the damping ratio ζ varies between 0.02 and 0.1. The approach to design when parameters are subject to variation is to select nominal ("worst-case") values for the parameters and to construct the design so that no possible value of the parameters can cause the specifications to be missed. In the present case, we select the nominal model to have $\omega_n = 1$ (the lowest value) and $\zeta = 0.02$ (also the lowest value). These correspond to the parameter values $k = 0.091$ and $d = 0.0036$, and with $J_1 = 1$ and $J_2 = 0.1$, the nominal equations are

$$\dot{\mathbf{x}} = \begin{bmatrix} 0 & 1 & 0 & 0 \\ -0.91 & -0.036 & 0.91 & 0.036 \\ 0 & 0 & 0 & 1 \\ 0.091 & 0.0036 & -0.091 & -0.0036 \end{bmatrix} \mathbf{x} + \begin{bmatrix} 0 \\ 0 \\ 0 \\ 1 \end{bmatrix} u,$$

$$y = [\,1 \quad 0 \quad 0 \quad 0\,]\mathbf{x}. \tag{7.5}$$

The corresponding transfer function is

$$G(s) = \frac{0.036(s + 25)}{s^2(s + 0.02 \pm j1)}. \tag{7.6}$$

When a trial design is completed the computer simulation will be run with the range of parameter values to ensure that the design has adequate robustness to tolerate these changes.

The dynamic performance specifications will be met if the closed-loop poles have an $\omega_n = 0.7$ rad/s and a closed-loop damping ratio of 0.5; these correspond to an open-loop crossover frequency of $\omega_c \approx 0.5$ rad/s and a phase margin of about $50°$.

FIGURE 7.3
Root locus for $KG(s)$.

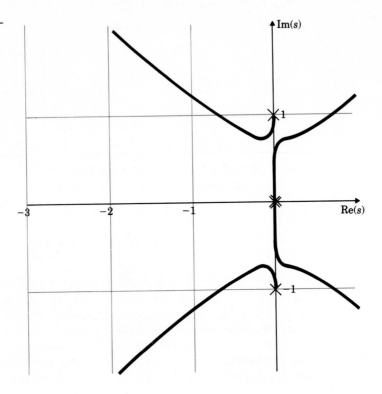

STEP 5: *Try a lead-lag or PID controller.* The proportional gain root locus for the nominal plant is drawn in Fig. 7.3, and the Bode plot is given in Fig. 7.4.

In order to illustrate some important aspect of compensation design, we will imagine that the designer first ignores the resonance and generates a design for the rigid body alone. Taking the process transfer function to be $1/s^2$, the feedback to be position plus derivative (star tracker plus rate gyro) with transfer function $D(s) = K(sT + 1)$, and the response objective to be $\omega_n = 0.7$ rad/s and $\zeta = 0.5$, a suitable controller is

$$D_1(s) = 0.5(1.4s + 1). \qquad (7.7)$$

The root locus is shown in Fig. 7.5 and the Bode plot in Fig. 7.6. From these plots it is seen that the low-frequency poles are reasonable but the system would be unstable because of the resonance.* One simple action at this point is to reduce our

---

*If we were to use the method described in Section 4.6.4 for nonlinear systems, we would expect the control to saturate and produce a limit cycle at about 1 rad/s, behavior that would rapidly deplete the control gas supply.

FIGURE 7.4
Open-loop frequency
response $KG(s)$ for
$K = 0.5$: (a) magnitude
and (b) phase.

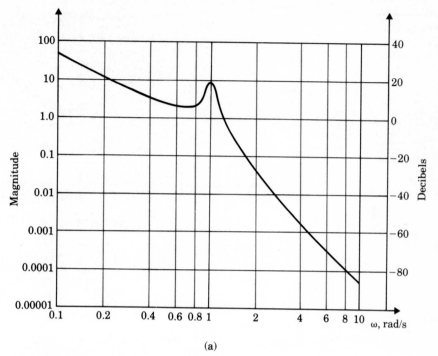

(a)

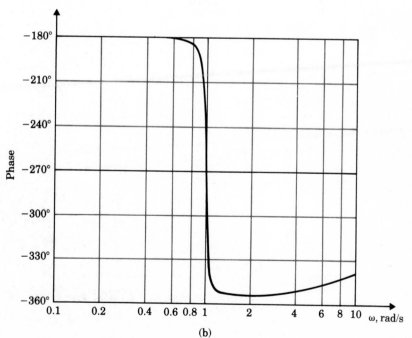

(b)

FIGURE 7.4
Open-loop frequency
response $KG(s)$ for
$K = 0.5$: (a) magnitude
and (b) phase.

FIGURE 7.5
Root locus for
$KD_1(s)G(s)$.

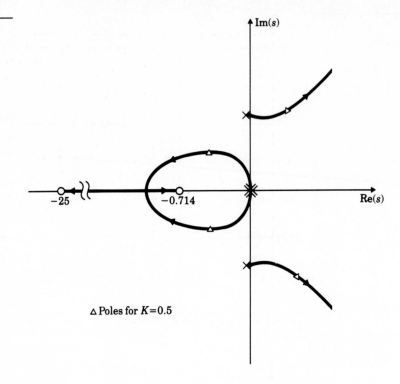

△ Poles for $K=0.5$

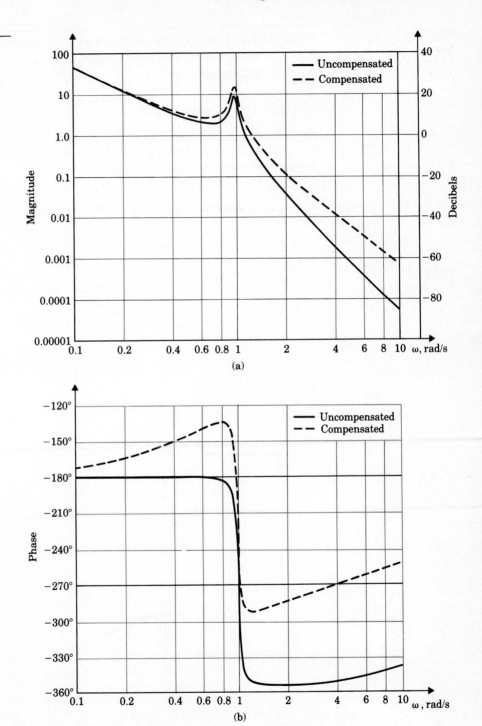

FIGURE 7.6
Frequency response
for $KD_1(s)G(s)$:
(a) magnitude and
(b) phase.

expectations with respect to bandwidth and slow the system down by lowering the gain until the system is stable. With so little damping, we must really go slowly. A bit of experimentation leads to

$$D_2(s) = 0.001(30s + 1), \qquad (7.8)$$

for which the root locus is drawn in Fig. 7.7 and the Bode plot in Fig. 7.8. From the Bode plot we see we have a gain margin of 50° with this design, but a crossover frequency of only $\omega = 0.04$ rad/s.

FIGURE 7.7
Root locus for
$KD_2(s)G(s)$.

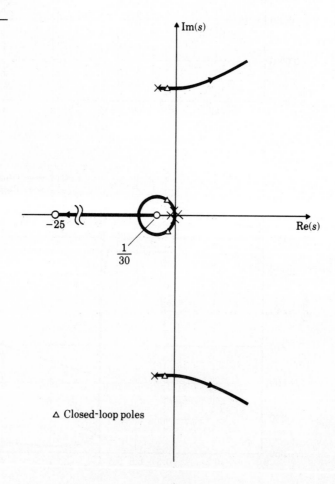

△ Closed-loop poles

FIGURE 7.8
Frequency response for $D_2(s)G(s)$: (a) magnitude and (b) phase.

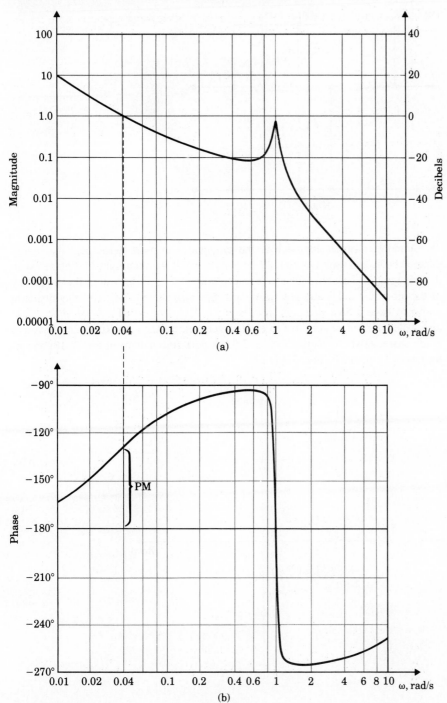

(a)

(b)

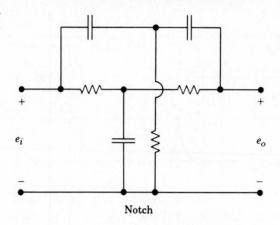

Notch

A more common approach to the problem is to place zeros near the lightly damped poles and use them to hold these poles back from the right-half plane. Such a compensation has a frequency response with very low gain near the frequency of the offending poles and has reasonable gain elsewhere. The frequency response seems to have a dent or notch in it, and the device is called a *notch filter*. (It is also called a *band reject filter* in electric network theory.) An RC circuit having a notch characteristic is shown in Fig. 7.9, its pole-zero pattern in Fig. 7.10, and its frequency response in Fig. 7.11.

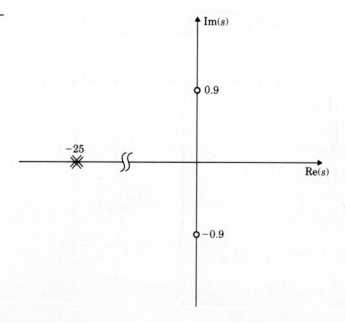

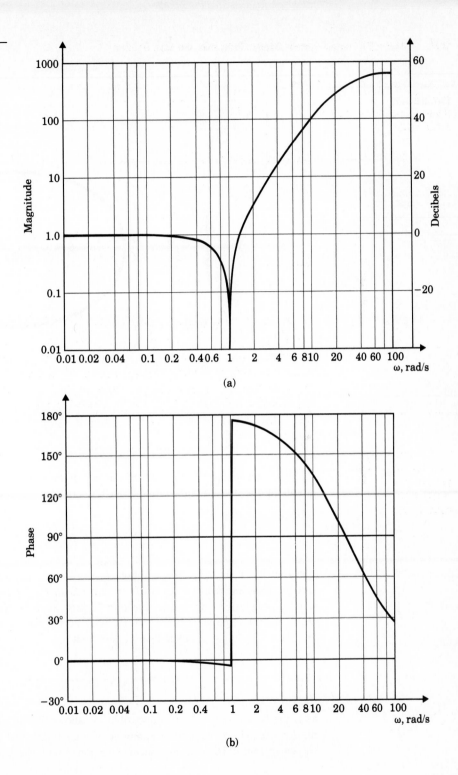

FIGURE 7.11
Frequency response
of a notch filter:
(a) magnitude and
(b) phase.

(a)

(b)

FIGURE 7.12
Root locus for
$KD_3(s)G(s)$.

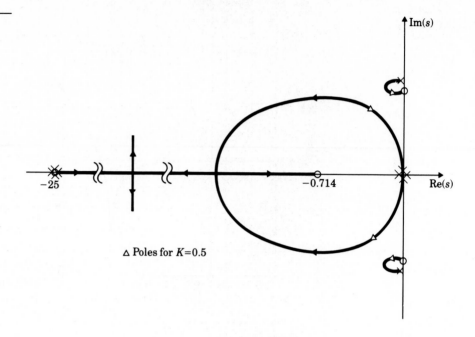

△ Poles for $K=0.5$

With this idea we begin with the compensation of Eq. (7.7) and add the notch, producing the revised compensator transfer function,

$$D_3(s) = 0.5(1.4s + 1)\frac{(s/0.9)^2 + 1}{[(s/25) + 1]^2}. \qquad (7.9)$$

The root locus for this case is shown in Fig. 7.12, the Bode plot in Fig. 7.13, and the transient recovery to $\theta_2(0) = 0.2$ rad in Fig. 7.14. This seems to be a satisfactory design and worth checking further for robustness; a more complete check will be done later, but a root locus using the compensator of Eq. (7.9) and the plant with $\omega_n = 2$ rather than 1, such that

$$\hat{G}(s) = \frac{(s/50) + 1}{s^2[(s^2/4) + 0.025 + 1]}, \qquad (7.10)$$

as if the boom were as stiff as possible, as shown in Fig. 7.15. Notice that now the low-frequency poles have damping of only 0.25. The frequency response is shown in Fig. 7.16, and the transient response with the varied parameter values

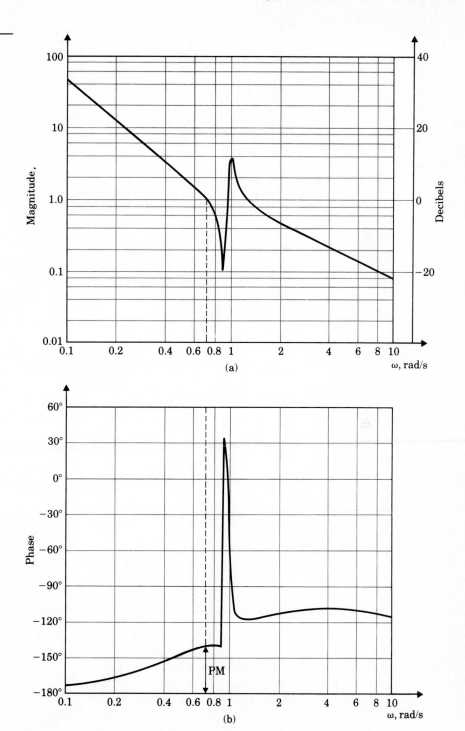

Frequency response
for $KD_3(s)G(s)$:
(a) magnitude and
(b) phase.

FIGURE 7.14
Transient recovery for $D_3(s)$
and $G(s)$: $\theta_2(0) = 0.2$ rad.

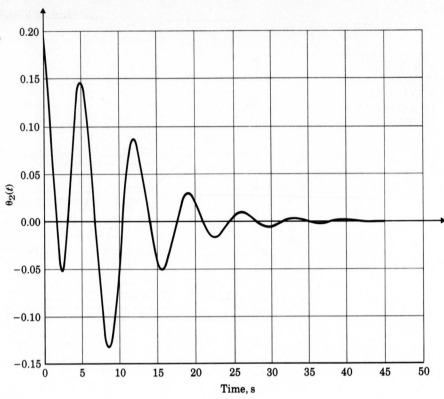

FIGURE 7.15
Root locus for $KD_3(s)\hat{G}(s)$.

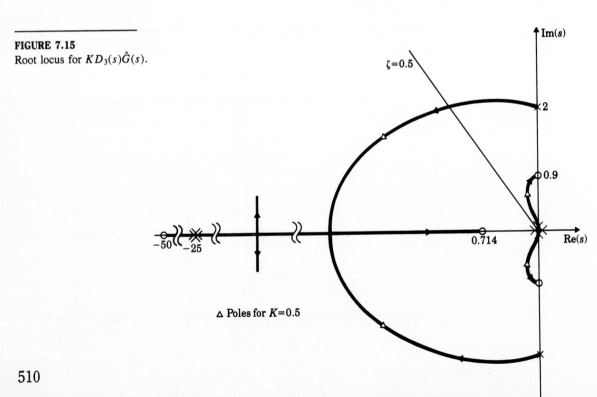

510

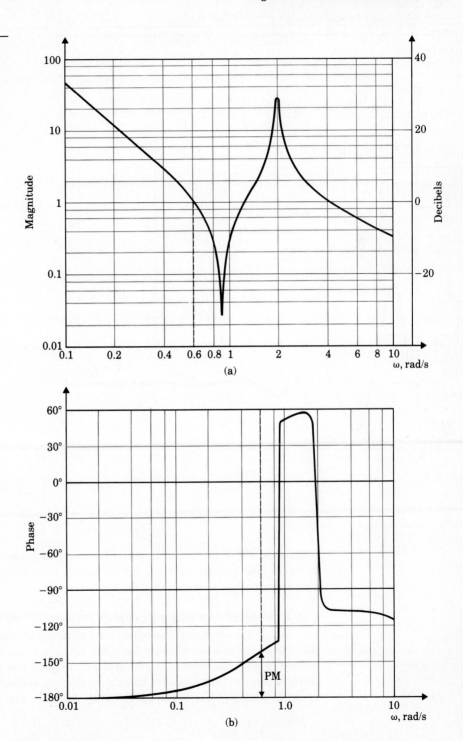

FIGURE 7.16
Frequency response
of $KD_3(s)\hat{G}(s)$:
(a) magnitude and
(b) phase.

FIGURE 7.17
Transient recovery for $D_3(s)$ and $\hat{G}(s)$: $\theta_2(0) = 0.2$ rad.

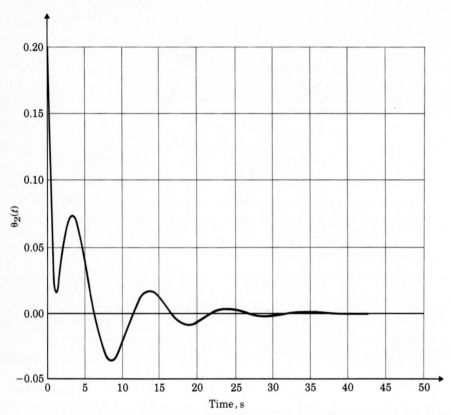

is drawn in Fig. 7.17. We could take a few more iterations of trial and error with the notch filter and rate feedback, but the system is complex enough that a look at state-space designs seems reasonable. We go to step 6.

STEP 6: *Try an optimal design using pole placement.* Using the state-variable formulation of the equations of motion, we can devise a controller that will place the closed-loop poles in arbitrary locations. Of course, used without thought, the method of pole placement can also result in a design that requires unreasonable levels of control effort or is very sensitive to changes in the plant transfer function. Guidelines for pole placement are given in Chapter 6; one approach that is often successful is to use optimal pole locations as found via the symmetric root locus. For the problem at hand, the symmetric root locus is drawn in Fig. 7.18. From this locus, to obtain a bandwidth around 0.7 rad/s we select closed-loop control poles at $-0.6 \pm j0.4$ and $-0.25 \pm j1.1$ and closed-loop estimator poles from the same locus at $-0.84 \pm j2.2$ and $-2.04 \pm j0.92$. These are not selected to be very fast in recognition of the noise in the position measurement.

If we select $\alpha_c(s)$, as discussed previously, from the symmetric root locus, the control gain is

$$\mathbf{K}_c = [-0.595 \quad 0.275 \quad 1.32 \quad 1.66], \tag{7.11}$$

FIGURE 7.18
Symmetric root locus for
the satellite.

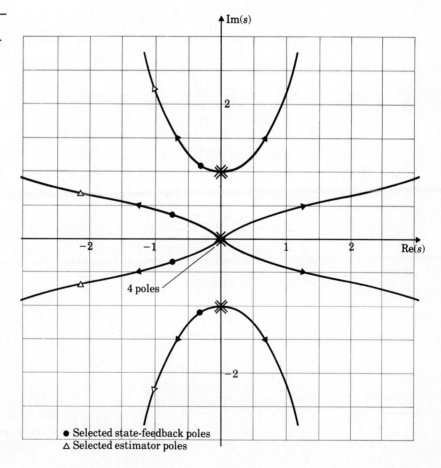

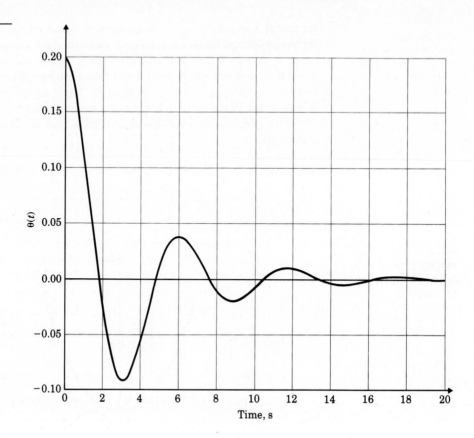

FIGURE 7.19
Transient recovery:
$\theta(0) = 0.2$ rad.

and the transient recovery for $\theta(0) = 0.2$ rad is shown in Fig. 7.19. The estimator characteristic polynomial, taken from the symmetric root locus, leads to an estimator (filter) gain of

$$\mathbf{L} = \begin{bmatrix} 5.73 \\ 16.42 \\ 33.22 \\ 29.98 \end{bmatrix}. \tag{7.12}$$

If we combine the control and estimation as presented in Chapter 6, the compensator transfer function that results is

$$D_4(s) = \frac{-94.8(s + 0.27)(s + 0.0543 \pm j0.894)}{(s + 0.676 \pm j2.88)(s + 3.04 \pm j1.53)} \tag{7.13}$$

The frequency response of this compensator is shown in Fig. 7.20. Notice that pole

FIGURE 7.20
Frequency response of
optimal compensator
$D_4(s)$: (a) magnitude and
(b) phase.

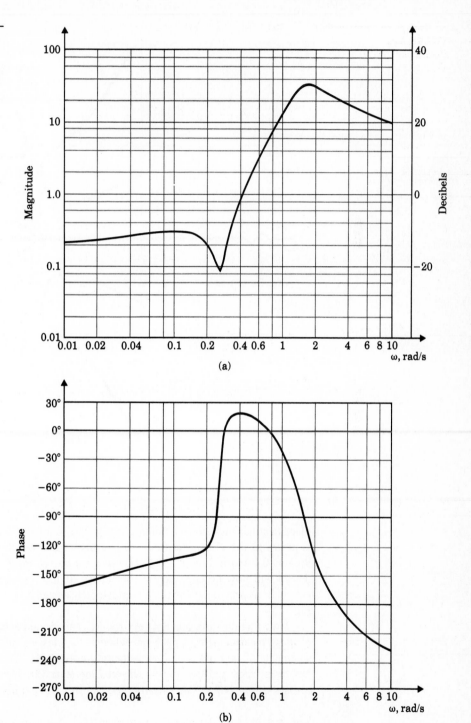

(a)

(b)

FIGURE 7.21
Root locus for
$D_4(s)G(s)$.

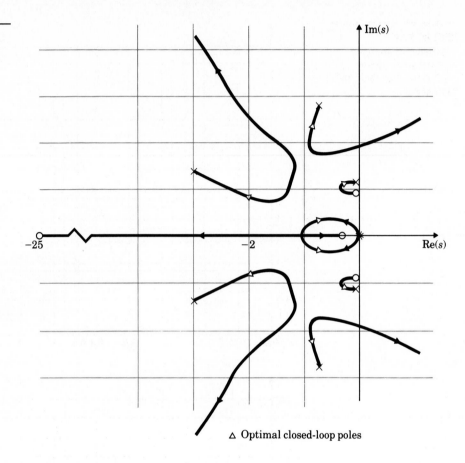

△ Optimal closed-loop poles

placement has introduced a notch directly. The root locus for this combined system is drawn in Fig. 7.21, the Bode plot in Fig. 7.22, and the recovery transient for $\theta(0) = 0.2$ rad in Fig. 7.23. Notice that the design meets the specification. This design will now be checked for robustness. If the resonant modes are shifted to $\omega_n = 2$, as was done for the PD-plus-notch design, we obtain the root locus of Fig. 7.24.

Note that the design is not very robust. This design is based on the assumption of noisy measurement of position, whereas the classical design assumes that measurements of both position and velocity are available. The classical compensator has three zeros and two poles, whereas the state-variable compensator has three zeros and four poles. The extra two poles of the state-variable design make it less

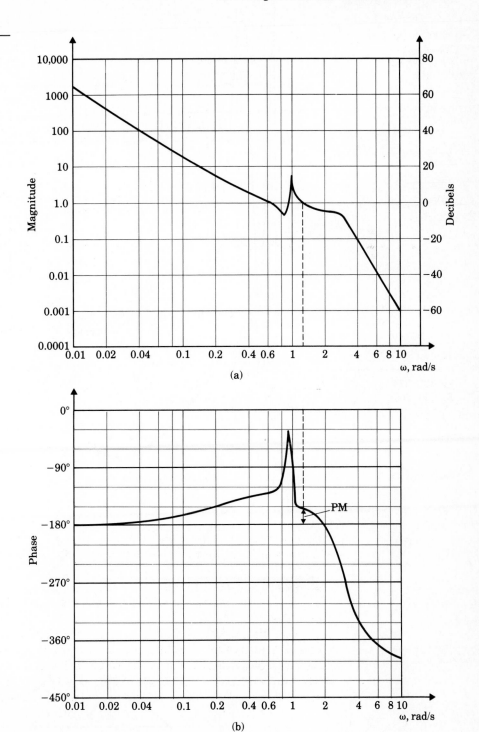

FIGURE 7.22
Frequency response of
compensated system
$D_4(s)G(s)$:
(a) magnitude and
(b) phase.

(a)

(b)

FIGURE 7.23
Transient recovery with
estimator: $\theta(0) = 0.2$ rad.

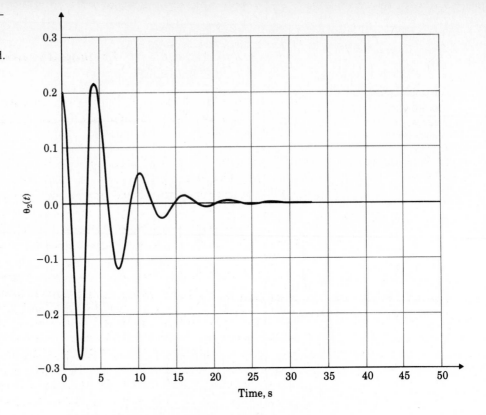

FIGURE 7.24
Root locus for
$D_4(s)\hat{G}(s)$.

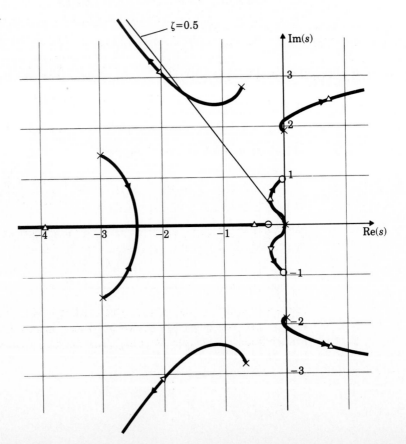

robust (see root locus of Fig. 7.24). We would expect that comparable results could be obtained if a reduced estimator is designed assuming a measurement of velocity as well as position.

STEP 7: *Simulate the design and compare the alternatives.* At this point we have two apparently satisfactory designs, with differing complexities and different robustness properties. We reconsider the original assumptions in the light of our trial designs. Both design methods have been strongly influenced by the presence of the lightly damped resonant mode caused by the coupled masses. However, as we saw in Chapter 2, the transfer function is a strong function of the fact that the actuator is on one body and the sensor is on the other (not collocated). Suppose we can satisfy the purposes of the mission by locating the sensor on the *same* mass that holds the actuator—that is, with a *collocated* actuator and sensor. If the actuators and sensors are collocated, the situation is drastically different! The system has zeros close to the flexible modes, and all the vibration modes can be stabilized by using negative *rate feedback*. This is easily seen from the root locus diagram typical of such systems shown in Fig. 7.25. Since control of position is typically desired, then direct rate feedback is not sufficient because position is unobtainable with a rate sensor. Position feedback with a lead compensator or combined position and rate feedback can be used to stabilize all the vibration and rigid-body modes. Consider the transfer function of the satellite with collocated actuator and sensor (to measure θ_1) for which the state matrices are

$$\mathbf{F} = \begin{bmatrix} 0 & 1 & 0 & 0 \\ -0.91 & -0.036 & 0.91 & 0.036 \\ 0 & 0 & 0 & 1 \\ 0.091 & 0.0036 & -0.091 & -0.0036 \end{bmatrix}, \quad \mathbf{G} = \begin{bmatrix} 0 \\ 0 \\ 0 \\ 1 \end{bmatrix},$$

$$\mathbf{H} = [0 \quad 0 \quad 1 \quad 0]. \tag{7.14}$$

The transfer function of the system is

$$G_{co}(s) = H(sI - F)^{-1}G = \frac{(s + 0.018 \pm j0.954)}{s^2(s + 0.02 \pm j1)}. \tag{7.15}$$

Notice the presence of a pair of zeros in the vicinity of the complex conjugate poles. If we now use just position-plus-rate feedback,

$$D_5(s) = 0.5(1.4s + 1), \tag{7.16}$$

then the system will not only be stabilized, but will have satisfactory response, because the resonant poles tend to cancel the complex conjugate zeros. Note that a notch filter is no longer necessary since the system provides its own notch. The

FIGURE 7.25
Simple control logic with collocated actuator and sensor: (a) rate feedback and (b) position feedback with lead compensator.

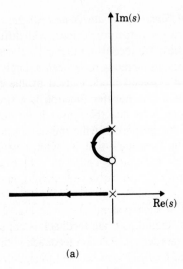

(a)

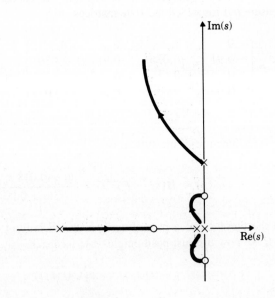

(b)

FIGURE 7.26
Root locus for $D_5(s)G_{co}(s)$.

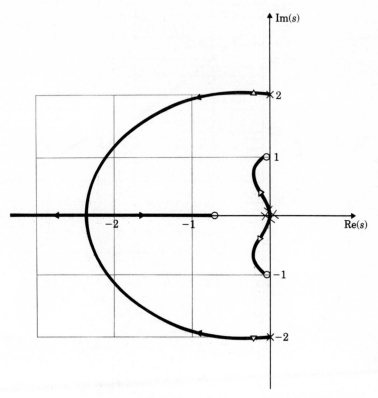

root locus of the system is shown in Fig. 7.26, and the frequency response is seen in Fig. 7.27. The result is a very simple, robust design achieved by moving the sensor from a noncollocated position to one collocated with the actuator. The result illustrates the very important point that, in achieving good feedback control, considerations of sensor location as well as technology and other features of the physical problem are typically more important than compensator transfer-function design.

Figure 7.28 shows the stability robustness measure (Eq. 5.67) for the satellite attitude control where the unstructured (multiplicative) bound on model error is described by

$$l_0(s) = \frac{10s^2}{s^2 + 8s + 64},$$

FIGURE 7.27
Frequency response
for $D_5(s)G_{co}(s)$:
(a) magnitude and
(b) phase.

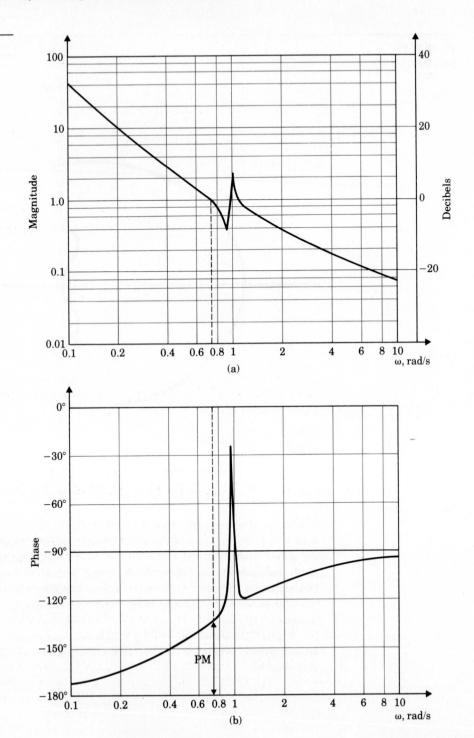

FIGURE 7.27
Frequency response
for $D_5(s)G_{co}(s)$:
(a) magnitude and
(b) phase.

FIGURE 7.28
Stability robustness
measure for satellite.

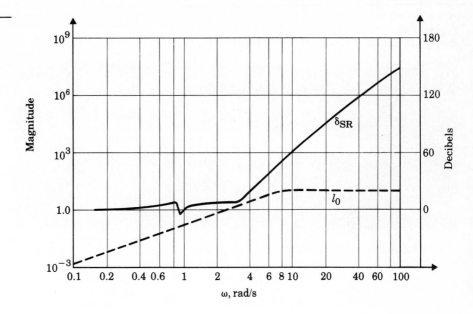

and the compensator is given by Eq. (7.16). Since the stability robustness measure remains above the bound on model error at all frequencies, the system is guaranteed to be stable for all models in the defined class.

7.3 Lateral and Longitudinal Control of a Boeing 747

The Boeing 747 is a large, wide-body jet transport airplane. A picture of a Boeing 747 is shown in Fig. 7.29, and a schematic with the relevant coordinates is shown in Fig. 7.30. The linearized equations of (rigid-body) motion for the Boeing 747 are of eighth-order but are separated into two fourth-order sets representing the longitudinal (u, w, q in Fig. 7.30) and lateral motion (β, r, p in Fig. 7.30). The longitudinal motion consists of axial, vertical, and pitching motion (see Example 6.4), while the lateral motion consists of rolling, yawing, and lateral movement. The elevator control surfaces and the throttle affect the longitudinal motion, whereas the aileron and rudder primarily affect lateral-directional motion. For a straight and wings-level reference condition, the coupling of lateral motion into longitudinal motion can usually be ignored and the equations of motion are treated as two decoupled fourth-order sets for designing the control, or *stability augmentation*,

FIGURE 7.29
Boeing 747 aircraft. *(Courtesy Boeing Commercial Airplane Co.)*

FIGURE 7.30
Definition of aircraft coordinates.

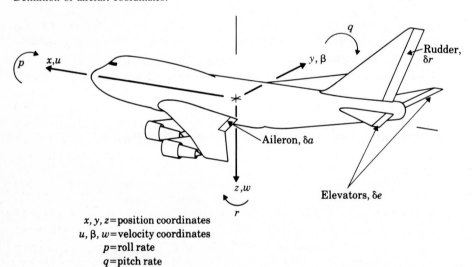

x, y, z = position coordinates
u, β, w = velocity coordinates
p = roll rate
q = pitch rate
r = yaw rate

for the aircraft. We will discuss the design of a stability-augmentation system for the lateral dynamics, called a *yaw damper,* and certain aspects of the autopilot having influence over the longitudinal behavior.

7.3.1 Yaw Damper

STEP 1: *Understand the process and its performance specifications.* Swept-wing aircraft have a natural tendency to be lightly damped in one of the lateral modes of motion. At typical commercial-aircraft cruising speeds and altitudes, this mode is sufficiently difficult for pilots to control that virtually every swept-wing aircraft has a feedback system to help the pilot. The specification of our control system is, therefore, to modify the natural dynamics so that the plane is easier for the pilot to fly. Studies have shown that pilots like natural frequencies $\lesssim 1$ rad/s with a damping ratio of $\zeta \gtrsim 0.5$. Aircraft with dynamics that violate these guidelines are generally considered fatiguing to fly and highly undesirable. Our system specifications, therefore, are to achieve lateral dynamics that meet these root constraints.

STEP 2: *Select a sensor.* The easiest measurement to take of aircraft motion is that of angular rate. Velocities can also be measured using pitot tubes and wind-vane devices, but these are noisier and less reliable for stabilization. Two angular rates— roll and yaw—partake in the lateral motion. Study of the lightly damped lateral mode indicates that it is primarily a yawing phenomenon; thus measurement of the yaw rate is a logical starting point for the design. Up until the early 1980s, the measurement had been made with a gyroscope, essentially a small, fast-spinning rotor that can yield an electric output proportional to the angular yaw rate of the aircraft. Since the early 1980s, most new aircraft systems are relying on a laser device (called a *ring-laser gyroscope*) for the measurement. Here, two laser beams traverse a closed path (often a triangle) in opposite directions. As the triangular device rotates, the apparent frequency of the two beams shifts, and this frequency shift is measured, producing a measure of rotational rate. These devices have fewer moving parts and are more reliable at less cost than the spinning-rotor variety of gyroscope.

STEP 3: *Select an actuator.* Two aerodynamic surfaces typically influence the lateral aircraft motion: the rudder and the ailerons (see Fig. 7.30). The lightly damped yaw mode that is being stabilized by the yaw damper is most affected by the rudder. Therefore, use of that single control input is a logical starting point for the design. Hydraulic devices are universally employed in large aircraft to provide the force that moves the aerodynamic surfaces. No other kind of device has been developed to provide the combination of high force, high speed, and light weight that is desired for the actuation of the controlling aerodynamic surfaces. On the other hand, low-speed flaps, which are frequently used to augment lift during take-off and landing, are typically actuated by an electric motor with a worm gear.

FIGURE 7.31
Root locus for yaw damper with direct feedback.

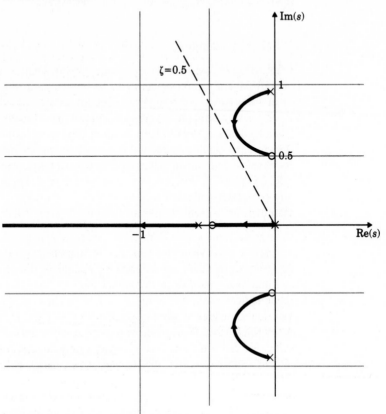

STEP 4: *Make a linear model.* The lateral perturbation equations (Heffley and Jewell, 1972) of motion in horizontal flight at 40,000 ft and forward speed $U = 774$ ft/s (Mach 0.8) are*

$$
\begin{bmatrix} \dot{\beta} \\ \dot{r} \\ \dot{p} \\ \dot{\phi} \end{bmatrix} = \begin{bmatrix} -0.0558 & -0.9968 & 0.0802 & 0.0415 \\ 0.598 & -0.115 & -0.0318 & 0 \\ -3.05 & 0.388 & -0.4650 & 0 \\ 0 & 0.0805 & 1 & 0 \end{bmatrix} \begin{bmatrix} \beta \\ r \\ p \\ \phi \end{bmatrix} + \begin{bmatrix} 0.00729 \\ -0.475 \\ 0.153 \\ 0 \end{bmatrix} \delta r
$$

$$
y = \begin{bmatrix} 0 & 1 & 0 & 0 \end{bmatrix} \begin{bmatrix} \beta \\ r \\ p \\ \phi \end{bmatrix}. \tag{7.17}
$$

The transfer function is

$$
G(s) = \frac{r(s)}{\delta r(s)} = \frac{-4.75(s + 0.498)(s + 0.012 \pm j0.488)}{(s + 0.0073)(s + 0.563)(s + 0.033 \pm j0.947)} \tag{7.18}
$$

so that the system has two stable real poles, and a pair of stable complex poles referred to as the "Dutch roll," the name describing the motions of a person skating on frozen canals in Holland. The stable real poles are referred to as the *spiral mode* ($s_1 = -0.0073$) and the *roll mode* ($s_2 = -0.563$). From looking at the natural roots, we see that the offending mode that needs repair for good pilot handling is the Dutch roll, or the roots at $s = -0.033 \pm j0.95$. The roots have an acceptable frequency, but their damping of $\zeta \cong 0.03$ is far short of the desired $\zeta \cong 0.5$.

STEP 5: *Try a lead-lag PID design.* As a first crack at the design, proportional feedback of the yaw rate to the rudder will be considered. The root locus versus the gain of this feedback is shown in Fig. 7.31, and its frequency response is shown in Fig. 7.32. The figures show that a damping of $\zeta \cong 0.45$ is achievable.

In practice, however, it is found that this simple feedback creates an objectionable quality during a steady turn, since it provides a steady rudder input opposite the pilot's input due to the steady yaw rate. Control engineers know from experience that this dilemma can be solved by the addition of a *washout circuit* in the feedback. The washout circuit has the feature that it passes only transient inputs

---

*The unit for β and ϕ are radians, and for r and p the units are radians per second.

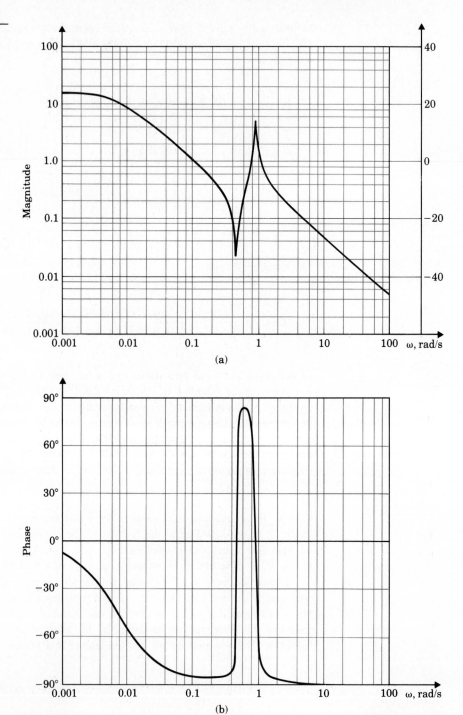

FIGURE 7.33
Yaw damper: (a) block
diagram and (b) block
diagram for analysis.

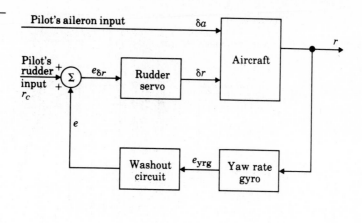

(a)

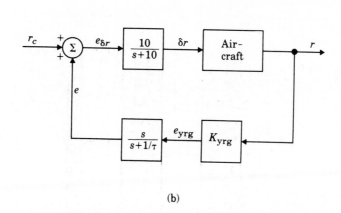

(b)

and "washes out" steady and low-frequency inputs. Figure 7.33 shows a block diagram of the yaw damper with the washout included.

The transfer function of the washout circuit is

$$H(s) = \frac{e(s)}{e_{yrg}(s)} = \frac{s}{s + (1/\tau)}.$$

The frequency response of the washout circuit is shown in Fig. 7.34(a and b) for $\tau = 3$ s, and its response to a unit step input is shown in Fig. 7.34(c). The root locus of the system with washout circuit is shown in Fig. 7.35. As seen from the root locus, with the addition of the washout very little increase in the damping

FIGURE 7.34
Frequency response of the washout circuit: (a) magnitude, (b) phase, and (c) step response.

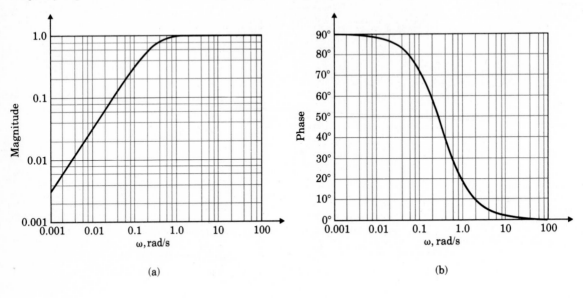

(a)   (b)

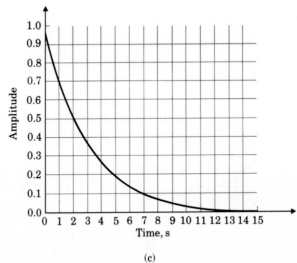

(c)

FIGURE 7.35
Root locus with washout circuit, $\tau = 3$.

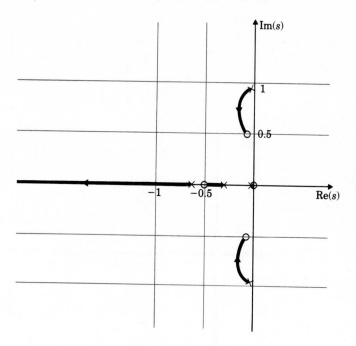

ratio can be obtained. The rudder servo represents the actuator dynamics

$$\frac{\delta r(s)}{e_{\delta r}(s)} = \frac{10}{s + 10},$$

which is fast compared with the dynamics of the system. The root locus of the system with washout and actuator dynamics is shown in Fig. 7.36. The associated frequency responses of the system are shown in Fig. 7.37. The response of the closed-loop system to an initial condition of $\beta_0 = 1°$ is shown in Fig. 7.38 for a root-locus gain of unity. The root locus indicates that the maximum achievable damping with this feedback scheme is approximately $\zeta = 0.3$. Although this a considerable improvement over the original aircraft, it is not as good as originally desired.

FIGURE 7.36
Root locus with washout circuit, $\tau = 3$, and actuator.

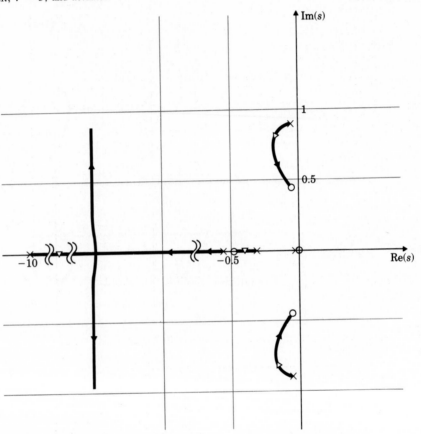

FIGURE 7.37
Frequency response of yaw damper, including washout and actuator: (a) magnitude and (b) phase.

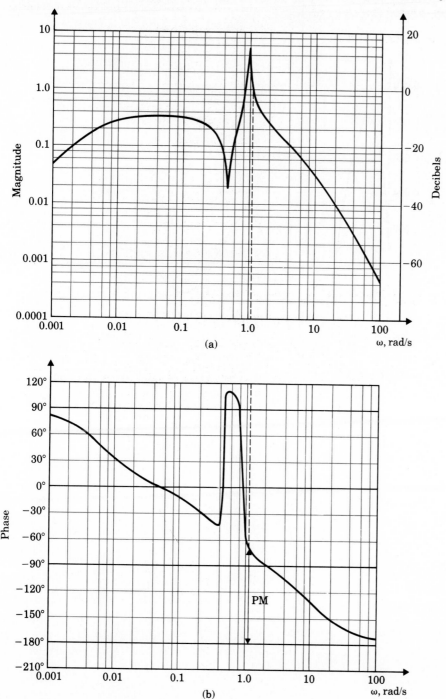

(a)

(b)

FIGURE 7.38
Initial condition response for $\beta_0 = 1°$.

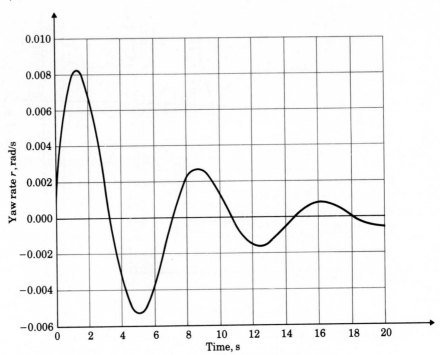

STEP 6: *Try an optimal design using pole placement.* If we augment the dynamic model of the system by adding the actuator and washout, we obtain the state-variable model:

$$
\begin{bmatrix} \dot{x}_A \\ \dot{\beta} \\ \dot{r} \\ \dot{p} \\ \dot{\phi} \\ \dot{x}_{\omega_0} \end{bmatrix}
=
\left[
\begin{array}{c:cccc:c}
-10 & 0 & 0 & 0 & 0 & 0 \\ \hdashline
0.00729 & -0.0558 & -0.997 & 0.0802 & 0.0415 & 0 \\
-0.475 & 0.598 & -0.1150 & -0.0318 & 0 & 0 \\
0.153 & -3.05 & 0.388 & -0.465 & 0 & 0 \\
0 & 0 & 0.0805 & 1 & 0 & 0 \\ \hdashline
0 & 0 & 1 & 0 & 0 & -0.333
\end{array}
\right]
$$

$$
\cdot
\begin{bmatrix} x_A \\ \hdashline \beta \\ r \\ p \\ \phi \\ \hdashline x_{\omega_0} \end{bmatrix}
+
\begin{bmatrix} 10 \\ \hdashline 0 \\ 0 \\ 0 \\ 0 \\ \hdashline 0 \end{bmatrix} e_{\delta r},
$$

$$
e = [\,0 \mid 0 \;\; 1 \;\; 0 \;\; 0 \mid -0.333\,]
\begin{bmatrix} x_A \\ \hdashline \beta \\ r \\ p \\ \phi \\ \hdashline x_{\omega_0} \end{bmatrix},
$$

where $e_{\delta r}$ is the input to the actuator and e is the output of the washout. The symmetric root locus for the augmented system is as shown in Fig. 7.39. If we select the state feedback poles from the symmetric root locus so that the complex roots have maximum damping ($\zeta = 0.4$), we find

$$
\alpha_c(s) = (s + 0.0051)(s + 0.468)(s + 0.279 \pm j0.628)(s + 1.106)(s + 9.89),
$$

then the state feedback gain is computed to be

$$
\mathbf{K} = [\,0.1059 \mid -0.191 \;\; -2.32 \;\; 0.0992 \;\; 0.0370 \mid 0.486\,].
$$

FIGURE 7.39
Symmetric root locus of lateral dynamics, including washout filter and actuator.

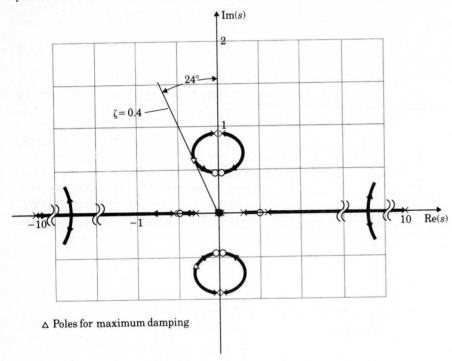

△ Poles for maximum damping

Note that the third entry in $\mathbf{K}$ is larger than the others so that the feedback of all four states is essentially the same as proportional feedback of r. This is also evident from the similarity of the root locus in Fig. 7.31 and the SRL of Fig. 7.39. If we select the estimator poles to be five times faster than the controller poles, that is,

$$\alpha_e(s) = (s + 0.0253)(s + 2.34)(s + 1.39 \pm j3.14)(s + 5.53)(s + 49.5),$$

then the estimator gain is found to be

$$\mathbf{L} = \begin{bmatrix} 250 \\ -2044 \\ -5158 \\ -24843 \\ -40113 \\ -15624 \end{bmatrix}.$$

The compensator transfer function from Eq. (6.113) is

$$D_c(s) = \frac{-844(s + 10.0)(s - 1.04)(s + 0.974 \pm j0.559)(s + 0.0230)}{(s + 0.0272)(s + 0.837 \pm j0.671)(s + 4.07 \pm j10.1)(s + 51.3)}.$$

$$(7.19)$$

Figure 7.40 shows the response of the system to an initial condition of $\beta_0 = 1°$. It is clear from the root locus that the damping can be improved by the SRL approach, and this is borne out by the reduced oscillatory behavior in the transient response of the system. However, this improvement has come at a considerable price. Note that the order of the compensator has increased from 1 in the original design (Fig. 7.33) to 6 in the design obtained using the controller/estimator/SRL approach.

Aircraft yaw dampers in use today generally employ a proportional feedback of yaw rate to rudder through a washout or through minor modifications to this design. The improved performance achievable with an optimal design approach utilizing full-state feedback and estimation is not judged worth the increase in complexity.

FIGURE 7.40
Initial condition response for $\beta_0 = 1°$, $\hat{x}_0 = 0$.

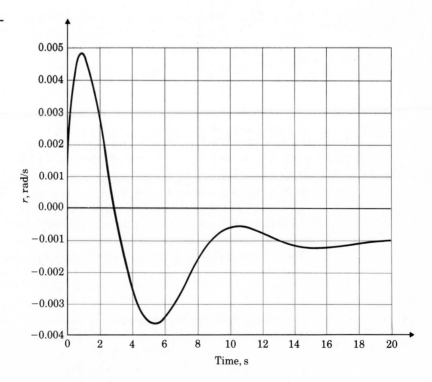

Perhaps a more fruitful approach to improvement of the design would be to add the aileron surface as a control variable along with the rudder. However, this approach, too, has not been considered worth its complexity.

STEPS 7 AND 8: *Verify the design.* Linear models of aircraft motion are reasonably accurate as long as the motion remains close to the reference condition and the actuators do not saturate. But since actuators are sized for safety so that they handle large transients, their saturation is very rare. The linear analysis-based design is, therefore, reasonably accurate, and we will not pursue a nonlinear simulation or further design verification.

7.3.2 Altitude-Hold Autopilot

STEP 1: *Understand the process and its performance specifications.* One of the pilot's many tasks is to hold a specific altitude. As an aid to keeping aircraft from colliding, those craft on an easterly path are required to be on an odd multiple of 1000 ft, while those on a westerly path are required to be on an even multiple of 1000 ft. It is therefore of some concern to the pilot that the altitude be held to within a few hundred feet. A well-trained attentive pilot can easily accomplish this task to within ±50 ft, and this kind of tolerance is what the air-traffic controllers expect pilots to maintain. Since this task requires the pilot to be fairly diligent, sophisticated aircraft often have an "altitude-hold autopilot" to perform the task. This system is fundamentally different from the yaw damper in that its role is to replace the pilot for certain periods of time, while the yaw damper's role is to help the pilot fly. Dynamic specifications, therefore, need not be such that pilots like the craft's "feel"; instead, the design should provide the kind of ride that pilots and passengers like. The damping ratio should still be in the vicinity of $\zeta \cong 0.5$, but, for a smooth ride, the natural frequency should be much slower than 1 rad/s.

STEP 2: *Select a sensor.* Clearly needed is a device to measure altitude, a task most easily done by measuring the atmospheric pressure. Almost from the time of the first Wright brothers' flight, this basic idea has been used in a device called a *barometric altimeter*. Before autopilots, the device consisted of a bellows whose free end was connected to a needle that directly indicated altitude on a dial. The same idea is used today, but the pressure is sensed electrically.

Because altitude control is essentially four integrations from the controlling elevator input, stabilization of the feedback loop cannot be accomplished by simple proportional feedback. Therefore pitch rate q is also used as a stabilizing feedback; it is measured by a gyroscope or ring-laser gyro identical to that used for yaw-rate measurement. Further stabilization from pitch-angle feedback is also helpful. It is obtained from either a ring-laser gyro inertial reference system or a "rate-integrating gyro," a device similar to the rate gyro, but structured differently so that its outputs are proportional to the angle of the aircraft's pitch (θ) and roll angles (ϕ).

STEP 3: *Select an actuator.* The aerodynamic surface typically used for pitch control on most aircraft is the elevator, δe. It is located on the horizontal tail, well removed from the aircraft's center of gravity so that its force produces an angular pitch rate and thus angle, which act to change the lift from the wing. In some high-performance aircraft, there are "direct-lift" control devices on the wing or perhaps small *canard* surfaces forward of the wing, which produce vertical forces on the aircraft much faster than elevators are able to. However, for purposes of our altitude hold, we will only consider the typical case of an elevator surface on the tail.

As for the rudder, hydraulic actuators are the preferred device to move the elevator surface.

STEP 4: *Make a linear model.* The longitudinal perturbation equations of motion for the Boeing 747 in horizontal flight, $U = 830$ ft/s at 20,000 ft (Mach 0.8) with a weight of 637,000 lb, are

$$
\begin{bmatrix} \dot{u} \\ \dot{w} \\ \dot{q} \\ \dot{\theta} \\ \dot{h} \end{bmatrix} = \begin{bmatrix} -0.00643 & 0.0263 & 0 & -32.2 & 0 \\ -0.0941 & -0.624 & 820 & 0 & 0 \\ -0.000222 & -0.00153 & -0.668 & 0 & 0 \\ 0 & 0 & 1 & 0 & 0 \\ 0 & -1 & 0 & 830 & 0 \end{bmatrix} \begin{bmatrix} u \\ w \\ q \\ \theta \\ h \end{bmatrix} + \begin{bmatrix} 0 \\ -32.7 \\ -2.08 \\ 0 \\ 0 \end{bmatrix} \delta e
$$

(7.20)

where the desired output for an altitude-hold autopilot is

$$
h = \begin{bmatrix} 0 & 0 & 0 & 0 & 1 \end{bmatrix} \begin{bmatrix} u \\ w \\ q \\ \theta \\ h \end{bmatrix}.
$$

(7.21)

The system has two pairs of stable complex poles and a root at $s = 0$. The complex pair at $-0.003 \pm j0.0098$ are referred to as the *phugoid mode** and the poles $-0.6463 \pm j1.1211$ are the *short-period modes.*

---

*The name was adopted by F. W. Lanchester (1908), who was the first to study the dynamic stability of aircraft analytically. It is apparently an incorrect version of a Greek word.

STEP 5: *Try a lead-lag or PID controller.* As a first step in the design, it is typically helpful to use an *inner-loop* feedback of q to δe so as to improve the damping of the short-period mode of the aircraft (see Fig. 7.41). The transfer function from δe to q is

$$\frac{q(s)}{\delta e(s)} = \frac{-2.08s(s + 0.0105)(s + 0.596)}{(s + 0.003 \pm j0.0098)(s + 0.646 \pm j1.21)} \tag{7.22}$$

The root locus for q feedback using Eq. (7.22) is as shown in Fig. 7.42. Since k_q is the root-locus parameter, the system matrix (Eq. 7.20) is now modified as follows:

$$\mathbf{F}_q = \mathbf{F} + k_q \mathbf{G} \mathbf{H}_q, \tag{7.23}$$

where $\mathbf{F}$ and $\mathbf{G}$ are defined in Eq. (7.20) and $\mathbf{H}_q = [0 \quad 0 \quad 1 \quad 0 \quad 0]$. One may pick a suitable gain k_q from the root locus. The selection procedure is the same one discussed in Chapter 4. (Recall the continuation locus and tachometer feedback example in Section 4.6.) If we choose $k_q = 1$, then the closed-loop poles are located at $-0.0039 \pm j0.0067, -1.683 \pm j0.277$ on the root locus and

$$\mathbf{F}_q = \begin{bmatrix} -0.00643 & 0.0263 & 0 & -32.2 & 0 \\ -0.0941 & -0.624 & 787 & 0 & 0 \\ -0.000222 & -0.00153 & -2.75 & 0 & 0 \\ 0 & 0 & 1 & 0 & 0 \\ 0 & -1 & 0 & 830 & 0 \end{bmatrix}.$$

Note that only the third column of $\mathbf{F}_q$ is different from $\mathbf{F}$. To add further capability in the autopilot, the pitch angle of the aircraft is typically also controlled. This is

FIGURE 7.41
Altitude-hold feedback system.

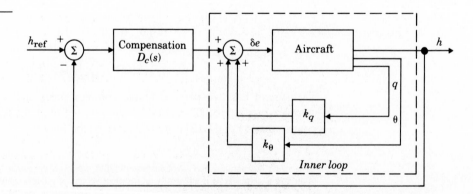

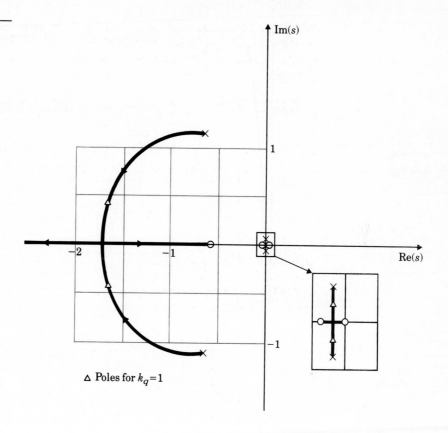

$\triangle$ Poles for $k_q=1$

achieved by feeding back also the pitch angle (θ) to elevator control. If we select*

$$\mathbf{K}_{\theta q} = [0 \quad 0 \quad -0.8 \quad -6 \quad 0]$$

in order to feedback θ and q, then the system matrix is

$$\mathbf{F}_{\theta q} = \mathbf{F}_q - \mathbf{G}\mathbf{K}_{\theta q} = \begin{bmatrix} -0.0064 & 0.0263 & 0 & -32.2 & 0 \\ -0.0941 & -0.624 & 761 & -196.2 & 0 \\ -0.002 & -0.0015 & -4.41 & -12.48 & 0 \\ 0 & 0 & 1 & 0 & 0 \\ 0 & -1 & 0 & 830 & 0 \end{bmatrix}, \quad (7.24)$$

with poles at $s = 0$, $-2.25 \pm j2.99$, -0.531, -0.0105. So far, the *inner loop* of the aircraft has been stabilized significantly. The uncontrolled aircraft has a

---

*A design based on SRL would suggest feeding back θ and q at this point.

FIGURE 7.43
Response of an attitude-
hold autopilot to a
command in θ.

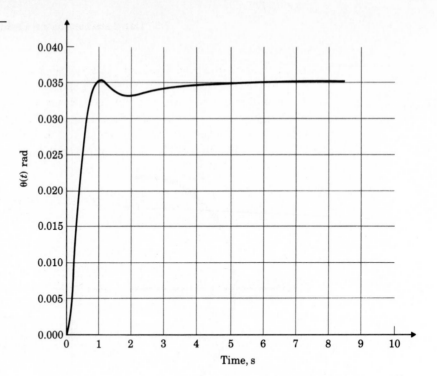

FIGURE 7.44
0° root locus with h
feedback only.

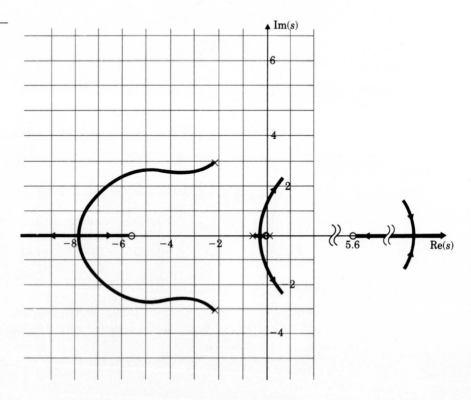

natural tendency to return to level attitude as evidenced by the phugoid roots at $s = -0.003 \pm j0.0098$. This inner stabilization is necessary to enable an outer h and $\dot{h}$ feedback to be successful; furthermore, the θ and q feedbacks can be used by themselves in an attitude-hold mode of an autopilot, in which a pilot wishes to control θ directly through input command. Figure 7.43 shows the response of an attitude-hold autopilot to a $2°$ (0.035 rad) step command in θ. The transfer function of the system is now

$$\frac{h(s)}{\delta e(s)} = \frac{32.7(s + 0.0045)(s + 5.64)(s - 5.61)}{s(s + 2.25 \pm j2.99)(s + 0.0105)(s + 0.0531)} \qquad (7.25)$$

A sketch of the root locus (Fig. 7.44) shows that feedback of altitude by itself does not yield an acceptable design. However, we may also feed back the altitude rate for stabilization. The root locus of the system with h and $\dot{h}$ feedback is shown in Fig. 7.45. The ratio of $\dot{h}$ to h found to be best after some iteration

FIGURE 7.45
$0°$ root locus with h and $\dot{h}$ feedback versus K_h.

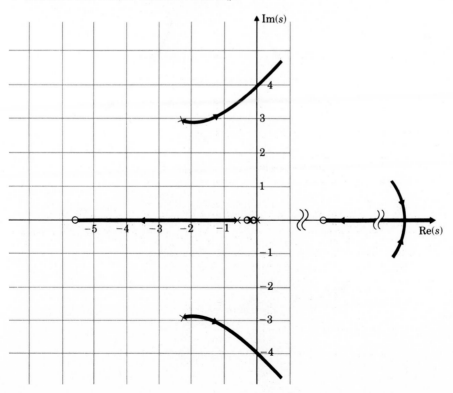

FIGURE 7.46
Symmetric root locus for altitude-hold design.

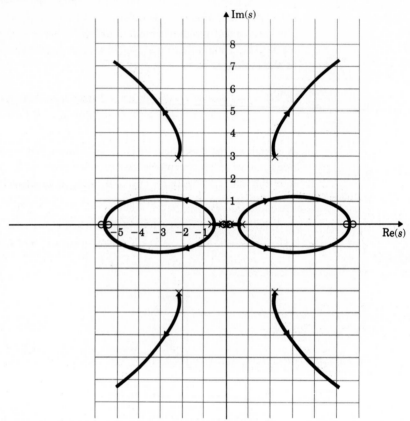

is 10:1,

$$G_e(s) = K_h(s + 0.1).$$

that is, we have placed a new zero.

The final design is the result of iterations between the q, θ, $\dot{h}$, and h feedback gains. Although this trial design was successful, use of the SRL approach has promise to expedite the process.

STEP 6: *Do an optimal design.* The symmetric root locus of the system is as shown in Fig. 7.46. If we choose the closed-loop poles at

$$\alpha_c(s) = (s + 0.0045)(s + 0.145)(s + 0.513)(s + 2.25 \pm j2.98),$$

then the required feedback gain is

$$\mathbf{K} = [\,-0.0011 \quad 0.0016 \quad -1.831 \quad -7.6113 \quad -0.001\,].$$

Since w is proportional to angle of attack α, state feedback would require an angle-of-attack sensor. Such sensors are expensive and sometimes unreliable, hence an estimator would be required (see Chapter 6). The step response of the system to a 100-ft step command in h is shown in Fig. 7.47, and the associated control effort

FIGURE 7.47
Step-response of altitude-hold autopilot.

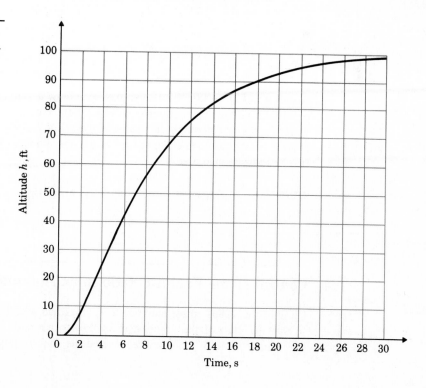

FIGURE 7.48
Control effort for 100-ft
step command in altitude.

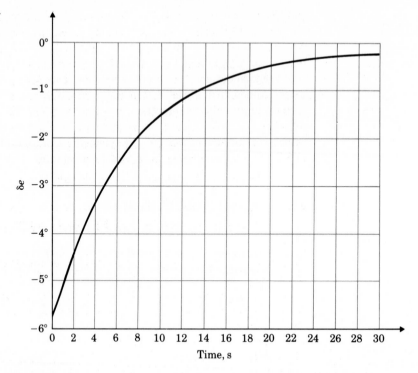

is shown in Fig. 7.48. This design has been done assuming that the linear model is valid for altitude changes under consideration. Simulations should be performed to verify this or determine the extent of the validity of the linear model. In order to achieve large altitude changes such as the one shown in Fig. 7.47, engine thrust may need to be increased.

STEPS 7 AND 8: *Verify the design.* See Section 7.3.1.

7.4 Control of the Fuel-Air Ratio in an Automotive Engine

Up until the 1980s, most automobile engines had a carburetor to meter the fuel so that the ratio of the gasoline mass flow to air mass flow (F/A) remained in the vicinity of 1:15. This device metered the fuel by relying on a pressure drop produced by the air flowing through a venturi. The device has performed adequately for the last 100 years in terms of keeping the engine running satisfactorily, but has historically allowed excursions of up to 20% in the F/A. After the implementation of exhaust-pollution regulations, these F/A excursions became unacceptable. During the 1970s, automobile companies improved the design and manufacturing process of the carburetors so that they became more accurate and delivered an accuracy in the vicinity of 3 to 5%. Through a combination of factors, this improved F/A accuracy helped lower exhaust-pollution levels. However, the carburetors were still open-loop devices in that no measurement was made of the F/A that entered the engine for subsequent feedback into the carburetor.

During the 1980s, all manufacturers turned to feedback control systems to provide a much-improved level of F/A accuracy, an action made necessary by the decreasing levels of allowable exhaust pollutants.

As for the other examples, we will discuss each step in the design of this system.

STEP 1: *Understand the process and its performance.* The method chosen to meet the exhaust standards has been to use a catalytic converter that simultaneously ox-idizes excess levels of exhaust carbon monoxide (CO) and unburned hydrocarbons (HC) and reduces excess levels of the oxides of nitrogen (NO and NO_2 or NO_x). This device is usually referred to as a "three-way catalyst" because of its effect on all three pollutants. This catalyst is ineffective when the F/A is more than 1% different from the stoichiometric level of 1:14.7; therefore, a feedback control system is required to maintain the F/A within 1% of that desired level. The system is depicted in Fig. 7.49. The dynamic phenomena that affect the relation between the sensed F/A output from the exhaust and the fuel-metering command in the intake manifold are (1) intake fuel and air mixing, (2) cycle delays due to the piston strokes in the engine, and (3) the time required for the exhaust to travel from the

FIGURE 7.49
F/A feedback control system.

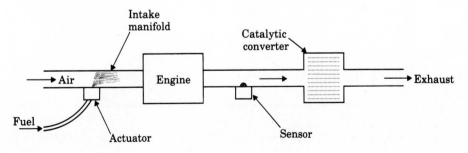

engine to the sensor. All these effects are strongly dependent on the speed and load of the engine. For example, engine speeds will vary from 600 to 6000 rpm. The result of these variations is that the time delays in the system that will affect the feedback control system behavior will also vary by at least 10:1, depending on the operating condition. The system undergoes transients as the driver demands more or less power through changes in the accelerator pedal, with the changes taking place over fractions of a second. Ideally, the feedback control system should be able to keep up with these transients.

STEP 2: *Select a sensor.* The discovery and development of the exhaust sensor was the key technological step that made this concept of exhaust-emission reduction possible. The device is made of zirconia; it is placed in the exhaust stream and yields a voltage related to the oxygen content of the gases. The F/A is uniquely related to the oxygen level. The voltage of the sensor is highly nonlinear with respect to F/A (Fig. 7.50), with almost all the change in voltage occurring precisely at the F/A where the feedback system must operate for effective performance of the catalyst. Therefore the gain of the sensor will be very high when the F/A is at the desired point (1:14.7), but will fall off considerably for F/A excursions away from 1:14.7.

Although other sensors are under development for possible use in F/A feedback control, no other sensor has demonstrated the capability to perform adequately. All

FIGURE 7.50
Exhaust sensor output.

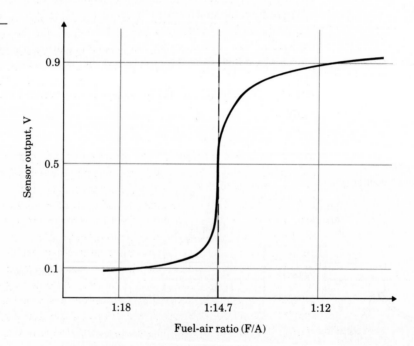

manufacturers are currently using zirconia sensors in their feedback control systems in production automobiles.

STEP 3: *Select an actuator.* Fuel metering can be accomplished by a carburetor or by fuel injection. To implement a feedback F/A system, the capability of adjusting the fuel metering electrically is required, since the sensor used provides an electric output.

Carburetors have been designed to provide this capability by including adjustable orifices that modify the primary fuel flow in response to the electric error signal. Most designs accomplish this by the use of on-off solenoid valves that modulate the pressure applied to a bellows in the carburetor. The bellows moves a needle to adjust the orifice size. The advantage of this approach appears to be that it is the cheapest of the alternatives.

Since fuel-injection systems are typically electrical by nature, they can be used to perform the fuel adjustment for F/A feedback by simply including the capability of summing the feedback signal in the computer code. In some cases, fuel injectors are placed at the inlet to every cylinder (called *multipoint injection*), and in some cases there is one large injector upstream (called *single-point* or *throttle body injection*). With the growing popularity of microprocessor engine control, single-point fuel injection is becoming much more popular, apparently because its cost is approaching that of the carburetor. Differences in performance between the carburetor and single-point injection are not major, since they both introduce fuel at approximately the same location. Multipoint injection does offer improved performance because the fuel is introduced much closer to the engine, with better distribution to the cylinders. Being closer reduces the time delays and thus yields better engine response. Because of a significant cost penalty, however, multipoint injection is typically only used on relatively expensive, high-performance automobiles.

STEP 4: *Make a linear model.* The sensor nonlinearity shown in Fig. 7.50 is severe enough that design effort based on a linearized model of it should be used with caution. Figure 7.51 shows a block diagram of the system, with the sensor shown to have a gain K_s. The two time constants indicated for the inlet manifold dynamics represent fast fuel flow in the form of vapor or droplets and slow fuel flow in the form of a liquid film on the manifold walls. The time delay is the sum of the delay caused by the four piston strokes from the intake process until the exhaust process and then the time required for the exhaust to travel from the engine itself to the sensor located perhaps 1 ft or so away. A sensor lag is also included in the process to account for the mixing that occurs in the exhaust manifold. Although the time constants and delay time change considerably, primarily as a function of engine load and speed, we will examine the design at a specific point where the

FIGURE 7.51
Block diagram of an F/A
control system.

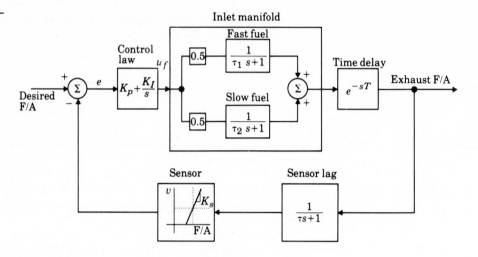

values are

$$\tau_1 = 0.02 \text{ s},$$
$$\tau_2 = 1 \text{ s},$$
$$T = 0.2 \text{ s},$$
$$\tau = 0.1 \text{ s}.$$

STEP 5: *Try a lead-lag or PID controller.* Given the tight error specifications and the wide variations in the required fuel command u_f due to varying engine-operating conditions, an integral control term is mandatory. Thus any required u_f can be provided when the error signal $e = 0$. The addition of a proportional term, although not often used, allows for an increase (doubling) in the bandwidth without degrading steady-state characteristics. In this example, we will use a control law that is proportional plus integral. The output from the control law is a voltage that drives the injector pulse-former to give a fuel pulse whose length is proportional to the voltage. The controller transfer function can be written as

$$D_c(s) = K_p + \frac{K_I}{s} = \frac{K_p}{s}(s + z), \qquad (7.26)$$

where

$$z = \frac{K_I}{K_p}$$

and z can be chosen as desired.

First, let's assume that the sensor is linear and can be represented by a gain K_s. Then z can be chosen for good stability and good response of the system. Figure 7.52 shows the frequency response of the system $K_s K_p = 1.0$ and $z = 0.3$, while

FIGURE 7.52
Frequency response of an F/A controller: (a) magnitude and (b) phase.

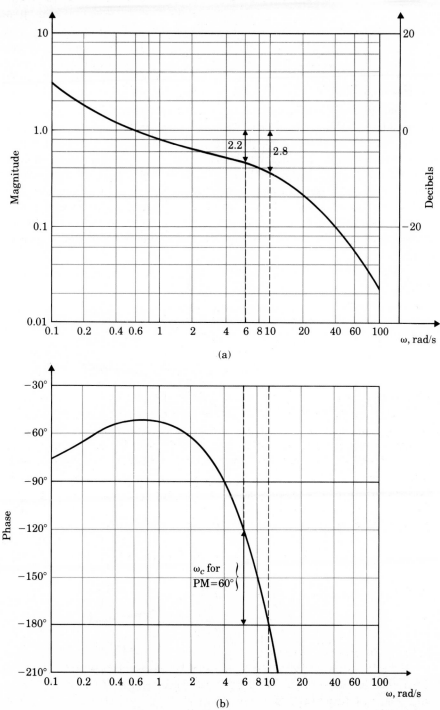

(a)

(b)

Fig. 7.53 shows a root locus of the system versus $K_s K_p$ with $z = 0.3$. Both analyses show that the system becomes unstable for $K_s K_p \cong 2.8$. Figure 7.52 shows that to achieve a phase margin of approximately 60°, the gain $K_s K_p$ should be $\cong 2.2$. The figure also shows that this produces a crossover frequency of 6.0 rad/s ($\approx$ 1 Hz). The root locus in Fig. 7.53 verifies that this candidate design will achieve acceptable damping ($\zeta \cong 0.5$).

Although this linear analysis shows that acceptable stability at a reasonable bandwidth ($\approx$ 1 Hz) can be achieved with a PI controller, a look at the nonlinear sensor characteristics (Fig. 7.50) shows that this, indeed, may not be achievable. Note that the slope of the sensor output is extremely high near the desired setpoint, thus producing a very high value of K_s. Therefore lower values of the controller gain K_p need to be used to maintain the overall $K_s K_p$ of 2.2 when including the effect of the high sensor gain. On the other hand, a value of K_p low enough to yield a stable system at F/A = 1:14.7 will yield sluggish response to transient errors that deviate much from the setpoint, because the effective sensor gain will be reduced

FIGURE 7.53
Root locus of an F/A controller.

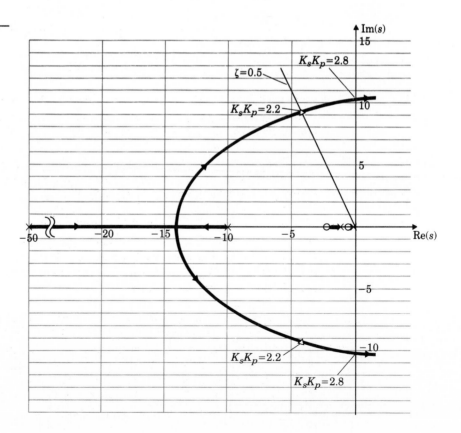

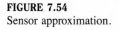

FIGURE 7.54
Sensor approximation.

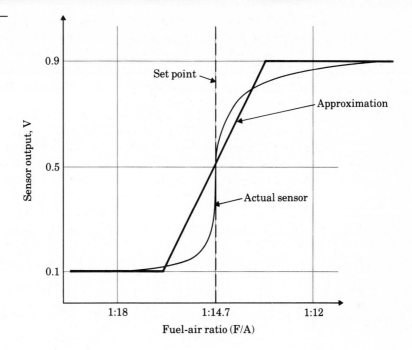

substantially. It is therefore necessary to account for the sensor nonlinearity in order to obtain satisfactory response characteristics of the system for anything other than minute disturbances about the setpoint. A first approximation to the sensor is shown in Fig. 7.54. Since the sensor gain at the setpoint is still quite different between the actual sensor and its approximation, this approximation will yield erroneous conclusions regarding stability about the setpoint; however, it will be useful in a simulation to determine response to initial conditions.

STEP 6: *Try an optimal controller.* The response of this system is dominated by the sensor nonlinearity, and any fine tuning of the control needs to account for that feature. Furthermore, the system dynamics are relatively simple, and it is unlikely that an optimal design approach will yield any improvement over the PI controller used. This step will therefore be omitted.

STEP 7: *Simulate design with nonlinearities.* Figure 7.55(a) is a plot of the system error with the approximate sensor of Fig. 7.54 and $K_p K_s = 2.0$. The slow response is apparent with 12.5 s before the error comes out of saturation and a time constant of almost 5 s once the linear region is reached. In real automobiles, these systems are operated with much higher gains and to show these effects, a simulation with $K_p K_s = 6.0$ is plotted in Fig. 7.55(b) and (c). At this gain, the linear system is unstable and the signals initially grow, as may be seen in the figures for time before 5 s.

FIGURE 7.55

System response with nonlinear sensor approximation.

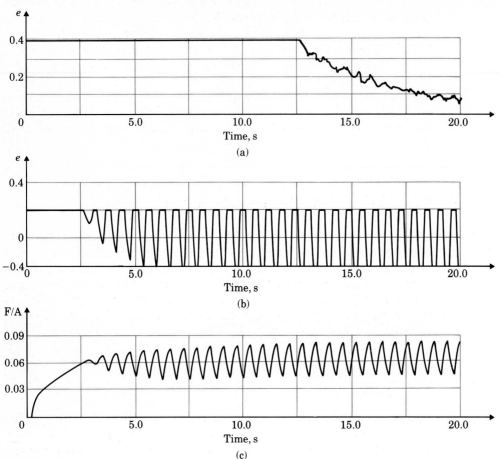

(a)

(b)

(c)

The growth is halted by the fact that as the input to the sensor nonlinearity gets large, the *effective* gain of the sensor decreases due to the saturation, and, in the limit, a limit cycle is reached having a frequency corresponding to the point where the root locus crosses the imaginary axis and having an amplitude such that the total effective gain of $K_p K_{s,eq} = 2.8$. As was described in Section 4.6.4, the effective gain of a saturation for moderately large inputs can be approximated by $4N/\pi A$ where N is the saturation level and A is the amplitude of the input signal. Here, $N = 0.4$ and, if $K_p = 0.1$, then we must have $K_{s,eq} = 28$. Thus we predict an input signal amplitude of $A = (4)(0.4)/(\pi)(28) = 0.018$. This value is closely verified by the plot of Fig. 7.55(c), the input to the nonlinearity in this case. The frequency of oscillation is also nearly 10.1 rad/s, as predicted by the root locus in Fig. 7.53.

In the actual implementation of feedback F/A controllers in automobile engines, sensor degradation over thousands of miles of use is of primary concern in order to meet the exhaust-pollution standards over the government-mandated 50,000 mi. In order to reduce the sensitivity of the average setpoint to changes in the sensor output characteristics, manufacturers typically modify the design discussed here. One approach is to feed the sensor output into relay function [Fig. 4.26(b)], thus completely eliminating any dependency on the sensor gain at the setpoint. The frequency of the limit cycle is then solely determined by controller constants and engine characteristics. Average steady-state F/A accuracy is also improved.

7.5   Control of a Digital Tape Transport

STEP 1: *Understand the process and its performance specifications.* Most high-performance tape transports are designed with a small capstan to pull the tape past the read/write heads with the take-up reels turned by DC motors. The capstan motion is isolated from the reels by vacuum columns that provide constant tension to the tape at the heads. This structure permits separate design of capstan, vacuum, and reel controls. A sketch of the system is shown in Fig. 7.56, and a schematic of the capstan control is drawn in Fig. 7.57. The objective of the capstan-control system is to control the speed and the tension of the tape at the read/write head. The tape is to be controlled at speeds up to 200 in/s, with start-up as fast as possible subject to keeping the tension below 12 N at all times to prevent permanent distortion of the tape.

STEP 2: *Select a sensor.* The sensor selection is easy in this case, being a DC tachometer to measure idler speed. Alternatives would be an AC generator, which would give a sinusoidal voltage signal whose amplitude is proportional to speed. This output voltage needs additional electronic signal processing such as rectifying and filtering to provide a suitable signal for the controller. Also possible is a digital shaft encoder that provides a digital reading of shaft position. Such devices have low inertia and low friction and are often used in digital control. Using digital control is increasingly attractive for motion-control problems; however, in the present case the need to convert the digital signal to analog form leads us to favor the DC tachometer.

STEP 3: *Select an actuator.* There are many choices for motion-control actuators, including hydraulic, pneumatic, AC electric, DC electric, and stepping motors. Hydraulic and pneumatic actuators require an auxiliary pump but can provide high forces and torques in a lightweight package. They are common in aircraft control-surface servos and chemical-process valve controls. However, for our purpose the

FIGURE 7.56
Typical tape-drive
mechanism: Reference
track 1 is nearest the
front (observer) end.

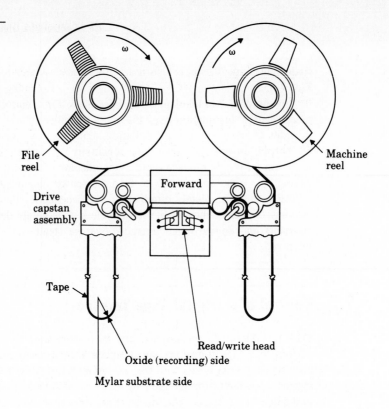

File reel

Machine reel

Drive capstan assembly

Forward

Tape

Read/write head

Oxide (recording) side

Mylar substrate side

FIGURE 7.57
Model of a tape drive.

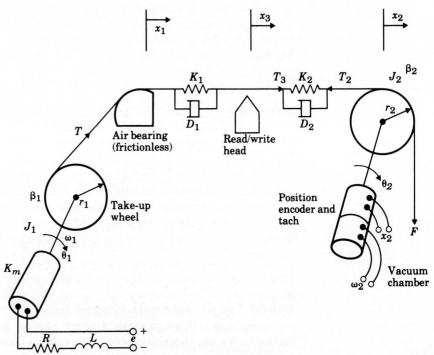

electric motors are the best choice, since they are reliable, inexpensive, and easy to maintain. Of these, the DC motor has the best low-speed acceleration and low-ripple-speed characteristics and is the machine of choice here.

STEP 4: *Make a linear model.* The system is in static equilibrium when $T_0 = F$ and $K_m i_0 = r_i T_0$. We define the variables as deviations from this equilibrium.
The equations of motion of the system are given by the laws of mechanics:

$$J_1 \frac{d\omega_1}{dt} + \beta_1 \omega_1 + r_1 T = K_m i, \tag{7.27}$$

$$\dot{x}_1 = r_1 \omega_1, \tag{7.28}$$

$$L \frac{di}{dt} + Ri + K_m \omega_1 = e, \tag{7.29}$$

$$\dot{x}_2 = r_2 \omega_2, \tag{7.30}$$

$$J_2 \frac{d\omega_2}{dt} + \beta_2 \omega_2 = r_2 T_2, \tag{7.31}$$

$$T = K_1(x_3 - x_1) + D_1(\dot{x}_3 - \dot{x}_1), \tag{7.32}$$

$$T_2 = K_2(x_2 - x_3) + D_2(\dot{x}_2 - \dot{x}_3), \tag{7.33}$$

$$x_1 = r_1 \theta_1, \tag{7.34}$$

$$x_2 = r_2 \theta_2, \tag{7.35}$$

$$x_3 = \frac{x_1 + x_2}{2}, \tag{7.36}$$

where

J_1 = combined inertia of wheel, plus take-up motor, 4×10^{-5} kg $\cdot$ m^2,

β_1 = viscous friction at take-up wheel, 0.01 N $\cdot$ m $\cdot$ s,

r_1 = wheel radius (take-up), 0.02m,

K_m = motor torque constant, 0.03V $\cdot$ s,

J_2 = inertia of idler, 10^{-5} kg $\cdot$ m^2,

β_2 = viscous friction at wheel, 0.01 N $\cdot$ m $\cdot$ s,

r_2 = radius of tape on idler, 0.02 m,

$D_{1,2}$ = damping in tape-stretch motion, 20 N/m/s,

$K_{1,2}$ = spring constant in tape-stretch motion, 4×10^4 N/m,

ω_1 = speed of drive wheel, $\dot{\theta}_1$, rad/s,

ω_2 = output speed measured by tachometer output $\dot{\theta}_2$, rad/s

T = tape tension, $T_2 = T_3$ at read-write head, N,

e = applied voltage, V,

R = armature resistance, 1Ω,

L = armature inductance, 10^{-3} H,

θ_1 = angular displacement of capstan,

θ_2 = tachometer shaft angle,

F_1 = constant force (tape tension from vacuum column), 6 N,

x_3 = position of tape at head, m,

$\dot{x}_3$ = velocity of tape at head, m/s.

The equations of the system are fifth-order and are already of the form

$$\dot{\mathbf{x}} = \mathbf{F}\mathbf{x} + \mathbf{G}u, \tag{7.37}$$

$$y = \mathbf{H}\mathbf{x}, \tag{7.38}$$

where

$$\mathbf{x} = [\, x_1 \quad \omega_1 \quad x_2 \quad \omega_2 \quad i \,]^{\mathrm{T}},$$

$$u = e.$$

There are two outputs of interest, tape position at the head (x_3) and tape tension T. For tape position, we have

$$x_3 = \frac{x_1 + x_2}{2}.$$

Thus for position at the head,

$$\mathbf{H}_3 = [\, 0.5 \quad 0 \quad 0.5 \quad 0 \quad 0 \,]. \tag{7.39}$$

The tension is given by Eq. (7.32), which we can write as

$$
\begin{aligned}
T &= K_1\left(\frac{x_1 + x_2}{2} - x_1\right) + D_1\left(\frac{r_1\omega_1 + r_2\omega_2}{2} - r_1\omega_1\right) \\
&= \frac{K_1}{2}(x_2 - x_1) + \frac{D_1}{2}(r_2\omega_2 - r_1\omega_1).
\end{aligned}
\tag{7.40}
$$

If we substitute the numbers into these equations, we find very large and very small values, indicating the need for time and amplitude scaling. Ideally, we would like the elements in $\mathbf{F}$, $\mathbf{G}$, and $\mathbf{H}$ to be in the range 0.1 to 10, if possible. We write the equations in the state order with the numbers and use x_1' and x_2' for the unscaled positions:

$$\dot{x}_1' = 0.02\omega_1,$$

$$4 \times 10^{-5}\dot{\omega}_1 = -400x_1' - 0.014\omega_1 + 400x_2' + 0.004\omega_2 + 0.03i,$$

$$\dot{x}_2' = 0.02\omega_2,$$

$$10^{-5}\dot{\omega}_2 = 400x_1' + 0.004\omega_1 - 400x_2' - 0.014\omega_2,$$

$$10^{-3}\frac{di}{dt} = -0.03\omega_1 - i + e. \tag{7.41}$$

After some experimentation, we decide to measure time in milliseconds, so $\tau = 10^3 t$, and to measure position in units of 10^{-5} m, so we take $x_1 = 10^5 x_1'$ and $x_2 = 10^5 x_2'$. With these changes, the equations are

$$\frac{dx_1}{d\tau} = 2\omega_1,$$

$$\frac{d\omega_1}{d\tau} = -0.1x_1 - 0.35\omega_1 + 0.1x_2 + 0.1\omega_2 + 0.75i,$$

$$\frac{dx_2}{d\tau} = 2\omega_2,$$

$$\frac{d\omega_2}{d\tau} = 0.4x_1 + 0.4\omega_1 - 0.4x_2 - 1.4\omega_2,$$

$$\frac{di}{d\tau} = -0.03\omega_1 - i + e. \tag{7.42}$$

These equations are described by the matrices:

$$\mathbf{F} = \begin{bmatrix} 0 & 2 & 0 & 0 & 0 \\ -0.1 & -0.35 & 0.1 & 0.1 & 0.75 \\ 0 & 0 & 0 & 2 & 0 \\ 0.4 & 0.4 & -0.4 & -1.4 & 0 \\ 0 & -0.03 & 0 & 0 & -1 \end{bmatrix}; \quad \mathbf{G} = \begin{bmatrix} 0 \\ 0 \\ 0 \\ 0 \\ 1 \end{bmatrix};$$

$$\mathbf{H} = [0 \quad 0 \quad 1 \quad 0 \quad 0] \qquad x_2 \text{ output;}$$

$$\mathbf{H}_3 = [0.5 \quad 0 \quad 0.5 \quad 0 \quad 0] \qquad x_3 \text{ output;}$$

$$\mathbf{H}_{T_2} = [-0.2 \quad -0.2 \quad 0.2 \quad 0.2 \quad 0] \qquad \text{tension output.}$$

The transfer function to measure position x_2 is

$$G(s) = \frac{X_2(s)}{E_1(s)} = \frac{0.6(s + 2)}{s^5 + 2.75s^4 + 3.22s^3 + 1.88s^2 + 0.418s}$$

$$= \frac{0.6(s + 2)}{s(s + 0.507)(s + 0.968)(s + 0.637 \pm j0.667)}. \tag{7.43}$$

STEP 5: *Try a lead-lag or PID controller.* The uncompensated root locus versus the proportional feedback gain and the open-loop frequency response of the system are shown in Figs. 7.58 and 7.59, respectively. It is desired to use feedback control to achieve a settling time of about 12 ms and a negligible overshoot in response to a maximum velocity command, defined as one causing no more than 6-N peak tension.

The specifications call for an ω_n of about 1 rad/ms, and, with a pole-zero excess (that is, the difference between number of poles and zeros) of 4, we will require substantial lead. One way to design a compensator is to use derivative-(tachometer)-plus-lead-network feedback. If we place both the lead network zero

FIGURE 7.58
Uncompensated root locus for a tape drive.

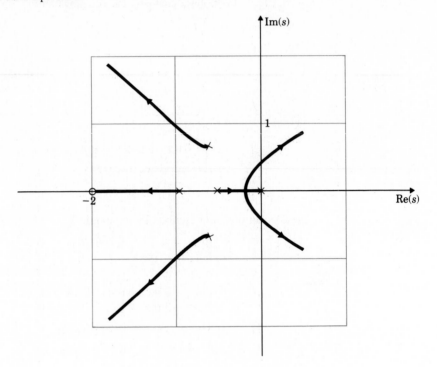

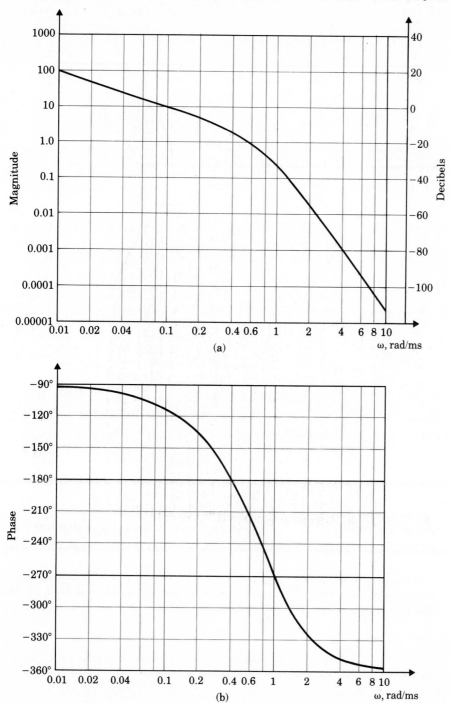

(a)

(b)

FIGURE 7.60
Compensated root locus
for a tape drive.

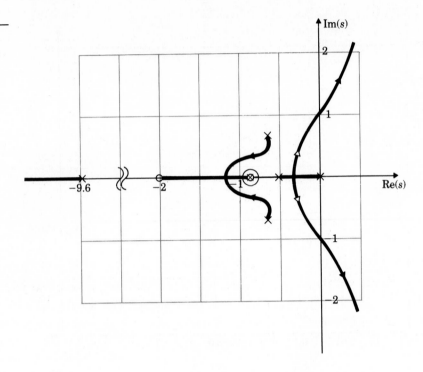

and the zero due to the tachometer* at -0.968 and place the lead network pole at -9.68, then the compensator is

$$D_c(s) = K_c \frac{(s + 0.968)^2}{s + 9.68}. \tag{7.44}$$

The root locus of the system is as shown in Fig. 7.60. The ramp response of the system is as shown in Fig. 7.61(a) and the associated tension is as shown in Fig. 7.61(b). The frequency response of the compensated system is seen in Fig. 7.62.

STEP 6: *An optimal design using pole placement.* We now consider designing a compensator using the state-variable approach. The symmetric root locus of the system is shown in Fig. 7.63. If we select the closed-loop poles of the system to

---

*Pole-zero cancellation is usually avoided because of sensitivity problems but may be acceptable if the canceled pole is sufficiently stable. However, unstable pole-zero cancellation is absolutely forbidden.

FIGURE 7.61
(a) Ramp response with
$D_c(s)$ compensation;
(b) tension for ramp
response.

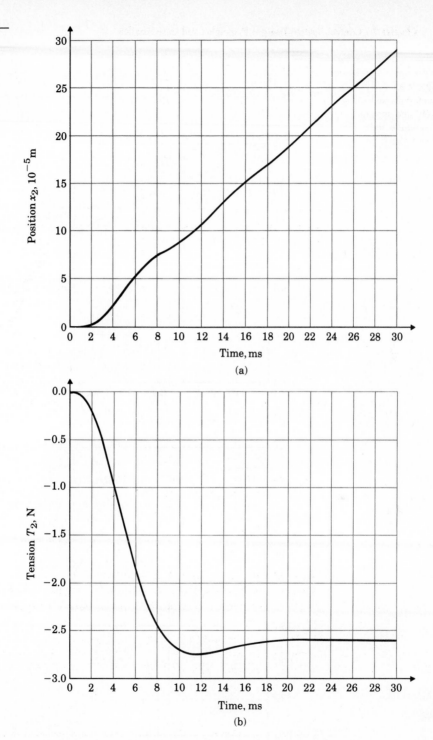

(a)

(b)

FIGURE 7.62
Frequency response of
compensated system:
(a) magnitude and
(b) phase.

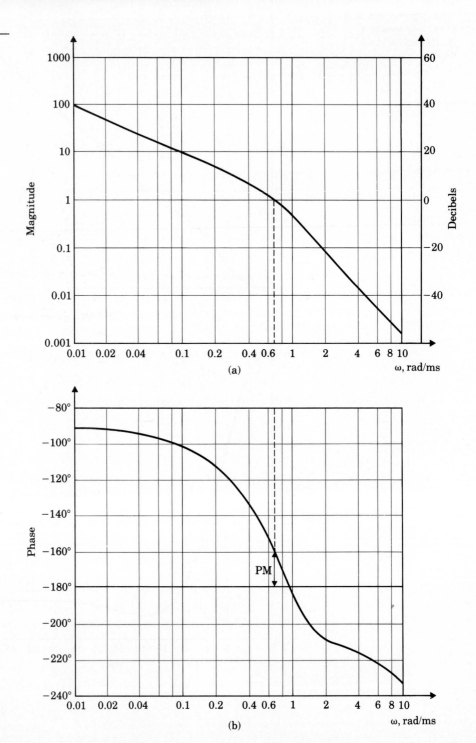

(a)

(b)

FIGURE 7.63
Symmetric root locus for tape drive.

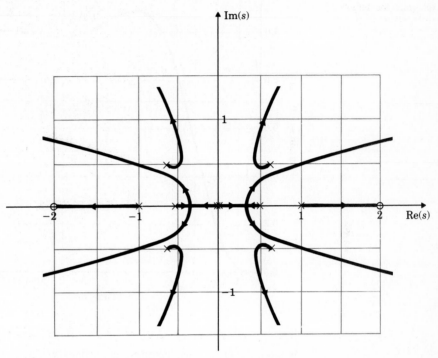

FIGURE 7.64
Step response for
full-state feedback.

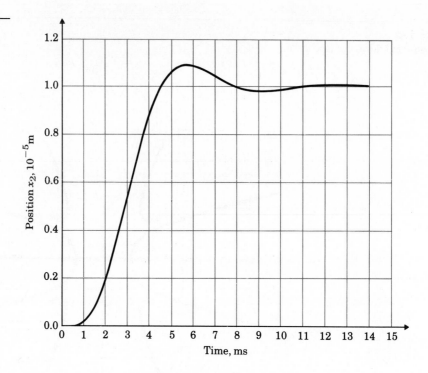

be on the SRL at a bandwidth of about 1 rad/ms,

$$\alpha_c(s) = (s + 0.451 \pm j0.937)(s + 0.947 \pm j0.581)(s + 1.16),$$

then the required state feedback gain is

$$\mathbf{K} = [\,0.802 \quad 2.58 \quad 0.489 \quad 0.964 \quad 1.21\,].$$

The step response of the system is shown in Fig. 7.64, where the reference input is multiplied by the inverse of the closed-loop DC gain; that is,

$$u = -\mathbf{K}\mathbf{x} + 1.29r,$$

so as to obtain zero steady-state error to a unit step. The associated control effort is shown in Fig. 7.65(a) and the tension in Fig. 7.65(b). The ramp response of the system and the associated control effort are shown in Fig. 7.66(a) and (b), respectively. If measurement of all the states is not available and an estimator is to be designed based on position measurement alone, such that its poles are four times faster than the control, then

$$\alpha_e(s) = (s + 1.80 \pm j3.75)(s + 3.79 \pm j2.32)(s + 4.64),$$

FIGURE 7.65
Full-state feedback:
(a) control effort and
(b) tension.

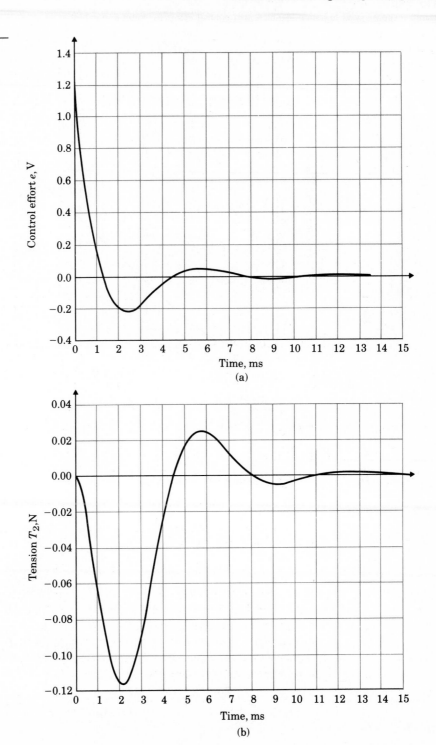

(a)

(b)

FIGURE 7.66
(a) Ramp response with full-state feedback; (b) control effort for ramp command.

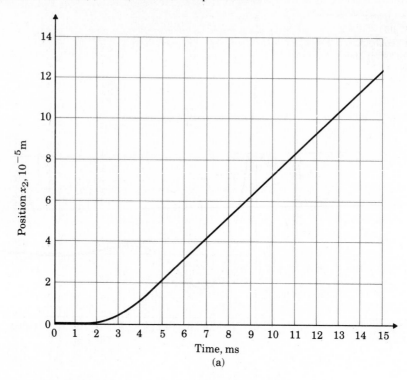

(a)

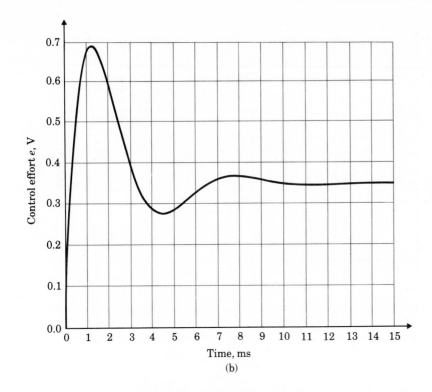

Time, ms
(b)

and the estimator gain is

$$\mathbf{L} = \begin{bmatrix} 403.9 \\ 50.6 \\ 13.1 \\ 38.6 \\ 1166.2 \end{bmatrix}.$$

The compensator transfer function is then given by

$$D_c(s) = \frac{-1905(s + 0.640 \pm j0.797)(s + 0.984 \pm j0.247)}{(s + 2.38)(s + 1.60 \pm j4.73)(s + 5.73 \pm j2.64)}, \qquad (7.45)$$

with frequency response shown in Fig. 7.67, which is essentially a *lead* network. The frequency response of the complete compensated system is as shown in Fig. 7.68.

The root locus of the compensated system is as shown in Fig. 7.69. Figure 7.70 shows a transient response of the tension and position, including the effect of the estimator.

STEPS 7 AND 8: *Verify the design.* At this point the system should be simulated with all nonlinearities included, followed by a prototype test.

569

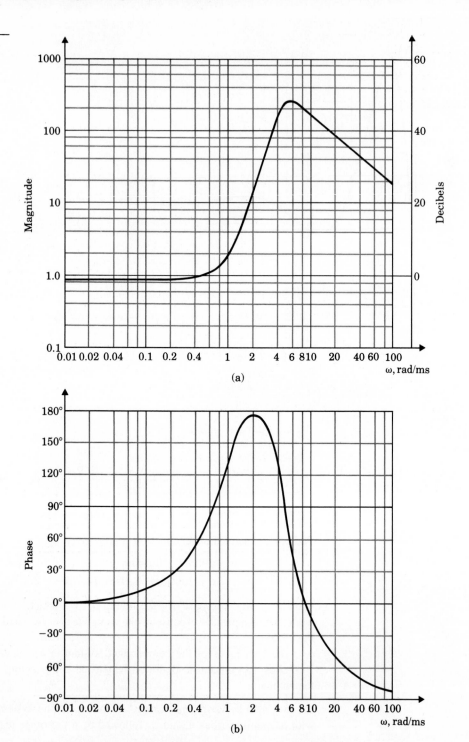

FIGURE 7.68
Frequency response of
compensated system:
(a) magnitude and
(b) phase.

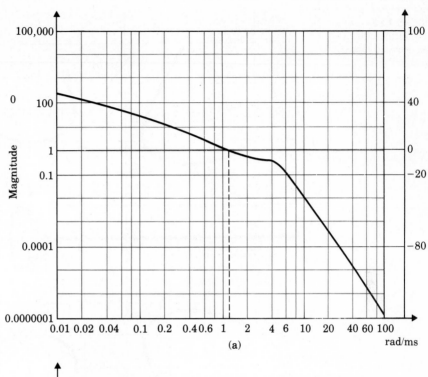

(a)

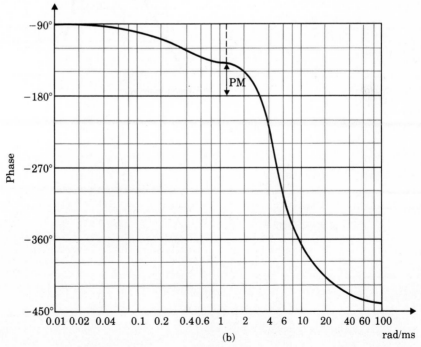

(b)

FIGURE 7.69
Root locus for a
compensated tape
drive.

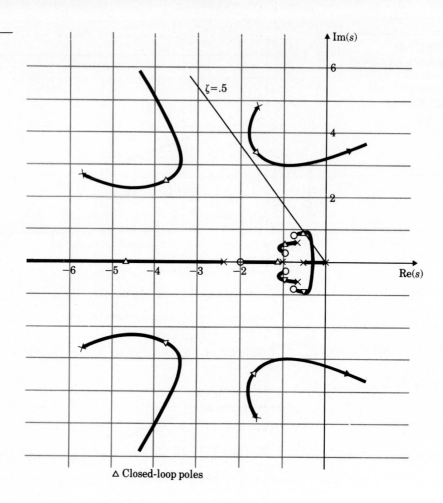

△ Closed-loop poles

FIGURE 7.70
Initial conditions $x_{30} =$
1, $\hat{x}_0 = 0$: (a) tension
and (b) output.

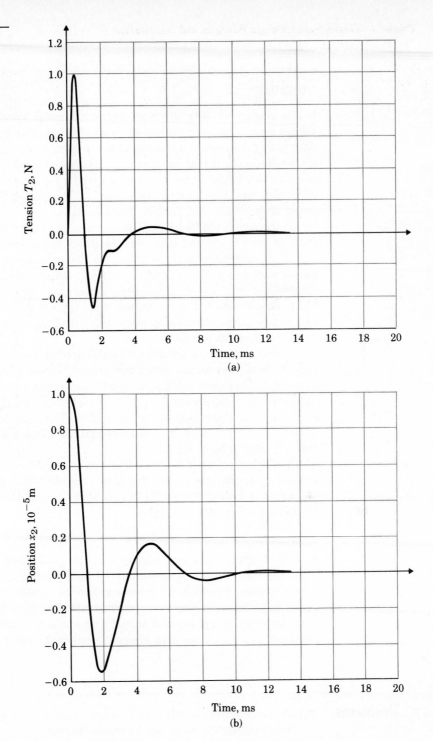

(a)

(b)

573

Summary

In this chapter we have given an outline of control-system design intended to enlarge the perception of the factors involved in obtaining a satisfactory controller. After listing the steps in a possible design sequence, we have presented four case studies illustrating the method. In the first study we looked at the control of satellite attitude. The dynamic compensation involved a lightly damped resonance, and a notch filter was introduced as one design aid. We also saw that an SRL approach would automatically add a similar notch in this case.

In the second case we looked at the control of a well-known commercial aircraft. In the yaw-damper case, we saw that a single feedback path produced a marginally acceptable result. Refining the design using the state-space SRL approach improved damping from $\zeta = 0.3$ to 0.4 but increased the order of the compensator from 1 to 6. In the altitude-hold case, we saw that a classical multiloop closure approach was difficult to design because of the high interaction between the loops, necessitating many design iterations. The SRL approach helped considerably in this process.

In the third case, a control for fuel/air ratio in the automobile was studied. Here we needed to control with a pure time delay and a highly nonlinear sensor. In practice, this system is left to oscillate constantly in a limit cycle because of the vast range of conditions under which it must perform well. The final case involved a tape drive, characteristic of motion control. The main features here were that more than one lead network is required and more than one output (velocity and tension) needs to be monitored.

In all these cases it became clear that the model used for design does not completely describe the true system to be controlled. In order to illustrate one of the analytical tools developed for this situation, an example of stability robustness determination has been included.

We leave this chapter with the observation that practical control designs require substantial calculations. The tools we have prescribed—the root locus and the Bode frequency response—can be used with great effectiveness for small problems; however, the state-space method of pole placement and the calculation of time responses, especially for nonlinear systems, require the aid of a computer. Fortunately, the development of computer aids is progressing rapidly, and we believe that every control engineer should have access to a CAD system. The curves developed in this chapter were primarily computed using CTRL-C, a product of Systems Control Technology, Inc.

Problems

7.1 Of the three types of PID control, which one is the most effective in reducing the error resulting from a constant disturbance? Explain.

7.2 Is there a greater change of instability when the sensor in a feedback control system is not collocated with the actuator? Explain.

7.3 Consider the plant $G(s) = 1/s^3$. What is the effect of adding a lead compensator,

$$D_c(s) = \frac{s + a}{s + b} \qquad a < b,$$

on the phase margin of this system?

7.4 Consider the closed-loop system shown in Fig. 7.71,

a) What is the phase margin if $K = 70,000$?
b) What is the gain margin if $K = 70,000$?
c) What value of K will yield a phase margin of $\sim 70°$?
d) What value of K will yield a phase margin of $\sim 0°$?
e) Sketch a root locus of the system, being careful of the locus location at $s = j\omega$ and determine from your locus what value of K lies on the stability boundary.
f) If the disturbance d is a constant and $K = 10,000$, what is the maximum allowable value of d if Y is to remain less than 0.1? (Assume $r = 0$.)
g) If it is required to allow larger values of d than the value you obtained in **(f)**, but with the same error constraint, discuss what steps you could take to alleviate the problem.

FIGURE 7.71

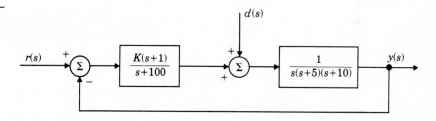

7.5 Consider the system shown in Fig. 7.72, which represents an attitude-rate control for an aircraft.

a) Design a compensator so that the dominant poles are at $-2 \pm j2$.
b) Sketch the Bode plot of your design and select compensation so that the crossover is at least $2\sqrt{2}$ rad/s and PM $\geq 50°$.
c) Sketch the root-locus of your design, finding $\omega_n > 2\sqrt{2}$ and $\zeta \geq 0.5$.

FIGURE 7.72

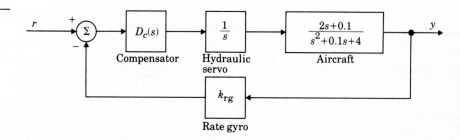

7.6 Consider the system in Fig. 7.73. Which of the following claims are true? Explain in detail.

FIGURE 7.73

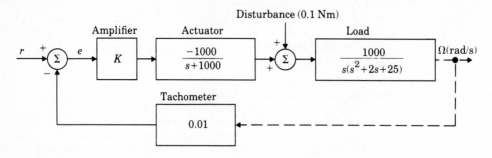

a) The actuator natural behavior (the pole at 1000 rad/s) must be included in an analysis to evaluate a usable maximum gain for which the control system is stable.

b) The value of K must be negative for a stable system.

c) For a certain value of K the control system will oscillate at a frequency within 1 rad/s of 5 rad/s.

d) The system is unstable if $K < -10$, that is, if $|K| > 10$.

e) If K must be *negative* for stability, the control system cannot counteract a *positive* disturbance.

f) A positive constant disturbance will speed up the load, thereby making the final value of e negative.

g) With only a positive constant command input r, e must have a final value greater than zero.

h) For $K = -1$ the loop is stable and for the disturbance shown there will be an error in speed, the magnitude of which in the steady state is less than 5 rad/s.

7.7 The stick balancer is shown in Fig. 7.74.

a) Using root-locus techniques, design a compensation that will place the dominant roots at $s = -5 \pm j5$ (that is, $\omega_n = 7$ rad/s, $\zeta = 0.707$).

b) Use Bode techniques to design a compensation to meet the following specifications:

$$\theta_{ss} < 0.001 \text{ for } T_d = 1, \text{ a constant,}$$

$$\text{Phase margin} \geq 50°,$$

$$\text{Closed} - \text{loop bandwidth} \approx 7 \text{ rad/s.}$$

FIGURE 7.74

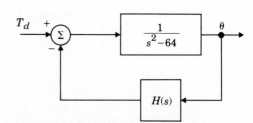

7.8 Consider the system in Fig. 7.75.

a) Suppose $G(s) = 2500K/[s(s + 25)]$.

1) Design a lead compensator so that the phase margin of the system is more than 45°; the steady-state error due to a ramp should be less than or equal to 0.01.
2) Design a lead compensator so that the overshoot is less than 25% and the settling time is less than 0.1 s (assume 1%).

b)
$$G(s) = \frac{K}{s(1 + 0.1s)(1 + 0.2s)}$$

and the performance specifications are $K_v = 100$, with a phase margin $\geq 40°$.

1) Is the lead compensation effective for this system? Find a lag compensator for the given specifications and plot the root locus of the compensated system.
2) Design a lag compensator such that the peak overshoot is less than 20% and $K_v = 100$.

c) 1) Repeat [**b(1)**] using a lead-lag compensator.
2) Find the root-locus of the compensated system and compare the findings with those of [**b(1)**].

FIGURE 7.75

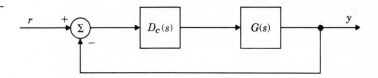

7.9 Consider the open-loop system

$$G(s) = \frac{10}{s[1 + (s/10)]}.$$

a) Sketch a Bode plot for this system. The design specifications are, for *low frequency*, $K_v = 100$ and PM = 45°. Sinusoidal inputs of up to 1 rad/s should be reproduced with $\leq 2\%$ error. For *high frequency*, sinusoidal inputs greater than 100 rad/s should be attenuated at the output to $\leq 5\%$ of their input value. The block diagram sketched in Fig. 7.76 may be useful for conceptual purposes.

b) First pick the open-loop gain K so that $K_v = 100$.

FIGURE 7.76

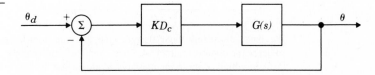

c) Show that a *sufficient* condition for meeting the specifications on the sinusoidal inputs is that the magnitude plot lie outside the shaded regions in Fig. 7.77. Recall $\theta/\theta_d = KG/(1 + KG)$ and $e/\theta_d = 1/1 + KG$.

d) Explain why introducing a lead network alone cannot meet the design specifications.

e) Explain why a lag network alone cannot meet the design specifications.

f) Develop a full design using a lead-lag compensator that meets all the design specifications. Remember, you are not allowed to alter the previously chosen open-loop gain.

FIGURE 7.77

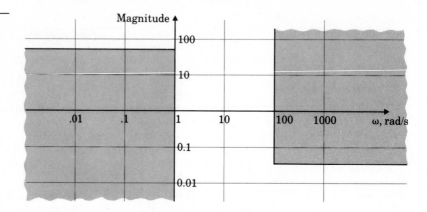

7.10 Consider the system in Fig. 7.78, where

$$G(s) = \frac{300}{s(s + 0.225)(s + 4)(s + 180)}.$$

It is desired to have an overall system satisfying the following specifications: (1) The steady-state error due to a step function is zero; (2) the phase margin is 55° and the gain margin is at least 6 dB; and (3) the gain crossover frequency is not smaller than that of the uncompensated plant.

a) What kind of compensation should be used and why?

b) Design a suitable compensator $D_c(s)$ to meet the specifications.

FIGURE 7.78

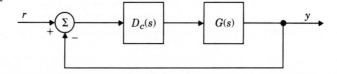

7.11 We have discussed three design methods: the root-locus method of Evans, the frequency-response method of Bode, and the state-variable-pole-placement method. Explain which of these methods is *best* described by the following statements. If you feel more than one method fits a given statement equally well, say so and explain why.

a) This method is readily used when the plant description must be obtained from experimental data.

b) This method permits most direct control over dynamic response characteristics such as rise time, percent overshoot, and settling time.

c) This method is most easily reduced to a computer-design algorithm.

d) This method permits most direct control over the steady-state error constants of K_p or K_v.

e) This method is most likely to lead to the *least-complex* controller capable of meeting the dynamic and static accuracy specifications.

f) This method allows the design to guarantee that the final design will be unconditionally stable.

g) This method can be used without modification for plants that include transportation lag terms such as

$$G(s) = \frac{e^{-2s}}{(s + 3)^2}$$

7.12 In design by the method of Bode we used lead and lag networks. For a type I system indicate the effects of these compensation networks on each of the design performance specifications listed below. Indicate the effects as "an increase," "substantially unchanged," or "a decrease" and use the second-order plant $G(s) = K/[s(s + 1)]$ to illustrate your conclusions.

a) K_v
b) Phase margin
c) Closed-loop bandwidth
d) Percent overshoot
e) Settling time

7.13 *Altitude control of a hot-air balloon.* The equations of vertical motion of a hot-air balloon (Fig. 7.79) linearized about vertical equilibrium are

$$\delta\dot{T} + \frac{1}{\tau_1}\delta T = \delta q,$$

$$\tau_2\ddot{z} + \dot{z} = a\delta T + w,$$

where

δT = deviation of hot-air temperature from temperature that gives buoyant force

z = altitude of balloon,

δq = deviation in burner heating rate from equilibrium rate, divided by thermal capacity of hot air,

w = vertical wind velocity.

An altitude-hold autopilot is to be designed for a balloon whose parameters are

$$\tau_1 = 250 \text{ s}, \qquad \tau_2 = 25 \text{ s}, \qquad a = 0.3 \text{m/s} - (°C)^{-1}.$$

Only altitude is sensed, so the control law is of the form

$$\delta q(s) = D(s)[z_d(s) - z(s)],$$

where z_d is the desired (commanded) altitude.

FIGURE 7.79

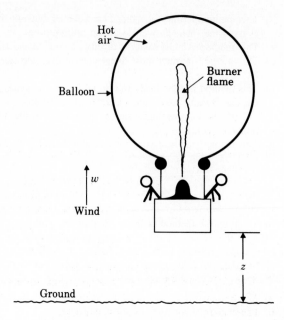

a) Sketch a root locus of the closed-loop eigenvalues versus gain K for a proportional-feedback controller, $\delta q = -K(z - z_d)$. Use Routh's criterion (or $s = j\omega$) to determine the value of gain above which the closed-loop system is unstable and the associated frequency at neutral stability.

b) Intuitively and from findings in (a), it is apparent that quite a bit of lead compensation is required to produce a satisfactory autopilot. Sketch a root locus of the closed-loop eigenvalues versus gain K for a double-lead compensator, $\delta q = -D(s)(z - z_d)$, where

$$-D(s) = K\left(\frac{s + 0.03}{s + 0.12}\right)^2.$$

c) Sketch Bode straight-line asymptote plots of *magnitude only* for the open-loop transfer functions of the proportional-feedback and the lead-compensated systems.

d) Select a gain K for the lead-compensated system to give a crossover frequency of 0.06 rad/s.

e) With the gains selected in (d), what is the steady-state error in altitude for a steady vertical wind of 1 m/s? (Be careful: first find the closed-loop transfer function from w to error.)

f) If the error in (e) is too large, how would you modify the compensation to give higher low-frequency gain? (Give a qualitative answer only.)

7.14 A satellite attitude control uses a reaction wheel to provide angular motion of the vehicle according to the equations:

$$\text{Satellite: } I\ddot{\phi} = T^c + T^{ex},$$
$$\text{Measurement: } \dot{Z} = \phi - aZ,$$
$$\text{Control: } T^c = -D(s)Z,$$

where

$$
\begin{aligned}
T^c &= \text{control torque,} \\
T^{ex} &= \text{external torque,} \\
\phi &= \text{angle to be controlled,} \\
Z &= \text{measurement from the sensor,} \\
I &= \text{satellite inertia, } 1000 \text{ kg/m}^2, \\
a &= \text{sensor constant, } 1 \text{ rad/s,} \\
D(s) &= \text{compensation.}
\end{aligned}
$$

a) Suppose $D(s) = K_0$, a constant. Draw the root locus with respect to K of the resulting closed-loop system.
b) For what range of K_0 is the resulting system stable?
c) Add a lead network with pole at $s = -1$ so that the closed-loop system has a natural frequency of $\omega_n = 0.04$ rad/s and a damping ratio of $\zeta = 0.5$.

$$
D(s) = K_1 \frac{s + z}{s + 1}.
$$

1) Where is the zero of the lead network ?
2) Draw the root locus of the compensated system and give the design value of K_1.

d) For what range of K_1 is the system stable?
e) What is the steady-state error to a constant T^{ex} with this design?
f) What is the "type" of this system with respect to rejection of T^{ex}?
g) Draw the Bode plot asymptotes of the *open-loop* system, with gain adjusted for the closed-loop value K_1 of (c). Use a scale of 1 in. per decade for magnitude and frequency and 1 in. per 90° for phase. Add the compensation of (c) and compute the *phase margin* of the result.
h) Select as the state ϕ, $\dot{\phi}$, and Z and write the state equations of the system. Select gains of K_ϕ and $K_{\dot\phi}$ to locate the control poles at $s = -0.02 \pm j0.02 \sqrt{3}$.

7.15 A standard control problem is sketched in Fig. 7.80, with three alternative solutions. The signal w is the plant noise and may be analyzed as if it were a step; the signal v is the sensor noise and may be analyzed as if it contained power to very high frequencies.

a) Compute the values of the parameters K_1, a, K_2, K_T, K_3, d, and K_D so that

$$
\frac{Y(s)}{R(s)} = \frac{16}{s^2 + 4s + 16}
$$

in each case. Note that in system III a pole is to be placed at $s = -4$.
b) Complete the following table:

System	K_v	$\left.\frac{y}{w}\right|_{s=0}$	$\left.\frac{y}{v}\right|_{s\to\infty}$
I			
II			
III			

Express the last entries as A/s^k to show how fast noise from v is attenuated at high frequencies.

c) Rank the three designs as to

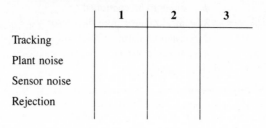

	1	2	3
Tracking			
Plant noise			
Sensor noise			
Rejection			

FIGURE 7.80

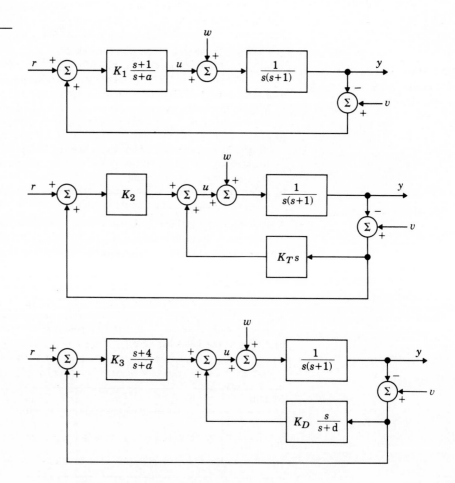

7.16 The equations of motion for a cart-stick balancer are

$$\dot{x} = \begin{bmatrix} 0 & 1 & 0 \\ 31.33 & 0 & 0.016 \\ -31.33 & 0 & -0.216 \end{bmatrix} x + \begin{bmatrix} 0 \\ -0.649 \\ 8.649 \end{bmatrix} u,$$

$$y = [10 \quad 0 \quad 0]x.$$

a) Compute the transfer function and determine the poles and zeros.
b) Determine the feedback gain K to move the poles of the system at third-order ITAE locations with $\omega_n = 4$ rad/s.
c) Determine the estimator gain L needed to place the estimator error poles at $-10, -10, -10$.
d) Determine the compensator for the design in (**b**) and (**c**). Draw a root locus of the system.
e) Suppose we use a reduced-order estimator with poles at $-10, -10$. What is the required estimator gain?
f) Determine the compensator for the reduced-order estimator. Draw a root locus of the system.
g) Compute the frequency responses of the two compensators.

7.17 The lateral-directional perturbation equations for a Boeing 747 on landing approach in horizontal flight at sea level, with $U = 221$ ft/s (Mach 0.198) weight of 564,000 lb are

$$\begin{bmatrix} \dot{\beta} \\ \dot{r} \\ \dot{p} \\ \dot{\phi} \end{bmatrix} = \begin{bmatrix} -0.0890 & -0.989 & 0.1478 & 0.1441 \\ 0.168 & -0.217 & -0.166 & 0 \\ -1.33 & 0.327 & -0.975 & 0 \\ 0 & 0.149 & 1 & 0 \end{bmatrix} \begin{bmatrix} \beta \\ r \\ p \\ \phi \end{bmatrix} + \begin{bmatrix} 0.0148 \\ -.151 \\ 0.0636 \\ 0 \end{bmatrix} \delta r,$$

$$y = [0 \quad 1 \quad 0 \quad 0] \begin{bmatrix} \beta \\ r \\ p \\ \phi \end{bmatrix}.$$

The transfer function is

$$G(s) = \frac{R(s)}{\delta R(s)} = \frac{-0.151(s + 105)(s + 0.0328 \pm j0.414)}{(s + 1.109)(s + 0.0425)(s + 0.646 \pm j0.731)}.$$

Draw the uncompensated root locus and frequency response of the system. What would be a satisfactory classical controller? Try a state-variable design approach by drawing a symmetric root locus for the system. Choose the closed-loop poles of the system on the symmetric root locus at

$$\alpha_c(s) = (s + 1.12)(s + 0.165)(s + 0.162 \pm j0.681),$$

and choose the estimator poles to be five times faster at

$$\alpha_e(s) = (s + 5.58)(s + 0.825)(s + 0.812 \pm j3.40).$$

Compute the compensator transfer function. Discuss the properties of the robustness of the system with respect to parameter variations and unmodeled dynamics. Note the similarity to the design with the flight conditions discussed in this chapter. What does this suggest about providing a continuous (nonlinear) control throughout the operating envelope?

7.18 Design and construct a device to keep a ball centered on a freely swinging beam. An example of such a device is shown in Fig.7.81. It uses coils surrounding permanent magnets as the actuator to move the beam, solar cells to sense the ball position, and a hall-effect device to sense the beam position. You should research other possible actuators and sensors as part of your design effort. Compare the quality of the control achievable for ball feedback only versus multiple loop feedback of both ball and beam position.

FIGURE 7.81
Ball balancer design example.

□8

Digital

Control

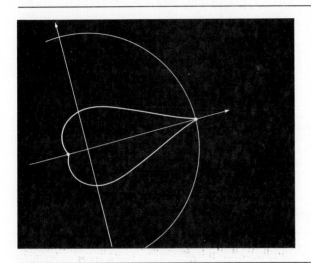

8.1 Introduction

The intent of this chapter is to provide the basic tools for the design of a control system that is to be implemented using a computer or microprocessor. The design methods are applicable to any type of computer (from microprocessor to large-scale computer); however, the effects of small word size and slow sample rates take on a more important role when a microprocessor is used. For a more complete discussion of this subject, see Franklin, Powell, and Workman (1990).

A typical topology of the type of system to be considered is shown in Fig. 8.1. There are two different approaches for the design of digital algorithms.

1. *Continuous design and digitization.* Perform a continuous design, then digitize the resulting compensation.

2. *Direct digital design.* Digitize the plant model, then perform a design using discrete analysis methods.

Both methods will be covered and their advantages and disadvantages discussed.

585

FIGURE 8.1
Basic control-system
block diagram.

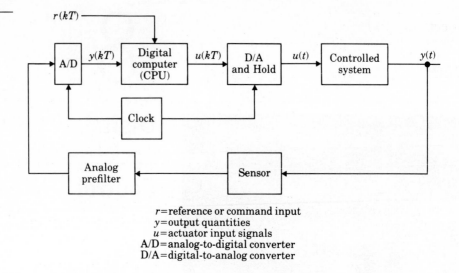

r=reference or command input
y=output quantities
u=actuator input signals
A/D=analog-to-digital converter
D/A=digital-to-analog converter

An actual design process is often a combination of the two methods. A first iteration to a digital design can be obtained using a discretization of a continuous design. Then the result is tuned up using a direct digital analysis and design. No matter what the process used, a prudent designer typically follows up with a simulation of the digitally controlled system, including all sampling and calculation delays, to validate the system performance predicted by linear analysis.

8.2 Theoretical Background

8.2.1 z Transform

In the analysis of continuous systems, we use the Laplace transform, which is defined by

$$L\{f(t)\} = F(s) = \int_0^\infty f(t)e^{-st}\,dt, \tag{8.1}$$

which leads directly to the important property that

$$L\{\dot{f}(t)\} = sF(s). \tag{8.2}$$

This relation enables us to easily find the transfer function of a linear continuous system, given the differential equation of that system.

FIGURE 8.2
A continuous and
sampled version
of signal f.

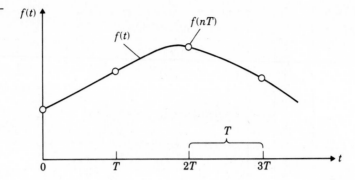

For discrete systems, a very similar procedure is available. The *z transform* is defined by

$$Z\{f(n)\} = F(z) = \sum_{n=0}^{\infty} f(n)z^{-n},\tag{8.3}$$

where $f(n)$ is the sampled version, as shown in Fig. 8.2, and $n = 1, 2, 3, \ldots$ refers to discrete sample times $t_1, t_2, t_3, \ldots$. This leads directly to a property analogous to Eq. (8.2), namely, that

$$Z\{f(n-1)\} = z^{-1}F(z),\tag{8.4}$$

where $n - 1$ refers to the value $f(n)$ one sample cycle in the past. This relation allows us to easily find the transfer function of a discrete system, given the difference equations of that system. For example, the general second-order difference equation,

$$y(n) = -a_1 y(n-1) - a_2 y(n-2) + b_0 u(n) + b_1 u(n-1) + b_2 u(n-2),\tag{8.5}$$

can be converted to the z transform of $y(n)$, $u(n)$, and so on by invoking Eq. (8.4) once or twice to arrive at

$$Y(z) = (-a_1 z^{-1} - a_2 z^{-2})Y(z) + (b_0 + b_1 z^{-1} + b_2 z^{-2})U(z),\tag{8.6}$$

which results in the transfer function

$$\frac{Y(z)}{U(z)} = \frac{b_0 + b_1 z^{-1} b_2 z^{-2}}{1 + a_1 z^{-1} a_2 z^{-2}}\tag{8.7}$$

8.2.2 *z* Transform Inversion

Table 8.1 relates simple discrete time functions to their z transforms and gives the Laplace transforms for the same time functions.

TABLE 8.1
Laplace Transforms and z Transforms of Simple Discrete Time Functions

$\mathcal{F}(s)$ is the Laplace transform of $f(t)$ and $F(z)$ is the z transform of $f(nT)$. Unless otherwise noted, $f(t) = 0$ if $t < 0$.

Number	$\mathcal{F}(s)$	$f(nT)$	$F(z)$
1		$1, n = 0; 0, n \neq 0$	1
2		$1, n = k; 0, n \neq k$	z^{-k}
3	$\dfrac{1}{s}$	$1(nT)$	$\dfrac{z}{z-1}$
4	$\dfrac{1}{s^2}$	nT	$\dfrac{Tz}{(z-1)^2}$
5	$\dfrac{1}{s^3}$	$\dfrac{1}{2!}(nT)^2$	$\dfrac{T^2}{2}\dfrac{z(z+1)}{(z-1)^3}$
6	$\dfrac{1}{s^4}$	$\dfrac{1}{3!}(nT)^3$	$\dfrac{T^3}{6}\dfrac{z(z^2+4z+1)}{(z-1)^4}$
7	$\dfrac{1}{s^m}$	$\displaystyle\lim_{a\to 0}\dfrac{(-1)^{m-1}}{(m-1)!}\dfrac{\partial^{m-1}}{\partial a^{m-1}}e^{-anT}$	$\displaystyle\lim_{a\to 0}\dfrac{(-1)^{m-1}}{(m-1)!}\dfrac{\partial^{m-1}}{\partial a^{m-1}}\dfrac{z}{z-e^{-aT}}$
8	$\dfrac{1}{s+a}$	e^{-anT}	$\dfrac{z}{z-e^{-aT}}$
9	$\dfrac{a}{(s+a)^2}$	nTe^{-anT}	$\dfrac{Tze^{-aT}}{(z-e^{-aT})^2}$
10	$\dfrac{1}{(s+a)^3}$	$\dfrac{1}{2}(nT)^2e^{-anT}$	$\dfrac{T^2}{2}\dfrac{e^{-aT}z(z+e^{-aT})}{(z-e^{-aT})^3}$
11	$\dfrac{1}{(s+a)^m}$	$\dfrac{(-1)^{m-1}}{(m-1)!}\dfrac{\partial^{m-1}}{\partial a^{m-1}}e^{-anT}$	$\dfrac{(-1)^{m-1}}{(m-1)!}\dfrac{\partial^{m-1}}{\partial a^{m-1}}\dfrac{z}{z-e^{-aT}}$

	$F(s)$	$f(nT)$	$F(z)$
12	$\dfrac{1}{s(s+a)}$	$1 - e^{-anT}$	$\dfrac{z(1-e^{-aT})}{(z-1)(z-e^{-aT})}$
13	$\dfrac{a}{s^2(s+a)}$	$\dfrac{1}{a}(anT - 1 + e^{-anT})$	$\dfrac{z[(aT-1+e^{-aT})z + (1-e^{-aT}-aTe^{-aT})]}{a(z-1)^2(z-e^{-aT})}$
14	$\dfrac{b-a}{(s+a)(s+b)}$	$e^{-anT} - e^{-bnT}$	$\dfrac{(e^{-aT}-e^{-bT})z}{(z-e^{-aT})(z-e^{-bT})}$
15	$\dfrac{s}{(s+a)^2}$	$(1-anT)e^{-anT}$	$\dfrac{z[z - e^{-aT}(1+aT)]}{(z-e^{-aT})^2}$
16	$\dfrac{a^2}{s(s+a)^2}$	$1 - e^{-anT}(1+anT)$	$\dfrac{z[z(1-e^{-aT}-aTe^{-aT}) + e^{-2aT} - e^{-aT} + aTe^{-aT}]}{(z-1)(z-e^{-aT})^2}$
17	$\dfrac{(b-a)s}{(s+a)(s+b)}$	$be^{-bnT} - ae^{-anT}$	$\dfrac{z[z(b-a)-(be^{-aT}-ae^{-bT})]}{(z-e^{-aT})(z-e^{-bT})}$
18	$\dfrac{a}{s^2+a^2}$	$\sin anT$	$\dfrac{z\sin aT}{z^2 - (2\cos aT)z + 1}$
19	$\dfrac{s}{s^2+a^2}$	$\cos anT$	$\dfrac{z(z-\cos aT)}{z^2 - (2\cos aT)z + 1}$
20	$\dfrac{s+a}{(s+a)^2+b^2}$	$e^{-anT}\cos bnT$	$\dfrac{z(z - e^{-aT}\cos bT)}{z^2 - 2e^{-aT}(\cos bT)z + e^{-2aT}}$
21	$\dfrac{b}{(s+a)^2+b^2}$	$e^{-anT}\sin bnT$	$\dfrac{ze^{-aT}\sin bT}{z^2 - 2e^{-aT}(\cos bT)z + e^{-2aT}}$
22	$\dfrac{a^2+b^2}{s[(s+a)^2+b^2]}$	$1 - e^{-anT}\left(\cos bnT + \dfrac{a}{b}\sin bnT\right)$	$\dfrac{z(Az + B)}{(z-1)[z^2 - 2e^{-aT}(\cos bT)z + e^{-2aT}]}$, where

$$A = 1 - e^{-aT}\cos bT - \frac{a}{b}e^{-aT}\sin bT$$

$$B = e^{-2aT} + \frac{a}{b}e^{-aT}\sin bT - e^{-aT}\cos bT$$

Given a general z transform, one can expand it into a sum of elementary terms using partial fraction expansion and find the resulting time series from the table. Again, these procedures are exactly the same as those used for continuous systems.

A z transform inversion technique that has no continuous counterpart is called *long division*. Given a z transform

$$Y(z) = \frac{N(z)}{D(z)}, \tag{8.8}$$

one simply divides the denominator into the numerator using long division. The result is a polynomial (perhaps infinite) in z, from which the time series can be found by using Eq. (8.3).

For example, a first-order system described by the difference equation

$$y(n) = ky(n-1) + u(n) \tag{8.9}$$

yields

$$\frac{Y(z)}{U(z)} = \frac{1}{1 - kz^{-1}}. \tag{8.10}$$

For an impulsive input defined by

$$u(0) = 1,$$
$$u(n) = 0 \qquad n \neq 0,$$

the z transform is

$$U(z) = 1$$

and

$$Y(z) = \frac{1}{1 - kz^{-1}}. \tag{8.11}$$

Therefore, to find the time series, we use long division as follows:

$$
\require{enclose}
\begin{array}{r}
1 + kz^{-1} + k^2z^{-2} + k^3z^{-3} + \cdots \\[2pt]
1 - kz^{-1} \enclose{longdiv}{1 } \\[2pt]
\underline{1 - kz^{-1}} \\[2pt]
kz^{-1} + 0 \\[2pt]
\underline{kz^{-1} - k^2z^{-2}} \\[2pt]
k^2z^{-2} + 0 \\[2pt]
\underline{kz^2z^{-2} - k^3z^{-3}} \\[2pt]
k^3z^{-3} \\[2pt]
\ddots
\end{array}
$$

to yield the infinite series for $Y(z)$:

$$Y(z) = 1 + kz^{-1} + k^2z^{-2} + k^3z^{-3} + \cdots,$$

which means that (see Eq. 8.3) the sampled time history of y is

$$y(0) = 1,$$
$$y(1) = k,$$
$$y(2) = k^2,$$
$$\vdots$$
$$y(n) = k^n.$$

8.2.3 Relationship between s and z

For continuous systems, we saw in Chapter 2 that certain behavior results from different pole locations in the s plane: oscillatory behavior for poles near the imaginary axis, exponential decay for poles on the negative real axis, and unstable behavior for poles with a positive real part. The same kind of association is also useful to designers for discrete systems. Consider the continuous signal

$$f(t) = e^{-at} \qquad t > 0, \tag{8.12}$$

which has the Laplace transform

$$F(s) = \frac{1}{s + a} \tag{8.13}$$

and corresponds to a pole at $s = -a$. The z transform of $f(nT)$ is

$$F(z) = Z\{e^{-anT}\},$$

which can be seen from Table 8.1 to be

$$F(z) = \frac{1}{1 - e^{-aT}z^{-1}}, \tag{8.14}$$

which corresponds to a pole at $z = e^{-aT}$. This means that a pole at $s = -a$ in the s plane corresponds to a pole at $z = e^{-aT}$ in the discrete domain. This is true in general, and the equivalent characteristics in the z plane are related to those in the s plane by the expression

$$z = e^{sT} \tag{8.15}$$

where T is the sample period. Table 8.1 also includes the Laplace transforms, demonstrating the $z = e^{sT}$ relationship in the denominators of the table entries.

Figure 8.3 shows the mapping of lines of constant damping ζ and natural frequency ω_n from the s plane to the upper half of the z plane, using Eq. (8.15). The mapping has several important features:

1. The stability boundary is the unit circle $|z| = 1$.
2. The small vicinity around $z = +1$ is essentially identical to the vicinity around $s = 0$.

FIGURE 8.3
Natural frequency and damping loci in z plane. (The lower half is the mirror image of the half shown.)

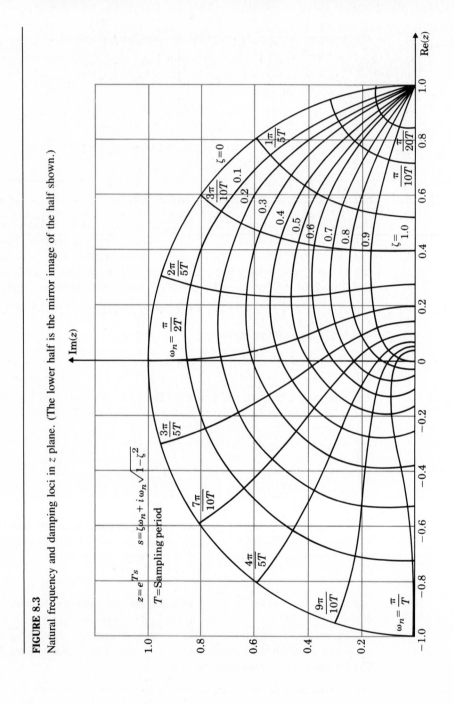

$z = e^{Ts}$ $s = \zeta\omega_n + i\omega_n\sqrt{1-\zeta^2}$

T = Sampling period

3. The z plane locations give response information normalized to the sample rate, rather than to time, as in the s plane.

4. The negative real z axis always represents a frequency of $\omega_s/2$, where $\omega_s = 2\pi/T =$ sample rate.

5. Vertical lines in the left-half s plane (constant real part or time constant) map into circles within the unit circle of the z plane.

6. Horizontal lines in the s plane (constant imaginary part or frequency) map into radial lines in the z plane.

7. There is no location in the z plane that represents frequencies greater than $\omega_s/2$. Physically, this is because one must sample at least twice as fast as a signal's frequency to represent it digitally. Mathematically, it is because of the nature of the trigonometric functions imbedded in Eq. (8.15).

8.2.4 Final-Value Theorem

The final-value theorem for continuous systems discussed in Section 2.7.1,

$$\lim_{t \to \infty} x(t) = \lim_{s \to 0} sX(s), \tag{8.16}$$

is often used to find steady-state system errors or steady-state gains of portions of a control system. The analog for discrete systems is obtained by noting that a continuous steady response is denoted by $X(s) = A/s$ and leads to the multiplication by s in Eq. (8.16). Therefore, since the steady response for discrete systems is

$$X(z) = \frac{A}{1 - z^{-1}},$$

the discrete final-value theorem is

$$\lim_{n \to \infty} x(n) = \lim_{z \to 1} (1 - z^{-1})X(z). \tag{8.17}$$

For example, to find the DC gain of the transfer function

$$G(z) = \frac{X(z)}{U(z)} = \frac{0.58(1 + z)}{z + 0.16},$$

let $u(n) = 1$ for $n \geq 0$, so that

$$U(z) = \frac{1}{1 - z^{-1}}$$

and

$$X(z) = \frac{0.58(1 + z)}{(1 - z^{-1})(z + 0.16)}.$$

Applying the final-value theorem yields

$$x(\infty) = \lim_{z \to 1} \left[\frac{0.58(1 + z)}{z + 0.16} \right] = 1,$$

and therefore the DC gain of $G(z)$ is unity. In general, we see that to find the DC gain of any stable transfer function (all poles within the unit circle), we simply substitute $z = 1$ and compute the resulting gain.

Since the gain of a system should not change whether represented continuously or discretely, this calculation is an excellent aid in finding a discrete controller that best matches a continuous controller. It is also a good check on the calculations associated with determining the discrete model of a system.

8.3   Continuous Design

The first part of this design procedure should already be familiar to the reader, since it has been the subject of the first seven chapters of the book. We carry out this part of the design as before, with no changes required to represent the fact that the control will eventually be implemented digitally. The second part of the procedure is to digitize the resulting compensation.

8.3.1 Digitization Procedures

The problem is: Given a continuous compensation $D(s)$, find the best equivalent discrete compensation $D(z)$. Or more exactly, given a $D(s)$ from the control system shown in Fig. 8.4, find the best digital implementation of that compensation. A digital implementation requires that y be sampled at some sample rate and that the computer output samples be smoothed in some manner so as to provide a continuous u. For ease of hardware design, the smoothing operation is almost always a simple hold (or zero-order hold, ZOH), which is shown in Fig. 8.5.

Therefore we can restate the problem as: Find the best $D(z)$ in the digital implementation shown in Fig. 8.6 to match a desired $D(s)$. It is important to note at the outset that there is no exact solution to this problem because $D(s)$ responds to the complete time history of $x(t)$, whereas $D(z)$ has access only to the samples

FIGURE 8.4
A continuous control system.

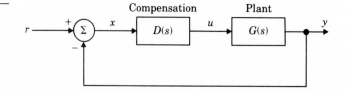

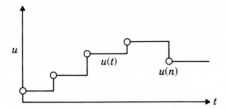

$x(n)$. In a sense, the various digitization approximations simply make different assumptions about what happens to $x(t)$ between the sample points.

Tustin's Method

One digitization method is to approach the problem as one of numerical integration. Suppose

$$\frac{U(s)}{X(s)} = D(s) = \frac{1}{s},$$

that is, pure integration. Therefore

$$u(nT) = \int_0^{mT-T} x(t)\,dt + \int_{nT-T}^{T} x(t)\,dt,$$

which can be rewritten as

$$u(nT) = u(nT - T) + \text{area under } x(t) \text{ over last } T, \qquad (8.18)$$

where T is the sample period. $u(nT)$ is usually written $u(n)$ for short, and the task at each step is to use trapezoidal integration, that is, to approximate $x(t)$ by a straight line between the two samples (Fig. 8.7). Therefore Eq. (8.18) becomes

$$u(nT) = u(nT - T) + \frac{T}{2}[x(nT - T) + x(nT)], \qquad (8.19)$$

or, taking the z transform,

$$\frac{U(z)}{X(z)} = \frac{T}{2}\frac{1 + z^{-1}}{1 - z^{-1}} = \frac{1}{(2/T)[(1 - z^{-1})/(1 + z^{-1})]}. \qquad (8.20)$$

For

$$D(s) = \frac{a}{s + a}$$

FIGURE 8.7
Trapezoidal integration.

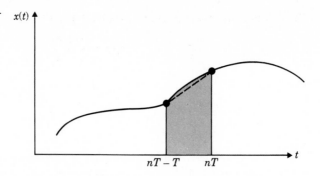

application of the same integration approximation yields

$$D(z) = \frac{a}{(2/T)[(1 - z^{-1})/(1 + z^{-1})] + a}.$$

In fact, the substitution

$$s = \frac{2}{T} \frac{1 - z^{-1}}{1 + z^{-1}} \qquad (8.21)$$

in any $D(s)$ yields a $D(z)$ based on the trapezoidal integration formula. This is called *Tustin's,* or the *bilinear approximation.*

Matched Pole-Zero Method (MPZ)

Another digitization method, called the *matched pole-zero,* is found by extrapolation of the relation between the s and z planes stated in Eq. (8.15). If we take the z transform of a sampled $x(t)$, then the poles of $X(z)$ are related to the poles of $X(s)$ according to $z = e^{sT}$. The idea of the matched pole-zero technique is to apply $z = e^{sT}$ to the poles and zeros of a transfer function, even though it doesn't strictly apply to transfer functions or even to the zeros of a time sequence. Like all the transfer function digitization methods, this method is an approximation; here the approximation is motivated partly by the fact that $z = e^{sT}$ is the correct s-to-z transformation for the poles of a time sequence and partly by the minimal amount of algebra required to determine the digitized transfer function.

Since physical systems often have more poles than zeros, it is also useful to arbitrarily add zeros to $D(z)$ at $z = -1$—that is, a $(1 + z^{-1})$ term—which causes an averaging of the current and past input values, as in the trapezoidal integration (Tustin's) method. The gain is selected so that the low-frequency gains of $D(s)$ and $D(z)$ match one another. The method can be summarized as follows:

1. Map poles and zeros according to $z = e^{sT}$.
2. Add $1 + z^{-1}$, $(1 + z^{-1})^2$, or the like if the numerator is of a lower order than the denominator.
3. Match DC or low-frequency gain.

For example, the matched pole-zero approximation of

$$D(s) = \frac{s + a}{s + b}$$

is

$$D(z) = k\frac{z - e^{-aT}}{z - e^{-bT}}, \qquad (8.22)$$

where

$$k = \frac{a}{b}\frac{1 - e^{-bT}}{1 - e^{-aT}},$$

and for

$$D(s) = \frac{s + a}{s(s + b)} \Rightarrow D(z) = k\frac{(z + 1)(z - e^{-aT})}{(z - 1)(z - e^{-bT})}, \qquad (8.23)$$

where matching gain at low frequency yields

$$k = \frac{a}{2b}\frac{1 - e^{-bT}}{1 - e^{-aT}}.$$

In both digitization methods, the fact that an equal power of z appears in the numerator and denominator of $D(z)$ implies that the difference equation output at time n will require a sample of the input at time n. For example, the $D(z)$ in Eq. (8.22) can be written

$$\frac{U(z)}{X(z)} = D(z) = k\frac{1 - \alpha z^{-1}}{1 - \beta z^{-1}},$$

where $\alpha = e^{-aT}$ and $\beta = e^{-bT}$, which by inspection can be seen to result in the difference equation

$$u(n) = \beta u(n - 1) + k[x(n) - \alpha x(n - 1)]. \qquad (8.24)$$

It is impossible to sample $x(n)$, compute $u(n)$, and then output $u(n)$ all in zero elapsed time; therefore Eq. (8.24) is technically impossible to implement. However, if the equation is simple enough or the computer is fast enough, a slight delay between the $x(n)$ sample and the $u(n)$ output will have a negligible effect on the actual response of the system compared with that expected from the original design. A rule of thumb would be to keep the delay on the order of one-twentieth of the system rise time.

Modified Matched Pole-Zero Method (MMPZ)

The $D(z)$ in Eq. (8.23) would also result in $u(n)$ being dependent on $x(n)$, the input at the same time point. If the structure of the computer hardware prohibits this relation or if the computations are particularly lengthy, it may be desirable to derive a $D(z)$ that has one less power of z in the numerator than in the denominator, so that the computer output $u(n)$ would only require input from the previous time, that is, $x(n-1)$. To do this, we simply modify step 2 in the matched pole-zero procedure. The second example,

$$D(s) = \frac{s+a}{s(s+b)},$$

would then become

$$D(z) = k\frac{z - e^{-aT}}{(z-1)(z - e^{-bT})},$$

where

$$k = \frac{a}{b}\frac{1 - e^{-bT}}{1 - e^{-aT}},$$

which results in

$$u(n) = (1 + e^{-bT})u(n-1) - e^{-bT}u(n-2) + k[x(n-1) - e^{-aT}x(n-2)].$$

In this equation an entire sample period is available to perform the calculation and to output $u(n)$. A discrete analysis of this controller would therefore more accurately explain the behavior of the actual system. However, because the controller is using data that is one cycle old, this controller will typically not perform as well as the MPZ in terms of the deviations of the desired system output.

Method Comparison

A numerical comparison of the magnitude of the frequency response for a first-order compensator is made in Fig. 8.8 for the three approximation techniques at two different sample rates. The results of the $D(z)$ computations used to draw Fig. 8.8 are shown in Table 8.2.

FIGURE 8.8
A comparison of discrete approximations.

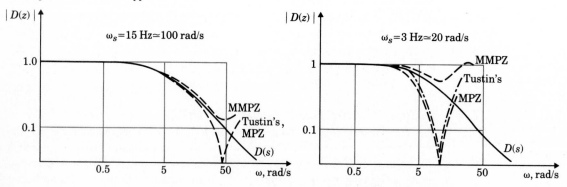

TABLE 8.2
Digital Approximations:
$D(z)$ **for** $D(s) = 5/(s+5)$

Method	$\omega_s = 15\text{Hz}$	$\omega_s = 3\text{Hz}$
Matched pole-zero (MPZ)	$0.143\dfrac{z+1}{z-0.715}$	$0.405\dfrac{z+1}{z-0.189}$
Modified MPZ (MMPZ)	$0.285\dfrac{1}{z-0.715}$	$0.811\dfrac{1}{z-0.189}$
Tustin's	$0.143\dfrac{z+1}{z-0.713}$	$0.454\dfrac{z+1}{z-0.0914}$

The figure shows that all the approximations are quite good at frequencies below about one-fourth the sample rate, $\omega_s/4$. If $\omega_s/4$ is sufficiently larger than the filter break frequency—that is, if the sampling is fast enough—the break characteristics are accurately reproduced. Tustin's and the MPZ methods show a notch at $\omega_s/2$ because of their zero term, $z + 1$. Other than the large difference at $\omega_s/2$, which is typically outside the range of interest, the three methods have similar accuracies. Since the MPZ techniques require much simpler algebra than Tustin's, they are preferred if the designer is performing the calculations by hand.

8.3.2 Design Example

For a $1/s^2$ plant, we wish to design a digital controller to have a closed-loop natural frequency $\omega_n \approx 0.3$ rad/s and $\zeta = 0.7$. The first step is to find the proper $D(s)$ defined in Fig. 8.9. The specifications can be met with

$$D(s) = k\frac{s+a}{s+b} \tag{8.25}$$

where

$a = 0.2,$
$b = 2.0,$
$k = 0.81,$

as can be verified by the root locus in Fig. 8.10. To digitize this $D(s)$, we first need to select a sample rate. For a system with $\omega_n = 0.3$ rad/s, a very "safe"

FIGURE 8.9
A continuous-design statement.

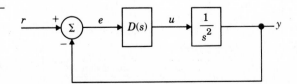

FIGURE 8.10
s plane locus versus *k*.

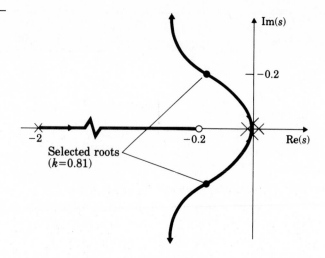

sample rate would be a factor of 20 faster than ω_n, yielding

$$\omega_s = 0.3 \times 20 = 6 \text{ rad/s, or 1 Hz.}$$

Thus let's pick $T = 1$ s. The matched pole-zero digitization of Eq. (8.25) is given by Eq. (8.22) and yields

$$D(z) = 0.389 \frac{z - 0.82}{z - 0.135} \tag{8.26}$$

or

$$D(z) = \frac{0.389 - 0.319z^{-1}}{1 - 0.135z^{-1}},$$

FIGURE 8.11
A digital control system.

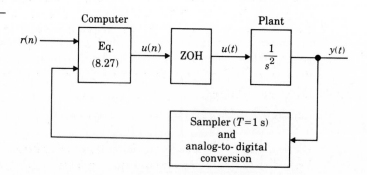

which leads to

$$u(n) = 0.135u(n - 1) + 0.389e(n) - 0.319e(n - 1), \qquad (8.27)$$

where

$$e(n) = r(n) - y(n),$$

and completes the digital algorithm design. The complete digital system is shown in Fig. 8.11.

8.3.3 Applicability Limits of Method

If an exact discrete analysis or a simulation of a system were performed and the digitization determined for a wide range of rates, the system would be unstable for sample rates slower than approximately $5 \times \omega_n$, and the damping would be substantially degraded for sample rates slower than $10 \times \omega_n$. At sample rates on the order of $20 \times \omega_n$ (or $20 \times$ bandwidth for more complex systems), this design method will yield reasonable results, and can be used with confidence for sample rates of $30 \times$ bandwidth or higher.

Basically, the errors come about because the technique ignores the lagging effect of the ZOH. An approximate method to account for this is to assume that the transfer function of the ZOH is

$$G_{\text{ZOH}}(s) = \frac{2/T}{s + (2/T)}. \qquad (8.28)$$

This is based on the idea that, on the average, the hold delays the signal by $T/2$, and the above is a first-order lag with a time constant of $T/2$ and a DC gain of 1. We could therefore patch up the original $D(s)$ design by inserting this $G_{\text{ZOH}}(s)$ into the original plant model and finding the $D(s)$ that yields satisfactory response.

One of the advantages of using the continuous design method, however, is that the sample rate need not be selected until after the basic feedback design is completed. Use of Eq. (8.28) eliminates this advantage, although it does partially alleviate a primary disadvantage, the approximate nature of the method.

8.4 Discrete Design

8.4.1 Analysis Tools

The first step in performing a control design or analysis of a system with some discrete elements is to find the discrete transfer function of the continuous portion.

For a system similar to that shown in Fig. 8.1, we wish to find the transfer function between $u(kT)$ and $y(kT)$. Unlike the case discussed in the previous section, there is an exact discrete equivalent for this system because the ZOH precisely describes what happens between samples, and the output $y(n)$ is dependent only on the input at the sample times $u(n)$.

For a plant described by a $G(s)$ and preceded by a ZOH, the discrete transfer function is

$$G(z) = (1 - z^{-1})Z \left\{ \frac{G(s)}{s} \right\}, \qquad (8.29)$$

where $Z\{F(s)\}$ is the z transform of the time series whose Laplace transform is $F(s)$, that is, on the same line in Table 8.1. The formula has the term $G(s)/s$ because the control comes in as a step input during each sample period. The term $1 - z^{-1}$ is there because a one-sample duration step can be thought of as an infinite duration step followed by a negative step one cycle delayed. This formula (Eq. 8.29) allows us to replace the mixed (continuous and discrete) system shown in Fig. 8.12(a) with a pure discrete equivalent system shown in Fig. 8.12(b).

The analysis and design of discrete systems is very similar to the analysis and design of continuous systems; in fact, all the same rules apply. The closed-loop transfer function of Fig. 8.12(b) is obtained using the same rules of block-diagram reduction; that is,

$$\frac{Y(z)}{R(z)} = \frac{DG}{1 + DG}. \qquad (8.30)$$

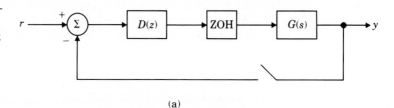

(a)

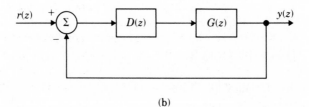

(b)

Since we'd like to find the characteristic behavior of the closed-loop system, we wish to find the factors of the denominator of Eq. (8.30), that is, find the roots of the characteristic equation

$$1 + D(z)G(z) = 0. \tag{8.31}$$

The root-locus techniques used in continuous systems to find roots of a polynomial in s apply equally well here for the polynomial in z. The rules apply directly without modification; however, the interpretation of the results is quite different, as we saw in Fig. 8.3. A major difference is that the stability boundary is now the unit circle instead of the imaginary axis.

A simple example of the discrete design tools discussed so far should help fix ideas. Suppose $G(s)$ in Fig. 8.12(a) is

$$G(s) = \frac{a}{s + a}. \tag{8.32}$$

It follows from Eq. (8.29) that

$$G(z) = (1 - z^{-1})Z\left[\frac{a}{s(s + a)}\right]$$

$$= (1 - z^{-1})\left[\frac{(1 - e^{-aT})z^{-1}}{(1 - z^{-1})(1 - e^{-aT}z^{-1})}\right]$$

$$= \frac{1 - \alpha}{z - \alpha}$$

where

$$\alpha = e^{-aT}.$$

To analyze the performance of a closed-loop proportional control law, that is, $D(z) = k$, we use standard root-locus rules. The result is shown in Fig. 8.13(a), and for comparison the root locus for a continuous controller is shown in Fig. 8.13(b). In contrast to the continuous case, which remains stable for all values of k, the discrete case becomes oscillatory with a decreasing damping ratio as z goes from 0 to -1 and eventually becomes unstable. This instability is due to the lagging effect of the ZOH, which is properly accounted for in the discrete analysis.

FIGURE 8.13
(a) z plane root locus;
(b) s plane root locus.

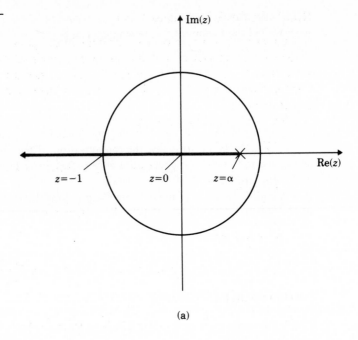

(a)

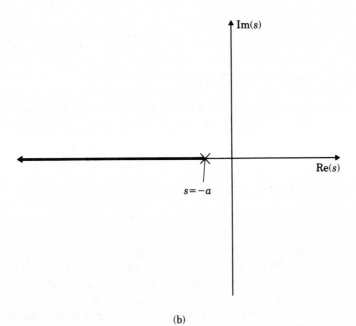

(b)

8.4.2 Feedback Properties

In continuous systems we typically start the design process by using proportional, derivative, or integral control laws or combinations of these, sometimes with a lag included. The same ideas are used in discrete designs directly, or perhaps the $D(z)$ that results from the digitization of a continuously designed $D(s)$ is used as a starting point.

The discrete control laws are

Proportional:

$$u(n) = k_p e(n) \Rightarrow D(z) = k_p; \tag{8.33}$$

Derivative:

$$u(n) = k_D[e(n) - e(n-1)],$$

which has the z transform

$$D(z) = k_D(1 - z^{-1}) = k_D \frac{z-1}{z}; \tag{8.34}$$

Integral:

$$u(n) = u(n-1) + k_I e(n),$$

which has the z transform

$$D(z) = \frac{k_I}{1 - z^{-1}} = \frac{k_I z}{z-1}. \tag{8.35}$$

8.4.3 Design Example

For an example, let's use the same problem that we used for the continuous design: the $1/s^2$ plant. Using Eq. (8.29), we have

$$G(z) = \frac{T^2}{2} \frac{z+1}{(z-1)^2}, \tag{8.36}$$

which becomes, with $T = 1$ s,

$$G(z) = \frac{1}{2} \frac{z+1}{(z-1)^2}. \tag{8.37}$$

Proportional feedback in the continuous case yields pure oscillatory motion, and in the discrete case we should expect even worse results. The root locus in Fig. 8.14 verifies this. For very low values of k (very low frequencies compared to the sample rate) the locus is tangent to the unit circle ($\zeta \cong 0$ and pure oscillatory motion), thus matching the behaviour of the proportional continuous design. For higher values of k, the locus diverges into the unstable region because of the effect of the ZOH and sampling.

FIGURE 8.14
z plane for a $1/s^2$ plant.

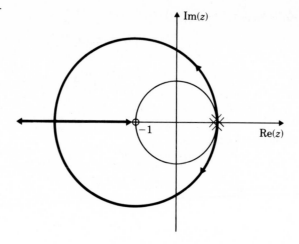

To compensate for this, let's add a velocity term to the control law, or

$$u(z) = k[1 + \gamma(1 - z^{-1})]e(z), \tag{8.38}$$

which yields

$$D(z) = k(1 + \gamma)\frac{z - [\gamma/(1 + \gamma)]}{z}. \tag{8.39}$$

Now the task is to find the values of γ and k that yield good performance. When we did this design previously, we wanted $\omega_n = 0.3$ rad/s and $\zeta = 0.7$. Figure 8.3 indicates that this s plane root location maps into a z plane location of

$$z = 0.8 \pm 0.2j.$$

FIGURE 8.15
Compensated z plane
locus for the $1/s^2$
example.

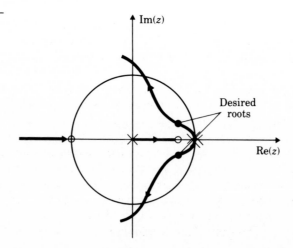

Desired
roots

Figure 8.15 shows that for $\gamma = 4$ and $k = 0.08$, or

$$D(z) = 0.4\frac{z - 0.8}{z},\qquad(8.40)$$

the roots are at the desired location. Normally, it is not particularly advantageous to match specific z plane root locations; rather it is only necessary to pick k and γ to obtain acceptable z plane roots, a much easier task. In this example, we want to match a specific location only so that we can compare the result with the previous design.

The control law that results is

$$u(z) = 0.08[1 + 4(1 - z^{-1})]$$

or

$$u(n) = 0.4e(n) - 0.32e(n - 1),\qquad(8.41)$$

which is very similar to the control law equation (8.27) obtained previously.

8.4.4 Design Comparison

The controller designed using purely discrete methods, Eq. (8.41), basically differs from the continuously designed controller, Eq. (8.27), only by the absence of the $u(n - 1)$ term. The $u(n - 1)$ term in Eq. (8.27) results from the lag term, $(s + b)$, in the compensation, Eq. (8.25), which is typically included in analog controllers for noise attenuation and because of the difficulty in building pure analog differentiators. Some equivalent lag in discrete design naturally appears as a pole at $z = 0$ (see Fig. 8.15) and represents the one sample delay in computing the derivative by a first difference. For more noise attenuation, the pole could be moved to the right of $z = 0$, thus resulting in less derivative action and more smoothing, the same trade-off that exists in continuous control design.

Other than the $u(n - 1)$ term, the two controllers represented by Eqs. (8.27) and (8.41) are very similar. This similarity results because the sample rate is fairly fast compared with ω_n; that is, $\omega_s \approx 20 \times \omega_n$. As the sample rate decreases, the numerical values in the compensations become increasingly different. For the discrete design, the actual system response follows that indicated by the z plane root locations, while the continuously designed system response diverges from that indicated by the s plane root locations.

As a general rule, discrete design should be used if sampling frequency is slower than $10 \times \omega_n$. At the very least, a continuous design with slow sampling ($\omega_s < 10\omega_n$) should be verified by a discrete analysis or simulation, and the compensation adjusted if necessary. A simulation of a digital control system is a good idea in any case. If the simulation properly accounts for all delays and possibly asynchronous behavior of different modules, it may expose instabilities that are impossible to detect using continuous or discrete linear analysis.

8.5   State-Space Design Methods

We have seen in Chapter 6 that a linear, constant-coefficient continuous system can be represented by a set of first-order matrix differential equations:

$$\dot{\mathbf{x}} = \mathbf{F}\mathbf{x} + \mathbf{G}u + \mathbf{G}_1 w, \tag{8.42}$$

where u is the control input to the system and w is a disturbance input. The output equation can be expressed as

$$y = \mathbf{H}\mathbf{x} + Ju, \tag{8.43}$$

and the solution to the state equation is

$$\mathbf{x}(t) = e^{\mathbf{F}(t-t_0)}\mathbf{x}(t_0) + \int_{t_0}^{t} e^{\mathbf{F}(t-\tau)}\mathbf{G}u(\tau)\,d\tau. \tag{8.44}$$

It is possible to use this solution to obtain a discrete state-space representation of the system. We wish to use this solution over one sample period to obtain a difference equation; hence we juggle the notation a bit (let $t = nT + T$ and $t_0 = nT$) and arrive at a particular version of Eq. (8.44):

$$\mathbf{x}(nT + T) = e^{\mathbf{F}T}\mathbf{x}(nT) + \int_{nT}^{nT+T} e^{\mathbf{F}(nT+T-\tau)}\mathbf{G}u(\tau)\,d\tau. \tag{8.45}$$

This result is not dependent on the type of hold, since u is specified in terms of its continuous time history $u(\tau)$ over the sample interval. A common and typically valid assumption is that of a zero-order hold (ZOH) with no delay, that is,

$$u(\tau) = u(nT), \qquad nT \le \tau < nT + T.$$

To facilitate the solution of Eq. (8.45) for a ZOH with no delay, let

$$\eta = nT + T - \tau.$$

Then we have

$$\mathbf{x}(nT + T) = e^{\mathbf{F}T}\mathbf{x}(nT) + \int_{0}^{T} e^{\mathbf{F}\eta}\,d\eta\,\mathbf{G}u(nT). \tag{8.46}$$

If we define

$$\mathbf{\Phi} = e^{\mathbf{F}T} \tag{8.47}$$

and

$$\mathbf{\Gamma} = \int_{0}^{T} e^{\mathbf{F}\eta}\,d\eta\,\mathbf{G}, \tag{8.48}$$

Eqs. (8.42) and (8.43) reduce to difference equations in standard form:

$$\mathbf{x}(n + 1) = \boldsymbol{\Phi}\mathbf{x}(n) + \boldsymbol{\Gamma}u(n) + \boldsymbol{\Gamma}_1 w(n),$$

$$y(n) = \mathbf{H}\mathbf{x}(n) + Ju(n), \tag{8.49}$$

where we include the effect of a disturbance w. If w is a constant, then $\boldsymbol{\Gamma}_1$ is given by Eq. (8.48), with $\mathbf{G}$ replaced by $\mathbf{G}_1$. If w is an impulse, then $\boldsymbol{\Gamma}_1 = \mathbf{G}_1$ and $w(n) = w_0 \delta_0(n)$.* The $\boldsymbol{\Phi}$ series expansion,

$$\boldsymbol{\Phi} = e^{\mathbf{F}T} = \mathbf{I} + \mathbf{F}T + \frac{\mathbf{F}^2 T^2}{2!} + \frac{\mathbf{F}^3 T^3}{3!} + \cdots,$$

can also be written

$$\boldsymbol{\Phi} = \mathbf{I} + \mathbf{F}T\boldsymbol{\Psi}, \tag{8.50}$$

where

$$\boldsymbol{\Psi} = \mathbf{I} + \frac{\mathbf{F}T}{2!} + \frac{\mathbf{F}^2 T^2}{3!} + \cdots.$$

The $\boldsymbol{\Gamma}$ integral in Eq. (8.48) can be evaluated term by term to give

$$\boldsymbol{\Gamma} = \sum_{k=0}^{\infty} \frac{\mathbf{F}^k T^{k+1}}{(k+1)!}\mathbf{G}$$

$$= \sum_{k=0}^{\infty} \frac{\mathbf{F}^k T^k}{(k+1)!}T\mathbf{G}$$

$$= \boldsymbol{\Psi}T\mathbf{G} \tag{8.51}$$

We evaluate $\boldsymbol{\Psi}$ by a series in the form

$$\boldsymbol{\Psi} \approx \mathbf{I} + \frac{\mathbf{F}T}{2}\left\{ \mathbf{I} + \frac{\mathbf{F}T}{3}\left[\cdots \frac{\mathbf{F}T}{N-1}\left(\mathbf{I} + \frac{\mathbf{F}T}{N} \right) \right] \cdots \right\}, \tag{8.52}$$

which has better numerical properties than the direct series of powers. We then find $\boldsymbol{\Gamma}$ from Eq. (8.51) and $\boldsymbol{\Phi}$ from Eq. (8.50). For a discussion of various methods of numerical determination of $\boldsymbol{\Phi}$ and $\boldsymbol{\Gamma}$, refer to Franklin, Powell, and Workman (1990) and Moler and van Loan (1978).

To compare this method of representing the plant with the discrete transfer function, we can take the z transform of Eq. (8.49), with $w = J = 0$, and obtain

$$[z\mathbf{I} - \boldsymbol{\Phi}]\mathbf{X}(z) = \boldsymbol{\Gamma}U(z), \tag{8.53}$$

$$Y(z) = \mathbf{H}\mathbf{X}(z). \tag{8.54}$$

---

*δ_0 is the discrete, or Kronecker, delta, which is zero except at $n = 0$, where it is 1.0.

Therefore

$$\frac{Y(z)}{U(z)} = \mathbf{H}[z\mathbf{I} - \mathbf{\Phi}]^{-1}\mathbf{\Gamma}. \tag{8.55}$$

For the satellite attitude control example discussed in Section 2.2, the $\mathbf{\Phi}$ and $\mathbf{\Gamma}$ matrices are easy to calculate using Eqs. (8.50) and (8.51) and the following values for $\mathbf{F}$ and $\mathbf{G}$:

$$\mathbf{F} = \begin{bmatrix} 0 & 1 \\ 0 & 0 \end{bmatrix}, \qquad \mathbf{G} = \begin{bmatrix} 0 \\ 1 \end{bmatrix}.$$

Since $\mathbf{F}^2 = 0$ in this case, $\mathbf{\Phi}$ has only two nonzero terms,

$$\mathbf{\Phi} = \mathbf{I} + \mathbf{F}T$$

$$= \begin{bmatrix} 1 & 0 \\ 0 & 1 \end{bmatrix} + \begin{bmatrix} 0 & 1 \\ 0 & 0 \end{bmatrix} T = \begin{bmatrix} 1 & T \\ 0 & 1 \end{bmatrix};$$

$$\mathbf{\Gamma} = \left[\mathbf{I}T + \mathbf{F}\frac{T^2}{2!} \right] \mathbf{G}$$

$$= \left\{ \begin{bmatrix} T & 0 \\ 0 & T \end{bmatrix} + \begin{bmatrix} 0 & 1 \\ 0 & 0 \end{bmatrix} \frac{T^2}{2} \right\} \begin{bmatrix} 0 \\ 1 \end{bmatrix} = \begin{bmatrix} T^2/2 \\ T \end{bmatrix}.$$

Hence, using Eq. (8.55), we obtain

$$\frac{Y(z)}{U(z)} = \begin{bmatrix} 1 & 0 \end{bmatrix} \left\{ z \begin{bmatrix} 1 & 0 \\ 0 & 1 \end{bmatrix} - \begin{bmatrix} 1 & T \\ 0 & 1 \end{bmatrix} \right\}^{-1} \begin{bmatrix} T^2/2 \\ T \end{bmatrix}$$

$$= \frac{T^2}{2} \frac{z+1}{(z-1)^2},$$

which is the same result that would have been obtained using the z transform tables. Note that to compute Y/U we find that the denominator is the determinant $\det[z\mathbf{I} - \mathbf{\Phi}]$, which comes from the matrix inverse in Eq. (8.55). This determinant is the characteristic polynomial of the transfer function, and the zeros of the determinant are the poles of the plant. We have two poles at $z = 1$ in this case, corresponding to two integrations in this plant's equations of motion.

We can further explore the question of poles and zeros and the state-space description by considering again the transform formulas, Eqs. (8.53) and (8.54). An interpretation of transfer-function poles from the perspective of the corresponding difference equation is that a pole is a value of z such that the equation has a nontrivial solution when the forcing input is zero. From Eq. (8.53), this implies that the linear equations

$$[z\mathbf{I} - \mathbf{\Phi}]\mathbf{X}(z) = [0]$$

have a nontrivial solution. From matrix algebra the well-known requirement for this is that $\det[z\mathbf{I} - \mathbf{\Phi}] = 0$. In the present case,

$$\det[z\mathbf{I} - \mathbf{\Phi}] = \det\left\{\begin{bmatrix} z & 0 \\ 0 & z \end{bmatrix} - \begin{bmatrix} 1 & T \\ 0 & 1 \end{bmatrix}\right\}$$

$$= \det\begin{bmatrix} z-1 & -T \\ 0 & z-1 \end{bmatrix}$$

$$= (z-1)^2 = 0,$$

which is the characteristic equation, as we have seen.

Along the same line of reasoning, a system zero is a value of z such that the system output is zero even with a nonzero state and input combination. Thus we are able to find a nontrivial solution for $\mathbf{X}(z_0)$ and $U(z_0)$ such that if $Y(z_0)$ is identically zero, then z_0 is a *zero* of the system. Combining Eqs. (8.53) and (8.54), we must satisfy the requirement

$$\begin{bmatrix} z\mathbf{I} - \mathbf{\Phi} & -\mathbf{\Gamma} \\ \mathbf{H} & 0 \end{bmatrix}\begin{bmatrix} \mathbf{X}(z) \\ u(z) \end{bmatrix} = [0]. \tag{8.56}$$

Once more the condition for the existence of nontrivial solutions is that the determinant of the square coefficient system matrix be zero. For the satellite example,

$$\det\begin{bmatrix} z-1 & -T & -T^2/2 \\ 0 & z-1 & -T \\ 1 & 0 & 0 \end{bmatrix} = \det\begin{bmatrix} -T & -T^2/2 \\ z-1 & -T \end{bmatrix}$$

$$= T^2 + \frac{T^2}{2}(z-1)$$

$$= \frac{T^2}{2}z + \frac{T^2}{2}$$

$$= \frac{T^2}{2}(z+1).$$

Thus we have a single zero at $z = -1$, as we have seen from the transfer function.

Much of the algebra for state-space control design is the same as for the continuous time case discussed in Chapter 6. The poles of a discrete system can be moved to desirable locations by linear state-variable feedback:

$$u = -\mathbf{K}\mathbf{x} \tag{8.57}$$

such that

$$\det[z\mathbf{I} - \mathbf{\Phi} + \mathbf{\Gamma}\mathbf{K}] = a_c(z), \tag{8.58}$$

provided the system is controllable. The system is controllable if the controllability matrix

$$\mathscr{C} = [\mathbf{\Gamma} \quad \mathbf{\Phi}\mathbf{\Gamma} \quad \mathbf{\Phi}^2\mathbf{\Gamma} \cdots \mathbf{\Phi}^{n-1}\mathbf{\Gamma}] \tag{8.59}$$

is full-rank.

A discrete full-order estimator has the form

$$\hat{\mathbf{x}}(n + 1) = \mathbf{\Phi}\hat{\mathbf{x}} + \mathbf{\Gamma}u(n) + \mathbf{L}[y(n) - \mathbf{H}\hat{\mathbf{x}}(n)], \qquad (8.60)$$

where $\hat{\mathbf{x}}$ is the state estimate. The error equation

$$\tilde{\mathbf{x}}(n + 1) = (\mathbf{\Phi} - \mathbf{LH})\tilde{\mathbf{x}}(n) \qquad (8.61)$$

can be given arbitrary dynamics $\alpha_e(z)$, provided that the system is observable, which requires that the observability matrix

$$\mathbb{O} = \begin{bmatrix} \mathbf{H} \\ \mathbf{H\Phi} \\ \mathbf{H\Phi}^2 \\ \vdots \\ \mathbf{H\Phi}^{n-1} \end{bmatrix} \qquad (8.62)$$

be full-rank.

Similar to the continuous time case, if the open-loop transfer function is

$$\frac{Y(z)}{U(z)} = \frac{b(z)}{a(z)}, \qquad (8.63)$$

then a state-space compensator can be designed such that

$$\frac{Y(z)}{R(z)} = \frac{\gamma(z)b(z)}{\alpha_c(z)\alpha_e(z)}, \qquad (8.64)$$

where r is the reference input. The polynomial $\alpha_c(z)$ is selected by the designer and results in a control gain $\mathbf{K}$ so that $\det[z\mathbf{I} - \mathbf{\Phi} + \mathbf{\Gamma K}] = \alpha_c(z)$ using exactly the same methods that were discussed in Chapter 6 for continuous systems. Similarly, the polynomial $\alpha_e(z)$ is selected by the designer and results in an estimator gain $\mathbf{L}$, so that $\det[z\mathbf{I} - \mathbf{\Phi} + \mathbf{LH}] = \alpha_e(z)$. If the estimator is structured according to Fig. 6.44(a), the system zeros $\gamma(z)$ will be identical to the estimator poles $\alpha_e(z)$, thus removing the estimator response from the closed-loop system response. However, if desired, the polynomial $\gamma(z)$ can be selected arbitrarily by the designer with suitable feedforward of the reference input. The reader is referred to Franklin, Powell, and Workman (1990) for details.

8.6   Hardware Characteristics

8.6.1 Analog-to-Digital (A/D) Converters

A/D converters are devices that convert a voltage level from a sensor to a digital word usable by the computer. At the most basic level, all digital words are binary numbers consisting of many bits that are set either to 1 or 0. Therefore the task of the A/D at each sample time is to convert a voltage level to the correct bit pattern and often to hold that pattern until the next sample time.

Of the many A/D conversion techniques that exist, the most common are based on counting schemes or successive approximation schemes. In counting methods, the input voltage may be converted to a train of pulses whose frequency is proportional to the voltage level. The pulses are then counted over a fixed period using a binary counter, thus resulting in a binary representation of the voltage level. A variation on this scheme is to start the count simultaneously with a linear (in time) voltage and to stop the count when the voltage reaches the magnitude of the input voltage to be converted.

The successive approximation technique tends to be much faster than the counting methods. It is based on successively comparing the input voltage to reference levels representing the various bits in the digital word. The input voltage is first compared with a reference that is half the maximum. If the input voltage is greater, the most significant bit is set, and the signal is then compared with a reference that is ¾ the maximum to determine the next bit, and so on. These converters require one clock cycle to set each bit, so they would need n cycles for an n-bit converter. At the same clock rate, a counter-based converter might require as many as 2^n cycles, which would usually be much slower.

For either technique, it will take longer to perform the conversion for a higher number of bits. Not surprisingly, the price of A/Ds goes up with both speed and bit size. For example, an eight-bit (resolution of 0.4%) A/D with the minimal performance capability of a 100 μs conversion time sold for a modest $3 in 1990, whereas faster models with a 50 ns conversion time sold for approximately $40.

If more than one channel of data needs to be sampled and converted to digital words, it is usually accomplished by use of a multiplexer rather than by multiple A/Ds. The multiplexer sequentially connects the A/D into the channel being sampled.

8.6.2 Digital-to-Analog (D/A) Converters

D/A converters are used to convert the digital words from the computer to a voltage level for driving actuators or perhaps a recording device such as an oscilloscope or stripchart recorder. The basic idea is that the binary bits are used to cause switches (electronic gates) to open or close, thus routing the electric current through an appropriate network of resistors so that the correct voltage level is generated. Since no counting or iteration is required for D/As, they tend to be much faster than A/D converters. In fact, a D/A is a component in an A/D based on successive approximation.

8.6.3 Analog Prefilters

This device is often placed between the sensor and the A/D. Its function is to reduce the higher-frequency noise components in the analog signal so as to prevent the noise from being switched to a lower frequency by the sampling process (called *aliasing* or *folding*).

FIGURE 8.16
An example of aliasing.

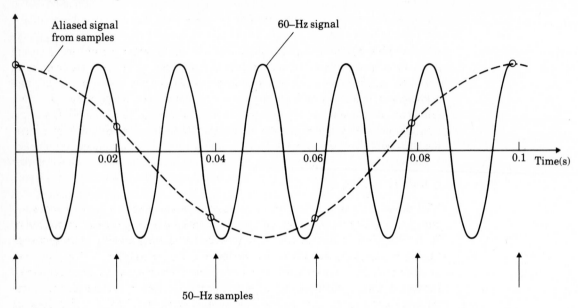

An example of aliasing is shown in Fig. 8.16, where a 60-Hz oscillatory signal is being sampled at 50 Hz. The figure shows the result from the samples as a 10-Hz signal and also shows the mechanism by which the frequency of the signal is aliased from 60 to 10 Hz. Aliasing will occur any time the sample rate is not at least twice as fast as any of the frequencies in the signal being sampled. Therefore, to prevent aliasing of a 60-Hz signal, the sample rate would have to be faster than 120 Hz, clearly much higher than the 50-Hz rate in the figure.

This phenomenon is one of the consequences of the sampling theorem of Nyquist and Shannon, which basically states that, for the signal to be accurately reconstructed from the samples, it must have no frequency component greater than half the sample rate ($\omega_s/2$). Another consequence is that the highest frequency that can be represented by discrete samples is $\omega_s/2$, an idea that has been discussed in Section 8.2.3.

The consequence of aliasing on a digital control system can be substantial. In a continuous system, noise components with a frequency much higher than the control-system bandwidth normally have a small effect because the system won't respond at the high frequency. However, in a digital system the frequency of the noise can potentially be aliased down in frequency to the vicinity of the system bandwidth; therefore, the system will respond, thus causing the noise to appear on the system's output.

The solution is to place an analog prefilter before the sampler. In many cases a simple first-order low-pass filter will do; that is,

$$H_p(s) = \frac{a}{s + a},$$

where the breakpoint a is selected lower than $\omega_s/2$ so that any noise present with frequencies greater than $\omega_s/2$ is attenuated by the prefilter. The lower the breakpoint frequency selected, the more the noise above $\omega_s/2$ is attenuated. However, too low a breakpoint may reduce the control system's bandwidth. The prefilter does not eliminate the aliasing; but by judicious choice of the prefilter breakpoint and the sample rate, the designer has the ability to reduce the magnitude of the aliased noise to some acceptable level.

8.6.4 The Computer

This is the unit that does all the computations. Most computers used today for control are built around an integrated circuit called a microprocessor that contains most of the functions needed. It could be housed in a desktop-size workstation or PC, or possibly in a shoebox-size dedicated microcomputer. The relatively low cost of microprocessor technology is making its use very attractive and has accounted for the large increase in the use of digital control systems that has occurred in the 1980s. The computer consists of a *central processor unit,* or *CPU,* which does the computations and provides the system logic, a clock to synchronize the system, memory modules for data and instruction storage, and a power supply to provide the various required voltages. The memory modules come in three basic varieties:

1. *ROM (read-only memory)* is the least expensive, but, once manufactured, its contents cannot be changed. Typically most of the memory in products manufactured in quantity is ROM.

2. *RAM (random-access memory)* is the most expensive, but its values can be changed by the CPU. It is required to store only the values that will be changed during the control process and typically represents only a small fraction of the total memory of a developed product.

3. *PROM (programmable read-only memory)* is a ROM whose values can be changed by a technician using a special device. It is typically used during product development to enable the designer to try different algorithms and parameter values.

Microprocessors for control applications generally come with a digital word size of either 8, 16, or 32 bits, although some have been available with 12 bits. Larger word sizes give better accuracy, but at an increase in cost. The most economical solution is often to use an 8-bit microprocessor, but to use two digital words to store one value *(double precision)* in the areas of the controller that are critical to the system accuracy.

8.7   Word-Size Effects

A numerical value can be represented with only limited precision in a digital computer. For fixed-point arithmetic, the resolution is 0.4% of full range for 8 bits and 0.1% for 10 bits; resolution drops by a factor of 2 for each additional bit. The effect of this quantization shows up in A/D conversion, multiplication truncation, and parameter storage errors. If the computer uses floating-point arithmetic, the resolution of the multiplication and parameter storage changes with the magnitude of the number being stored, the resolution affecting only the mantissa, while the exponent essentially continually adjusts the full scale.

8.7.1 Random Effects

As long as a system has varying inputs or disturbances, A/D errors and multiplication errors act in a random manner on the system and essentially produce noise at the output of the system. The output noise due to a particular noise source (A/D or multiplication) has a mean value of

$$\bar{n}_o = H_{\mathrm{DC}} \bar{n}_I, \tag{8.65}$$

where

H_{DC} = DC gain of transfer function between noise source and output,

$\bar{n}_I$ = mean value of noise source.

The mean value of the noise will be zero for a round-off process but has a value

$$\bar{n}_I = \frac{q}{2},$$

where q is the resolution level for a truncation process. Although most A/Ds round off, producing no mean error, some truncate and do produce an error. The total noise effect is the sum of all noise sources.

The variance of the output noise σ_o is always nonzero, irrespective of whether the process truncates or rounds. The value of the output variance is most easily found by solving the discrete Lyapunov equation

$$\mathbf{R}_x = \mathbf{\Phi}\mathbf{R}_x\mathbf{\Phi}^{\mathrm{T}} + \mathbf{\Gamma}\mathbf{\Gamma}^{\mathrm{T}}\sigma_I^2, \tag{8.66}$$

where

$\sigma_o^2 = \mathbf{H}\mathbf{R}_x\mathbf{H}^{\mathrm{T}}$,

σ_o = output noise rms,

$\mathbf{R}_x$ = state covariance matrix,

σ_I^2 = input noise variance,

$\mathbf{\Phi}, \mathbf{H}$ = system description matrices from Section 8.5,

$\mathbf{\Gamma}$ = noise input matrix.

The input noise variance has magnitude

$$\sigma_I^2 = \frac{q^2}{12} \tag{8.67}$$

for either round-off or truncation.

The evaluation of Eq. (8.66) is usually done using computer tools as described in Franklin, Powell, and Workman (1990). Carrying out the calculations for a system with a small-word-size computer (8 bits or less) will generally show that the noise response of the system becomes more sensitive as the sampling rate increases and can be a design issue for very fast sample rates. However, for 16- or 32-bit computers, the onset of significant noise errors occurs at such a high sample rate that quantization noise is not a factor in design.

The sensitivity of a system to A/D errors can be partly alleviated by adding lag in the digital controller or by adding more bits to the A/D converter. Different structures of the digital controller have no effect.

On the other hand, multiplication errors can be reduced substantially for high-order (greater than second-order) controllers by proper structuring of a given control transfer function. For example, a second-order transfer function with real roots,

$$D(z) = \frac{U(z)}{E(z)} = \frac{z - 0.8}{(z - 0.2)(z - 0.3)}, \tag{8.68}$$

can be implemented in a *direct* manner, yielding

$$u(n) = 0.5u(n - 1) - 0.06u(n - 2) + e(n) - 0.8e(n - 1), \tag{8.69}$$

or can be implemented in a *parallel* manner that results from a partial fraction expansion of Eq. (8.68). The result, shown in Fig. 8.17, yields

$$x_1(n) = 0.2x_1(n - 1) + 6e(n),$$
$$x_2(n) = 0.3x_2(n - 1) + 5e(n),$$
$$u(n) = x_1(n) - x_2(n). \tag{8.70}$$

Note that the transfer functions to the output from the multiplications in Eq. (8.69) are substantially different from those in Eq. (8.70). It is also possible to implement the $D(z)$ with a *cascade* factorization, which would be two first-order blocks arranged serially for this example.

FIGURE 8.17
Parallel implementation.

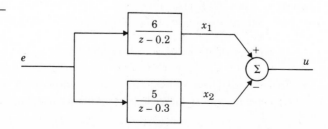

Either cascade or parallel implementations would be preferred to the direct implementation for digital controllers with a small word size because of their substantially reduced response to quantization errors.

8.7.2 Systematic Effects

Parameters such as the numerical values in Eqs. (8.69) and (8.70), if in error, will change the dynamic behavior of a system. In a high-order controller with a small word size and a direct implementation, a very small percentage error in a stored parameter can result in substantial root-location changes and can sometimes cause instability. In the Apollo command module, a 14-bit word size for parameter storage would have resulted in instability if a direct implementation had been used in the sixth-order compensator! These effects are amplified when there are two compensator poles close together (or repeated) or when there are fast sample rates, when all poles tend to clump around $z = +1$ and are close to one another. These effects can be reduced by using larger word sizes, parallel or cascade implementations, double-precision parameter storage, and slower sample rates.

Under conditions of constant disturbances and input commands, multiplication errors can also cause systematic errors. Typically the result is a steady-state error or possibly a limit cycle. The steady-state error results from the deadband that exists in any digital controller. The magnitude of the deadband is proportional to the DC gain from the multiplication error to the output and to the resolution level. Stable limit cycles occur only for controllers with lightly damped poles. These effects are also reduced by larger word sizes, parallel or cascade implementations, and slower sample rates.

8.8 Sample-Rate Selection

The selection of the best sample rate for a digital control system is the result of a compromise among many factors. The basic motivation to lower the sample rate ω_s is cost. A decrease in sample rate means more time is available for the control calculations; hence slower computers can be used for a given control function or more control capability can be achieved from a given computer. Either way, the cost per function is lowered. For systems with A/D converters, less demand on conversion speed will also lower cost. These economic arguments indicate that the best engineering choice is the slowest possible sample rate that still meets all performance specifications.

Factors that could provide a lower limit to the acceptable sample rate are

1. Tracking effectiveness, as measured by closed-loop bandwidth or by time-response requirements, such as rise time and settling time

2. Regulation effectiveness, as measured by the error response to random plant disturbances

3. Resonances in the plant that require a sampling rate twice their natural frequency

4. Sensitivity to plant-parameter variations

5. Error due to measurement noise and the associated prefilter design methods.

A fictitious limit on the sample rate occurs when using continuous-design techniques. The inherent approximation in the method may give rise to system instabilities as the sample rate is lowered, and this can lead the designer to conclude that a lower limit on ω_s has been reached when, in fact, the proper conclusion is that the approximations are invalid; the solution is not to sample faster but to refine the design with a direct digital-design method.

8.8.1 Tracking Effectiveness

An absolute lower bound to the sample rate is set by a specification to track a command input with a certain frequency (the system bandwidth). The *sampling theorem* states that in order to reconstruct an unknown, band-limited, continuous signal from samples of that signal, one must sample at least twice as fast as the highest frequency contained in the signal. Therefore, in order for a closed-loop system to track an input at a certain frequency, it must absolutely have a sample rate twice as fast as that frequency; as a result, ω_s must be about ten times the system bandwidth ($\omega_s \Rightarrow 10 \times \omega_{BW}$). We also saw from the z plane mapping, $z = e^{sT}$, that the highest frequency that can be represented by a discrete system is $\omega_s/2$, supporting this conclusion.

It is important to note the distinction between the closed-loop bandwidth ω_{BW} and the highest frequencies in the open-loop plant dynamics, since these two frequencies can be quite different. For example, closed-loop bandwidths could be an order of magnitude *less* than open-loop modes of resonances for some control problems. Information concerning the state of the plant resonances for purposes of control can be extracted from sampling the output without satisfying the sampling theorem because some a priori knowledge is available (even though imprecise) concerning these dynamics, and the system is *not* required to track these frequencies. Thus, a priori knowledge of the dynamic model of the plant can be included in the compensation in the form of a notch filter.

The "closed-loop bandwidth" limitation provides the fundamental lower bound on the sample rate. In practice, however, the theoretical lower bound of sampling at twice the bandwidth of the reference input signal would not be judged sufficient in terms of the quality of the desired time responses. For a system with a rise time on the order of 1 s, thus yielding a closed-loop bandwidth on the order of 0.5 Hz, it is not unreasonable to insist on a sample rate of 5 to 20 Hz, which is a factor of 10 to 40 times ω_{BW}. This is necessary in order to reduce the delay between a command and the system response to the command and to smooth the system output to the control steps coming out of the ZOH. In other words, most digital control systems have sample rates that are considerably faster than the lower limit imposed by the sampling theorem.

8.8.2 Disturbance Rejection

Disturbance rejection is an important aspect—if not the most important one—of any control system. Disturbances enter a system with various frequency characteristics ranging from steps to white noise. For the purpose of sample-rate selection, the higher-frequency random disturbances are the most influential.

The ability of the control system to reject disturbances with a good continuous controller represents a lower bound on the error response that can be hoped for when implementing the controller digitally. In fact, some degradation over the continuous design must occur because the sampled values are slightly out of date at all times except at precisely the sampling instants. However, if the sample rate is very fast compared with the frequencies contained in the noisy disturbance, no appreciable loss should be expected from the digital system as compared with the continuous controller. At the other extreme, if the sample time is very long compared with the characteristic frequencies of the noise, the response of the system because of noise is essentially the same as the response if there were no control at all. The selection of a sample rate will place the response somewhere in between these two extremes, and thus the impact of the sample rate on the disturbance rejection of the system may be very influential to the designer in selecting the sample rate.

Although the best choice of sample rate in terms of the ω_{BW} multiple is dependent on the frequency characteristics of the noise and the degree to which random disturbance rejection is important to the quality of the controller, sample rates on the order of 20 to 50 times ω_{BW} are typical.

8.8.3 Plant Resonances

In some cases, a plant may have open loop resonances that are considerably faster than the closed-loop bandwidth of the system. It is sometimes necessary to sample at least two times faster than these resonances. Systems where the resonance is very lightly damped and where the controller is adding damping to the resonant mode are particularly susceptible to this sensitivity. A determination of whether this limitation needs to be adhered to can be determined by performing a disturbance noise analysis using the Lyapunov equation as described for quantization noise in Eq. (8.66) or by simulating the digitally controlled system including random disturbance inputs.

8.8.4 Parameter Sensitivity

Any control design relies to some extent on knowledge of the parameters representing plant dynamics. Discrete systems exhibit an increasing sensitivity to parameter errors for a decreasing ω_s when the sample period becomes comparable to the period of any of the open-loop vehicle dynamics. For systems with all plant dynamics in the vicinity of the closed-loop bandwidth or slower, root-location changes due to parameter errors will not likely be a constraining factor unless the parameter error is quite large. However, for systems with a structural (or other) resonance stabilized by a notch filter, imperfect knowledge of the plant resonance characteristics will lead to changes in the system roots, and possibly to instabilities. This sensitivity to plant parameter increases as the sample rate decreases and could limit how slow the sample rate is in some cases. Typically, however, some other factor limits the sample rate, and at worst, some effort needs to be made in the controller design to minimize system sensitivity.

8.8.5 Control-System Design Methodology

Prefilter Design

Digital control systems with analog sensors typically include an analog prefilter between the sensor and the sampler or an A/D converter as an antialiasing device. The prefilters are low-pass, and the simplest transfer function is

$$H_p(s) = \frac{a}{s + a} \tag{8.71}$$

so that the noise above the prefilter breakpoint a is attenuated. The design goal is to provide enough attenuation at half the sample rate ($\omega_s/2$) so that the noise above $\omega_s/2$, when aliased into lower frequencies by the sampler, will not be detrimental to the control-system performance.

A conservative design procedure is to select the breakpoint and ω_s sufficiently higher than the system bandwidth so that the phase lag from the prefilter does not significantly alter the system stability, and thus the prefilter can be ignored in the basic control-system design. Furthermore, for a good reduction in the high-frequency noise at $\omega_s/2$, the sample rate is selected about 5 or 10 times higher than the prefilter breakpoint. The implication of this prefilter design procedure is that sample rates need to be on the order of 30 to 100 times faster than the system bandwidth. If done this way, the prefilter design procedure is likely to provide the lower bound on the selection of the sample rate.

An alternative design procedure is to allow significant phase lag from the prefilter at the system bandwidth and thus to require that the control design be carried out with the analog prefilter characteristics included. This procedure allows the designer to use very low sample rates, but at the expense of increased complexity in the original design, since the prefilter must be included in the plant transfer function. If this procedure is used and low prefilter breakpoints are allowed, the effect of sample rate on sensor noise is small and essentially places no limits on the sample rate.

Asynchronous Modules

As noted in the previous section, divorcing the prefilter design from the control-law design may result in the necessity for a faster sample rate than otherwise. This same result may exhibit itself for other types of modularization. For example, a smart sensor with its own computer running asynchronously from the primary control computer will not be amenable to direct digital design, because the overall system transfer function depends on the phasing between the smart sensor and the primary digital controller. This situation is similar to that of the digitization errors discussed in Section 8.3 and will therefore require sample rates on the order of $30 \times \omega_{BW}$.

Summary

Most control systems today are implemented using a digital computer. This chapter has briefly described the steps that must be taken to transfer the compensations designed throughout the book into a form suitable for programming into a digital computer. The effects of sample rate and computer word size have also been discussed.

Problems

8.1 The z transform of a discrete time filter $h(k)$ is

$$H(z) = \frac{1 + (1/2)z^{-1}}{[1 - (1/2)z^{-1}][1 + (1/3)z^{-1}]}.$$

a) Let $u(k)$ and $y(k)$ be the discrete input and output of this filter. Find a difference equation relating $u(k)$ and $y(k)$.
b) Find the natural frequency and damping of the filter's poles.
c) Is the filter stable?

8.2 Use the z transform to solve the difference equation

$$y(k) - 3y(k-1) + 2y(k-2) = 2u(k-1) - 2u(k-2),$$
$$u(k) = k \quad k \geq 0$$
$$= 0 \quad k < 0,$$
$$y(k) = 0 \quad k < 0.$$

8.3 The one-sided z transform is defined as

$$F(z) = \sum_0^\infty f(k)z^{-k}.$$

a) Show that the one-sided transform of $f(k+1)$ is $Z\{f(k+1)\} = zF(z) - zf(0)$.
b) Use the one-sided transform to solve for the transforms of the Fibonacci numbers generated by the difference equation $u_{k+2} = u_{k+1} + u_k$. Let $u_0 = u_1 = 1$. *Hint:* You will need to compute the transform of $f(k+2)$.
c) Compute the location of the poles of the transform of the Fibonacci numbers.
d) Compute the inverse transform of the numbers.
e) Show that if u_k is the kth Fibonacci number, then the ratio u_{k+1}/u_k will go to $(1 + \sqrt{5})/2$, the "golden ratio" of the Greeks.

8.4 a) The following transfer function is a lead network designed to add about $60°$ lead at $\omega_1 = 3$ rad/s:

$$H(s) = \frac{s+1}{0.1s+1}.$$

For each of the following design methods compute and plot in the z plane the pole and zero locations and compute the amount of phase lead given by the network at $z_1 = e^{j\omega_1 T}$. Let $T = 0.25$ s.

1) Tustin's rule
2) Pole-zero mapping

b) Plot on log-log paper over the frequency range $\omega = 0.1 \to \omega = 100$ rad/s the amplitude Bode plots of each of the above equivalents and compare with $H(s)$. *Hint:* Amplitude Bode plots are given by

$$|H(z)| = |H(e^{j\omega T})|$$

8.5 a) The following transfer function is a lag network designed to introduce gain attenuation of a factor of 10 (20 dB) at $\omega = 3$:

$$H(s) = \frac{10s+1}{100s+1}.$$

For each of the following design methods compute and plot in the z plane the pole-zero pattern of the resulting discrete equivalent and give the gain attenuation at $z_1 = e^{j\omega_1 T}$. Let $T = 0.25$ s.

1) Tustin's rule
2) Pole-zero mapping

b) For each case computed, plot the Bode amplitude curves over the range $\omega_1 = 0.01 \to \omega_h = 10$ rad/s.

8.6 Consider the linear equation $\mathbf{Ax} = \mathbf{b}$, where $\mathbf{A}$ is an $n \times n$ matrix. One way of solving for $\mathbf{x}$, given $\mathbf{b}$, is by the following algorithm (a discrete-time system):

$$\mathbf{x}(k+1) = (\mathbf{I} + c\mathbf{A})\mathbf{x}(k) - c\mathbf{b},$$

where c is a scalar to be chosen.

a) Show that the solution of $\mathbf{Ax} = \mathbf{b}$ is the equilibrium point $\mathbf{x}^*$ of the discrete-time system. An equilibrium point $\mathbf{x}^*$ of a discrete-time system $\mathbf{x}(k + 1) = \mathbf{f}[\mathbf{x}(k)]$ is given by $\mathbf{x}^* = \mathbf{f}(\mathbf{x}^*)$.

b) Consider the error $\mathbf{e}(k) = \mathbf{x}(k) - \mathbf{x}^*$. Write the linear equation that relates the error $\mathbf{e}(k + 1)$ to $\mathbf{e}(k)$.

c) Suppose $|1 + c\lambda_i(\mathbf{A})| < 1$, $i = 1, \ldots, n$. Show that starting from any initial guess $\mathbf{x}_0$, the algorithm converges to $\mathbf{x}^*$. $\lambda_i(\mathbf{A})$ denotes the ith eigenvalue of $\mathbf{A}$. *Hint:* For any matrix $\mathbf{B}$, $\lambda_i(\mathbf{I} + \mathbf{B}) = 1 + \lambda_i(\mathbf{B})$.

8.7 Consider the configuration for an autopilot, shown in Fig. 8.18, where

$$G(s) = \frac{40(s + 2)}{(s + 10)(s^2 - 1.4)}.$$

a) Find the transfer function $G(z)$ for $T = 1$ assuming it is preceded by ZOH.

b) Find the ranges of K such that the closed-loop system is stable.

c) Draw the root locus.

d) Compare the results in (b) to the case where an analog controller is used; that is, where the sampling switch is always closed. Which system has a bigger allowable K?

8.8 Write a computer program to compute $\boldsymbol{\Phi}$ and $\boldsymbol{\Gamma}$ from $\mathbf{F}$, $\mathbf{G}$, and sample period T. Use your program to compute $\boldsymbol{\Phi}$ and $\boldsymbol{\Gamma}$ if

a) $\mathbf{F} = \begin{bmatrix} -1 & 0 \\ 0 & -2 \end{bmatrix}$, $\quad \mathbf{G} = \begin{bmatrix} 1 \\ 1 \end{bmatrix}$, $\quad T = 0.2$ s.

b) $\mathbf{F} = \begin{bmatrix} -3 & -2 \\ 1 & 0 \end{bmatrix}$, $\quad \mathbf{G} = \begin{bmatrix} 1 \\ 0 \end{bmatrix}$, $\quad T = 0.2$ s.

8.9 a) For

$$\boldsymbol{\Phi} = \begin{bmatrix} 1 & T \\ 0 & 1 \end{bmatrix}, \qquad \boldsymbol{\Gamma} = \begin{bmatrix} T^2/2 \\ T \end{bmatrix},$$

find a transform matrix $\mathbf{T}$ such that, if $\mathbf{x} = \mathbf{Tw}$, the equations in $\mathbf{w}$ will be in control canonical form.

b) Compute $\mathbf{K}_w$, the gain, such that, if $u = -\mathbf{K}_w \mathbf{w}$, the characteristic equation will be $a_c(z) = z^2 - 1.6z + 0.7$.

c) Use $\mathbf{T}$ from (a) to compute $\mathbf{K}_x$, the gain in the $\mathbf{x}$ states.

8.10 Consider a control realization with transfer function $1/s^2$ and piecewise constant input of the form

$$u(t) = u(i\Delta), \qquad i\Delta \le t \le (i + 1)\Delta.$$

FIGURE 8.18

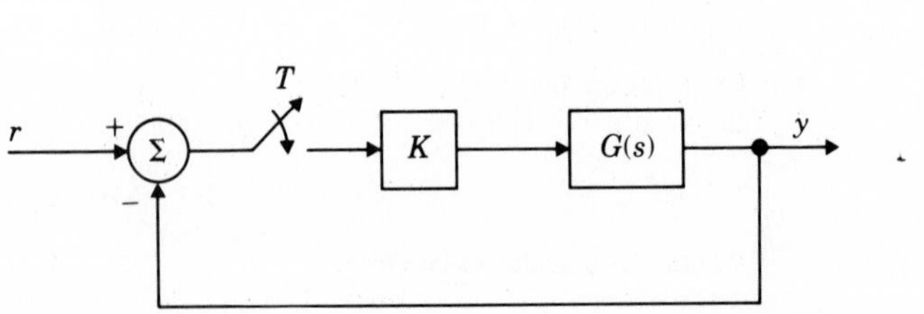

a) Show that if we restrict attention to the time instants $i\Delta, i = 0, 1, 2, \ldots,$ the resulting so-called sampled-data system can be described by the equations

$$\begin{bmatrix} x_1(i+1) \\ x_2(i+1) \end{bmatrix} = \begin{bmatrix} 1 & 0 \\ \Delta & 1 \end{bmatrix} \begin{bmatrix} x_1(i) \\ x_2(i) \end{bmatrix} + \begin{bmatrix} \Delta \\ \Delta^2/2 \end{bmatrix} u(i),$$

$$y(i) = [0 \quad 1][x_i(i) \quad x_2(i)]^T.$$

b) Design a second-order observer that will always reduce the error in an estimate of the initial state vector to zero in time 2Δ or less.

c) Is it possible to estimate the initial state exactly with a first-order observer? Justify your answer.

8.11 *Single-axis satellite attitude control.* Satellites often require attitude control for proper orientation of antennas and sensors with respect to the earth. Figure 2.5 shows a communication satellite with a three-axis attitude-control system. To gain insight into the three-axis problem, we often consider one axis at a time. Figure 8.19 depicts the case where motion is allowed only about an axis perpendicular to the page. The equations of motion of the system are given by

$$I\ddot{\theta} = M_C + M_D,$$

where I is the moment of inertia of the satellite about its mass center, M_C is the control torque applied by the thrustors, M_D is the sum of the disturbance torques, and θ is the angle of the satellite axis with respect to an "inertial" reference. The inertial reference must have no angular acceleration. Normalizing, we define

$$u = \frac{M_C}{I}, \qquad w_d = \frac{M_D}{I}$$

and obtain

$$\ddot{\theta} = u + w_d.$$

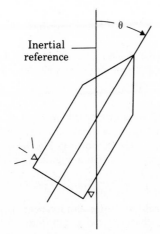

FIGURE 8.19
Satellite control schematic.

Inertial reference

Taking the Laplace transform,

$$\theta(s) = \frac{1}{s^2}[u(s) + w_d(s)],$$

which becomes, with no disturbance,

$$\frac{\theta(s)}{u(s)} = \frac{1}{s^2} = G_1(s).$$

In the discrete case with u being applied through a zero-order hold, we can use the methods of Chapter 8 to obtain the discrete transfer function

$$G_1(z) = \frac{\theta(z)}{u(z)} = \frac{T^2}{2}\frac{z+1}{(z-1)^2}.$$

a) Sketch the root locus of this system with proportional feedback. What is the type of this system?

b) Add a lead network to your locus so that the dominant poles are at $\zeta = 0.5$ and $\omega_n = 3\pi/10T$.

c) What is the feedback gain if $T = 1$ s? If $T = 2$ s?

d) Plot the closed-loop step response for $T = 1$ s.

8.12 Compute Φ by changing states so that the system matrix is diagonal.

a) Compute $e^{\mathbf{A}T}$, where

$$\mathbf{A} = \begin{bmatrix} -1 & 0 \\ 0 & -2 \end{bmatrix},$$

using the infinite series.

b) Show that if $\mathbf{F} = \mathbf{T}^{-1}\mathbf{A}\mathbf{T}$ for some nonsingular transformation matrix $\mathbf{T}$, then $e^{\mathbf{F}T} = \mathbf{T}^{-1}e^{\mathbf{A}T}\mathbf{T}$.

c) Show that if

$$\mathbf{F} = \begin{bmatrix} -3 & 1 \\ -2 & 0 \end{bmatrix},$$

there exists a $\mathbf{T}$ such that $\mathbf{T}^{-1}\mathbf{A}\mathbf{T} = \mathbf{F}$. *Hint:* Write $\mathbf{T}^{-1}\mathbf{A} = \mathbf{F}\mathbf{T}^{-1}$, assume four unknowns for the elements of $\mathbf{T}^{-1}$, and solve. For more general interest, write $\mathbf{T}^{-1}$ as a matrix of columns and show that the columns are eigenvectors of $\mathbf{F}$.

d) Compute $e^{\mathbf{F}T}$.

8.13 It is possible to suspend a mass of magnetic material by means of an electromagnet whose current is controlled by the position of the mass (Woodson and Melcher, 1968). A schematic of a possible setup is shown in Fig. 8.20, and a photo of a working system at Stanford University is shown in Fig. 2.27. The equations of motion are

$$m\ddot{x} = -mg + f(x, \mathbf{I}),$$

where the force on the ball due to the electromagnet is given by $f(x, \mathbf{I})$. At equilibrium the magnet force balances the gravity force. Suppose we call the current

FIGURE 8.20

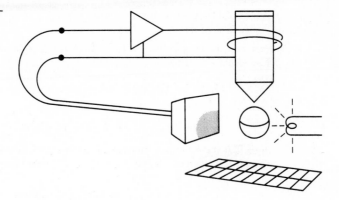

there $\mathbf{I}_0$. If we write $\mathbf{I} = \mathbf{I}_0 + i$, expand f about $x = 0$ and $\mathbf{I} = \mathbf{I}_0$, and neglect higher-order terms, we obtain

$$m\ddot{x} = k_1 x + k_2 i.$$

Reasonable values are $m = 0.02$ kg, $k_1 = 20$ N/m, and $k_2 = 0.4$ N/A.

a) Compute the transfer function from i to x and draw the (continuous) root locus for simple feedback $i = -Kx$.

b) Let the sample period be 0.02 s and compute the plant discrete transfer function when used with a zero-order hold.

c) Design a digital control for the magnetic levitation to meet the specifications $t_r \leq 0.1$ s, $t_s \leq 0.4$ s, and percent overshoot $\leq 20\%$.

d) Plot a root locus of your design versus k_1 and discuss the possibility of balancing balls of various masses.

e) Plot a step response of your design to an initial disturbance displacement on the ball and show both x and the control current i. If the sensor can measure x over a range of only ± 0.25 cm and the amplifier can provide a current of only 1 A, what is the *maximum* displacement possible for control, neglecting the nonlinear terms in $f(x, \mathbf{I})$?

8.14 In Problem 8.13 we described an experiment in magnetic levitation whose equation of motion reduced to

$$\ddot{x} = 1000x + 20u.$$

Let the sampling time be 0.01 s.

a) Use the pole placement technique to design this system to meet the specifications that settling time be less than 0.25 s and the overshoot to an initial offset in x be less than 20%.

b) Plot step responses of x, $\tilde{x}$, and u for an initial x displacement.

c) Plot the root locus for changes in the plant gain, and mark the design pole locations.

d) Introduce a command reference with feed-forward so that the estimate of $\dot{x}$ is not forced by r. Measure or compute the frequency response from r to system error $r - x$ and give the highest frequency for which the error amplitude is less than 20% of the command amplitude.

8.15 *A servomechanism for antenna azimuth control.* It is desired to control the elevation of an antenna designed to track a satellite (Fig. 8.21). The antenna and drive parts have a moment of inertia J and damping B, arising to some extent from bearing and aerodynamic friction, but mostly from the back emf of the DC-drive motor (Fig. 8.22). The equations of motion are

$$J\ddot{\theta} + B\dot{\theta} = T_c + T_d,$$

where T_c is the net torque from the drive motor and T_d is the disturbance torque due to wind. If we define

$$a = \frac{B}{J}, \qquad u = \frac{T_c}{B}, \qquad w_d = \frac{T_d}{B},$$

the equations reduce to

$$\frac{1}{a}\ddot{\theta} + \dot{\theta} = u + w_d.$$

Transformed, they are

$$\theta(s) = \frac{1}{s[(s/a) + 1]}[u(s) + w_d(s)],$$

or, with no disturbances,

$$\frac{\theta(s)}{u(s)} = \frac{1}{s[(s/a) + 1]} = G_2(s).$$

FIGURE 8.21
Satellite tracking antenna. *(Courtesy Ford Aerospace and Communications Corp.)*

FIGURE 8.22
Schematic diagram of
antenna.

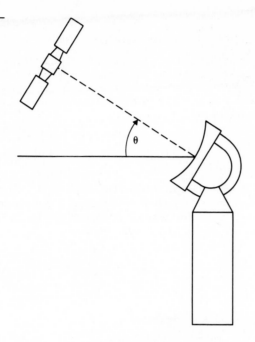

The discrete case, with $u(n)$ applied through a zero-order hold, yields

$$G_2(z) = \frac{\theta(z)}{u(z)} = K\frac{z + b}{(z - 1)(z - e^{-aT})},$$

where

$$K = \frac{aT - 1 + e^{-aT}}{a}, \qquad b = \frac{1 - e^{-aT} - aTe^{-aT}}{aT - 1 + e^{-aT}}.$$

a) Write the equations in state form with $x_1 = \dot{y}$. Give the matrices $\mathbf{F}$, $\mathbf{G}$, and $\mathbf{H}$. Let $a = 0.1$.

b) Letting $T = 1$, design $\mathbf{K}$ for equivalent poles at $s = -(1/2) \pm j\left(\sqrt{3}/2\right)$. Plot the step response of the resulting design.

c) Design an estimator with $\mathbf{L}$ selected so that $a_e(z) = z^2$; that is, poles are at the origin.

d) Use the estimated states for computing the control and introduce the reference input so as to leave the state estimate undisturbed. Plot the step response from this reference input and from a wind gust (step) disturbance.

e) Plot the root locus of the closed-loop system with respect to the plant gain and mark the locations of the closed-loop poles.

8.16 *Tank fluid temperature control.* The temperature of a tank of fluid with a constant flow rate in and out is to be controlled by adjusting the temperature of the incoming fluid. The temperature of the incoming fluid is controlled by a mixing valve that adjusts the relative amounts of hot and cold supplies of the fluid (see Fig. 8.23).

FIGURE 8.23
Tank temperature control.

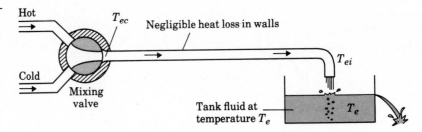

Because of the distance between the valve and the point of discharge into the tank, there is a time delay between the application of a change in the mixing valve and the discharge of the flow with the changed temperature into the tank. The differential equation governing the tank temperature is

$$\dot{T}_e = \frac{1}{cM}(q_{in} - q_{out}),$$

where

$$
\begin{aligned}
T_e &= \text{tank temperature,} \\
c &= \text{specific heat of the fluid,} \\
M &= \text{fluid mass contained in the tank,} \\
q_{in} &= c\dot{m}_{in}T_{ei}, \\
q_{out} &= c\dot{m}_{out}T_e, \\
\dot{m} &= \text{mass flow rate } (\dot{m}_{in} = \dot{m}_{out}), \\
T_{ei} &= \text{temperature of fluid entering tank.}
\end{aligned}
$$

The temperature at the input to the tank at time t, however, is the control temperature τ_d seconds in the past, which may be expressed as

$$T_{ei}(t) = T_{ec}(t - \tau_d),$$

where

$$
\begin{aligned}
\tau_d &= \text{delay time,} \\
T_{ec} &= \text{temperature of fluid immediately after the control valve and directly} \\
&\quad \text{controllable by the valve.}
\end{aligned}
$$

Combining constants, we obtain

$$\dot{T}_e(t) + aT_e(t) = aT_{ec}(t - \tau_d),$$

where

$$a = \frac{\dot{m}}{M},$$

which, transformed, becomes

$$\frac{T_e(s)}{T_{ec}(s)} = \frac{e^{-\tau_d s}}{(s/a) + 1} = G_3(s).$$

To form the discrete transfer function of G_3 preceded by a zero-order hold, we must compute

$$G_3(z) = Z\left\{ \frac{1 - e^{-sT}}{s} \frac{e^{-\tau_d s}}{(s/a) + 1} \right\}.$$

We assume that $\tau_d = lT - mT$, $0 \le m < 1$. Then

$$G_3(z) = Z\left\{ \frac{1 - e^{-sT}}{s} \frac{e^{\ell sT} e^{msT}}{(s/a) + 1} \right\}$$

$$= (1 - z^{-1})z^{-\ell} Z\left\{ \frac{e^{msT}}{s(s/a) + 1} \right\}$$

$$= (1 - z^{-1})z^{-1} Z\left\{ \frac{e^{msT}}{s} - \frac{e^{msT}}{s + a} \right\}$$

$$= \frac{z - 1}{z} \frac{1}{z^{\ell}} Z\{1(t + mT) - e^{-a(t + mT)}1(t + mT)\}$$

$$= \frac{z - 1}{z} \frac{1}{z^{\ell}} \left(\frac{z}{z - 1} - \frac{e^{-amT} z}{z - e^{-aT}} \right)$$

$$= \frac{1}{z^{\ell}} \frac{(1 - e^{-amT})z + e^{-amT} - e^{-aT}}{z - e^{-aT}}$$

$$= \frac{1 - e^{-mT}}{z^{\ell}} \frac{z + \alpha}{z - e^{-aT}};$$

$$\alpha = \frac{e^{-amT} - e^{-aT}}{1 - e^{amT}}.$$

It is easy to see that the zero location, $-\alpha$, varies from $\alpha = \infty$ at $m = 0$ to $\alpha = 0$ as $m \to 1$ and that $G_3(1) = 1.0$ for all a, m, and l. For the specific values of $\tau_d = 1.5$, $T = 1$, and $a = 1$, the transfer function, with $l = 2$ and $m = 1/2$, reduces to

$$G_3(z) = \frac{z + 0.6065}{z^2(z - 0.3679)}.$$

a) Write the equations in state-space form.
b) Design a feedback gain such that $\alpha_c(z) = z^3$.
c) Design an estimator such that $\alpha_e(z) = z^3$.
d) Plot the root locus of the system with respect to the plant gain.
e) Determine the step response of the system.

Appendix A: Laplace Transforms

Laplace transforms can be used to study the complete response characteristics of feedback systems, including the transient response. This is in contrast to the use of Fourier transforms in which the steady-state response is the main concern. The Laplace transform of $f(t)$ denoted by $\mathscr{L}\{f(t)\} = F(s)$ is a function of the complex variable $s = \sigma + j\omega$, where

$$F(s) \triangleq \int_{0-}^{\infty} f(t)e^{-st}\,dt \tag{A.1}$$

and is referred to as the *unilateral* (or *one-sided*) *Laplace transform.** A function $f(t)$ will have a Laplace transform if it is of exponential order, which means if there exists a real number σ such that

$$\lim_{t \to \infty} \left| f(t)e^{-\sigma t} \right| = 0. \tag{A.2}$$

For example, ae^{bt} is of exponential order, whereas e^{t^2} is not. If $F(s)$ exists for some $s_0 = \sigma_0 + j\omega_0$, then it exists for all s such that

$$\mathrm{Re}(s) \geq \sigma_0. \tag{A.3}$$

The smallest value of σ_0 for which $F(s)$ exists is called the *abscissa of convergence*, and the region to the right of $\mathrm{Re}(s) \geq \sigma_0$ is called the *region of convergence*.

---

*Bilateral (or two-sided) Laplace transforms and the so-called $\mathscr{L}_+$ transforms, in which the lower value of integral is 0+, also arise elsewhere.

Typically, two-sided Laplace transforms exist for a specified range,

$$\alpha < \mathrm{Re}(s) < \beta,$$

which defines the strip of convergence. Some properties of Laplace transforms are listed in Table A.1 and some Laplace transform pairs are listed in Table A.2. For a thorough study of Laplace transforms and extensive tables, the reader is referred to Churchill (1972), Campbell and Foster (1948), and, for the two-sided transform, Van der Pol and Bremmer (1955).

TABLE A.1
Properties of Laplace Transforms

Number	Laplace transform	Time function	Comment
—	$F(s)$	$f(t)$	Transform pair
1	$\alpha F_1(s) + \beta F_2(s)$	$\alpha f_1(t) + \beta f_2(t)$	Superposition
2	$F(s)e^{-s\lambda}$	$f(t - \lambda)$	Time delay $(\lambda \geq 0)$
3	$\dfrac{1}{|a|}F\left(\dfrac{s}{a}\right)$	$f(at)$	Time scaling
4	$F(s + a)$	$e^{-at}f(t)$	Shift in frequency
5	$s^m F(s) - s^{m-1}f(0)$ $- s^{m-2}\overset{\circ}{f}(0) - \cdots - f^{(m-1)}(0)$	$f^{(m)}(t)$	Differentiation
6	$\dfrac{1}{s}F(s) + \dfrac{[\overset{\circ}{f}(t)dt\,|0}{s}$	$\int f(\zeta)d\zeta$	Integration
7	$F_1(s)F_2(s)$	$f_1(t) * f_2(t)$	Convolution
8	$\lim\limits_{s\to\infty} sF(s)$	$f(0+)$	Initial-value theorem
9	$\lim\limits_{s\to 0} sF(s)$	$\lim\limits_{t\to\infty} f(t)$	Final-value theorem
10	$\dfrac{1}{2\pi j}\displaystyle\int_{c-j\infty}^{c+j\infty} F_1(\zeta)F_2(s - \zeta)d\zeta$	$f_1(t)f_2f(t)$	Time product
11	$-\dfrac{d}{ds}F(s)$	$tf(t)$	Multiplication by time

TABLE A.2
Table of Laplace Transforms

Number	$F(s)$	$f(t), \quad t \geq 0$
1	1	$\delta(t)$
2	$1/s$	$1(t)$
3	$1/s^2$	t
4	$2!/s^3$	t^2
5	$3!/s^4$	t^3
6	$m!/s^{m+1}$	t^m
7	$1/(s + a)$	e^{-at}
8	$1/(s + a)^2$	te^{-at}
9	$1/(s + a)^3$	$\dfrac{1}{2!}t^2 e^{-at}$
10	$1/(s + a)^m$	$\dfrac{1}{(m-1)!}t^{m-1} e^{-at}$
11	$\dfrac{a}{s(s + a)}$	$1 - e^{-at}$
12	$\dfrac{a}{s^2(s + a)}$	$\dfrac{1}{a}(at - 1 + e^{-at})$
13	$\dfrac{b - a}{(s + a)(s + b)}$	$e^{-at} - e^{-bt}$
14	$\dfrac{s}{(s + a)^2}$	$(1 - at)e^{-at}$
15	$\dfrac{a^2}{s(s + a)^2}$	$1 - e^{-at}(1 + at)$
16	$\dfrac{(b - a)s}{(s + a)(s + b)}$	$be^{-bt} - ae^{-at}$
17	$a/(s^2 + a^2)$	$\sin at$
18	$s/(s^2 + a^2)$	$\cos at$
19	$\dfrac{s + a}{(s + a)^2 + b^2}$	$e^{-at}\cos bt$
20	$\dfrac{b}{(s + a)^2 + b^2}$	$e^{-at}\sin bt$
21	$\dfrac{a^2 + b^2}{s\left[(s + a)^2 + b^2\right]}$	$1 - e^{-at}\left(\cos bt + \dfrac{a}{b}\sin bt\right)$

Appendix B: A Review of

Complex Variables

This appendix is a brief summary of some results on complex variable theory with emphasis on the facts needed in control theory. For a comprehensive study of basic complex variable theory the reader is referred to standard textbooks such as Churchill et al.(1976).

B.1 The Definition of a Complex Number

The complex numbers are distinguished from purely real numbers in that they also contain the *imaginary operator,* which we shall denote by j. By definition

$$j^2 = -1 \quad \text{or} \quad j = \sqrt{-1}. \tag{B.1}$$

A complex number may be defined as

$$A = \sigma + j\omega \tag{B.2}$$

where σ is the real part and ω is the imaginary part:

$$\sigma = \text{Re}\{A\} \qquad \omega = \text{Im}\{A\}. \tag{B.3}$$

Note that the imaginary part of A is itself a real number. We may represent the complex number A in the cartesian coordinate system as shown in Fig. B.1(a). The complex plane is shown and A is represented by a single point. An alternative representation is in the polar form. The complex number is represented by a vector with length r and an angle θ where the angle is in radians and is measured from

FIGURE B.1

(a) *Cartesian* representation of a complex number; (b) polar representation of a complex number.

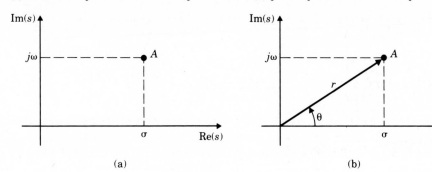

(a) (b)

the positive real axis in the counter-clockwise direction as shown in Fig. B.1(b). The complex number, A, in polar form is denoted by

$$A = |A| \angle \arg(A) = r \angle \theta = re^{j\theta} \qquad 0 \le \theta < 2\pi \qquad \text{(B.4)}$$

$$r = |A| = \sqrt{(\sigma^2 + \omega^2)} \qquad \text{(B.5)}$$

where the number r is the length of the vector representing A and is called the magnitude or modulus or absolute value of A and

$$\tan(\theta) = \frac{\omega}{\sigma} \qquad \text{(B.6)}$$

The *conjugate* of A is defined as

$$A^* = \sigma - j\omega \qquad \text{(B.7)}$$

therefore,

$$(A^*)^* = A \qquad \text{(B.8)}$$

$$(A_1 \pm A_2)^* = A_1{}^* \pm A_2{}^* \qquad \text{(B.9)}$$

$$\left(\frac{A_1}{A_2}\right)^* = \frac{A_1{}^*}{A_2{}^*} \qquad \text{(B.10)}$$

$$\text{Re}(A) = \frac{(A + A^*)}{2} \qquad \text{Im}(A) = \frac{(A - A^*)}{2j} \qquad \text{(B.11)}$$

and

$$AA^* = (|A|)^2. \qquad \text{(B.12)}$$

B.2   Algebraic Manipulations

The rules for .addition, multiplication, and division are as expected. Let

$$A_1 = \sigma_1 + j\omega_1 \qquad A_2 = \sigma_2 + j\omega_2 \qquad \text{(B.13)}$$

then

$$A_1 + A_2 = (\sigma_1 + j\omega_1) + (\sigma_2 + j\omega_2) = (\sigma_1 + \sigma_2) + j(\omega_1 + \omega_2). \qquad \text{(B.14)}$$

Complex numbers may be added or subtracted graphically. Each complex number is represented by a vector extended from the origin. The sum is obtained by adding the two vectors by constructing a parallelogram as shown in Fig. B.2(a). Alternatively, starting at the tail of one vector one may draw a parallel vector to the other.

$$A_1 A_2 = (\sigma_1 + j\omega_1)(\sigma_2 + j\omega_2)$$
$$= (\sigma_1\sigma_2 - \omega_1\omega_2) + j(\omega_1\sigma_2 + \sigma_1\omega_2). \qquad \text{(B.15)}$$

The product of two complex numbers may be done graphically using polar representations as shown in Fig. B.2(b). The division of two complex numbers is carried

FIGURE B.2
(a) Addition of complex numbers; (b) multiplication of complex numbers; (c) division of complex numbers.

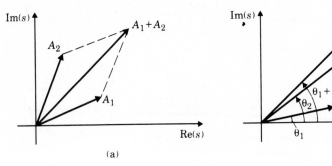

(a)

(b)

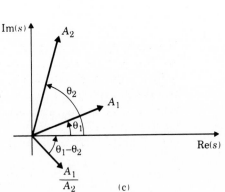

(c)

out by *rationalization*. This means that both the numerator and denominator in the ratio are multiplied by the conjugate of the denominator:

$$\frac{A_1}{A_2} = \frac{A_1 A_2{}^*}{A_2 A_2{}^*} \tag{B.16}$$

$$= \frac{(\sigma_1\sigma_2 + \omega_1\omega_2) + j(\omega_1\sigma_2 - \sigma_1\omega_2)}{(\sigma_2^2 + \omega_2^2)} \tag{B.17}$$

From Eq. (B.4) it follows that

$$A^{-1} = \frac{1}{r}e^{-j\theta} \qquad r \neq 0. \tag{B.18}$$

Also if $A_1 = r_1 e^{j\theta_1}$ and $A_2 = r_2 e^{j\theta_2}$ then

$$A_1 A_2 = r_1 r_2 e^{j(\theta_1 + \theta_2)} \tag{B.19}$$

and

$$\frac{A_1}{A_2} = \frac{r_1}{r_2} e^{j(\theta_1 - \theta_2)} \qquad r_2 \neq 0. \tag{B.20}$$

The division of complex numbers may be carried out graphically in polar coordinates as shown in Fig. B.2(c).

B.3   Graphical Evaluation of Magnitude and Phase

Consider the transfer function

$$\mathbf{G}(s) = \frac{\prod_{i=1}^{i=m}(s + z_i)}{\prod_{i=1}^{i=n}(s + p_i)}. \tag{B.21}$$

The value of the transfer function for sinusoidal inputs is found by replacing s by $j\omega$. The gain and phase are given by $\mathbf{G}(j\omega)$ and may be determined analytically or by a graphical procedure. Consider the pole-zero configuration for such a $\mathbf{G}(s)$ and a point $s_0 = j\omega_0$ on the imaginary axis as shown in Fig. B.3. Also consider the vectors drawn from the poles and the zero to s_0. The magnitude of the transfer function evaluated at $s_0 = j\omega_0$ is simply the ratio of the distance from the zero to the product of all the distances from the poles:

$$|\mathbf{G}(j\omega_0)| = \frac{r_1}{r_2 r_3 r_4}. \tag{B.22}$$

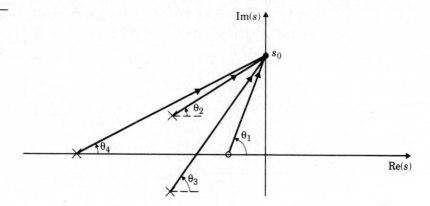

The phase is given by the sum of the angles from the zero minus the sum of the angles from the poles:

$$\arg\{\mathbf{G}(j\omega_0)\} = \angle\mathbf{G}(j\omega_0) = \theta_1 - (\theta_2 + \theta_3 + \theta_4). \qquad \text{(B.23)}$$

This may be explained as follows. The term $s + z_1$ is a vector addition of its two components. We may determine this equivalently as $s - (-z_1)$, which amounts to translation of the vector $s + z_1$ starting at $-z_1$ as shown in Fig. B.4. This means that a vector drawn from the zero location to s_0 is equivalent to $s + z_1$. The same reasoning applies to the poles. We reflect p_1, p_2, and p_3 about the origin to obtain the pole locations. Then the vectors drawn from $-p_1, -p_2$, and $-p_3$ to s_0 are the same as the vectors in the denominator represented in polar coordinates. Note that this method is not limited to the evaluation on the imaginary axis. The same method may be used for s_0 at other places in the complex plane.

FIGURE B.4
Illustration of graphical
computation of $(s + z_1)$.

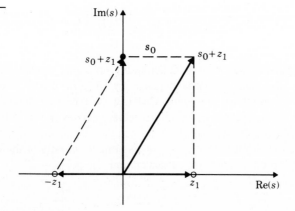

B.4   Differentiation and Integration

The usual rules apply to complex differentiation. Let $\mathbf{G}(s)$ be differentiable with respect to s then

$$\frac{d\mathbf{G}(s)}{ds} = \frac{d\operatorname{Re}(\mathbf{G}(s))}{ds} + j\frac{d\operatorname{Im}(\mathbf{G}(s))}{ds}. \tag{B.24}$$

The standard rules also apply in the integration except that the constant of integration is a complex constant:

$$\int \mathbf{G}(s)ds = \int \operatorname{Re}(\mathbf{G}(s))ds + j\int \operatorname{Im}(\mathbf{G}(s))ds + c \tag{B.25}$$

where c is a complex constant.

B.5   Euler's Relations

Let us now derive an important relationship involving the complex exponential. Define

$$A = \cos(\theta) + j\sin(\theta) \tag{B.26}$$

where θ is in radians. Then

$$\frac{dA}{d\theta} = -\sin(\theta) + j\cos(\theta) = j^2\sin(\theta) + j\cos(\theta)$$

$$= j(\cos(\theta) + j(\sin(\theta)) = jA. \tag{B.27}$$

Collecting the terms involving A we obtain

$$\frac{dA}{A} = j\ d\theta. \tag{B.28}$$

Integrating both sides of (B.28) yields

$$\ln(A) = j\theta + c \tag{B.29}$$

where c is a constant of integration. If we let $\theta = 0$ in (B.29) we find that $c = 0$ or

$$A = e^{j\theta} = \cos(\theta) + j\sin(\theta). \tag{B.30}$$

Similarly,

$$A^* = e^{-j\theta} = \cos(\theta) - j\sin(\theta). \tag{B.31}$$

From (B.30) and (B.31) it follows that

$$\cos(\theta) = \frac{(e^{j\theta} + e^{-j\theta})}{2} \tag{B.32}$$

and

$$\sin(\theta) = \frac{(e^{j\theta} - e^{-j\theta})}{2j}. \tag{B.33}$$

B.6 Analytic Functions

Assume that $\mathbf{G}$ is a complex valued function defined in the complex plane. Let s_0 be in the domain of $\mathbf{G}$, which is assumed to be finite in some disk centered at s_0. Thus $\mathbf{G}(s)$ is defined not only at s_0 but also at all points in the disk centered at s_0. The function $\mathbf{G}$ is said to be *analytic* if its derivative exists at s_0 and at each point in the neighborhood of s_0.

B.7 Cauchy's Theorem

A *contour* is a piecewise smooth arc that consists of a number of smooth arcs joined together. A simple closed contour is a contour that does not intersect itself and ends on itself. Let C be a closed contour and let $\mathbf{G}$ be analytic inside and on C as shown in Fig. B.5(a). Cauchy's theorem states that

$$\oint_C \mathbf{G}(s)ds = 0. \tag{B.34}$$

A corollary to this theorem is the following. Let C_1 and C_2 be two paths connecting the points A_1 and A_2 as in Fig. B.5(b), then

$$\int_{C_1} \mathbf{G}(s)ds = -\int_{C_2} \mathbf{G}(s)ds. \tag{B.35}$$

B.8 Singularities and Residues

If a function $\mathbf{G}(s)$ is not analytic at s_0 but is analytic at some point in every neighborhood of s_0, s_0 is said to be a *singularity* of $\mathbf{G}$. A singular point is said to be

FIGURE B.5
(a) A closed contour. (b) Two different paths between A_1 and A_2.

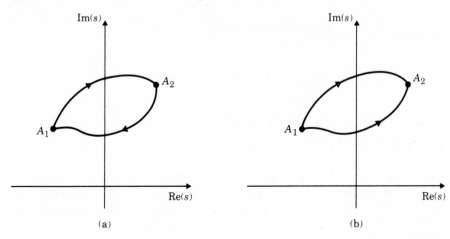

(a)   (b)

an *isolated singularity* if $G(s)$ is analytic everywhere else in the neighborhood of s_0 except at s_0. Let $G(s)$ be a *rational* function (i.e., ratio of polynomials). Then $G(s)$ is analytic except at the location of the poles (i.e., roots of the denominator). All singularities of rational algebraic functions are the pole locations. Let a rational $G(s)$ be analytic except at s_0. Then we may write $G(s)$ in its Laurent-series expansion

$$G(s) = \frac{A_{-n}}{(s - s_0)^n} + \cdots + \frac{A_{-1}}{(s - s_0)} + B_0 + B_1(s - s_0) + \cdots. \tag{B.36}$$

The coefficient A_{-1} is called the residue of $G(s)$ at s_0 and may be evaluated by Cauchy's integral formula as

$$A_{-1} = \frac{1}{2\pi j} \oint_C G(s)\,ds \tag{B.37}$$

where C denotes a closed arc within an analytic region centered at s_0 and contains no other singularity as shown in Fig. B.6. s_0 is not repeated with $n = 1$ and in this case,

$$A_{-1} = \text{Res}(G(s); s_0) = (s - s_0)G(s)\big|_{s = s_0}, \tag{B.38}$$

which is the familiar "cover up" method of computing residues.

FIGURE B.6
Contour around an
isolated singularity.

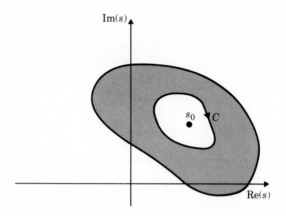

B.9 Residue Theorem

If the contour C contains l singularities, then (B.38) may be generalized to yield
Cauchy's residue theorem:

$$\frac{1}{2\pi j} \oint \mathbf{G}(s)ds = \sum_{i=1}^{i=l} \text{Res}(\mathbf{G}(s); s_i). \tag{B.39}$$

B.10 The Argument Principle

Before stating the argument principle, we need a preliminary result from which the
principle follows readily. Let $\mathbf{G}(s)$ be an analytic function on and inside a closed
contour C except for a finite number of poles inside C. Then for C described in a
positive sense (clockwise direction),

$$\frac{1}{2\pi j} \oint \frac{\mathbf{G}'(s)}{\mathbf{G}(s)} ds = N - P \tag{B.40}$$

or

$$\frac{1}{2\pi j} \oint d[\ln \mathbf{G}] = N - P \tag{B.41}$$

where N and P are the total number of zeros and poles of $\mathbf{G}$ inside C, respectively.
A pole or zero of multiplicity k is counted k times.

Proof: Let s_0 be a zero of **G** with multiplicity k. Then in some neighborhood of that point we may write **G**(s) as,

$$\mathbf{G}(s) = (s - s_0)^k f(s) \tag{B.42}$$

where $f(s)$ is analytic and $f(s_0) \neq 0$. If we differentiate (B.41) we obtain

$$\mathbf{G}'(s) = k(s - s_0)^{k-1} f(s) + (s - s_0)^k f'(s). \tag{B.43}$$

Now (B.40) may be rewritten as

$$\frac{\mathbf{G}'(s)}{\mathbf{G}(s)} = \frac{K}{(s - s_0)} + \frac{f'(s)}{f(s)}. \tag{B.44}$$

Therefore, $\mathbf{G}'(s)/\mathbf{G}(s)$ has a pole at $s = s_0$ with residue K. This analysis may be repeated for every zero. Hence, the sum of all the residues of $\mathbf{G}'(s)/\mathbf{G}(s)$ is the number of zeros of $\mathbf{G}(s)$ inside C. If s_0 is a pole with multiplicity l we may write

$$h(s) = (s - s_0)^l \mathbf{G}(s) \tag{B.45}$$

where $h(s)$ is analytic and $h(s_0) \neq 0$. Then (B.42) may be rewritten as

$$\mathbf{G}(s) = \frac{h(s)}{(s - s_0)^l}. \tag{B.46}$$

Differentiating (B.43) we obtain,

$$\mathbf{G}'(s) = \frac{h'(s)}{(s - s_0)} - \frac{l\,h(s)}{(s - s_0)^{l+1}} \tag{B.47}$$

so that

$$\frac{\mathbf{G}'(s)}{\mathbf{G}(s)} = \frac{-l}{(s - s_0)} + \frac{h'(s)}{h(s)}. \tag{B.48}$$

This analysis may be repeated for every pole. The result is that the sum of the residues of $\mathbf{G}'(s)/\mathbf{G}(s)$ at all the poles of $\mathbf{G}(s)$ is $-P$. We now state the argument principle.

Using (B.38):

$$\frac{1}{2\pi j} \oint_C d\,[\ln\mathbf{G}(s)] = (N - P) \tag{B.49}$$

where $d/ds \ln \mathbf{G}(s)$ was substituted for $\mathbf{G}'(s)/\mathbf{G}(s)$. If we write $\mathbf{G}(s)$ in polar form then,

$$\oint_\Gamma d\,[\ln\mathbf{G}(s)] = \oint_\Gamma d\,\big\{ [\ln|\mathbf{G}(s)| + j\arg(\ln\mathbf{G}(s))] \big\}$$

$$= \ln|\mathbf{G}(s)|\big|_{s=s_1}^{s=s_2} + j\,\arg[\mathbf{G}(s)]\big|_{s=s_1}^{s=s_2}. \tag{B.50}$$

Since Γ is a closed contour, the first term is zero but the second term is 2π times the net encirclements of the origin:

$$\frac{1}{2\pi j} \oint_{\Gamma} d\,[\ln \mathbf{G}(s)] = N - P. \qquad (B.51)$$

Intuitively, the argument principle may be stated as follows. Let $\mathbf{G}(s)$ be a rational function that is analytic except possibly at a finite number of points. Select an arbitrary contour in the s-plane so that $\mathbf{G}(s)$ is analytic at every point on the contour (the contour does not pass through any of the singularities). The corresponding mapping into the $\mathbf{G}(s)$ plane will encircle the origin a number of times, in the same sense as the s-plane contour, equal to the difference between the zeros and poles of $\mathbf{G}(s)$ encircled by the s-plane contour. For example, if the contour encircles a single zero, the mapping will encircle the origin once in the clockwise direction. Similarly, if the contour only encloses a single pole, the mapping will encircle the origin in the counter-clockwise direction. If the contour encircles no singularities, or if the contour encloses an equal number of poles and zeros, there will be no encirclement of the map of the origin. A contour evaluation of $\mathbf{G}(s)$ will encircle the origin if there is a nonzero net difference between the encircled singularities. The mapping is conformal as well, which means that the magnitude and sense of the angles between smooth arcs is preserved. Chapter 5 provides a more detailed intuitive treatment of the argument principle and its application to feedback control in the form of the Nyquist stability theorem.

Appendix C: Summary of Matrix Theory

We assume that the reader is already somewhat familiar with matrix theory and with the solution of linear systems of equations. However, a brief summary of matrix theory is presented here, with emphasis on the results needed in control theory. The reader is referred to Strang (1988) and Gantmacher (1959) for further study.

C.1 Matrix

An array of numbers arranged in rows and columns is referred to as a *matrix*. If **A** is a matrix with m rows and n columns, an $m \times n$ (read m by n) matrix, it is denoted by

$$A = \begin{bmatrix} a_{11} & a_{12} & \cdots & a_{1n} \\ a_{21} & a_{22} & \cdots & a_{2n} \\ \vdots & \vdots & & \vdots \\ a_{m1} & a_{m2} & \cdots & a_{mn} \end{bmatrix}, \tag{C.1}$$

where the entries a_{ij} are its elements. If $m = n$, then the matrix is square; otherwise it is rectangular. Sometimes a matrix is simply denoted by $\mathbf{A} = [a_{ij}]$. If $m = 1$ or $n = 1$, then the matrix reduces to a *row vector* or a *column vector*, respectively.

C.2 Elementary Operations on Matrices

If **A** and **B** are matrices of the same dimension, then their sum is defined by

$$\mathbf{C} = \mathbf{A} + \mathbf{B}, \tag{C.2}$$

647

where

$$c_{ij} = a_{ij} + b_{ij}; \tag{C.3}$$

that is, the addition is done element by element. It is easy to verify that

$$\mathbf{A} + \mathbf{B} = \mathbf{B} + \mathbf{A}, \tag{C.4}$$

$$(\mathbf{A} + \mathbf{B}) + \mathbf{C} = \mathbf{A} + (\mathbf{B} + \mathbf{C}). \tag{C.5}$$

Two matrices can be multiplied if they are compatible. Let $\mathbf{A} = m \times n$ and $\mathbf{B} = n \times p$, then the $m \times p$ matrix,

$$\mathbf{C} = \mathbf{AB}, \tag{C.6}$$

is the product of the two matrices, where

$$c_{ij} = \sum_{k=1}^{n} a_{ik} b_{kj}. \tag{C.7}$$

Matrix multiplication satisfies the associative law

$$\mathbf{A(BC)} = \mathbf{(AB)C} \tag{C.8}$$

but not the commutative law; that is, in general,

$$\mathbf{AB} \neq \mathbf{BA}. \tag{C.9}$$

C.3 Trace

The *trace* of a square matrix is the sum of its diagonal elements;

$$\text{trace } \mathbf{A} = \sum_{i=1}^{n} a_{ii}. \tag{C.10}$$

C.4 Transpose

The $n \times m$ matrix obtained by interchanging the rows and columns of $\mathbf{A}$ is called the *transpose of matrix* $\mathbf{A}$

$$A^{\mathrm{T}} = \begin{bmatrix} a_{11} & a_{21} & \cdots & a_{m1} \\ a_{12} & a_{22} & \cdots & \cdots \\ \vdots & \vdots & & \vdots \\ a_{1n} & a_{2n} & \cdots & a_{mn} \end{bmatrix}. \tag{C.11}$$

A matrix is said to be *symmetric* if

$$\mathbf{A}^T = \mathbf{A}. \tag{C.12}$$

It is easy to show that

$$(\mathbf{AB})^T = \mathbf{B}^T\mathbf{A}^T \tag{C.13}$$

$$(\mathbf{ABC})^T = \mathbf{C}^T\mathbf{B}^T\mathbf{A}^T \tag{C.14}$$

$$(\mathbf{A} + \mathbf{B})^T = \mathbf{A}^T + \mathbf{B}^T. \tag{C.15}$$

C.5 Matrix Inverse and Determinant

The *determinant* of a square matrix, denoted by "det," is defined by Laplace's expansion

$$\det \mathbf{A} = \sum_{j=1}^{n} a_{ij}\gamma_{ij} \qquad \text{for any } i = 1, 2, \ldots, n, \tag{C.16}$$

where γ_{ij} is called the *cofactor* and

$$\gamma_{ij} = -1^{i+j} \det M_{ij}, \tag{C.17}$$

where $\det M_{ij}$ is called a *minor*. The term M_{ij} is the same as the matrix $\mathbf{A}$ except that its ith row and jth column have been removed. Note that M_{ij} is always an $(n-1) \times (n-1)$ matrix, and the minors and cofactors are identical except possibly for a sign.

The *adjugate* of a matrix is the transpose of the matrix of its cofactors

$$\text{adj } \mathbf{A} = [\gamma_{ij}]^T. \tag{C.18}$$

It can be shown that $(\mathbf{A})(\text{adj}\mathbf{A}) = (\det \mathbf{A})\mathbf{I}$. If $\det \mathbf{A} \neq 0$, then the *inverse* of a matrix $\mathbf{A}$ is defined by

$$\mathbf{A}^{-1} = \frac{\text{adj } \mathbf{A}}{\det \mathbf{A}} \tag{C.19}$$

and has the property that

$$\mathbf{AA}^{-1} = \mathbf{A}^{-1}\mathbf{A} = \mathbf{I}, \tag{C.20}$$

where $\mathbf{I}$ is called the *identity matrix:*

$$\mathbf{I} = \begin{bmatrix} 1 & 0 & \cdots & \cdots & 0 \\ 0 & 1 & 0 & \cdots & 0 \\ \vdots & \vdots & \ddots & & \vdots \\ & & & & 0 \\ 0 & \cdots & \cdots & 0 & 1 \end{bmatrix}, \tag{C.21}$$

that is, with 1s along the diagonal and 0s elsewhere. Note that a matrix has an inverse—that is, it is *nonsingular* if its determinant is nonzero. The inverse of the product of two matrices is the product of the inverse of the matrices in reverse order:

$$(\mathbf{AB})^{-1} = \mathbf{B}^{-1}\mathbf{A}^{-1} \tag{C.22}$$

and

$$(\mathbf{ABC})^{-1} = \mathbf{C}^{-1}\mathbf{B}^{-1}\mathbf{A}^{-1}. \tag{C.23}$$

C.6   Properties of the Determinant

When dealing with determinants of matrices, the following elementary (row or column) operations are useful.

1. If any row (or column) of $\mathbf{A}$ is multiplied by a scalar α, the resulting matrix $\overline{\mathbf{A}}$ has the determinant

$$\det \overline{\mathbf{A}} = \alpha \det \mathbf{A}. \tag{C.24}$$

Hence

$$\det[\alpha\mathbf{A}] = \alpha^n \det \mathbf{A}. \tag{C.25}$$

2. If any two rows (or columns) of $\mathbf{A}$ are interchanged to obtain $\overline{\mathbf{A}}$, then

$$\det \overline{\mathbf{A}} = -\det \mathbf{A}. \tag{C.26}$$

3. If a multiple of a row (or column) of $\mathbf{A}$ is added to another to obtain $\overline{\mathbf{A}}$, then

$$\det \overline{\mathbf{A}} = \det \mathbf{A}. \tag{C.27}$$

It is also easy to show that

$$\det \mathbf{A} = \det \mathbf{A}^\mathrm{T} \tag{C.28}$$

and

$$\det[\mathbf{AB}] = \det \mathbf{A} \det \mathbf{B}. \tag{C.29}$$

Using the latter relation on Eq. (C.20) we have that

$$\det \mathbf{A} \det \mathbf{A}^{-1} = 1. \tag{C.30}$$

If $\mathbf{A}$ and $\mathbf{B}$ are square matrices, then the determinant of the block triangular matrix

$$\det \begin{bmatrix} \mathbf{A} & \mathbf{C} \\ 0 & \mathbf{B} \end{bmatrix} = \det \mathbf{A} \det \mathbf{B} \tag{C.31}$$

is the product of the determinants of the diagonal blocks.

If **A** is nonsingular, then

$$\det \begin{bmatrix} \mathbf{A} & \mathbf{B} \\ \mathbf{C} & \mathbf{D} \end{bmatrix} = \det \mathbf{A} \det \left[\mathbf{D} - \mathbf{CA}^{-1}\mathbf{B} \right]. \tag{C.32}$$

Using this identity, the transfer function of a scalar system can be written in a compact form:

$$G(s) = \mathbf{H}(s\mathbf{I} - \mathbf{F})^{-1}\mathbf{G} + J = \frac{\det \begin{bmatrix} s\mathbf{I} - \mathbf{F} & \mathbf{G} \\ -\mathbf{H} & J \end{bmatrix}}{\det[s\mathbf{I} - \mathbf{F}]}. \tag{C.33}$$

C.7   Inverse of Block Triangular Matrices

If **A** and **B** are square invertible matrices, then

$$\begin{bmatrix} \mathbf{A} & \mathbf{C} \\ 0 & \mathbf{B} \end{bmatrix}^{-1} = \begin{bmatrix} \mathbf{A}^{-1} & -\mathbf{A}^{-1}\mathbf{CB}^{-1} \\ 0 & \mathbf{B}^{-1} \end{bmatrix}. \tag{C.34}$$

C.8   Special Matrices

Some matrices have a special structure and are given a name. We have already defined the identity matrix, which has a special form. A *diagonal matrix* is one with nonzero elements along the main diagonal and zeroes elsewhere:

$$\mathbf{A} = \begin{bmatrix} a_{11} & & & & 0 \\ & a_{22} & & & \\ & & a_{33} & & \\ & & & \ddots & \\ 0 & & & & a_{nn} \end{bmatrix}. \tag{C.35}$$

A matrix is said to be *(upper) triangular,* if all the elements below the main diagonal are zero:

$$\mathbf{A} = \begin{bmatrix} a_{11} & a_{12} & \cdots & & a_{1n} \\ 0 & a_{22} & & & \cdot \\ \vdots & 0 & & & \cdot \\ 0 & \vdots & \ddots & \ddots & \cdot \\ 0 & 0 & 0 & & a_{nn} \end{bmatrix}. \tag{C.36}$$

The determinant of a diagonal or triangular matrix is simply the product of its diagonal elements.

A matrix is said to be in the *(upper) companion form* if it has the structure

$$
\mathbf{A}_c = \begin{bmatrix}
-a_1 & -a_2 & & \cdots & -a_n \\
1 & 0 & & \cdots & 0 \\
0 & 1 & 0 & \cdots & 0 \\
\vdots & & \ddots & & \vdots \\
0 & \cdots & \cdots & 1 & 0
\end{bmatrix}.
\tag{C.37}
$$

Note that all the information is contained in the first row. Other variants of this form are the lower left or right companion matrices. A *Vandermonde matrix* has the following structure

$$
\mathbf{A} = \begin{bmatrix}
1 & a_1 & a_1^2 & \cdots & a_1^{n-1} \\
1 & a_2 & a_2^2 & \cdots & a_2^{n-1} \\
\vdots & \vdots & \vdots & & \vdots \\
1 & a_n & a_n^2 & \cdots & a_n^{n-1}
\end{bmatrix}.
\tag{C.38}
$$

C.9 Rank

The *rank* of a matrix is the number of its linearly dependent rows or columns. If the rank of $\mathbf{A}$ is r, then all $(r + 1) \times (r + 1)$ submatrices of $\mathbf{A}$ are singular, and there is at least one $r \times r$ submatrix that is nonsingular. It is also true that

$$
\text{Row rank of } \mathbf{A} = \text{Column rank of } \mathbf{A}.
\tag{C.39}
$$

C.10 Characteristic Polynomial

The *characteristic polynomial* of a matrix $\mathbf{A}$ is defined by

$$
a(s) \triangleq \det[s\mathbf{I} - \mathbf{A}]
$$
$$
= s^n + a_1 s^{n-1} + \cdots + a_{n-1}s + a_n,
\tag{C.40}
$$

where the roots of the polynomial are referred to as *eigenvalues of* $\mathbf{A}$. We can write

$$
a(s) = (s - \lambda_1)(s - \lambda_2)\cdots(s - \lambda_n),
\tag{C.41}
$$

where $\{\lambda_i\}$ are the eigenvalues of $\mathbf{A}$. The characteristic polynomial of a companion matrix is

$$
a(s) = \det[s\mathbf{I} - \mathbf{A}_c] = s^n + a_1 s^{n-1} + \cdots + a_{n-1}s + a_n.
\tag{C.42}
$$

C.11 The Cayley-Hamilton Theorem

The Cayley-Hamilton theorem states that every square matrix $\mathbf{A}$ satisfies its characteristic equation. This means that if $\mathbf{A}$ is an $n \times n$ matrix with characteristic equation $a(s)$, then

$$a(\mathbf{A}) \triangleq \mathbf{A}^n + a_1\mathbf{A}^{n-1} + \cdots + a_{n-1}\mathbf{A} + a_n\mathbf{I} = 0. \tag{C.43}$$

C.12 Eigenvalues and Eigenvectors

If λ is a scalar and V is a nonzero vector, then

$$\mathbf{A}V = \lambda V \tag{C.44}$$

defines λ as the *eigenvalue* of the square matrix $\mathbf{A}$ and V is the associated (right) eigenvector. If we collect both terms on the right,

$$(\lambda\mathbf{I} - \mathbf{A})V = 0. \tag{C.45}$$

Since V is nonzero, then

$$\det[\lambda\mathbf{I} - \mathbf{A}] = 0 \tag{C.46}$$

so that λ is an eigenvalue of the matrix $\mathbf{A}$ as defined earlier. The normalization of the eigenvectors is arbitrary; that is, if V is an eigenvector so is αV. The eigenvectors are usually normalized to have unit length; that is $\|V\|^2 = V^\mathsf{T}V = 1$.

If W^T is a nonzero row vector such that

$$W^\mathsf{T}\mathbf{A} = \lambda W^\mathsf{T}, \tag{C.47}$$

then W is called a *left eigenvector of* $\mathbf{A}$. Note that we can write

$$\mathbf{A}^\mathsf{T}W = \lambda W \tag{C.48}$$

so that W is simply a *right eigenvector* of $\mathbf{A}^\mathsf{T}$.

C.13 Similarity Transformations

Consider the arbitrary nonsingular matrix $\mathbf{T}$ such that

$$\overline{\mathbf{A}} = \mathbf{T}^{-1}\mathbf{A}\mathbf{T}. \tag{C.49}$$

This kind of matrix operation is referred to as a *similarity transformation*. If $\mathbf{A}$ has a full set of eigenvectors, then $\mathbf{T}$ can be chosen to be the set of eigenvectors and

$\overline{\mathbf{A}}$ will be diagonal. Consider the set of equations in state-variable form:

$$\dot{\mathbf{x}} = \mathbf{Fx} + \mathbf{G}u. \tag{C.50}$$

If we make a change of variables

$$\mathbf{T}\xi = \mathbf{x}, \tag{C.51}$$

then

$$\mathbf{T}\dot{\xi} = \mathbf{FT}\xi + \mathbf{G}u,$$
$$\dot{\xi} = \mathbf{T}^{-1}\mathbf{FT}\xi + \mathbf{T}^{-1}\mathbf{G}u$$
$$= \overline{\mathbf{F}}\xi + \overline{\mathbf{G}}u, \tag{C.52}$$

where

$$\overline{\mathbf{F}} = \mathbf{T}^{-1}\mathbf{FT},$$
$$\overline{\mathbf{G}} = \mathbf{T}^{-1}\mathbf{G}. \tag{C.53}$$

The characteristic polynomial of $\overline{\mathbf{F}}$ is

$$\det[s\mathbf{I} - \overline{\mathbf{F}}] = \det[s\mathbf{I} - \mathbf{T}^{-1}\mathbf{FT}]$$
$$= \det[s\mathbf{T}^{-1}\mathbf{T} - \mathbf{T}^{-1}\mathbf{FT}]$$
$$= \det[\mathbf{T}^{-1}(s\mathbf{I} - \mathbf{F})\mathbf{T}]$$
$$= \det \mathbf{T}^{-1} \det[s\mathbf{I} - \mathbf{F}] \det \mathbf{T}. \tag{C.54}$$

Using Eq. (C.30) then,

$$\det[s\mathbf{I} - \overline{\mathbf{F}}] = \det[s\mathbf{I} - \mathbf{F}]. \tag{C.55}$$

From Eq. (C.55) one can see that $\overline{\mathbf{F}}$ and $\mathbf{F}$ both have the same characteristic polynomial, and we have the important result that a similarity transformation does not change the eigenvalues of a matrix. From Eq. (C.51) a new state made up of a linear combination of the old state has the same eigenvalues as the old set.

C.14   Matrix Exponential

Let $\mathbf{A}$ be a square matrix. The matrix exponential of $\mathbf{A}$ is defined as the series

$$e^{\mathbf{A}t} = \mathbf{I} + \mathbf{A}t + \mathbf{A}^2\mathbf{t}^2/2! + \mathbf{A}^3\mathbf{t}^3/3! + \cdots. \tag{C.56}$$

It can be shown that the series converges. If $\mathbf{A}$ is an $n \times n$ matrix, then $e^{\mathbf{A}t}$ is also an $n \times n$ matrix and can be differentiated:

$$\frac{d}{dt}e^{\mathbf{A}t} = \mathbf{A}e^{\mathbf{A}t}. \tag{C.57}$$

Other properties of matrix exponential are

$$e^{\mathbf{A}t_1} e^{\mathbf{A}t_2} = e^{\mathbf{A}(t_1+t_2)} \qquad (C.58)$$

and

$$e^{\mathbf{A}} e^{\mathbf{B}} \neq e^{\mathbf{B}} e^{\mathbf{A}} \qquad (C.59)$$

unless $\mathbf{A}$ and $\mathbf{B}$ commute, that is, $\mathbf{AB} = \mathbf{BA}$.

C.15   Fundamental Subspaces

The *range space* of $\mathbf{A}$ (or the column space of $\mathbf{A}$) is defined by the set of vectors $\mathbf{x}$;

$$\mathbf{x} = \mathbf{Ay} \qquad (C.60)$$

for some vector $\mathbf{y}$. The range space of $\mathbf{A}$ is denoted by $\mathcal{R}(\mathbf{A})$.

The *null space* of $\mathbf{A}$ is defined by the set of vectors $\mathbf{x}$ such that

$$\mathbf{Ax} = 0 \qquad (C.61)$$

and is denoted by $\mathcal{N}(\mathbf{A})$. If $\mathbf{x} \epsilon \mathcal{N}(\mathbf{A})$ and $\mathbf{y} \epsilon \mathcal{R}(\mathbf{A}^{\mathrm{T}})$, then $\mathbf{y}^{\mathrm{T}}\mathbf{x} = 0$; that is, every vector in the null space of $\mathbf{A}$ is orthogonal to every vector in the range space of $\mathbf{A}^{\mathrm{T}}$.

C.16   Singular Value Decomposition

The singular value decomposition (SVD) is one of the most useful tools in linear algebra. It has been widely used in control theory during the last decade. Let A be an $m \times n$ matrix. Then there always exist matrices $\mathbf{U}, \mathbf{S}, \mathbf{V}$ such that

$$\mathbf{A} = \mathbf{USV}^{\mathrm{T}}$$

where $\mathbf{U}$ and $\mathbf{V}$ are orthogonal matrices,

$$\mathbf{UU}^{\mathrm{T}} = \mathbf{I}, \qquad \mathbf{VV}^{\mathrm{T}} = \mathbf{I}$$

and $\mathbf{S}$ is a quasi-diagonal matrix with the singular values as its diagonals

$$\mathbf{S} = \begin{bmatrix} \Sigma & 0 \\ 0 & 0 \end{bmatrix}$$

where Σ is a diagonal matrix of nonzero singular values in descending order

$$\sigma_1 \geq \sigma_2 \geq \cdots \geq \sigma_r > 0.$$

The unique diagonals of $\mathbf{S}$ are called the *singular values*. The maximum singular value is denoted by $\overline{\sigma}(\mathbf{A})$ and the minimum singular value is denoted by $\underline{\sigma}(\mathbf{A})$. The rank of the matrix is the same as the number of nonzero singular values. The columns of $\mathbf{U}$ and $\mathbf{V}$,

$$\mathbf{U} = [u_1 u_2 \cdots u_m]$$
$$\mathbf{V} = [v_1 v_2 \cdots v_n]$$

are called the left and right *singular vectors*, respectively. Using SVD provides complete information about the fundamental subspaces associated with a matrix:

$$\mathcal{N}(\mathbf{A}) = span\{v_{r+1} \quad v_{r+2} \quad \cdots \quad v_n\}$$
$$\mathcal{R}(\mathbf{A}) = span\{u_1 \quad u_2 \quad \cdots \quad u_r\}$$
$$\mathcal{R}(\mathbf{A}^T) = span\{v_1 \quad v_2 \quad \cdots \quad v_r\}$$
$$\mathcal{N}(\mathbf{A}^T) = span\{u_{r+1} \quad u_{r+2} \quad \cdots \quad u_m\}$$

The norm $\|\mathbf{A}\|_2$ of the matrix is given by

$$\|\mathbf{A}\|_2 = \overline{\sigma}(\mathbf{A})$$

If $\mathbf{A}$ is a function of ω then the $\|\mathbf{A}\|_\infty$ is given by

$$\|\mathbf{A}\|_\infty = \max_\omega \overline{\sigma}(\mathbf{A})$$

C.17 Positive Definite Matrices

A matrix $\mathbf{A}$ is said to be *positive semidefinite* if,

$$\mathbf{x}^T \mathbf{A} \mathbf{x} \geq 0, \qquad \text{all } \mathbf{x}.$$

The matrix is said to be *positive definite* if, in the above, equality holds only for $\mathbf{x} = 0$. A symmetric matrix is positive definite if and only if all of its eigenvalues are positive. It is positive semidefinite if and only if all of its eigenvalues are nonnegative. An alternative method for determination of positive definiteness is by a test of minors of the matrix. A matrix is positive definite if all the leading principal minors are positive. It is positive semidefinite if they are all nonnegative.

References

Ackermann, J., "Der Entwurf Linearer Regelungssysteme im Zustandsraum," *Regelungs-technik und Prozessdatenverarbeitung*, **7**:297–300, 1972.

Airy, G. B., "On the Regulator of the Clock-Work for Effecting Uniform Movement of Equatorials," *Memoirs of the Royal Astronomical Society*, **11**:249–267, 1840.

Anderson, B. D. O., and J. B. Moore, *Linear Optimal Control*, Prentice-Hall, Englewood Cliffs, N. J., 1971.

Ashley, H., *Engineering Analysis of Flight Vehicles*, Addison-Wesley, Reading, Mass., 1974.

Åström, K. J., "Frequency Domain Properties of Otto Smith Regulators," *Int. J. Control*, **26**(2):307–314, 1977.

——, *Introduction to Stochastic Control Theory*, Academic Press, New York, 1970.

Bellman, R., *Dynamic Programming*, Princeton University Press, Princeton, N.J., 1957.

Bellman R., and R. Kalaba, eds., *Mathematical Trends in Control Theory*, Dover, New York, 1964.

Blakelock, J. H., *Automatic Control of Aircraft and Missiles*, John Wiley, New York, 1965.

Bode, H. W., "Feedback: The History of an Idea," given at the Conference on Circuits and Systems, New York, 1960, and reprinted in Bellman and Kalaba, eds., *Mathematical Trends in Control Theory*, Dover, 1964.

——, *Network Analysis and Feedback Amplifier Design*, Van Nostrand, New York, 1945.

Bryson, A. E., and Y. C. Ho, *Applied Optimal Control*, Blaisdell, Waltham, Mass., 1969.

Buckley, P. S., *Techniques of Process Control*, John Wiley, New York, 1964.

Butterworth, S., "On the Theory of Filter Amplifiers," *Wireless Engineering*, **7**:536–541, 1930.

Callender, A., D. R. Hartree, and A. Porter, "Time Lag in a Control System," *Philosoph. Trans. Royal Soc. London*, Cambridge University Press, London, 1936.

Campbell, D. P., *Process Dynamics*, John Wiley, New York, 1958.

Campbell, G. A., and R. N. Foster, *Fourier Integrals for Practical Applications,* Van Nostrand, New York, 1948.

Cannon, R. H., Jr., *Dynamics of Physical Systems,* McGraw-Hill, New York, 1967.

Churchill, R. V., *Operational Mathematics,* Third ed., McGraw-Hill, New York, 1972.

Churchill, R. V., J. W. Brown, and R. F. Verhey, *Complex Variables and Applications,* McGraw-Hill, New York, 1976.

Clark, R. N., *Introduction to Automatic Control Systems,* John Wiley, New York, 1962.

Craig, J., *Introduction to Robotics: Mechanics and Control,* Addison-Wesley, Reading, Mass., 1985.

Dorf, R.C., *Modern Control Systems,* Fifth ed., Addison-Wesley, Reading, Mass., 1989.

Doyle, J. C., and G. Stein, "Multivariable Feedback Design: Concepts for a Classical/Modern Synthesis," *IEEE Trans. Auto. Contr.,* **AC-26**(1):4–16, February 1981.

Electrocraft Corp., *DC Motor Speed Controls Servo Systems,* Fifth ed., 1600 Second St. South, Hopkins, Minn. 55343, 1980.

Elgerd, O. I., *Electric Energy Systems Theory,* McGraw-Hill, New York, 1982.

Elgerd, O. I., and W. C. Stephens, "Effect of Closed-Loop Transfer Function Pole and Zero Locations on the Transient Response of Linear Control Systems," *AIEE,* **42**:121–127, 1959.

Emami-Naeini, A., and G. F. Franklin, "Zero Assignment in the Multivariable Robust Servomechanism," *Proc. IEEE Conf. Dec. Contr.,* 891–893, December 1982.

Emami-Naeini, A., and P. Van Dooren, "Computation of Zeros of Linear Multivariable Systems," *Automatica,* **18**(4):415–430, 1982.

——, and ——, "On Computation of Transmission Zeros and Transfer Functions," *Proc. IEEE Conf. Dec. Contr.,* 51–55, 1982.

Etkin, B., *Dynamics of Atmospheric Flight,* John Wiley, New York, 1959.

Evans, W. R., "Graphical Analysis of Control Systems," *Trans. AIEE,* **67**:547–551, 1948.

Francis, B. A., and W. M. Wonham, "The Internal Model Principle of Control Theory," *Automatica,* **12**:457–465, 1976.

Francis, B. A., "A Course in H_∞ Control Theory," *Lecture Notes in Control and Information Sciences,* 88, Springer-Verlag, 1987.

Francis, J. G. F., "The QR Transformation: A Unitary Analogue to the LR Transformation, Parts I and II," *Comp. J.,* **4**:265–272, 332–345, 1961.

Franklin, G. F., and A. Emami-Naeini, "A New Formulation of the Multivariable Robust Servomechanism Problem," an internal report, ISL, Stanford University, Stanford, Calif., July 1983.

——, and ——, "Robust Servomechanism Design Applied to Control of Reel-to-Reel Digital Tape Transports," *Proc. Asilomar Conf.,* 108–113, 1981.

Franklin, G. F., J. D. Powell, and M. L. Workman, *Digital Control of Dynamic Systems,* Second ed., Addison-Wesley, Reading, Mass., 1990.

Fuller, A. T., "The Early Development of Control Theory," *Trans. ASME, J. Dyn. Syst. Meas. Contr.,* **98**:109–118, 224–235, 1976.

Gantmacher, F. R., *The Theory of Matrices,* Vols. I and II, Chelsea Publishing Co., New York, 1959.

Garbow, B. S., F. M. Boyle, J. J. Dongarra, and C. B. Moler, *Matrix Eigensystem Routines—EISPACK Guide Extension,* lecture notes in Computer Science 51, Springer, Berlin, 1977.

Gardner, M. F., and J. L. Barnes, *Transients in Linear Systems,* John Wiley, New York, 1942.

Gopinath, B., "On the Control of Linear Multiple Input-Output Systems," *Bell Sys. Tech. J.,* **50,** March 1971.

Graham, D., and R. C. Lathrop, "The Synthesis of Optimum Response: Criteria and Standard Forms," *Trans. AIEE,* **72**(Pt. 1):273–288, 1953.

Gunkel, T. L., III, and G. F. Franklin, "A General Solution for Linear Sampled Data Control," *Trans. ASME J. Basic Eng.,* **85-D**:197–201, 1963.

Halliday, D., and R. Resnick, *Fundamentals of Physics,* John Wiley, New York, 1970.

Hang, C. C., K. J. Åström, and W. K. Ho, "Refinements of the Ziegler-Nichols Tuning Formula," Report CI–90–1, National University of Singapore, Dept. of Electrical Engineering, April 1990.

Heffley, R. K., and W. F. Jewell, "Aircraft Handling Qualities," System Technology, Inc., Tech. Rept. 1004-1, Hawthorne, Calif., May 1972.

Hubbard, M., Jr., and J. D. Powell, "Closed-Loop Control of Internal Combustion Engine Exhaust Emissions," SUDAAR No. 473, Dept. of Aero/Astro, Stanford University, Stanford, Calif., February 1974.

Hurewicz, W., "Filters and Servo Systems with Pulsed Data," Chap. 5 in H. M. James, N. B. Nichols, and R. S. Phillips, *Theory of Servomechanisms,* Radiation Lab. Series, Vol. 25, McGraw-Hill, New York, 1947.

James, H. M., N. B. Nichols, and R. S. Phillips, *Theory of Servomechanisms,* Radiation Lab. Series, Vol. 25, McGraw-Hill, New York, 1947.

Johnson, R. C., Jr., A. S. Foss, G. F. Franklin, R. V. Monopoli, and G. Stein, "Toward Development of a Practical Benchmark Example for Adaptive Control," *IEEE Control Systems Magazine,* **1**(4):25–28, December 1981.

Kailath, T., *Linear Systems,* Prentice-Hall, Englewood Cliffs, N.J., 1980.

Kalman, R. E., "A New Approach to Linear Filtering and Prediction Problems," *ASME J. Basic Eng.,* **82**:34–45, 1960.

——, "On the General Theory of Control Systems," *Proc. First Int. Cong. Auto. Contr.,* Moscow, 1960.

Kalman, R., and J. E. Bertram, "Control System Analysis and Design vi the Second Method of Lyapunov. II. Discrete Systems," *Trans. ASME Ser. D.J. Basic Eng,* **82**:394–400, 1960.

Kalman R., Y. C. Ho, and K. S. Narendra, "Controllability of Linear Dynamical Systems," in *Contributions to Differential Equations,* **1**(2), John Wiley, New York, 1961.

Kane, T. R., and D. A. Levinson, *Dynamics; Theory and Application,* McGraw-Hill, New York, 1985.

Kautsky, J., N.K. Nichols, and P. Van Dooren, "Robust Pole Assignment in Linear State Feedback," *Int. J. Control,* **41**(5): 1129–1155, 1985.

Kochenburger, R. J., "A Frequency Response Method for Analyzing and Synthesizing Contactor Servomechanisms," *Trans. AIEE,* **69**:270–283, 1950.

Kuo, B. C., *Automatic Control Systems,* Fourth ed., Prentice-Hall, Englewood Cliffs, N.J., 1982.

———, ed., *Incremental Motion Control,* Vol. 2, *Step Motors and Control Systems,* SRL Publishing Co., Champaign, Ill., 1980.

———, ed., *Proc. Symp. Incremental Motion Cont. Systems and Devices, Pt. I: Step Motors and Controls,* University of Illinois, Urbana, 1972.

Lanchester, F. W., *Aerodometics,* Archibald Constable, London, 1908.

LaSalle, L. P., and S. Lefschetz, *Stability by Liapunov's Direct Method,* Academic Press, New York, 1961.

Laub, A. J., A. Linnemann, and M. Wette, "Algorithms and Software for Pole Assignment by State Feedback," *Proc. 2nd Symp. CACSD,* March 1985.

Lyapunov, A. M., "Problème général de la stabilité du mouvement," *Ann. Fac. Sci. Toulouse,* **9**:203–474, 1907. (Translation of the original paper published in 1893 in *Comm. Soc. Math. Kharkow* and reprinted as Vol. 17 in *Ann. Math Studies,* Princeton University Press, Princeton, N.J., 1949.)

Ljung, L., and T. Söderström, *Theory and Practice of Recursive Identification,* MIT Press, Cambridge, Mass. 1983.

Luenberger, D. G., "Observing the State of a Linear System," *IEEE Trans. Military Electr.,* **MIL-8**:74–80, 1964.

McRuer, D. T., I. Askenas, and D. Graham, *Aircraft Dynamics and Automatic Control,* Princeton University Press, Princeton, N.J., 1973.

Mason, S. J., "Feedback Theory: Some Properties of Signal Flow Graphs," *Proc. IRE,* **41**:1144–1156, 1953.

———, "Feedback Theory: Further Properties of Signal Flow Graphs," *Proc. IRE,* **44**: 920–926, 1956.

Maxwell, T. C., "On Governors," *Proc. Royal Soc. London,* **16**:270–283, 1868.

Mayr, O., *The Origins of Feedback Control,* MIT Press, Cambridge, Mass., 1970.

Mees, A.I., *Dynamics of Feedback Systems,* John Wiley, New York, 1981.

Minimis, G. S., and C. C. Paige, "An Algorithm for Pole Assignment of Time Invariant Systems," *Int. J. Contr.,* **35**(2):341–354, 1982.

Misra, P., and R. V. Patel, "Numerical Algorithms for Eigenvalue Assignment by Constant and Dynamic Output Feedback," *IEEE Trans. Auto. Control,* **AC–34** (6): 577–588, 1989.

Moler, C., and C. van Loan, "Nineteen Dubious Ways to Compute the Exponential of a Matrix," *SIAM Review,* **20**(4), 1978.

Nyquist, H., "Regeneration Theory," *Bell Sys. Tech. J.,* **11**:126–147, 1932.

Ogata, K., *Modern Control Engineering,* Prentice-Hall, Englewood Cliffs, N.J., 1970.

Pontryagin, L. S., V. G. Boltyanskii, R. V. Gamkrelidze, and E. F. Mishchenko, *The Mathematical Theory of Optimal Processes,* John Wiley, New York, 1962.

Ragazzini, J. R., R. H. Randall, and F. A. Russell, "Analysis of Problems in Dynamics by Electronic Circuits," *Proc. IRE,* **35**(5):442–452, May 1947.

Rosenbrock, H. H., *State Space and Multivariable Theory,* John Wiley, New York, 1970.

Routh, E. J., *Dynamics of a System of Rigid Bodies,* Macmillan & Co., London, 1905.

Safonov, M. G., A. J. Laub, and G. Hartmann, "Feedback Properties of Multivariable Systems: The Role and Use of Return Difference Matrix," *IEEE Trans. Auto Control*, **AC-26,** 47–65, 1981.

Schmitz, E. "Robotic Arm Control," Ph.D. thesis, Dept. of Aero/Astro, Stanford University, Stanford, Calif., 1985.

Sinha, N. K., and B. Kuszta, *Modeling and Identification of Dynamic Systems,* Van Nostrand Reinhold Co., New York, 1983.

Smith, O. J. M., *Feedback Control Systems,* McGraw-Hill, New York, 1958.

Smith, R. J., *Electronics: Circuits and Devices,* Second ed., John Wiley, New York, 1980.

Strang, G., *Linear Algebra and Its Applications,* Third ed., Harcourt Brace Jovanovich, New York, 1988.

Swift, J., *On Poetry: A Rhapsody,* 1733.

Thomson, W. T., *Theory of Vibration with Applications,* Second ed., Prentice-Hall, Englewood Cliffs, N.J., 1981.

Trankle, T. L., "Development of WMEC Tampa Maneuvering Model from Sea Trial Data," Systems Control Technology, Report No. MA-RD-760-87201, March 1987.

Truxal, J. G., *Control System Synthesis,* McGraw-Hill, New York, 1955.

Van der Pol, B., and H. Bremmer, *Operational Calculus,* Cambridge University Press, New York, 1955.

Van Dooren, P., A. Emami-Naeini, and L. Silverman, "Stable Extraction of the Kronecker Structure of Pencils," *Proc. IEEE Conf. Dec. Contr.,* 521–524, 1978.

Vidyasagar, M., *Nonlinear Systems Analysts,* Prentice-Hall, Englewood Cliffs, N.J., 1978.

Wiener, N., *The Extrapolation, Interpolation and Smoothing of Stationary Time Series,* John Wiley, New York, 1949.

——, "Generalized Harmonic Analysis," *Acta. Math.,* **55**:117, 1930.

Woodson, H. H., and J. R. Melcher, *Electromechanical Dynamics, Part I: Discrete Systems,* John Wiley, New York, 1968.

Zames, G., "On the Input-Output Stability of Time-Varying Nonlinear Feedback Systems— Part I: Conditions Derived Using Concepts of Loop Gain, Conicity and Positivity," *IEEE Trans. Auto. Contr.,* **AC-11**:465–476, 1966.

——, "On the Input-Output Stability of Time-Varying Nonlinear Feedback Systems—Part II: Conditions Involving Circles in the Frequency Plane and Sector Nonlinearities," *IEEE Trans. Auto. Contr.,* **AC-11**:228–238, 1966.

Ziegler, J. G., and N. B. Nichols, "Optimum Settings for Automatic Controllers," *Trans. ASME,* **64**:759–768, 1942.

——, and ——, "Process Lags in Automatic Control Circuits," *Trans. ASME,* **65**(5):433–444, July 1943.

Index

Table of Laplace Transforms

Number	$F(s)$	$f(t), t \geq 0$
1	1	$\delta(t)$
2	$1/s$	$1(t)$
3	$1/s^2$	t
4	$2!/s^3$	t^2
5	$3!/s^4$	t^3
6	$m!/s^{m+1}$	t^m
7	$1/(s + a)$	e^{-at}
8	$1/(s + a)^2$	te^{-at}
9	$1/(s + a)^3$	$\dfrac{1}{2!}t^2 e^{-at}$
10	$1/(s + a)^m$	$\dfrac{1}{(m-1)!}t^{m-1}e^{-at}$
11	$\dfrac{a}{s(s + a)}$	$1 - e^{-at}$
12	$\dfrac{a}{s^2(s + a)}$	$\dfrac{1}{a}(at - 1 + e^{-at})$
13	$\dfrac{b - a}{(s + a)(s + b)}$	$e^{-at} - e^{-bt}$
14	$\dfrac{s}{(s + a)^2}$	$(1 - at)e^{-at}$
15	$\dfrac{a^2}{s(s + a)^2}$	$1 - e^{-at}(1 + at)$
16	$\dfrac{(b - a)s}{(s + a)(s + b)}$	$be^{-bt} - ae^{-at}$
17	$a/(s^2 + a^2)$	$\sin at$
18	$s/(s^2 + a^2)$	$\cos at$
19	$\dfrac{s + a}{(s + a)^2 + b^2}$	$e^{-at}\cos bt$
20	$\dfrac{b}{(s + a)^2 + b^2}$	$e^{-at}\sin bt$
21	$\dfrac{a^2 + b^2}{s\left[(s + a)^2 + b^2\right]}$	$1 - e^{-at}\left(\cos bt + \dfrac{a}{b}\sin bt\right)$